DATE			

Analog Electronics Handbook

To my wife
To my son
for their endurance

Analog Electronics Handbook

T. H. COLLINS

PRENTICE HALL

New York • London • Toronto • Sydney • Tokyo

First published 1989 by
Prentice Hall International (UK) Ltd
66 Wood Lane End, Hemel Hempstead,
Hertfordshire, HP2 4RG
A division of
Simon & Schuster International Group

Printed and bound in Great Britain by
A. Wheaton & Co. Ltd, Exeter.

Library of Congress Cataloging in Publication Data

Collins, T. H.
Analog electronics handbook/T. H. Collins.
p. cm.
Includes index.
ISBN 0-13-033119-8
1. Analog electronic systems–Handbooks, manuals, etc.
I. Title. TK7825.C63 1988 87-26871

621.381–dc19

1 2 3 4 5 92 91 90 89 88

ISBN 0-13-033119-8

Contents

Preface

Analog Electronics is rapidly coming back into its own as the enormous vogue for digital electronics settles down.

The applications of analog electronics are widespread and many will never give way to digital systems simply because of the analog nature of most of the world around us.

It was with those considerations in mind that *Analog Electronics Handbook* was conceived. The book's aim is to provide a fairly comprehensive compendium of basic information on the topics which make up analog electronics and to present it in a practical and accessible form as a reference for engineers, students, teachers and researchers.

To cover all of these topics in depth, giving physical theories, derivations, formulas and so on would have resulted in an encyclopedic work of several volumes, inevitably of little use to people who need practical information quickly. For this reason extensive theories, derivations and calculations have been shortened or omitted, leaving concise but clear reviews of key concepts with ready-to-use facts, formulas, circuits, results and solutions to common problems. In many cases, worked examples are provided to illustrate the use of formulas in practical situations.

Since certain mathematical and physical concepts such as vector theory, magnetism, Laplace and Fourier transforms are essential to an understanding of analog electronics, these have been covered in the same way as more 'central' topics. However, the reader is assumed to have a working knowledge of differential and integral calculus and the principles of modern physics and network analysis.

Inevitably, there will be some subjects or items missing. For example, integrated circuit technology is expanding exponentially and new developments are announced almost daily, making it impossible for a book of this sort to keep up with them. Nevertheless, if *Analog Electronics Handbook* saves the reader the considerable effort and time required to go to a technical library and search through a number of books, it will have served its purpose.

Glossary

A_c	common mode gain of differential amplifier
A_d	differential gain of differential amplifier
A_f	closed loop gain
$A_{f,\infty}$	closed loop gain of ideal amplifier
A_r	loop gain
a_x, a_y, a_z	components of vector **a**
B	bandwidth
$\mathbf{B}$	magnetic induction
B_n	noise bandwidth
B_o	open loop bandwidth
B_f	closed loop bandwidth
B_r	retentivity
C	capacitor capacitance circulation
C_b	junction capacitance
$C_{cb'}$	Miller capacitance of bipolar transistor
C_d	diffusion capacitance
C_{DG}	drain–gate capacitance
C_i	input capacitance
CMRR	common mode rejection ratio
D	distortion factor
$\mathbf{D}$	displacement vector flux density
d	percentage distortion distance
$\mathbf{E}$	electric field strength
e	electromotive force induced voltage
F	force discrimination ratio noise factor
$F(s)$	Laplace transform of $f(t)$
$F(x)$	probability distribution function
$F(\omega)$	Fourier transform of $f(t)$
f	frequency
f_a	absolute error in a
f_{ap}	percentage error in a
f_{ar}	relative error in a
f_p	parallel resonance frequency
f_s	series resonance frequency
f_T	transition frequency
$f(x)$	probability density function
G	conductance shunt conductivity directive gain of antenna
GBP	gain-bandwidth product
g	gyration constant
g_m	transconductance
g_{max}	maximum transconductance
g_o	output admittance
H	rejection ratio CMRR
$\mathbf{H}$	magnetic field strength
H_c	coercive force
HF	high frequency
$H(j\omega)$	transfer function
h	height Planck's constant (4.32×10^{-15} eV s)
I	DC current
I_b	bias current
I_C	collector current (DC), diode current (DC)
I_D	drain current (DC)

I_{cbo}	leakage current (open emitter)
I_{ceo}	leakage current (open base)
I_{DSS}	max. drain current of JFET
I_E	emitter current (DC)
I_H	holding current
I_i	input current
I_m	amplitude of current i
I_o	output current
I_{os}	input offset current
I_p	primary peak current
I_s	signal current (DC)
	diode leakage current
	secondary peak current
I_{rms}	rms-value of current
I_Z	Zener current
i	AC current
$\mathbf{i}$	unit vector along x-axis
i_b	base current (AC)
i_c	collector current (AC)
i_d	drain current (AC)
i_g	gate current (AC)
i_n	noise current
i_o	output current (AC)
i_s	signal current (AC)
J	current density
$J_n(m)$	Bessel function
$\mathbf{j}$	unit vector along y-axis
k	Boltzmann's constant (1.38×10^{-23} J K^{-1})
$\mathbf{k}$	unit vector along z-axis
L	coefficient of self-inductance
LDR	light dependent resistor
l	length
l_e	effective length
M	magnetization
	coefficient of mutual inductance
MTI	moving target indication
MUF	maximum usable frequency
M_H	Hall constant
m_a	modulation factor
m_f	modulation index
	deviation ratio
NA	numerical aperture
NF	noise figure
N_w	spectral density of white noise
$N(\omega)$	spectral density
n	turns ratio
$\mathbf{n}$	unit normal vector
P	power
	polarization
	polarization density
	flux density
P_{av}	average power
P_{is}	power density of isotropic radiator
PPI	plan position indication
$\mathbf{p}$	electric dipole moment
p	instantaneous power
Q	charge
	quality factor
q	electron charge
R	resistance
	resistor
R_L	load resistance
R_m	reluctance
	magnetic resistance
R_r	radiation resistance
R_s	source resistance
$R(t, \tau)$	autocorrelation function
r_c	dynamic collector resistance
r_d	dynamic diode resistance
r_{DS}	channel resistance
$r_{DS,on}$	minimum channel resistance
r_e	dynamic emitter resistance
r_g	gyration resistance
r_i	input resistance
$r_{i,c}$	common mode input resistance
$r_{i,d}$	differential mode input resistance
$r_{i,f}$	input resistance of feedback amplifier
r_o	output resistance
$r_{o,f}$	output resistance of feedback amplifier
S	area
	slew rate
	stability factor
S_{eff}	effective area

$S_i(f)$	spectrum of current noise
S/H	sample-and-hold
S/N	signal-to-noise ratio
$S_v(f)$	spectrum of voltage noise
s	Laplace variable line regulation
T	absolute temperature in K torque
t	temperature in °C time
t_a	ambient temperature
t_j	junction temperature
V	DC voltage
V^+	positive supply voltage
V^-	negative supply voltage
V_{BB}	interbase voltage
V_{BO}	breakover voltage
VDR	Voltage Dependent Resistor
V_m	amplitude, peak value
V_p	primary (peak) voltage
V_P	pinch-off voltage
V_{pp}	peak-to-peak value
V_s	secondary (peak) voltage
VSWR	Voltage Standing Wave Ratio
V_z	Zener voltage
v	AC voltage
v_c	carrier voltage
v_d	drift velocity
v_g	group velocity
v_o	output signal voltage
v_p	phase velocity
v_s	signal voltage
W	energy
w	energy density
X_C	capacitive reactance
X_L	inductive reactance
$\bar{x}$	average value of variable x
y, Y	admittance
Z	impedance
Z_i	input impedance
Z_L	load impedance
Z_o	output impedance characteristic impedance of transmission line

α	temperature coefficient attenuation coefficient current amplification factor (emitter-collector)
α_r	current amplification factor (reverse)
α_f	current amplification factor (forward)
β	current amplification factor (base-collector)
γ	emitter efficiency relative detuning specific conductivity
Δ	Laplace operator
∇	nabla operator, del
δ	skin depth
ε	dielectric constant permittivity
ε_r	relative permittivity
ε_v	permittivity of vacuum
η	efficiency intrinsic stand-off ratio
η_a	antenna efficiency
θ	thermal resistance
λ	wavelength
μ	voltage amplification factor mobility permeability specific magnetic conductivity
μ'	voltage transfer ratio
μ_v	permeability of vacuum
μ_r	relative permeability
ρ	specific resistance correlation coefficient of two variables reflection coefficient
Σ	sum
σ_x	standard deviation of n data x
σ_{xy}	covariance of variables x and y
τ	time constant
Φ	magnetic flux
Φ_g	thermal generation rate
Φ_r	thermal recombination rate
ϕ	phase shift

$\phi(x,y,z)$	scalar field	ω_c	cutoff frequency
χ	electric susceptibility		carrier frequency
χ_m	magnetic susceptibility	ω_m	center frequency
ψ	flux		modulation frequency
	electric flux	ω_o	resonance frequency
ω	angular frequency		oscillation frequency

SYMBOLS

ideal source of DC voltage

ideal source of AC voltage

ideal source of DC current

ideal source of AC current

diode

Zener diode

tunnel diode

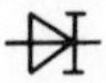

backward diode

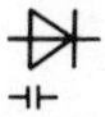

capacitance diode
varicap
varactor

light emitting diode

photodiode

Schottky diode

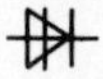

Shockley diode

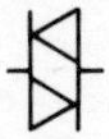

diac

thyristor

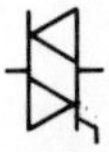

triac

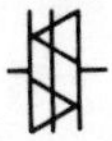

SBS

SCS

npn-transistor

pnp-transistor

Schottky transistor

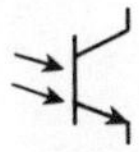

phototransistor

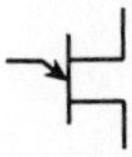

unijunction transistor

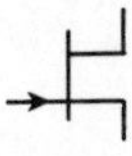

n-channel junction FET

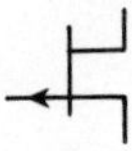

p-channel junction FET

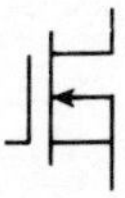

n-channel depletion FET

p-channel depletion FET

n-channel enhancement FET

p-channel enhancement FET

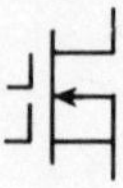

dual-gate n-channel FET

n-channel photoFET

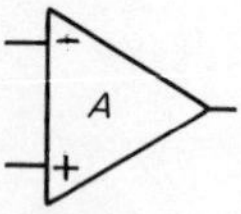

operational amplifier

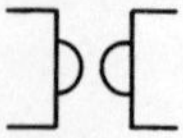

gyrator

crystal

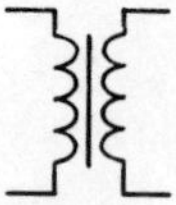

transformer

resistor impedance

NTC-thermistor

PTC-thermistor

VDR

LDR

variable resistor

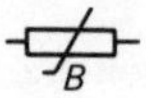

fluxistor

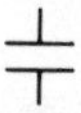

capacitor/capacitance

inductor/inductance

// in parallel with

common

Tables

Table 1 The Greek alphabet

alpha	A	α	nu	N	ν
beta	B	β	xi	Ξ	ξ
gamma	Γ	γ	omicron	O	o
delta	Δ	δ	pi	Π	π
epsilon	E	ϵ	rho	P	ρ
zeta	Z	ζ	sigma	Σ	σ
eta	H	η	tau	T	τ
theta	Θ	θ	upsilon	Υ	υ
iota	I	ι	phi	Φ	ϕ
kappa	K	κ	chi	X	χ
lambda	Λ	λ	psi	Ψ	ψ
mu	M	μ	omega	Ω	ω

Table 2 Decimal multiples and fractions

Prefix	*Symbol*	*Multiplier*
exa	E	10^{18}
peta	P	10^{15}
tera	T	10^{12}
giga	G	10^{9}
mega	M	10^{6}
kilo	k	10^{3}
milli	m	10^{-3}
micro	μ	10^{-6}
nano	n	10^{-9}
pico	p	10^{-12}
femto	f	10^{-15}
atto	a	10^{-18}
bronto	b	10^{-21}

Table 3 Functions of sums of angles

$$\sin(x \pm y) = \sin x \cos y \pm \cos x \sin y$$
$$\cos(x \pm y) = \cos x \cos y \mp \sin x \sin y$$
$$\sin(x + y) + \sin(x - y) = 2 \sin x \cos y$$
$$\cos(x + y) + \cos(x - y) = 2 \cos x \cos y$$
$$\sin(x + y) - \sin(x - y) = 2 \cos x \sin y$$
$$\cos(x + y) - \cos(x - y) = -2 \sin x \sin y$$
$$\tan(x + y) = \frac{\tan x + \tan y}{1 - \tan x \tan y}$$

Table 4 Hyperbolic functions

$$\sinh x = \frac{\exp x - \exp(-x)}{2} = x + \frac{x^3}{3!} + \frac{x^5}{5!} + \frac{x^7}{7!} + \cdots$$
$$\cosh x = \frac{\exp x + \exp(-x)}{2} = 1 + \frac{x^2}{2!} + \frac{x^4}{4!} + \frac{x^6}{6!} + \cdots$$
$$\tanh x = \frac{\sinh x}{\cosh x}$$
$$\coth x = \frac{\cosh x}{\sinh x} = \frac{1}{\tanh x}$$
$$\sinh(x \pm jy) = \sinh x \cos y \pm j \cosh x \sin y$$
$$\cosh(x \pm jy) = \cosh x \cos y \pm j \sinh x \sin y$$
$$\left.\begin{aligned}\cosh(jx) &= \tfrac{1}{2}(\exp jx + \exp(-jx)) = \cos x \\ \sinh(jx) &= \tfrac{1}{2}(\exp jx - \exp(-jx)) = j \sin x\end{aligned}\right\} \text{de Moivre's theorem}$$
$$\exp(\pm jx) = \cos x \pm j \sin x$$
$$\cosh x = \cos jx$$
$$j \sinh x = \sin jx$$
$$\tanh(x \pm jy) = \frac{\sinh 2x}{\cosh 2x + \cos 2y} \pm j \frac{\sin 2y}{\cosh 2x + \cos 2y}$$
$$\coth(x \pm jy) = \frac{\sinh 2x}{\cosh 2x - \cos 2y} \pm j \frac{\sin 2y}{\cosh 2x - \cos 2y}$$

Table 5 Laplace transforms

$f(t)$	$F(s)$
unit impulse $\delta(t)$	1
unit step $1(t)$	$\frac{1}{s}$
$\frac{t^{n-1}}{(n-1)!}\ (n=1,2,3,\ldots)$	$\frac{1}{s^n}$
$t^n \exp(-at)(n=0,1,2,3,\ldots)$	$\frac{n!}{(s+a)^{n+1}}$
$\frac{1}{b-a}(\exp(-at)-\exp(-bt))$	$\frac{1}{(s+a)(s+b)}$
$\frac{1}{b-a}(b\exp(-bt)-a\exp(-at))$	$\frac{s}{(s+a)(s+b)}$
$\frac{1}{ab}\left[1+\frac{1}{a-b}(b\exp(-at)-a\exp(-bt))\right]$	$\frac{1}{s(s+a)(s+b)}$
$\exp(-at)\sin\omega t$	$\frac{\omega}{(s+a)^2+\omega^2}$
$\exp(-at)\cos\omega t$	$\frac{s+a}{(s+a)^2+\omega^2}$
$\sinh at$	$\frac{a}{s^2-a^2}$
$\cosh at$	$\frac{s}{s^2-a^2}$
$\frac{1}{a^2}(at-1+\exp(-at))$	$\frac{1}{s^2(s+a)}$
$\frac{\omega_n}{\sqrt{1-\zeta^2}}\exp(-\zeta\omega_n t)\sin\omega_n\sqrt{1-\zeta^2}\,t$	$\frac{\omega_n^2}{s^2+2\zeta\omega_n s+\omega_n^2}$
$\frac{-1}{\sqrt{1-\zeta^2}}\exp(-\zeta\omega_n t)\sin(\omega_n\sqrt{1-\zeta^2}\,t-\phi)$ $\phi=\tan^{-1}\frac{\sqrt{1-\zeta^2}}{\zeta}$	$\frac{s}{s^2+2\zeta\omega_n s+\omega_n^2}$
$1-\frac{1}{\sqrt{1-\zeta^2}}\exp(-\zeta\omega_n t)\sin(\omega_n\sqrt{1-\zeta^2}\,t+\phi)$ $\phi=\tan^{-1}\frac{\sqrt{1-\zeta^2}}{\zeta}$	$\frac{\omega_n^2}{s(s^2+2\zeta\omega_n s+\omega_n^2)}$
$\frac{1}{\sqrt{\pi t}}$	$\frac{1}{\sqrt{s}}$
$J_n(at)$ (Bessel function)	$\frac{1}{\sqrt{s^2+a^2}}\left[\frac{s^2+a^2-s}{a}\right]^n$ $(n=0,1,2,3\ldots)$

Table 6 The electromagnetic spectrum

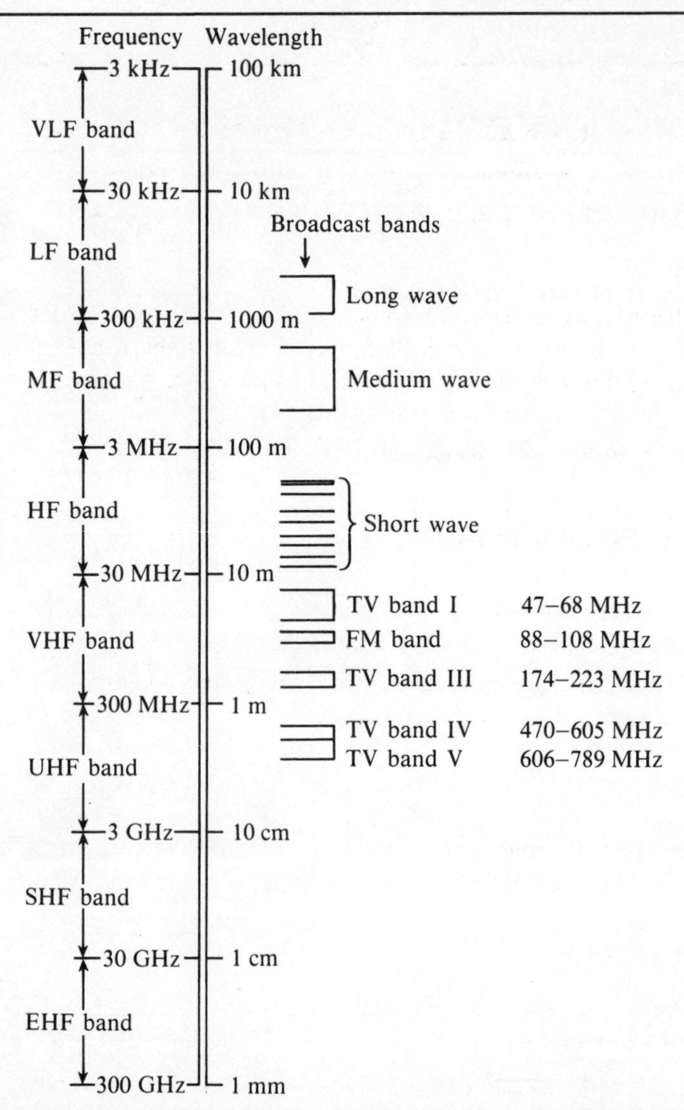

1
Vector Theory

Vector theory was developed around 1850. It soon proved to be an important link between various disciplines by providing a mathematical description of key concepts in mechanical engineering, aerodynamics and hydrodynamics, electricity and magnetism.

1.1 DEFINITIONS AND LAWS

Vector

A **vector** is a quantity which is characterized by both magnitude and direction. Examples are: force, velocity, electric field. Notation: $\boldsymbol{a}$. Magnitude of $\boldsymbol{a}$: a or $|\boldsymbol{a}|$.

Scalar

A **scalar** is a quantity which is characterized by magnitude only. Examples: mass, length, time, potential.

Linear operations

$$\boldsymbol{a} + \boldsymbol{b} = \boldsymbol{b} + \boldsymbol{a} \qquad \text{(commutative law)}$$

$$\boldsymbol{a} + (\boldsymbol{b} + \boldsymbol{c}) = (\boldsymbol{a} + \boldsymbol{b}) + \boldsymbol{c} \qquad \text{(associative law)}$$

$$(\lambda + \mu)\boldsymbol{a} = \lambda\boldsymbol{a} + \mu\boldsymbol{a} \quad (\lambda \text{ and } \mu \text{ are constants}) \qquad \text{(distributive law)}$$

$$\lambda(\boldsymbol{a} + \boldsymbol{b}) = \lambda\boldsymbol{a} + \lambda\boldsymbol{b} \qquad \text{(distributive law)}$$

The resultant vector of the sum of two vectors follows the **triangle law** (parallelogram) of vector addition (Fig. 1.1).

Representation

Usually a vector is represented in a right-handed coordinate system (Fig. 1.2). In such a system three **unit vectors** are assumed, having the directions of the positive x-, y- and z-axes. These unit vectors are $\boldsymbol{i}, \boldsymbol{j}$ and $\boldsymbol{k}$ respectively.

A vector $\boldsymbol{a}$ (Fig. 1.2) can be written as $\boldsymbol{a} = \boldsymbol{i}a_x + \boldsymbol{j}a_y + \boldsymbol{k}a_z$ where a_x, a_y and a_z are the **components** of the vector. Magnitude of $\boldsymbol{a}$: $a = \sqrt{a_x^2 + a_y^2 + a_z^2}$.

If the vector $\boldsymbol{r}$ is located between the origin O and a point $P(x_o, y_o, z_o)$ then $\boldsymbol{r} = \boldsymbol{i}x_o + \boldsymbol{j}y_o + \boldsymbol{k}z_o$ (Fig. 1.3). This vector is the **position vector** or **radius vector** from O to P. Notation: P($\boldsymbol{r}$).

Scalar field

If to each point $P(x, y, z)$ in a space there corresponds a number (scalar) $\phi(x, y, z)$, then $\phi(x, y, z)$ is called a **scalar field**. Examples are: air pressure on a weather map, electrostatic potential. A time-independent scalar field is called **stationary**.

Vector field

If to each point $P(x, y, z)$ in a space there corresponds a vector $\boldsymbol{a}(x, y, z)$, then $\boldsymbol{a}(x, y, z)$ is called a **vector field**. Examples are: electrostatic field, magnetic field. A time-independent vector field is called **stationary**.

1.2 MULTIPLICATION OF VECTORS

1.2.1 Scalar product (dot product)

The scalar product of two vectors $\boldsymbol{a}$ and $\boldsymbol{b}$, denoted by $\boldsymbol{a} \cdot \boldsymbol{b}$ is defined as the product of the magnitudes of $\boldsymbol{a}$ and $\boldsymbol{b}$ and the cosine of the angle θ between them. Thus

$$\boldsymbol{a} \cdot \boldsymbol{b} = ab \cos \theta$$

The scalar product is a scalar.

Laws of scalar products

$$\boldsymbol{a} \cdot \boldsymbol{b} = \boldsymbol{b} \cdot \boldsymbol{a} \qquad \text{(commutative law)}$$

$$\boldsymbol{a} \cdot (\boldsymbol{b} + \boldsymbol{c}) = \boldsymbol{a} \cdot \boldsymbol{b} + \boldsymbol{a} \cdot \boldsymbol{c} \qquad \text{(distributive law)}$$

$$\lambda(\boldsymbol{a} \cdot \boldsymbol{b}) = (\lambda \boldsymbol{a}) \cdot \boldsymbol{b} = \boldsymbol{a} \cdot (\lambda \boldsymbol{b}) \qquad (\lambda \text{ is a constant})$$

The **angle** between two vectors $\boldsymbol{a}$ and $\boldsymbol{b}$ is expressed as

$$\cos\theta = \frac{a_x b_x + a_y b_y + a_z b_z}{\sqrt{a_x{}^2 + a_y{}^2 + a_z{}^2}\sqrt{b_x{}^2 + b_y{}^2 + b_z{}^2}}$$

Example

If the vectors are located between the origin and the points $P(1, 2, 3)$ and $Q(0, 1, 0)$, then

$$\cos\theta = \frac{1(0) + 2(1) + 3(0)}{\sqrt{1^2 + 2^2 + 3^2}\sqrt{0^2 + 1^2 + 0^2}}$$
$$= \tfrac{1}{7}\sqrt{14}$$

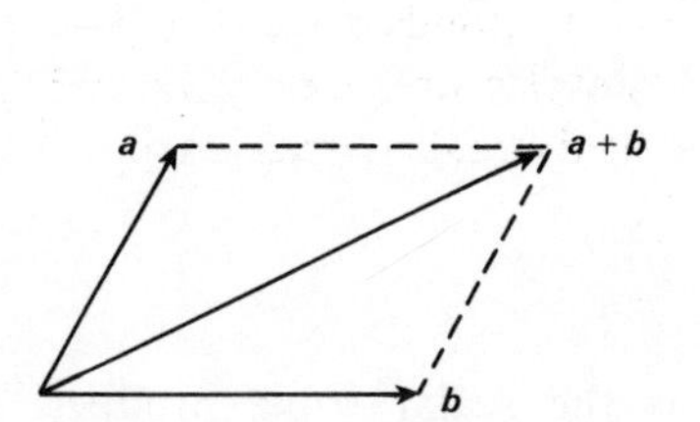

Fig. 1.1 Addition of two vectors

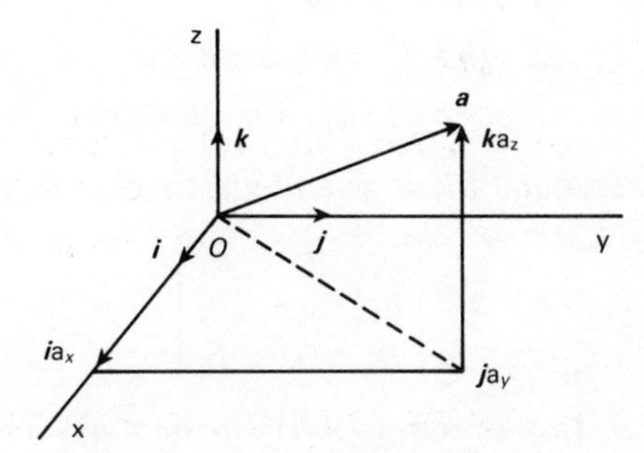

Fig. 1.2 Vector representation and unit vectors in cartesian coordinates

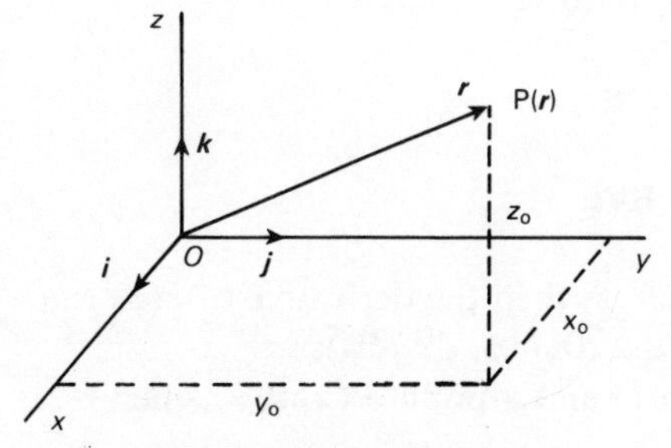

Fig. 1.3 Position vector or radius vector

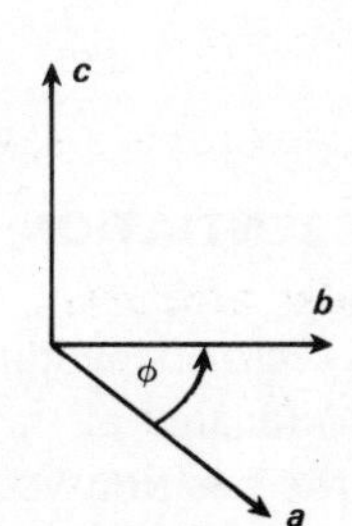

Fig. 1.4 Vector product of two vectors

1.2.2 Vector product (cross product)

The vector product of two vectors $\boldsymbol{a}$ and $\boldsymbol{b}$, deoted by $\boldsymbol{a} \times \boldsymbol{b}$ is a vector $\boldsymbol{c}$ with the following properties:

- the magnitude of $\boldsymbol{c}$ is the product of the magnitudes of $\boldsymbol{a}$ and $\boldsymbol{b}$ and the sine of the angle θ between them,
- the direction of $\boldsymbol{c}$ is perpendicular to the plane of $\boldsymbol{a}$ and $\boldsymbol{b}$ such that $\boldsymbol{a}$, $\boldsymbol{b}$ and $\boldsymbol{c}$ form a right-handed system (Fig. 1.4).

In symbols:

$$\boldsymbol{c} = \boldsymbol{a} \times \boldsymbol{b}$$

$$c = ab \sin \theta$$

The following laws for vector products apply:

$$\boldsymbol{a} \times \boldsymbol{b} = -\boldsymbol{b} \times \boldsymbol{a} \qquad \text{(anti-commutative law)}$$

$$\boldsymbol{a} \times (\boldsymbol{b} \times \boldsymbol{c}) = \boldsymbol{a} \times \boldsymbol{b} + \boldsymbol{a} \times \boldsymbol{c} \qquad \text{(distributive law)}$$

$$\lambda(\boldsymbol{a} \times \boldsymbol{b}) = (\lambda\boldsymbol{a}) \times \boldsymbol{b} = \boldsymbol{a} \times (\lambda\boldsymbol{b}) \qquad (\lambda = \text{constant})$$

If $\boldsymbol{a}$ and $\boldsymbol{b}$ are given by their components, then

$$\boldsymbol{a} \times \boldsymbol{b} = \begin{vmatrix} \boldsymbol{i} & \boldsymbol{j} & \boldsymbol{k} \\ a_x & a_y & a_z \\ b_x & b_y & b_z \end{vmatrix}$$

$\boldsymbol{a} \cdot (\boldsymbol{b} \times \boldsymbol{c}) = \boldsymbol{c} \cdot (\boldsymbol{a} \times \boldsymbol{b}) = \boldsymbol{b} \cdot (\boldsymbol{c} \times \boldsymbol{a})$. This is the scalar triple product. It equals the volume of a parallelepiped with edges $\boldsymbol{a}$, $\boldsymbol{b}$ and $\boldsymbol{c}$ (Fig. 1.5).

$$\boldsymbol{a} \cdot (\boldsymbol{a} \times \boldsymbol{b}) = 0$$

$$\boldsymbol{a} \times (\boldsymbol{b} \times \boldsymbol{c}) = \boldsymbol{b}(\boldsymbol{a} \cdot \boldsymbol{c}) - \boldsymbol{c}(\boldsymbol{a} \cdot \boldsymbol{b}) \qquad \text{(vector triple product)}$$

$$(\boldsymbol{a} \times \boldsymbol{b}) \times \boldsymbol{c} = \boldsymbol{b}(\boldsymbol{a} \cdot \boldsymbol{c}) - \boldsymbol{a}(\boldsymbol{b} \cdot \boldsymbol{c})$$

$$(\boldsymbol{a} \times \boldsymbol{b}) \cdot (\boldsymbol{c} \times \boldsymbol{d}) = (\boldsymbol{a} \cdot \boldsymbol{c})(\boldsymbol{b} \cdot \boldsymbol{d}) - (\boldsymbol{a} \cdot \boldsymbol{d})(\boldsymbol{b} \cdot \boldsymbol{c})$$

1.3 DIFFERENTIATION; SPACE CURVE

If $\boldsymbol{r}(\phi)$ is a vector depending on a scalar ϕ, then the derivative of $\boldsymbol{r}(\phi)$ is a vector $\mathrm{d}\boldsymbol{r}/\mathrm{d}\phi$ located at the point $(\mathrm{d}r_x/\mathrm{d}\phi, \mathrm{d}r_y/\mathrm{d}\phi, \mathrm{d}r_z/\mathrm{d}\phi)$.

If $\boldsymbol{r}$ is the position vector between O and a point $P(x, y, z)$, then

$$\boldsymbol{r}(\phi) = \boldsymbol{i}x(\phi) + \boldsymbol{j}y(\phi) + \boldsymbol{k}z(\phi)$$

As ϕ changes, the terminal point of $\boldsymbol{r}$ describes a **space curve** (Fig. 1.6) and $\mathrm{d}\boldsymbol{r}/\mathrm{d}\phi$ is a vector in the direction of the tangent of the space curve (Fig. 1.7).

The **unit tangent vector** $\boldsymbol{t}$ is a vector parallel to the tangent of the space curve and having unit length:

$$\boldsymbol{t}=\frac{\boldsymbol{i}\dfrac{\mathrm{d}x}{\mathrm{d}\phi}+\boldsymbol{j}\dfrac{\mathrm{d}y}{\mathrm{d}\phi}+\boldsymbol{k}\dfrac{\mathrm{d}z}{\mathrm{d}\phi}}{\sqrt{\left(\dfrac{\mathrm{d}x}{\mathrm{d}\phi}\right)^2+\left(\dfrac{\mathrm{d}y}{\mathrm{d}\phi}\right)^2+\left(\dfrac{\mathrm{d}z}{\mathrm{d}\phi}\right)^2}}$$

Example

If $\boldsymbol{r}(\phi)=\boldsymbol{i}\cos\phi+\boldsymbol{j}\sin\phi$, then the tangent vector $\boldsymbol{t}$ is

$$\boldsymbol{t}=\frac{-\boldsymbol{i}\sin\phi+\boldsymbol{j}\cos\phi}{\sqrt{\sin^2\phi+\cos^2\phi}}$$
$$=-\boldsymbol{i}\sin\phi+\boldsymbol{j}\cos\phi$$

If $\boldsymbol{r}$ is dependent on two scalars ϕ and ψ, then $\boldsymbol{r}(\phi,\psi)$ describes a **surface**.

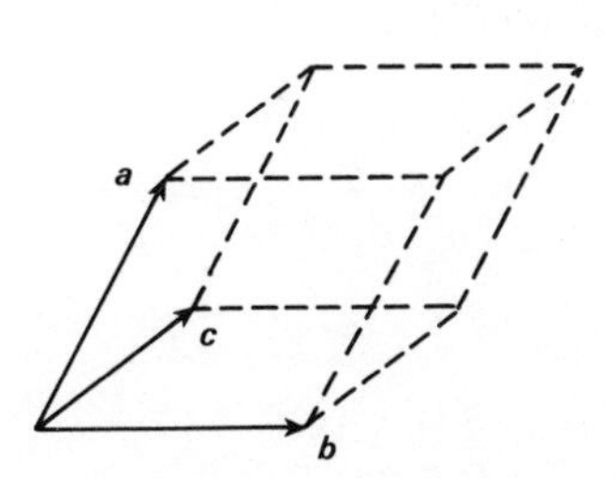

Fig. 1.5 The scalar triple product

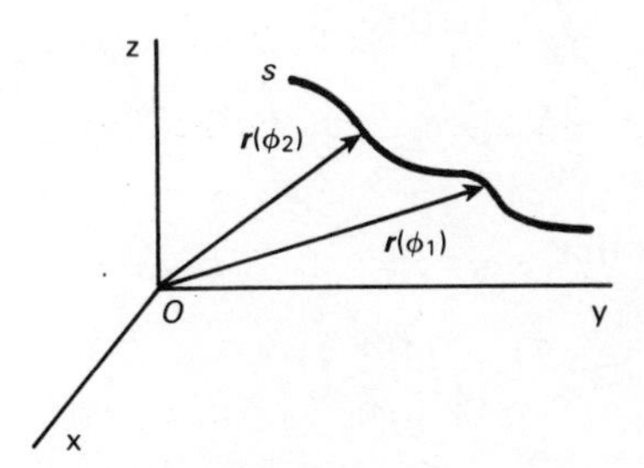

Fig. 1.6 Space curve described by a position vector $\boldsymbol{r}(\phi)$

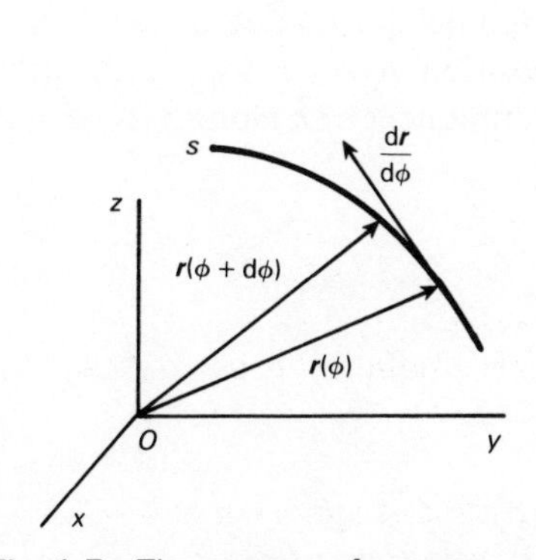

Fig. 1.7 The tangent of a space curve $\boldsymbol{r}(\phi)$

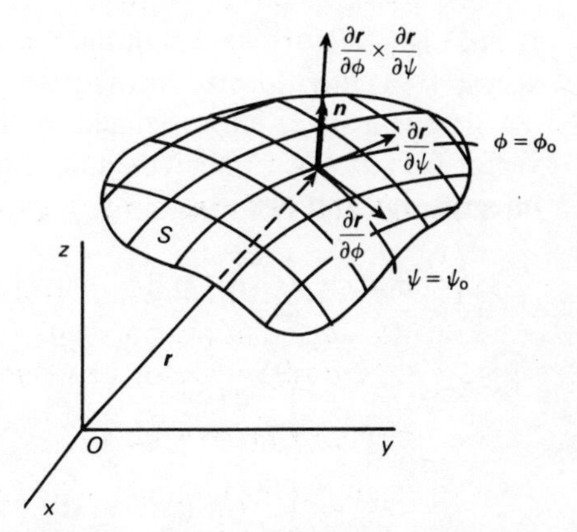

Fig. 1.8 The unit normal $\boldsymbol{n}$ to a point on a surface S

A vector normal to a point on a surface S is $\partial \boldsymbol{r}/\mathrm{d}\phi \times \partial \boldsymbol{r}/\mathrm{d}\psi$ (Fig. 1.8).

The **unit normal** $\boldsymbol{n}$ to a point on a surface S is the vector perpendicular to S and having unit length (Fig. 1.8):

$$\boldsymbol{n} = \frac{\dfrac{\partial \boldsymbol{r}}{\partial \phi} \times \dfrac{\partial \boldsymbol{r}}{\partial \psi}}{\left| \dfrac{\partial \boldsymbol{r}}{\partial \phi} \times \dfrac{\partial \boldsymbol{r}}{\partial \psi} \right|}$$

Example

A surface is given by the equation $2x + 3y + z = 0$. When taking $x = \phi$, $y = \psi$, it follows that $z = -2\phi - 3\psi$. The parameter equation of the surface is then

$$\begin{aligned} \boldsymbol{r}(\phi, \psi) &= \boldsymbol{i}\phi + \boldsymbol{j}\psi - \boldsymbol{k}(2\phi + 3\psi) \\ &= (\phi, \psi, -2\phi - 3\psi) \end{aligned}$$

$$\left.\begin{aligned} \frac{\partial \boldsymbol{r}}{\partial \phi} &= (1, 0, -2) \\ \frac{\partial \boldsymbol{r}}{\partial \psi} &= (0, 1, -3) \end{aligned}\right\} \frac{\partial \boldsymbol{r}}{\partial \phi} \times \frac{\partial \boldsymbol{r}}{\partial \psi} = (2, 3, 1)$$

so that

$$\boldsymbol{n} = \frac{1}{\sqrt{14}} (2, 3, 1)$$

1.4 INTEGRATION; CIRCULATION; FLUX

If $\boldsymbol{r}(\phi)$ is a vector depending on a scalar ϕ, then the integral of $\boldsymbol{r}(\phi)$ is a vector $\int \boldsymbol{r}(\phi)\, \mathrm{d}(\phi)$ located at the point $(\int r_1(\phi)\, \mathrm{d}\phi, \int r_2(\phi)\, \mathrm{d}\phi, \int r_3(\phi)\, \mathrm{d}\phi)$.

If s is a space curve defined by the position vector $\boldsymbol{r}(\phi)$, located in a vector field $\boldsymbol{a}(x, y, z)$, then the tangential component of $\boldsymbol{a}$ along s is the **line integral**. In symbols:

$$\begin{aligned} \int_{P_1}^{P_2} a_t\, \mathrm{d}r &= \int_{P_1}^{P_2} (\boldsymbol{a} \cdot \boldsymbol{t})\, \mathrm{d}r \\ &= \int_{P_1}^{P_2} \boldsymbol{a} \cdot \mathrm{d}\boldsymbol{r} \\ &= \int_{P_1}^{P_2} (a_x\, \mathrm{d}x + a_y\, \mathrm{d}y + a_z\, \mathrm{d}z) \end{aligned}$$

(Fig. 1.9).

Example

If $\boldsymbol{a}$ is defined as $\boldsymbol{a} = \boldsymbol{i}(3x^2 + 6y) - \boldsymbol{j}14yz + \boldsymbol{k}20xz^2$ and s as $x = \phi$, $y = \phi^2$, $z = \phi^3$ joining the points (0, 0, 0) and (1, 1, 1), then

$$\frac{\mathrm{d}\boldsymbol{r}}{\mathrm{d}\phi} = \boldsymbol{i} + \boldsymbol{j}2\phi + \boldsymbol{k}3\phi^2 \quad \text{or} \quad \mathrm{d}\boldsymbol{r} = (\boldsymbol{i} + \boldsymbol{j}2\phi + \boldsymbol{k}3\phi^2)\, \mathrm{d}\phi$$

Along s we have $\boldsymbol{a} = \boldsymbol{i}9\phi^2 - \boldsymbol{j}14\phi^5 + \boldsymbol{k}20\phi^7$ so that

$$\begin{aligned} \boldsymbol{a} \cdot \mathrm{d}\boldsymbol{r} &= (\boldsymbol{i}9\phi^2 - \boldsymbol{j}14\phi^5 + \boldsymbol{k}20\phi^7) \cdot (\boldsymbol{i} + \boldsymbol{j}2\phi + \boldsymbol{k}3\phi^2)\, \mathrm{d}\phi \\ &= (9\phi^2 - 28\phi^6 + 60\phi^9)\, \mathrm{d}\phi \end{aligned}$$

Finally

$$\int_s \boldsymbol{a} \cdot \mathrm{d}\boldsymbol{r} = \int_0^1 (9\phi^2 - 28\phi^6 + 60\phi^9)\, \mathrm{d}\phi = 5$$

If s is a closed curve the integral along s is denoted as

$$C = \oint_s \boldsymbol{a} \cdot \mathrm{d}\boldsymbol{r}$$

The integral is called the **circulation** of $\boldsymbol{a}$ around s. If, for example, $\boldsymbol{a}$ is a force, C is the work done in moving around s.

If a surface S and a vector field $\boldsymbol{a}(x, y, z)$ are given, the **surface integral** of $\boldsymbol{a}$ over S is

$$\iint_S \boldsymbol{a} \cdot \mathrm{d}\boldsymbol{S} = \iiint_S (\boldsymbol{a} \cdot \boldsymbol{n})\, \mathrm{d}S$$

where $\boldsymbol{n}$ is the unit normal to a point on $\mathrm{d}S$ and $\mathrm{d}\boldsymbol{S}$ is a vector with magnitude $\mathrm{d}S$ and whose direction is $\boldsymbol{n}$. Thus $\boldsymbol{a} \cdot \boldsymbol{n}$ is the magnitude of a normal to surface S (Fig. 1.10). This integral is called the **flux** of $\boldsymbol{a}$ over S.

If S is a closed surface the flux is denoted as

$$\psi = \oiint_S \boldsymbol{a} \cdot \mathrm{d}\boldsymbol{S} = \oiint_S (\boldsymbol{a} \cdot \boldsymbol{n})\, \mathrm{d}S$$

If $\phi(x, y, z)$ is a scalar field and v a volume, the **volume integral** of ϕ over v is

$$\iiint_v \phi(x, y, z)\, \mathrm{d}v$$

If $\boldsymbol{a}(x, y, z)$ is a vector field and v a volume, the volume integral of $\boldsymbol{a}$ over v is

$$\iiint_v \boldsymbol{a}(x, y, z)\, \mathrm{d}v$$

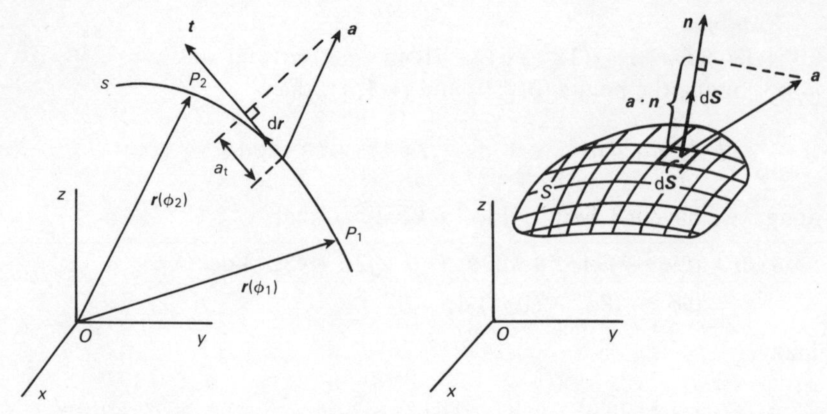

Fig. 1.9 Illustration of the line integral

Fig. 1.10 The normal to a surface S

which can be resolved into its components

$$\boldsymbol{i} \iiint_v a_x \, \mathrm{d}v + \boldsymbol{j} \iiint_v a_y \, \mathrm{d}v + \boldsymbol{k} \iiint_v a_z \, \mathrm{d}v$$

1.5 OPERATORS

In differentiating vector quantities it is customary to use the operators

$$\boldsymbol{\nabla} = \boldsymbol{i} \frac{\partial}{\partial x} + \boldsymbol{j} \frac{\partial}{\partial y} + \boldsymbol{k} \frac{\partial}{\partial z} \qquad \text{(nabla operator, also called 'del')}$$

$$\Delta = \boldsymbol{\nabla} \cdot \boldsymbol{\nabla} = \nabla^2$$

$$= \frac{\partial^2}{\partial x^2} + \frac{\partial^2}{\partial y^2} + \frac{\partial^2}{\partial z^2} \qquad \text{(Laplacian operator)}$$

1.6 GRADIENT

If $\phi(x, y, z)$ is a scalar field, the **gradient** of ϕ is defined as

$$\boldsymbol{\nabla} \phi = \text{grad } \phi$$

$$= \boldsymbol{i} \frac{\partial \phi}{\partial x} + \boldsymbol{j} \frac{\partial \phi}{\partial y} + \boldsymbol{k} \frac{\partial \phi}{\partial z}$$

Example

If ϕ is defined by $\phi = 3x^2y - y^3z^2$ then

$$\nabla\phi = \boldsymbol{i}\frac{\partial}{\partial x}(3x^2y - y^3z^2) + \boldsymbol{j}\frac{\partial}{\partial y}(3x^2y - y^3z^2) + \boldsymbol{k}\frac{\partial}{\partial z}(3x^2y - y^3z^2)$$

$$= \boldsymbol{i}(6xy) + \boldsymbol{j}(3x^2 - 3y^2z^2) + \boldsymbol{k}(-2y^3z)$$

At a point $P(1, -2, -1)$ the gradient is $(\nabla\phi)_P = \boldsymbol{i}(-12) + \boldsymbol{j}(-9) + \boldsymbol{k}(-16)$ and its magnitude $|\nabla\phi|_P = \sqrt{(12^2 + 9^2 + 16^2)} \simeq 22$.

The gradient of a scalar field is a vector field which describes direction, sense and magnitude of the rate of change of the scalar field.

Formulas involving gradients

$$\nabla(\phi_1 + \phi_2) = \nabla\phi_1 + \nabla\phi_2$$

$$\nabla(\lambda\phi) = \lambda\nabla\phi \quad (\lambda = \text{constant})$$

$$\nabla(\phi_1\phi_2) = \phi_1\nabla\phi_2 + \phi_2\nabla\phi_1$$

1.7 DIRECTIONAL DERIVATIVE OF A SCALAR

The directional derivative of a scalar ϕ in the direction of a vector $\boldsymbol{a}$ is defined as $\nabla\phi\cdot\boldsymbol{a}$ (Fig. 1.11). This is the rate of change of ϕ in the direction $\boldsymbol{a}$.

Example

If a scalar field is defined by $\phi(x, y, z) = x^2yz + 4xz^2$ and the direction of $\boldsymbol{a}$ is $2\boldsymbol{i} - \boldsymbol{j} - 2\boldsymbol{k}$, we find the directional derivative in a point $P(1, -2, -1)$ as follows:

$$\nabla\phi = \boldsymbol{i}(2xyz + 4z^2) + \boldsymbol{j}(x^2z) + \boldsymbol{k}(x^2y + 8xz)$$

$$(\nabla\phi)_P = 8\boldsymbol{i} - \boldsymbol{j} - 10\boldsymbol{k}$$

A unit vector in the given direction is

$$\frac{2\boldsymbol{i} - \boldsymbol{j} - 2\boldsymbol{k}}{\sqrt{2^2 + 1^2 + 2^2}} = \tfrac{2}{3}\boldsymbol{i} - \tfrac{1}{3}\boldsymbol{j} - \tfrac{2}{3}\boldsymbol{k}$$

The directional derivative is therefore

$$(8\boldsymbol{i} - \boldsymbol{j} - 10\boldsymbol{k})\cdot(\tfrac{2}{3}\boldsymbol{i} - \tfrac{1}{3}\boldsymbol{j} - \tfrac{2}{3}\boldsymbol{k}) = \tfrac{37}{3} = 12.3$$

The line integral $\int_P^Q \nabla\phi\cdot d\boldsymbol{r}$ is independent of the path joining P and Q. Its value is solely determined by the location of the points P and Q.

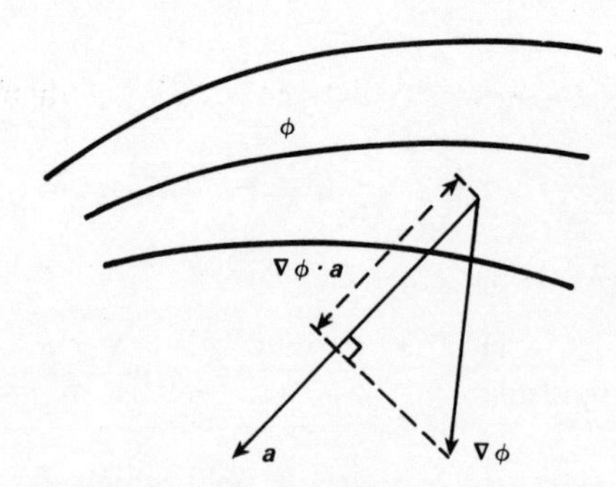

Fig. 1.11 The directional derivative of a scalar

If the path is an arbitrary closed curve s, the integral

$$\oint_s \nabla \phi \cdot \mathrm{d}\boldsymbol{r} = 0$$

1.8 DIVERGENCE

The **divergence** of a vector field $\boldsymbol{a}(x, y, z)$ is defined as

$$\begin{aligned}\nabla \cdot \boldsymbol{a} &= \operatorname{div} \boldsymbol{a} \\ &= \frac{\partial a_x}{\partial x} + \frac{\partial a_y}{\partial y} + \frac{\partial a_z}{\partial z}\end{aligned}$$

A vector field whose divergence is zero is called **solenoidal** or **sourceless.** Divergence describes the presence or absence of sources in a certain region of a vector field.

Formulas involving divergence

$$\nabla \cdot (\boldsymbol{a} + \boldsymbol{b}) = \nabla \cdot \boldsymbol{a} + \nabla \cdot \boldsymbol{b}$$

$$\nabla \cdot (\lambda \boldsymbol{a}) = \lambda(\nabla \cdot \boldsymbol{a}) \quad (\lambda = \text{constant})$$

$$\nabla \cdot (\phi \boldsymbol{a}) = (\nabla \phi) \cdot \boldsymbol{a} + \phi(\nabla \cdot \boldsymbol{a}) \quad (\phi = \text{scalar field})$$

Example 1

If $\boldsymbol{a}$ is defined by $\boldsymbol{a} = \boldsymbol{i}x^2z - \boldsymbol{j}2y^3z^2 + \boldsymbol{k}xy^2z$ then

$$\nabla \cdot \boldsymbol{a} = 2xz - 6y^2z^2 + xy^2$$

In a point $P(1, -1, 1)$ we have $(\nabla \cdot \boldsymbol{a})_P = -3$.

Example 2

If $\boldsymbol{a}$ is the vector field $\nabla \phi$ where $\phi(x, y, z) = 2x^3y^2z^4$ then

$$\nabla \cdot (\nabla \cdot \phi) = \Delta\phi = \frac{\partial^2\phi}{\partial x^2} + \frac{\partial^2\phi}{\partial y^2} + \frac{\partial^2\phi}{\partial z^2}$$

$$= 12xy^2z^4 + 4x^3z^4 + 24x^3y^2z^2$$

Example 3

If $\boldsymbol{a}$ is defined by $\boldsymbol{a} = \boldsymbol{i}x^2z - \boldsymbol{j}2y^3z^2 + \boldsymbol{k}xy^2z$ then

$$\nabla \cdot \boldsymbol{a} = 2xz - 6y^2z^2 + xy^2$$

Example 4

If $\boldsymbol{a}$ is defined by $\boldsymbol{a} = \boldsymbol{i}y^2z + \boldsymbol{j}x$ then

$$\nabla \cdot \boldsymbol{a} = 0$$

Thus $\boldsymbol{a}$ is a sourceless vector field.

1.9 ROTATION

The rotation of a vector field $\boldsymbol{a}$ is defined as

$$\nabla \times \boldsymbol{a} = \text{rot } \boldsymbol{a} = \text{curl } \boldsymbol{a} = \begin{vmatrix} \boldsymbol{i} & \boldsymbol{j} & \boldsymbol{k} \\ \frac{\partial}{\partial x} & \frac{\partial}{\partial y} & \frac{\partial}{\partial z} \\ a_x & a_y & a_z \end{vmatrix}$$

$$= \boldsymbol{i}\left(\frac{\partial a_z}{\partial y} - \frac{\partial a_y}{\partial z}\right) + \boldsymbol{j}\left(\frac{\partial a_x}{\partial z} - \frac{\partial a_z}{\partial x}\right) + \boldsymbol{k}\left(\frac{\partial a_y}{\partial x} - \frac{\partial a_x}{\partial y}\right)$$

Rotation describes the rotational properties of the field $\boldsymbol{a}$ (e.g. moving fluid). If $\nabla \times \boldsymbol{a} = 0$, the field is called a **conservative** or **irrotational** vector field.

A conservative vector field can be derived from a scalar field: $\boldsymbol{a} = \nabla \phi$ and the scalar field is called the **scalar potential.**

Conversely, if $\boldsymbol{a} = \nabla \phi$, then $\nabla \times \boldsymbol{a} = 0$.

Example 1

If $\boldsymbol{a}$ is defined by $\boldsymbol{a} = \boldsymbol{i}x^2y - \boldsymbol{j}2xz + \boldsymbol{k}2yz$ then

$$\nabla \times \boldsymbol{a} = \boldsymbol{i}(2x + 2z) - \boldsymbol{k}(x^2 + 2z)$$

This means that the field has curl.

Example 2
If $\boldsymbol{a}$ is defined by $\boldsymbol{a} = \boldsymbol{r}/r^2$ then $\nabla \times \boldsymbol{a} = 0$; the field is conservative.

Formulas involving rotation

$$\nabla \times (\boldsymbol{a} + \boldsymbol{b}) = \nabla \times \boldsymbol{a} + \nabla \times \boldsymbol{b}$$

$$\nabla \times (\lambda \boldsymbol{a}) = \lambda(\nabla \times \boldsymbol{a}) \quad (\lambda = \text{constant})$$

$$\nabla \times (\phi \boldsymbol{a}) = (\nabla \phi) \times \boldsymbol{a} + \phi(\nabla \times \boldsymbol{a}) \quad (\phi = \text{scalar})$$

$$\nabla \times (\nabla \times \boldsymbol{a}) = -\Delta \boldsymbol{a} + \nabla (\nabla \cdot \boldsymbol{a})$$

$$\nabla \cdot (\boldsymbol{a} \times \boldsymbol{b}) = \boldsymbol{b} \cdot (\nabla \times \boldsymbol{a}) - \boldsymbol{a} \cdot (\nabla \times \boldsymbol{b})$$

$$\nabla \times (\boldsymbol{a} \times \boldsymbol{b}) = (\boldsymbol{b} \cdot \nabla)\boldsymbol{a} - \boldsymbol{b} \cdot (\nabla \cdot \boldsymbol{a}) - (\boldsymbol{a} \cdot \nabla)\boldsymbol{b} + \boldsymbol{a} \cdot (\nabla \cdot \boldsymbol{b})$$

$$\nabla \cdot (\boldsymbol{a} \cdot \boldsymbol{b}) = \boldsymbol{b} \times (\nabla \times \boldsymbol{a}) + (\boldsymbol{b} \cdot \nabla)\boldsymbol{a} + \boldsymbol{a} \times (\nabla \times \boldsymbol{b}) + (\boldsymbol{a} \cdot \nabla) \cdot \boldsymbol{b}$$

$$\nabla \cdot (\nabla \phi) = \Delta \phi$$

$$\nabla \times (\nabla \phi) = 0$$

$$\nabla \cdot (\nabla \times \boldsymbol{a}) = 0$$

$$\nabla \times (\nabla \times \boldsymbol{a}) = \nabla \cdot (\nabla \cdot \boldsymbol{a}) - \Delta \boldsymbol{a}$$

$$\nabla \cdot (\phi \boldsymbol{a}) = \phi \nabla \cdot \boldsymbol{a} + \boldsymbol{a} \cdot \nabla \phi$$

1.10 DIVERGENCE THEOREM OF GAUSS (Green's theorem in space)

If v is the volume bounded by a closed surface S and $\boldsymbol{a}$ is a vector field, then

$$\iiint_v \nabla \cdot \boldsymbol{a} \, \mathrm{d}v = \oiint_S \boldsymbol{a} \cdot \boldsymbol{n} \, \mathrm{d}S$$

$$= \oiint_S \boldsymbol{a} \cdot \mathrm{d}\boldsymbol{S}$$

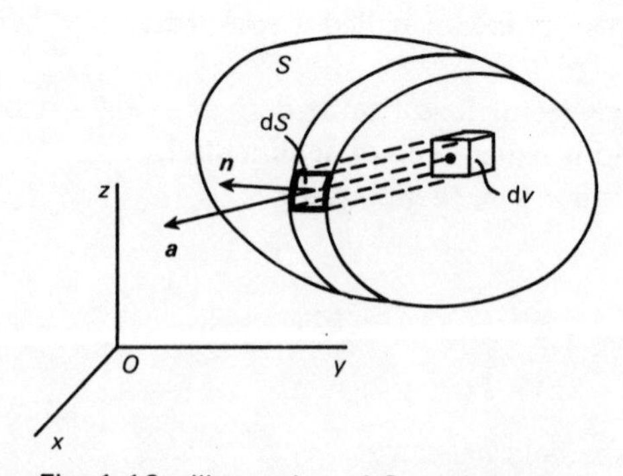

Fig. 1.12 Illustration of Gauss' theorem

(Fig. 1.12). Thus the integral of the divergence of $\boldsymbol{a}$ over v equals the surface integral of the normal component of $\boldsymbol{a}$ over S.

If field $\boldsymbol{a}$ is sourceless, $\nabla \cdot \boldsymbol{a} = 0$ so that $\oint\!\!\oint_S \boldsymbol{a} \cdot \mathrm{d}\boldsymbol{S} = 0$, that is the flux is zero.

1.11 ROTATION THEOREM OF STOKES

If S is a two-sided surface bounded by a closed curve s and $\boldsymbol{a}$ is a vector field, then

$$\oint_s \boldsymbol{a} \cdot \mathrm{d}\boldsymbol{s} = \iint_S (\nabla \times \boldsymbol{a}) \cdot \boldsymbol{n}\, \mathrm{d}S$$

$$= \iint_S (\nabla \times \boldsymbol{a}) \cdot \mathrm{d}\boldsymbol{S}$$

(Fig. 1.13); i.e., the line integral of the tangential component of $\boldsymbol{a}$ around a closed curve s equals the surface integral of the normal component of curl $\boldsymbol{a}$ taken over any surface S having s as its boundary.

Example 1

If $\boldsymbol{a}$ is defined by $\boldsymbol{a} = \boldsymbol{i}(2x - y) - \boldsymbol{j}(yz^2) - \boldsymbol{k}(y^2 z)$ and S is the upper half surface of the sphere $x^2 + y^2 + z^2 = 1$, its boundary s is described by $x^2 + y^2 + z^2 = 1$, $z = 0$. Then

$$\iint_S (\nabla \times \boldsymbol{a}) \cdot \boldsymbol{n}\, \mathrm{d}S = \oint_s \boldsymbol{a} \cdot \mathrm{d}\boldsymbol{s} = \pi$$

Example 2

If $\boldsymbol{a} = \boldsymbol{r}$ and S is the sphere $x^2 + y^2 + z^2 = 1$ then

$$\iint_S \boldsymbol{a} \cdot \boldsymbol{n}\, \mathrm{d}S = 4\pi$$

or

$$\iiint_v \nabla \cdot \boldsymbol{r}\, \mathrm{d}v = \iiint_v 3\, \mathrm{d}v$$

$$= 3\left(\frac{4}{3}\pi\right) = 4\pi$$

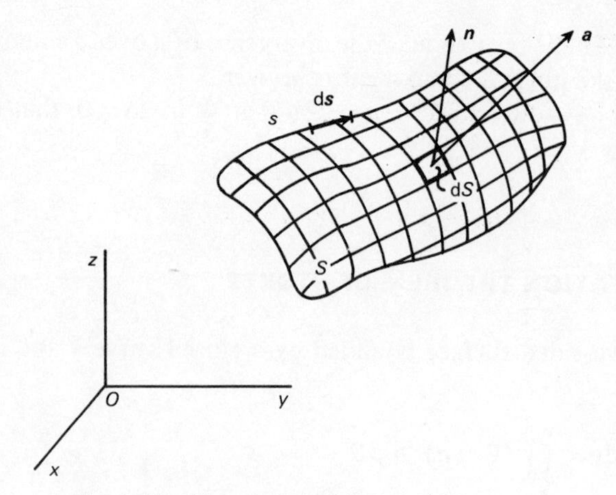

Fig. 1.13 Illustration of Stokes' theorem

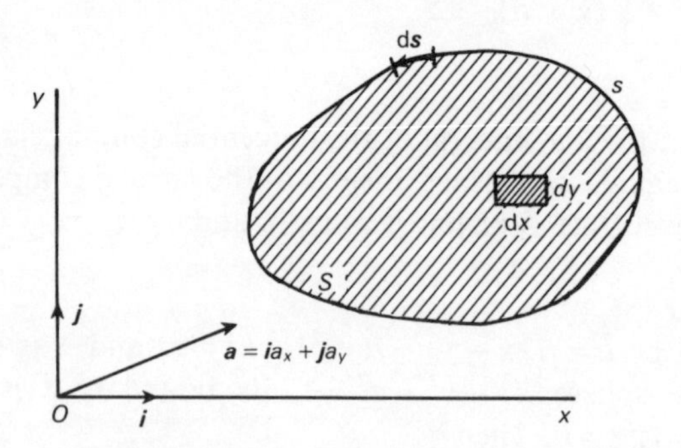

Fig. 1.14 Illustration of Green's theorem in the plane

1.12 GREEN'S THEOREM IN THE PLANE

If S is a closed region in the xy-plane bounded by the closed curve s and $\boldsymbol{a}$ is a vector field in the xy-plane (Fig. 1.14), then

$$\oint_s \boldsymbol{a} \cdot \mathrm{d}\boldsymbol{s} = \iint_S \left(\frac{\partial a_y}{\partial x} - \frac{\partial a_x}{\partial y} \right) \mathrm{d}x\, \mathrm{d}y$$

$$= \iint_S (\boldsymbol{\nabla} \times \boldsymbol{a}) \cdot \boldsymbol{k}\; \mathrm{d}x\, \mathrm{d}y$$

Green's theorem in the plane is a special case of Stokes' theorem while Gauss' theorem is a generalization of Green's theorem. It follows from

Green's theorem:

$$\oint_s \nabla \phi \cdot \mathrm{d}\boldsymbol{s} = 0$$

$$\oint_s \boldsymbol{a} \cdot \mathrm{d}\boldsymbol{s} = 0 \text{ if the vector field } \boldsymbol{a} \text{ is conservative}$$

2 Electricity

2.1 ELECTROSTATIC FORCES

Electrostatics is the study of stationary electric charges. They can be divided into two groups: positive and negative charges. Charges of one group repel each other while attracting members of the other group.

The interaction between charges is described by **Coulomb's law**. This states that the force exerted by two charges Q_1 and Q_2 on each other is (Fig. 2.1)

$$F_{12}(r) = \frac{1}{4\pi\varepsilon} \frac{Q_1 Q_2}{r_{12}^2} \frac{r_{12}}{r_{12}}$$

where

F is the force in newtons

Q_1, Q_2 are the charges in coulombs

r_{12} is the position vector from Q_1 to Q_2 in meters

$\varepsilon = \varepsilon_v \varepsilon_r$ is the **dielectric constant** or **permittivity** of the medium or dielectric

ε_v is the permittivity of vacuum ($\simeq 8.85 \times 10^{-12}$ F m$^{-1} \simeq 10^{-9}/36\pi$)

ε_r is the **relative permittivity** (in air $\varepsilon_r \simeq 1$).

The Coulomb force between two charges is independent of the presence of other charges in the vicinity.

For a system of charges $Q_1, Q_2, \ldots, Q_n$, their force on a point charge Q is (Fig. 2.2)

$$F(r) = \frac{Q}{4\pi\varepsilon} \sum_{i=1}^{n} \frac{Q_i}{r_i^2} \cdot \frac{r_i}{r_i}$$

where $F(r)$ is the vector sum of all forces and r_i is the position vector from Q_i to Q.

Example

Two negative charges Q_1 and Q_2 are located at (0, 0) and (0, 1) (meters), where $Q_1 = 2\ \mu C$, $Q_2 = 5\ \mu C$. Then

$$F = \frac{Q_1 Q_2}{4\pi\varepsilon r^2} j$$

$$= \frac{(-2 \times 10^{-6})(-5 \times 10^{-6})}{4\left(\frac{10^{-9}}{36}\right)1^2} j$$

$$= 0.09j \text{ newtons}$$

2.2 THE ELECTROSTATIC FIELD

The **electric field intensity** or **field strength** E at a point of a field is defined as the force exerted on a charge of 1 coulomb at that point.

The force on a point charge Q at rest in a field E is (Fig. 2.3)

$$F = QE$$

where E is expressed in $\mathrm{V\,m^{-1}}$.

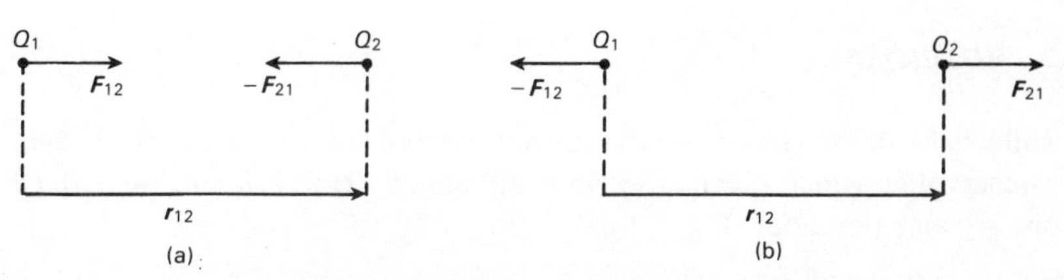

Fig. 2.1 (a) Force between two point charges of opposite polarities (b) Force between two point charges of equal polarities

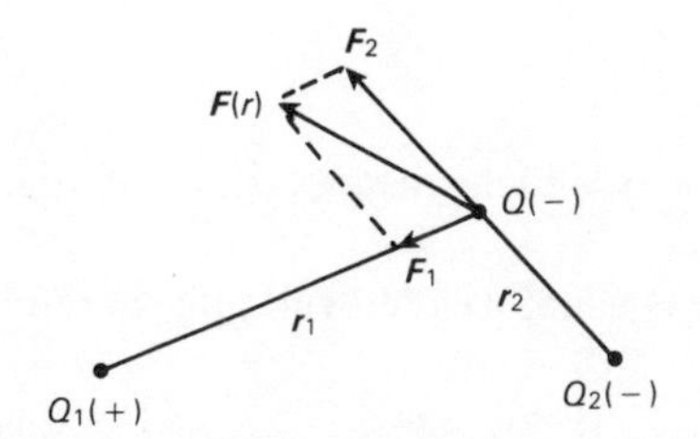

Fig. 2.2 Resultant force of three point charges

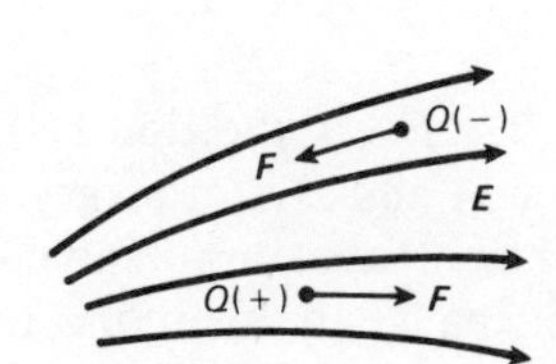

Fig. 2.3 Forces exerted by an electric field on point charges

Example 1
The field strength in a point A due to a point charge Q at a distance r is (Fig. 2.4)

$$E(r) = \frac{1}{4\pi\varepsilon}\frac{Q}{r^2}\frac{r}{r}$$

where r is the position vector from Q to A.

Example 2
The field in a point $A(r)$ due to a number of point charges $Q_1, Q_2, \ldots, Q_n$ is (Fig. 2.5(a))

$$E(r) = \frac{1}{4\pi\varepsilon}\sum_{i=1}^{n}\frac{Q_i(r - r_i)}{|\vec{r} - \vec{r_i}|^3}$$

where r_i is the position vector from Q_i to A.

Field lines are curves whose tangent, at every point, is in the direction of the field in that point. Figure 2.5(b) shows the field lines of a point charge, Figs. 2.5(c) and 2.5(d) show the field lines of two opposite and two positive charges respectively.

2.3 POTENTIAL

It follows from the law of conservation of energy that the electrostatic field is **conservative** which means (Sections 1.9 and 1.12) that it can be derived from a scalar potential (Fig. 2.6)

$$E = -\nabla\phi$$

The **potential** between two points A and B in a field E is defined as the work done when moving a positive charge of 1 coulomb from B to A (Fig. 2.7)

$$\phi_A - \phi_B = \phi_{AB}$$

$$= -\int_B^A E \cdot ds \text{ (volts)}$$

Since $\nabla \times E = 0$ (Section 1.9), $\oint_s E \cdot ds = 0$ which means that the path between A and B is arbitrary.

The potential or potential difference is usually called **voltage**; therefore we will change the symbol ϕ to V.

Example 1
The potential in a point A at a distance r_A from a point charge Q (Fig. 2.8) is

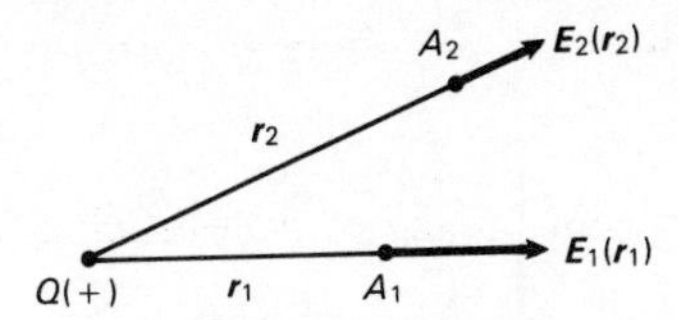

Fig. 2.4 Field vectors at two different points due to a point charge

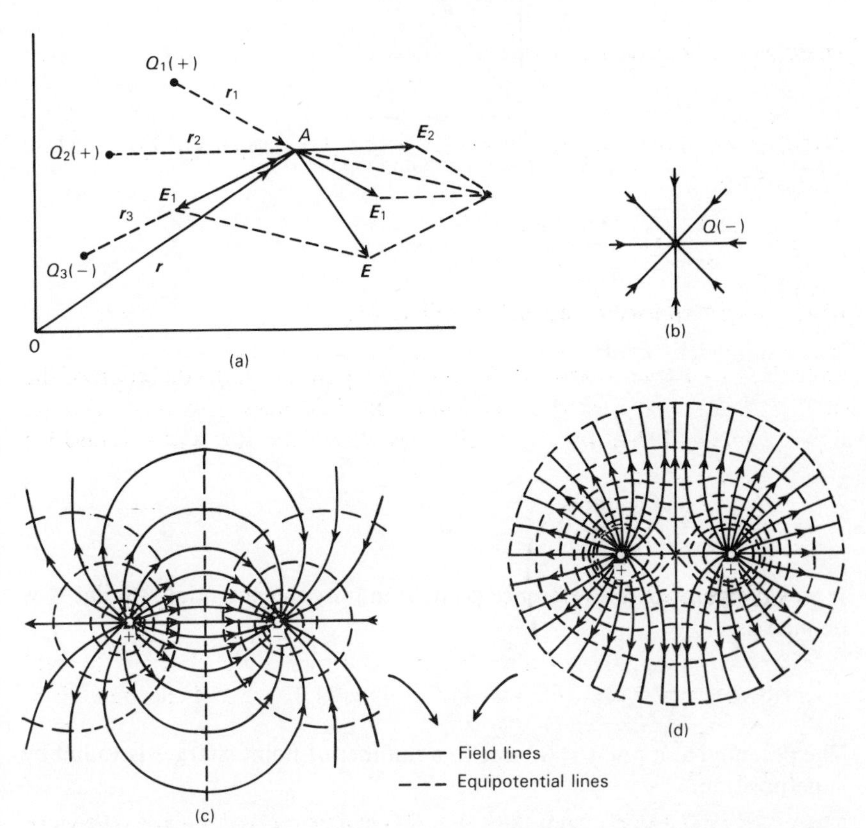

Fig. 2.5 (a) Resulting field vector at A due to a number of point charges (b) Field lines produced by a negative point charge (c) Field lines and equipotential lines due to two equal and opposite charges (d) Field lines and equipotential lines due to two equal positive charges

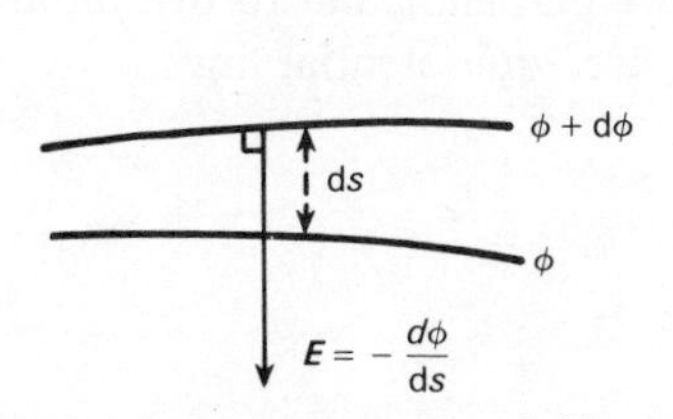

Fig. 2.6 Relation between field vector and scalar potential

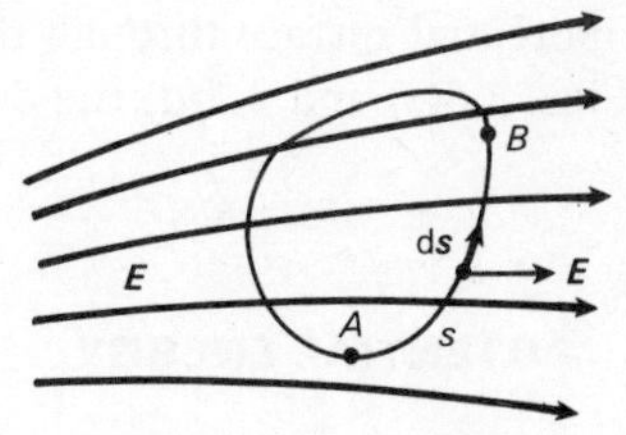

Fig. 2.7 The potential between two points in an electrostatic field

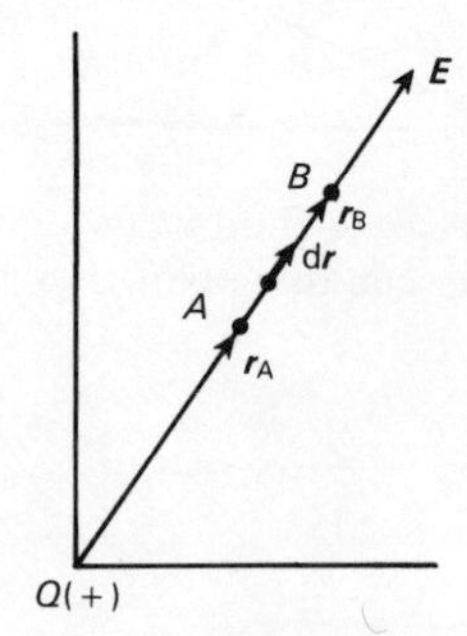

Fig. 2.8 The potential in a point $A(\boldsymbol{r}_A)$

given by

$$V_{AB} = -\int_B^A \boldsymbol{E}\cdot \mathrm{d}\boldsymbol{r}$$

$$= \frac{1}{4\pi\varepsilon}\int_A^B \frac{Q}{r^2}\,\mathrm{d}r$$

$$= \frac{-Q}{4\pi\varepsilon}\left(\frac{1}{r_B} - \frac{1}{r_A}\right)$$

If we take point B as a reference point at infinity the potential at point A is defined as

$$V(\boldsymbol{r}_A) = \frac{1}{4\pi\varepsilon}\frac{Q}{r_A}$$

The potential at a point $A(\boldsymbol{r})$ due to a number of point charges is found by superposition:

$$V(\boldsymbol{r}) = \sum_{i=1}^{n} V_i(\boldsymbol{r}_i)$$

An **equipotential surface** is a surface in an electrostatic field where all points have the same potential. A field line through a point and the equipotential surface through that point are perpendicular to one another. In Figs. 2.5(c) and 2.5(d) the dotted lines are equipotential lines.

2.4 POTENTIAL ENERGY

The work required to bring two charges Q_1 and Q_2 from an infinite distance to a distance r_{12} is

$$W = \frac{1}{4\pi\varepsilon} \int_{\infty}^{r_{12}} \frac{Q_1 Q_2}{r^2} (-\mathrm{d}r)$$

$$= \frac{1}{4\pi\varepsilon} \frac{Q_1 Q_2}{r_{12}}$$

In general, the work required to assemble an arrangement of n charges is

$$W = \frac{1}{2} \frac{1}{4\pi\varepsilon} \sum_{\substack{i,j=1 \\ i \neq j}}^{n} \frac{Q_i Q_j}{r_{ij}}$$

This is the **potential energy** of the system (not to be confused with potential).

2.5 THE DIPOLE

A dipole consists of two equal and opposite charges separated by a distance a (Fig. 2.9).

The potential at a point A due to a dipole at a distance $r \gg a$ (Fig. 2.10) is expressed by

$$V(\mathbf{r}) \simeq \frac{1}{4\pi\varepsilon} \frac{Q\boldsymbol{a} \cdot \boldsymbol{r}}{r^3}$$

$$= \frac{1}{4\pi\varepsilon} \frac{Qa \cos\theta}{r^2}$$

$$= \frac{1}{4\pi\varepsilon} \frac{\boldsymbol{p} \cdot \boldsymbol{r}}{r^3}$$

where $Q\mathbf{a} = \boldsymbol{p}$ is the **electric dipole moment** (direction from $-Q$ to $+Q$).

The field strength at a point $A(\boldsymbol{r})$ due to a dipole $(r \gg a)$ is easily described by its components in spherical coordinates (Figs. 2.11 and 2.12):

$$E_r = -\frac{\partial V}{\partial r} = \frac{1}{4\pi\varepsilon} \frac{2p\cos\theta}{r^3}$$

$$E_\theta = -\frac{\partial V}{\partial s} = -\frac{\partial V}{r\,\mathrm{d}\theta} = \frac{1}{4\pi\varepsilon} \frac{p\sin\theta}{r^3}$$

If a dipole is placed in a uniform electric field $\boldsymbol{E}$ a torque acts on the dipole (Fig. 2.13):

$$\boldsymbol{T} = \boldsymbol{p} \times \boldsymbol{E}$$

If the field is nonuniform an additional force $\boldsymbol{F}$ is exerted on the dipole;

$$\boldsymbol{F} = (\boldsymbol{p} \cdot \nabla) \cdot \boldsymbol{E}$$

The potential energy of the dipole is expressed by the simple formula

$$W_{\mathrm{p}} = -\boldsymbol{p} \cdot \boldsymbol{E}$$

2.6 THE ELECTROSTATIC FIELD IN DIELECTRIC MEDIA

If a dielectric medium is placed in an electric field (between the plates of a charged capacitor) the dipoles of the medium align themselves in the direction of the field, building up a net charge at the boundary (Fig. 2.14). The dipoles neutralize part of the charge, causing a decrease of voltage between the plates. If this voltage is held constant, more charge is transported to the plates. The total surface charge density $\sigma = Q/S = D$ has increased while the field $E = Vd$ is still the same. Thus the ratio $D/E = \varepsilon$ is a property of the medium, called **permittivity**.

$\boldsymbol{D}$ is a vector, called **displacement vector**. It is the charge which is displaced through a unit area normal to the field lines. Also $\boldsymbol{D}$ is the electric field inside the medium. In symbols we have

$$\boldsymbol{D} = \varepsilon \boldsymbol{E}$$

for an isotropic medium. Since $\varepsilon = \varepsilon_{\mathrm{v}}\varepsilon_{\mathrm{r}}$ we can write

$$\boldsymbol{D}_{\mathrm{v}} = \varepsilon_{\mathrm{v}} \boldsymbol{E}$$

for vacuum so that

$$\boldsymbol{D} = \varepsilon_{\mathrm{r}} \boldsymbol{D}_{\mathrm{v}}$$

and the total surface charge density is

$$\sigma = \varepsilon_{\mathrm{r}} \sigma_{\mathrm{v}}$$

where σ_{v} is the surface charge density of the free charges.

The effect of an electric field on a medium is called **polarization**. If we consider (Fig. 2.15) a polarized cylinder of cross section $\mathrm{d}S$, all charges inside the cylinder compensate, leaving two charges $+\mathrm{d}Q$ and $-\mathrm{d}Q$ at the ends. The cylinder can be considered as a dipole with moment $\mathrm{d}Q\boldsymbol{l}$ which is the sum of the molecular dipole moments Σp. The dipole moment per unit volume is thus

$$\frac{\Sigma p}{\mathrm{d}v} = \frac{\mathrm{d}Ql}{\mathrm{d}Sl} = \frac{dQ}{dS} = \sigma_{\mathrm{p}}$$

which is the surface charge density due to polarization. This is denoted as P, the **polarization** or **polarization density**, and it is equal to the dipole

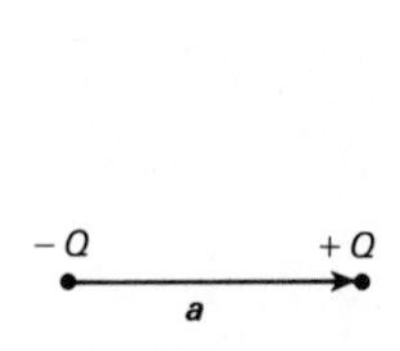

Fig. 2.9 Definition of a dipole

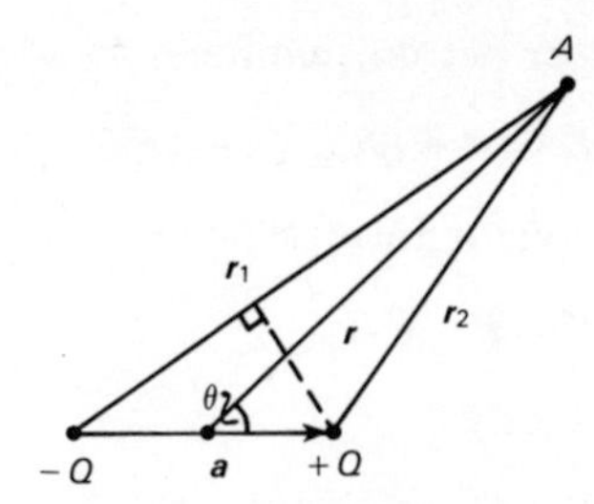

Fig. 2.10 Potential in a point A due to a dipole

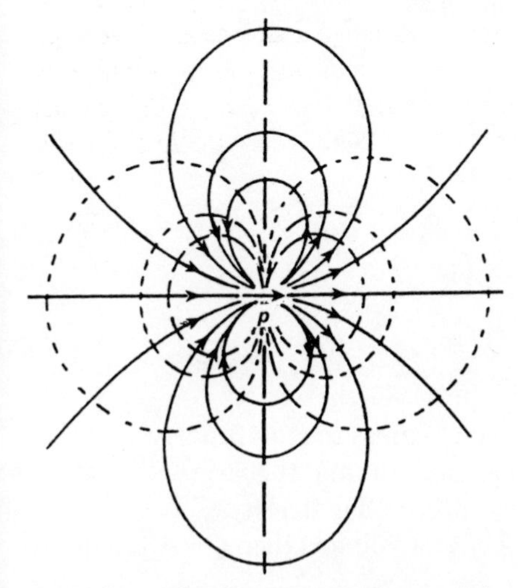

Fig. 2.11 Field lines and equipotential lines due to a dipole

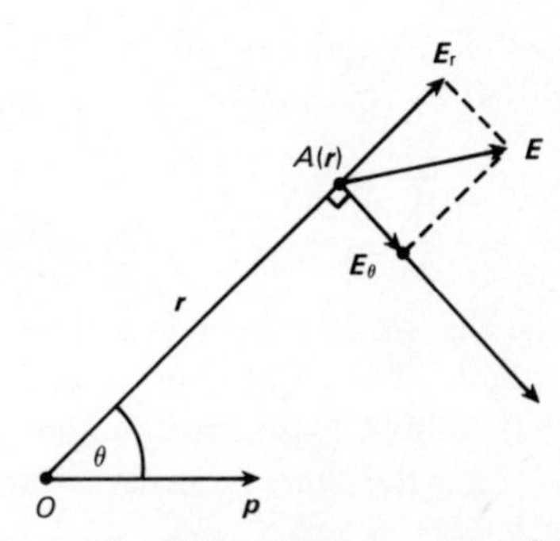

Fig. 2.12 Field vector at a point A due to a dipole

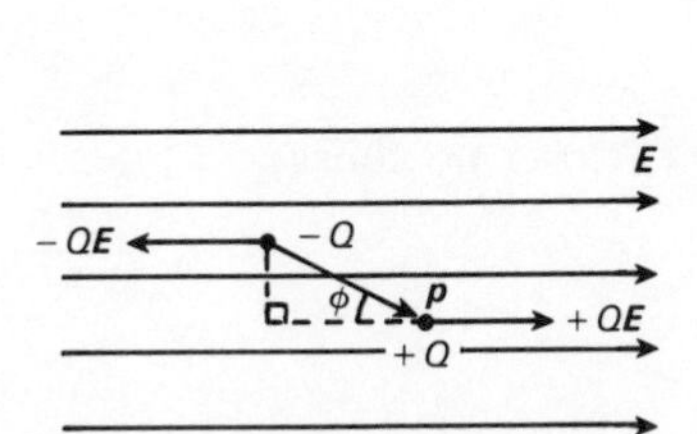

Fig. 2.13 Forces acting on a dipole in a uniform electrostatic field

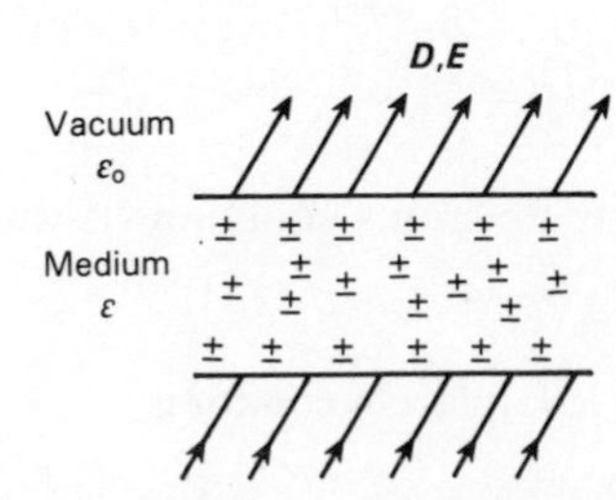

Fig. 2.14 Dipoles in a medium placed in an electrostatic field

moment per unit volume. Thus $\sigma = \sigma_p + \sigma_v$ or

$$D = P + D_v = P + \varepsilon_v E$$

or in vector form

$$\boldsymbol{D} = \boldsymbol{P} + \varepsilon_v \boldsymbol{E}$$

Conversely

$$\boldsymbol{P} = \boldsymbol{D} - \varepsilon_v \boldsymbol{E}$$

so that $\boldsymbol{P}$ is the difference between the electric fields of medium and vacuum. Also

$$\boldsymbol{P} = \varepsilon_v(\varepsilon_r - 1)\boldsymbol{E} = \varepsilon_v \chi \boldsymbol{E}$$

where $\chi = \varepsilon_r - 1$ is the **electric susceptibility** of the medium.

2.7 ELECTRIC FLUX

Gauss' law expresses the flux through a closed surface. If a charge Q is enclosed by this surface the law states that

$$\psi = \oiint_S \boldsymbol{D} \cdot \mathrm{d}\boldsymbol{S} = Q$$

or the electric flux through a closed surface equals the total charge Q where Q contains both free charges and polarization (bound) charges.

It follows from this law that $D = \mathrm{d}\psi/\mathrm{d}S =$ **flux density**.

If ρ is the charge density (charge per unit volume) then $Q = \iiint_v \rho \, \mathrm{d}v$ so that $\nabla \cdot \boldsymbol{D} = \rho$ and

$$\nabla \cdot \boldsymbol{E} = \frac{\rho}{\varepsilon}$$

This is Gauss' law in differential form.

Furthermore $\boldsymbol{E} = -\nabla V$ so that

$$\nabla^2 V = \Delta V = -\frac{\rho}{\varepsilon}$$

This is **Poisson's equation**. If the volume encloses no charge

$$\Delta V = 0$$

This is **Laplace's equation**.

Example 1

If Q is a point charge, the fields $\boldsymbol{D}$ and $\boldsymbol{E}$ at an arbitrary distance r from Q

can be found by placing Q in the center of a sphere with radius r (Fig. 2.16). Then

$$\oiint_S \boldsymbol{D} \cdot \mathrm{d}\boldsymbol{S} = D_r(r) \oiint_S \mathrm{d}S = D_r(r) 4\pi r^2 = Q$$

Thus

$$D_r(r) = \frac{Q}{4\pi r^2} \quad \text{and} \quad E_r(r) = \frac{Q}{4\pi\varepsilon r^2}$$

Example 2

A metal sphere of radius R is charged to a charge Q (Fig. 2.17). Then at a distance

$$r \leqslant R\colon\ D = 0 \text{ and } E = 0$$

$$r > R\colon\ E = \frac{Q}{4\pi\varepsilon r^2}$$

The potentials are

$$r \leqslant R\colon\ V(r) = \frac{Q}{4\pi\varepsilon R} \quad \text{(constant)}$$

$$r > R\colon\ V(r) = \frac{Q}{4\pi\varepsilon r}$$

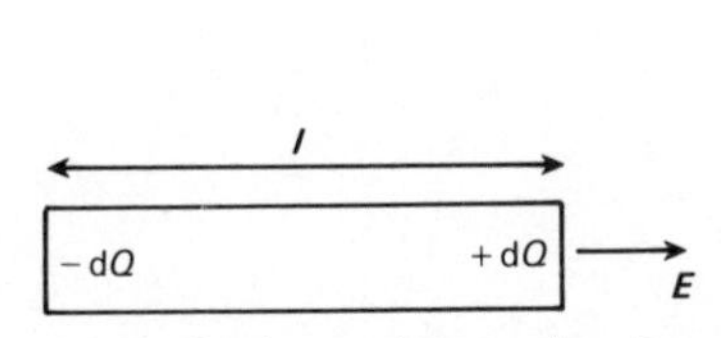

Fig. 2.15 Surface charge density or polarization

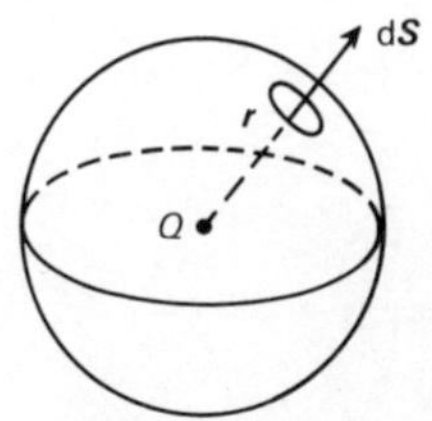

Fig. 2.16 Calculation of **D** and **E** at a point on a sphere around a charge Q

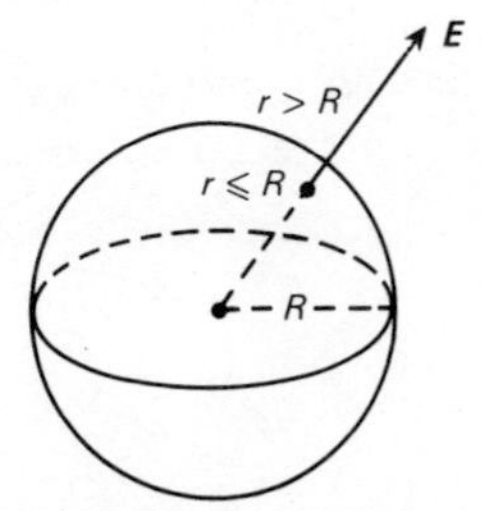

Fig. 2.17 Calculation of **D** and **E** due to a charged sphere

2.8 BOUNDARY BETWEEN CONDUCTOR AND DIELECTRIC

Inside a charged conductor (Fig. 2.18) $\mathbf{E} = \mathbf{D} = 0$ so that the charge is collected at the surface. This is an equipotential surface which means that the field vectors are normal to the surface.

Gauss' law applied to a surface element $\mathrm{d}S$ states that $\mathbf{D} \cdot \mathrm{d}\mathbf{S} = D_n \cdot \mathrm{d}S = \mathrm{d}Q$. If the surface charge density is σ we find by integration (Fig. 2.19)

$$\oiint_S \mathbf{D} \cdot \mathrm{d}\mathbf{S} = \oiint_S D_n \,\mathrm{d}S = Q = \sigma S = D_n S$$

Thus D_n is the surface charge density and the electric field at the surface is

$$E = E_n = \frac{\sigma}{\varepsilon}$$

2.9 BOUNDARY BETWEEN TWO DIELECTRICS

When taking a pillbox of infinitesimal height about a section of the plane separating the two dielectrics (Fig. 2.20(a)) and applying Gauss' law we find that the normal flux densities at the boundary are equal, thus

$$D_{n_1} = D_{n_2}$$

It follows from $\oint_s \mathbf{E} \cdot \mathrm{d}\mathbf{s} = 0$ that the tangential components of $\mathbf{E}$ are also equal:

$$E_{t_1} = E_{t_2}$$

(Fig. 2.20(b)).

2.10 CAPACITANCE

The **capacitance** of two charged conductors is defined as the ratio of the charge on the conductors and the potential difference (voltage) between them:

$$C = \frac{Q}{V}$$

Example 1

Consider the capacitance of a parallel-plate capacitor with two different dielectrics (Fig. 2.21). The $\mathbf{D}$ vector is $\mathbf{D} = D_n = \sigma = \varepsilon E = Q/S$ (Section 2.8).

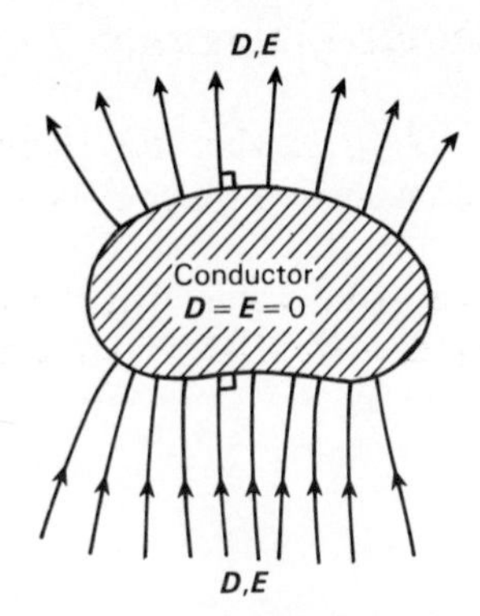

Fig. 2.18 Fields at the boundary of a conductor

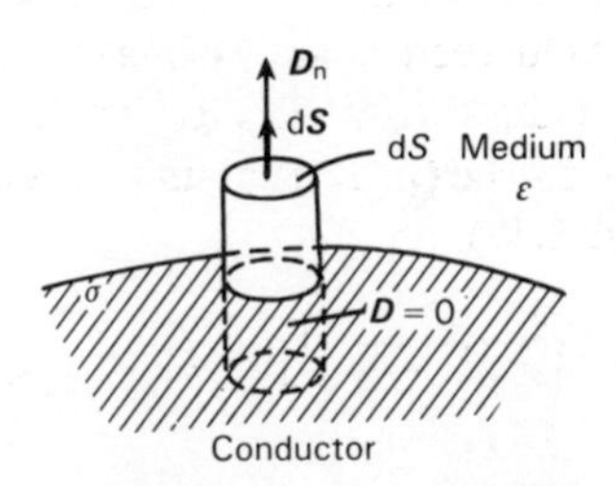

Fig. 2.19 Gauss' law applied to a surface element of a conductor

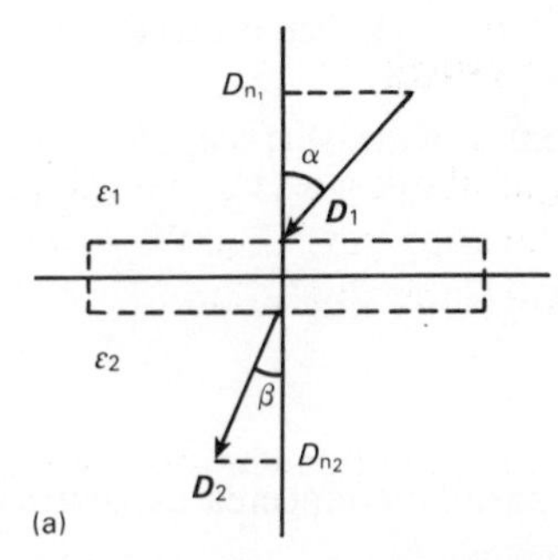

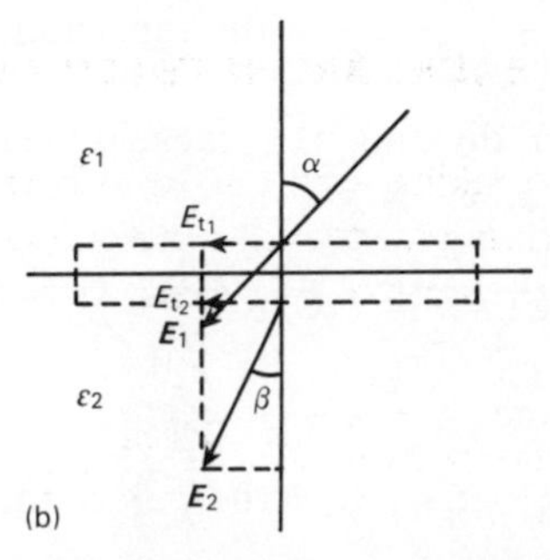

Fig. 2.20 (a) Behavior of $\boldsymbol{D}$ at the boundary of two dielectrics (b) Behavior of $\boldsymbol{E}$ at the boundary of two dielectrics

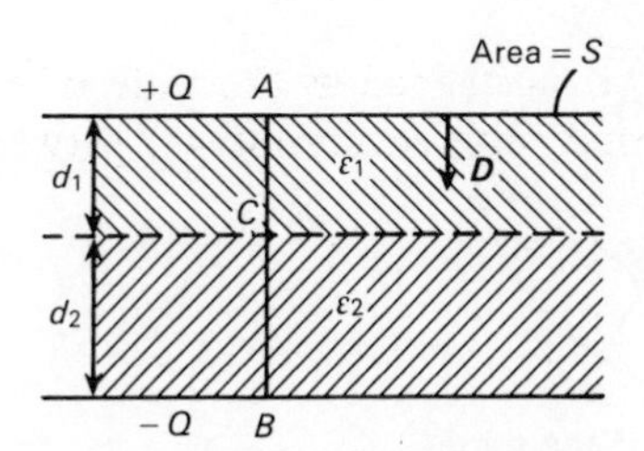

Fig. 2.21 Derivation of the capacitance of a parallel plate capacitor

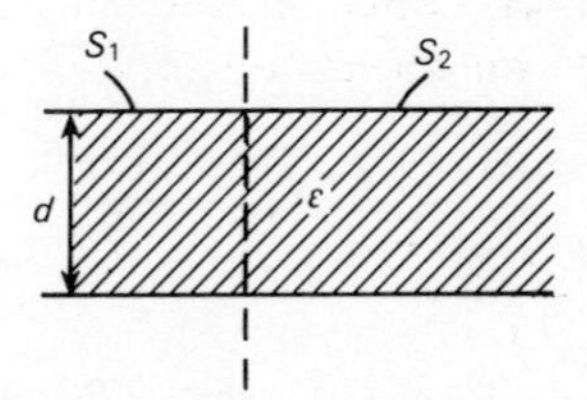

Fig. 2.22 Derivation of the capacitance of two parallel connected capacitors

$$V_{AB} = V_{AC} + V_{CB}$$

$$= \int_A^C E_1 \cdot ds + \int_C^B E_2 \cdot ds$$

$$= \frac{Q}{\varepsilon_1 S} d_1 + \frac{Q}{\varepsilon_2 S} d_2$$

so that

$$C = \frac{S}{\frac{d_1}{\varepsilon_1} + \frac{d_2}{\varepsilon_2}}$$

In the case where $\varepsilon_1 = \varepsilon_2 = \varepsilon$ and $d_1 + d_2 = d$ we have

$$C = \frac{\varepsilon S}{d}$$

the well-known formula for the parallel-plate capacitor. Also

$$\frac{1}{C} = \frac{1}{C_1} + \frac{1}{C_2} = \text{the capacitance of two series connected capacitors}$$

When dividing the capacitor into two parts with surfaces S_1 and S_2 (Fig. 2.22)

$$C_1 = \frac{\varepsilon S_1}{d}, \quad C_2 = \frac{\varepsilon S_2}{d}$$

but since $S = S_1 + S_2$

$$C = C_1 + C_2 = \text{the capacitance of two parallel connected capacitors}$$

Example 2

Consider a spherical capacitor. It follows from Section 2.7 that the

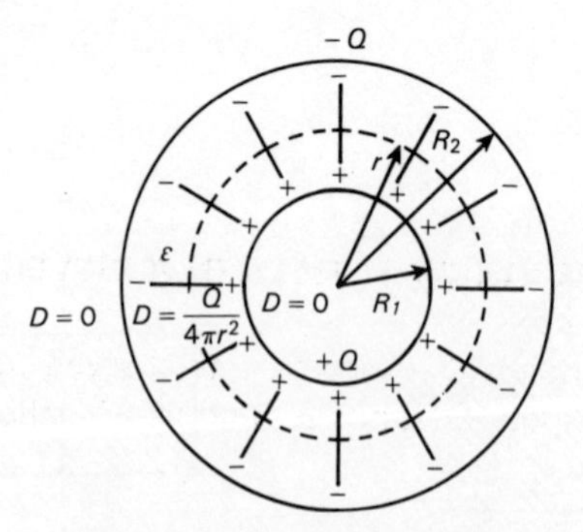

Fig. 2.23 Derivation of the capacitance of two spherical conductors

capacitance of a spherical capacitor with radius R is

$$C = 4\pi\varepsilon R$$

where ε is the permeability of the medium outside the sphere. If we have two concentric spheres with radii R_1 and R_2 and dielectric ε between them (Fig. 2.23)

$$D = \frac{Q}{4\pi r^2} \quad (R_1 < r < R_2)$$

$$V_{AB} = \int_A^B E \cdot \mathrm{d}r$$

$$= \frac{Q}{4\pi\varepsilon} \int_{R_1}^{R_2} \frac{\mathrm{d}r}{r^2}$$

$$= \frac{Q}{4\pi\varepsilon} \left(\frac{1}{R_1} - \frac{1}{R_2} \right)$$

so that

$$C = 4\pi\varepsilon \frac{R_1 R_2}{R_2 - R_1}$$

Example 3

Capacitance of a cylindrical capacitor (Fig. 2.24) is

$$\frac{C}{l} = \frac{2\pi\varepsilon}{\ln \dfrac{R_2}{R_1}}$$

Example 4

Capacitance of two parallel conductors (Fig. 2.25) is

$$\frac{C}{l} = \frac{\pi\varepsilon}{\ln (\mathrm{d}/R)}$$

2.11 ENERGY OF THE ELECTRIC FIELD IN A DIELECTRIC

A parallel-plane capacitor with dielectric ε is charged to a voltage V. A small charge dQ is transported from one conductor to the other. The work required is

$$\mathrm{d}W = V\,\mathrm{d}Q = \frac{Q}{C}\,\mathrm{d}Q$$

To charge the capacitor from the uncharged state to a charge Q requires

$$W = \int_0^Q \frac{Q}{C}\,\mathrm{d}Q = \frac{1}{2}\frac{Q^2}{C} = \tfrac{1}{2}QV = \tfrac{1}{2}CV^2$$

Since $C = \varepsilon Sd$, $E = Vd$ and the volume $v = Sd$, the energy density in the medium is

$$w = \frac{W}{v} = \frac{\frac{1}{2}CV^2}{Sd} = \tfrac{1}{2}\varepsilon E^2 = \tfrac{1}{2}DE \ (\mathrm{J\ m^{-3}})$$

Example

Two parallel plates are separated by a distance d; the field between the plates is $\boldsymbol{E}$. What is the attractive force between the plates?

Since work and force are related by $W = \boldsymbol{F} \cdot \boldsymbol{d}$ we find

$$F = \frac{W}{d} = \frac{Q^2}{2Cd} = \frac{Q^2}{2\varepsilon S} = \frac{\varepsilon E^2 S}{2}$$

2.12 CURRENT

Electric **current** is defined as the flow of charge carriers across a reference surface. If this flow is constant we can write

$$I = \frac{Q}{t}$$

A voltage across a conductor gives rise to an electric field $\boldsymbol{E}$ inside the conductor, causing the current to flow. If this field $\boldsymbol{E}$ is independent of time, the current is called **stationary**.

The **current density** $\boldsymbol{J}$ is the current flowing through a unit area so that the total current is equal to the integral of the current density vector taken over the surface:

$$I = \iint_S \boldsymbol{J} \cdot \mathrm{d}\boldsymbol{S}$$

In general, if the current is not stationary and S is a closed surface, the integral expresses the rate at which charge is leaving the volume enclosed.

If $\iiint_v \rho \, \mathrm{d}v$ is the total charge inside the volume we have

$$\iint_S \boldsymbol{J} \cdot \mathrm{d}\boldsymbol{S} = -\frac{\mathrm{d}}{\mathrm{d}t} \iiint_v \rho \, \mathrm{d}v$$

Applying Gauss' law leads to the equation

$$\operatorname{div} \boldsymbol{J} + \frac{\mathrm{d}\rho}{\mathrm{d}t} = 0$$

This is the **continuity equation**. Thus for a stationary current

$$\operatorname{div} \boldsymbol{J} = 0$$

In homogeneous solids $\boldsymbol{J}$ is proportional to $\boldsymbol{E}$, thus $\boldsymbol{J} = \gamma \boldsymbol{E}$ where γ is the **specific conductivity** of the solid (Fig. 2.26).

If S_1 and S_2 are two surfaces normal to the field $\boldsymbol{E}$ and having potentials V_1 and V_2 respectively, a charge transported from A to B requires work

$$\begin{aligned} V_A - V_B &= V_{AB} \\ &= \int_A^B \boldsymbol{E} \cdot \mathrm{d}\boldsymbol{l} \\ &= \int_A^B E \, \mathrm{d}l \\ &= \int_A^B \frac{J}{\gamma} \, \mathrm{d}l \end{aligned}$$

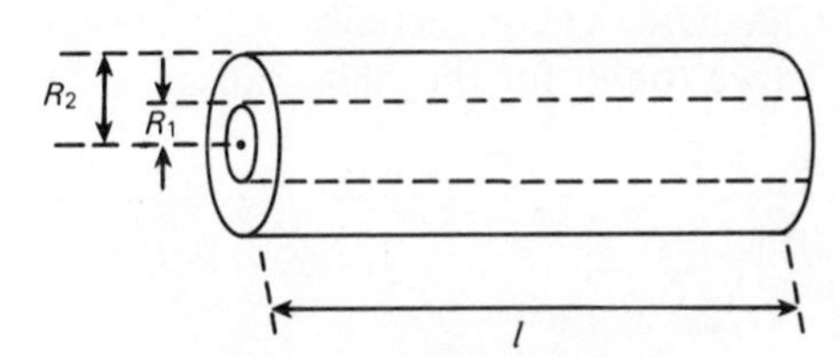

Fig. 2.24 Derivation of the capacitance of a cylindrical capacitor

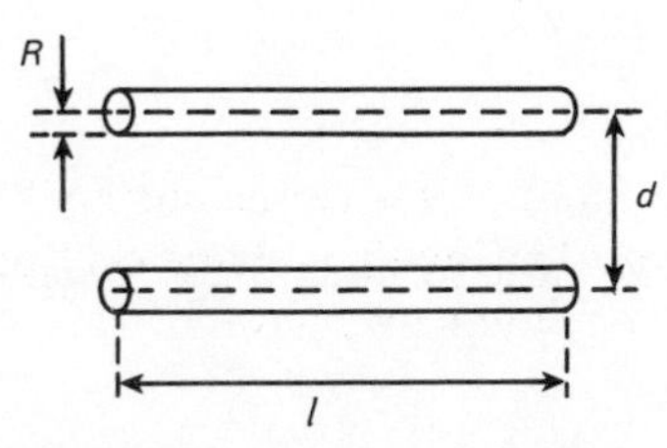

Fig. 2.25 Derivation of the capacitance of two parallel conductors

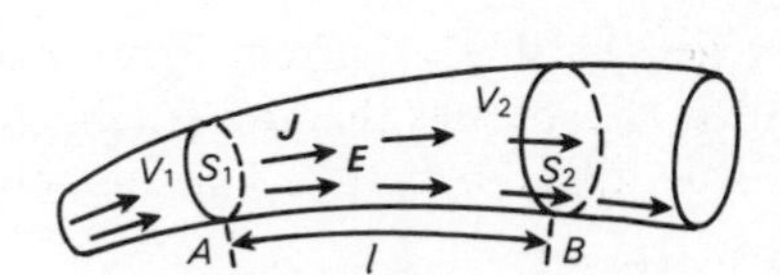

Fig. 2.26 The current density vector $\boldsymbol{J}$

Because $J = I/S$ we have

$$V_{AB} = \frac{I}{\gamma} \int_A^B \frac{dl}{S(l)}$$

If $S(l) = S$ (= constant)

$$V_{AB} = \frac{Il}{\gamma S} \quad \text{or} \quad V_{AB} = IR$$

This is **Ohm's law**. In this equation

$$R = \frac{l}{\gamma S} = \frac{\rho l}{S}$$

is the **resistance** between A_1 and A_2, $\rho = 1/\gamma$ is the **specific resistance** and $G = \gamma Sl$ is the **conductance**.

Example 1

The capacitance of a parallel-plate capacitor is $C = \varepsilon S/d$ so that

$$\frac{C}{G} = \frac{\varepsilon}{\gamma}$$

Example 2

Two concentric spheres with radii R_1 and R_2 have a medium of conductivity γ between them. Calculate the conductance.

In Section 2.10 we found for the capacitance

$$C = 4\pi\varepsilon \frac{R_1 R_2}{R_2 - R_1}$$

so that

$$G = 4\pi\gamma \frac{R_1 R_2}{R_2 - R_1}$$

If a current is not stationary, the momentary current is the charge dQ passing through a reference surface in a time dt:

$$i(t) = \frac{dQ}{dt}$$

When we substitute $Q = CV$ we arrive at the basic formula for a capacitor:

$$i(t) = C\frac{dV}{dt}$$

3
Magnetism

3.1 MAGNETIC INDUCTION

The interaction between electric charges was described by Coulomb, and the interaction between electric currents was described by Ampère who discovered that two parallel wires carrying stationary currents in the same direction are attracted to each other. When reversing the direction of one of the currents the wires repelled each other.

The force acting on a length l of one of the wires (Fig. 3.1) can be expressed by

$$\boldsymbol{F}_{12} = \mu \frac{I_1 I_2 l}{2\pi r} \boldsymbol{l}_{12} \text{ (newton)}$$

where $\boldsymbol{l}_{12}$ is the unit normal in the direction of r. This force is a magnetic force due to moving charges. Their interaction can be described by a magnetic field having a **magnetic induction *B***.

The constant μ is a property of the medium called **permeability**. Similar to permittivity, the permeability of a vacuum is denoted by $\mu_v (4\pi 10^{-7}\ \mathrm{H\,m^{-1}})$ and the **relative permeability** is defined by $\mu = \mu_v \mu_r$ (in air $\mu_r \simeq 1$).

Flux in an electrostatic field is defined as $\psi = \iint_S \boldsymbol{D} \cdot \mathrm{d}\boldsymbol{S}$. Accordingly, the **magnetic flux** Φ is defined as

$$\Phi = \iint_S \boldsymbol{B} \cdot \mathrm{d}\boldsymbol{S} \text{ (weber)}$$

It was found experimentally that if $\boldsymbol{B}$ is due to a stationary current,

$$\iint_S \boldsymbol{B} \cdot \mathrm{d}\boldsymbol{S} = 0$$

Since

$$\iint_S \boldsymbol{B} \cdot \mathrm{d}\boldsymbol{S} = \iiint_v (\boldsymbol{\nabla} \cdot \boldsymbol{B})\, \mathrm{d}v$$

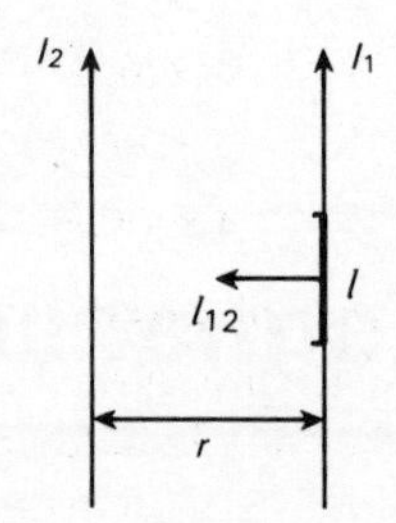

Fig. 3.1 Illustration of Ampère's law

it follows that

$$\nabla \cdot \boldsymbol{B} = 0$$

Thus the magnetic field has no divergence (there are no free magnetic charges) and magnetic flux lines are continuous and closed.

3.2 MAGNETIC FIELD INTENSITY

Since $\boldsymbol{B}$ depends upon the medium, then so do all derived magnetic quantities. Therefore a medium-independent variable $\boldsymbol{H}$ is defined as:

$$\boldsymbol{H} = \frac{\boldsymbol{B}}{\mu} \ (\mathrm{A\ m^{-1}})$$

$\boldsymbol{H}$ is the **magnetic field intensity** or **magnetic field strength.**

An atom can be considered as a small closed circuit. The current through the circuit produces a magnetic dipole field which can be described by a magnetic dipole moment. The sum of all dipole moments per unit volume is called the **magnetization** $\boldsymbol{M}$.

In a similar way to that of Section 2.6 we have

$$\boldsymbol{M} = \boldsymbol{B} - \mu_v \boldsymbol{H}$$

which is the difference of magnetic inductions of medium and vacuum, and

$$\boldsymbol{M} = \mu_v (\mu_r - 1)\boldsymbol{H} = \mu_v \chi_m \boldsymbol{H}$$

where $\chi_m = \mu_r - 1$ is the **magnetic susceptibility** of the medium.

3.3 MAGNETIC PROPERTIES OF MATERIALS

Materials can be divided into three classes: **diamagnetic**, **paramagnetic** and **ferromagnetic**.

$-10^{-6} < \chi_m < 0$ ($\mu_r < 1$) (diamagnetic, e.g. H_2O, Bi, Cu)

$10^{-6} < \chi_m < 10^{-4}$ ($\mu_r > 1$) (paramagnetic, e.g. O_2, Pt)

$\chi_m \gg 1$ ($\mu_r \gg 1$) (ferromagnetic, e.g. Fe, Ni, Co)

An electron revolving in an orbit around the nucleus of an atom causes a magnetic dipole field. If the atom is placed in an external magnetic field the electron will be accelerated or decelerated depending upon the directions of external field and dipole field. The total magnetic field strength will either increase or decrease.

Diamagnetic materials weaken the magnetic field, paramagnetic materials strengthen it. The majority of materials have diamagnetic properties.

In ferromagnetic material large numbers of atoms or molecules are grouped in domains (Weiss domains). In such a domain the dipole moments all point the same way. In an external magnetic field these domains line up and the material is magnetized.

When plotting B or M against H, a curve as shown in Fig. 3.2 is obtained showing that μ_r is not a constant. This is the **hysteresis curve**. For sufficiently large values of H, B remains almost constant; the material is **saturated** and all the domains have lined up in the direction of the applied field.

When H is reduced to zero, B has a nonzero value B_r called **retentivity** (permanent magnetization). To reduce B to zero, a value $H = H_c$ is required, called the **coercive force**.

Magnetic hard materials have values of $\mu_v H_c > 10^4$ A m^{-1}. They are used for permanent magnets.

3.4 AMPÈRE'S CIRCUITAL LAW

It was shown in Section 3.1 that

$$F_{12} = \left(\mu \frac{I_1}{2\pi r}\right) I_2 l$$

The factor $\mu I_1/(2\pi r)$ is the magnetic induction $B = F_{12}/(I_2 l)$ so that $H = I_1/(2\pi r)$.

When we take a current-carrying wire (Fig. 3.3) and calculate the magnetic field strength along a circular path we find

$$\oint_s \boldsymbol{H} \cdot \mathrm{d}s = \frac{B}{\mu} 2\pi r = I$$

In general, the integral of $\boldsymbol{H}$ along a closed curve equals the algebraic sum of all currents encircled by the curve:

$$\oint_s \boldsymbol{H}\cdot d\boldsymbol{s} = \sum_i I_i$$

This is **Ampère's circuital law.**

Since (Section 2.12)

$$I = \iint_S \boldsymbol{J}\cdot d\boldsymbol{S}$$

we have

$$\oint_s \boldsymbol{H}\cdot d\boldsymbol{s} = \iint_S \boldsymbol{J}\cdot d\boldsymbol{S}$$

$$= \iint_S (\nabla \times \boldsymbol{H})\cdot d\boldsymbol{S} \quad \text{(Stokes' theorem)}$$

so that

$$\nabla \times \boldsymbol{H} = \boldsymbol{J}$$

This is Ampère's circuital law in differential form.

In regions outside the current-carrying wire

$$\nabla \times \boldsymbol{H} = 0$$

Then $\boldsymbol{H}$ is derivable as the gradient of a magnetic potential ϕ (Section 1.9):

$$\boldsymbol{H} = -\nabla \phi$$

But $\nabla \cdot \boldsymbol{B} = \nabla \cdot \boldsymbol{H} = 0$ so that $\Delta\phi = 0$ which means that there exist no free magnetic charges.

3.5 BOUNDARY CONDITIONS

At the boundary of two media with different values of μ (Fig. 3.4) we find that the normal component of $\boldsymbol{B}$ is continuous since $\nabla \cdot \boldsymbol{B} = 0$ (Section 3.1). Thus

$$B_{n_1} = B_{n_2}$$

According to Section 3.4: $\nabla \times \boldsymbol{H} = 0$ so that the tangential component of $\boldsymbol{H}$ is continuous:

$$H_{t_1} = H_{t_2}$$

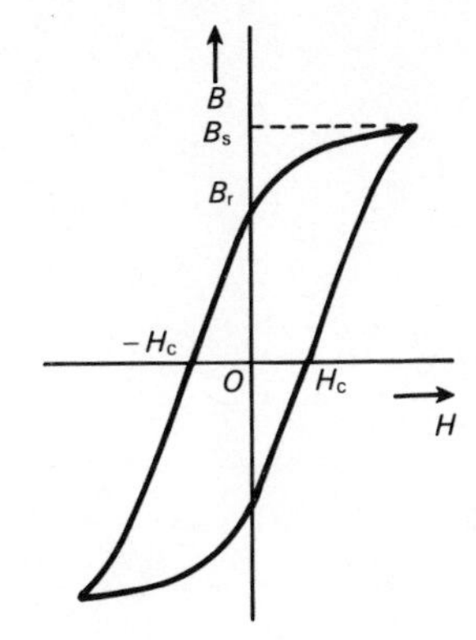

Fig. 3.2 The hysteresis curve

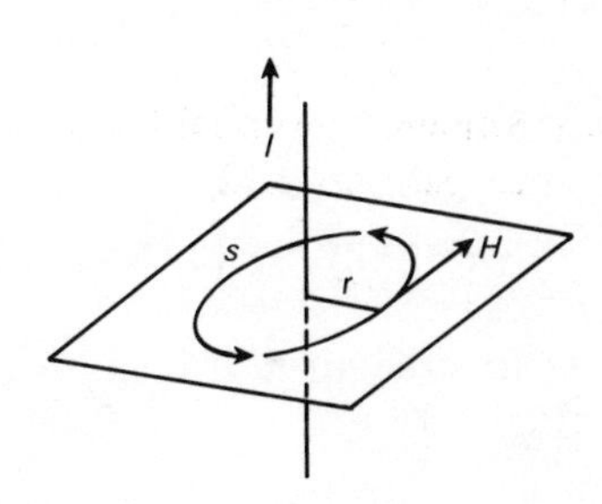

Fig. 3.3 Ampère's circuital law

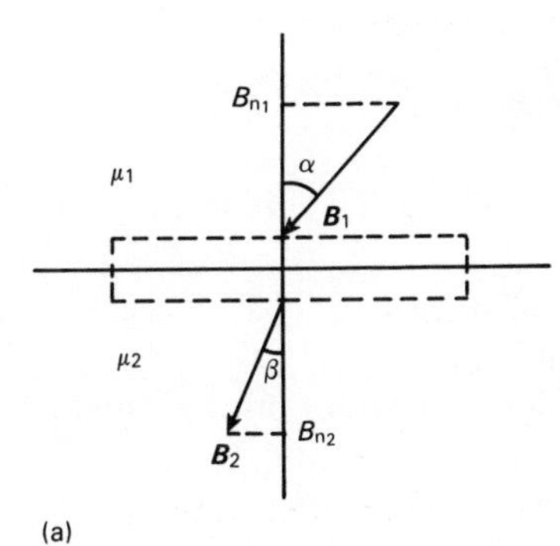

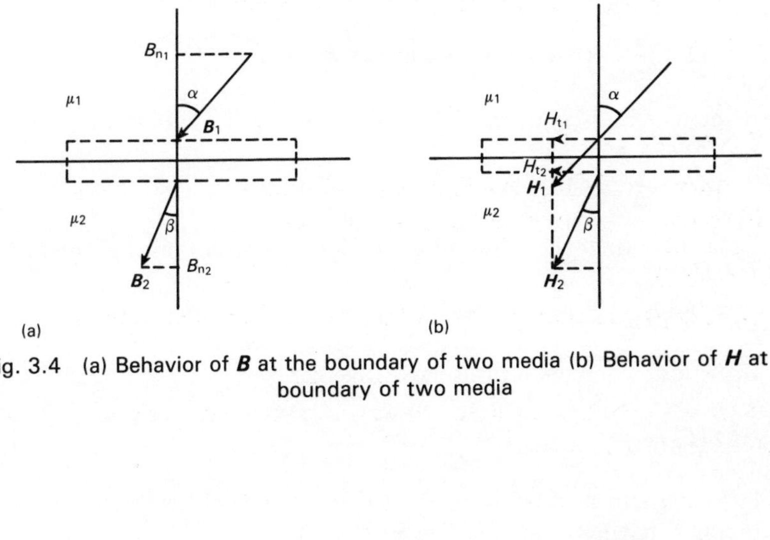

(a)

(b)

Fig. 3.4 (a) Behavior of ***B*** at the boundary of two media (b) Behavior of ***H*** at the boundary of two media

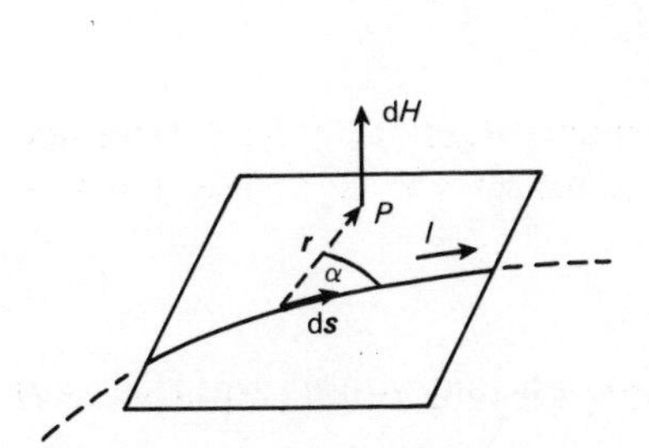

Fig. 3.5 Illustration of Laplace's formula

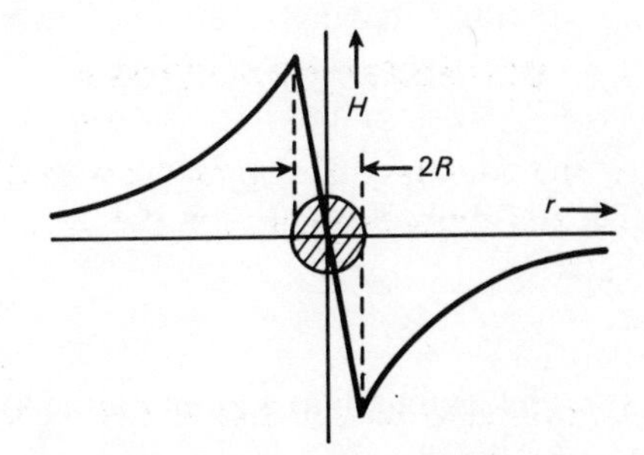

Fig. 3.6 The magnetic field due to a current carrying conductor

From Fig. 3.4 we find

$$\frac{\tan \alpha}{\tan \beta} = \frac{\mu_1}{\mu_2}$$

3.6 EXAMPLES OF MAGNETIC FIELDS

(a) Field strength in a point P due to a conductor carrying a current I (Fig. 3.5):

$$dH = \frac{I}{4\pi r^3} \, ds \times r \quad \text{(Laplace's formula)}$$

(b) Field strength at a distant r from an infinite long straight conductor (Fig. 3.6):

$$H = \frac{I}{2\pi r} \quad (r \geqslant R) \text{ (Biot and Savart law)}$$

If the wire is circular, the field strength inside the wire is

$$H = \frac{Ir}{2\pi R^2} \quad (r \leqslant R)$$

(c) Field strength at the center of a circular loop with radius R (Fig. 3.7):

$$H = \frac{I}{2R}$$

If the loop contains n turns close together,

$$H = \frac{nI}{2R}$$

(d) Field strength at a point on the axis of a solenoid with length l and radius R having n turns per meter (Fig. 3.8):

$$H = \frac{nI}{2l} (\cos \alpha_1 + \cos \alpha_2)$$

In a long solenoid ($l \gg R$)

$$H \simeq \frac{nI}{l}$$

(e) Field strength at a point on the axis of a toroid with n turns (Fig. 3.9):

$$H = \frac{nI}{2\pi R}$$

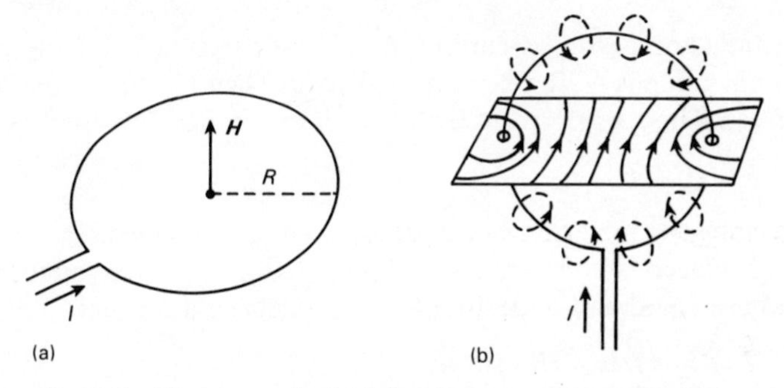

Fig. 3.7 The magnetic field due to a current carrying loop

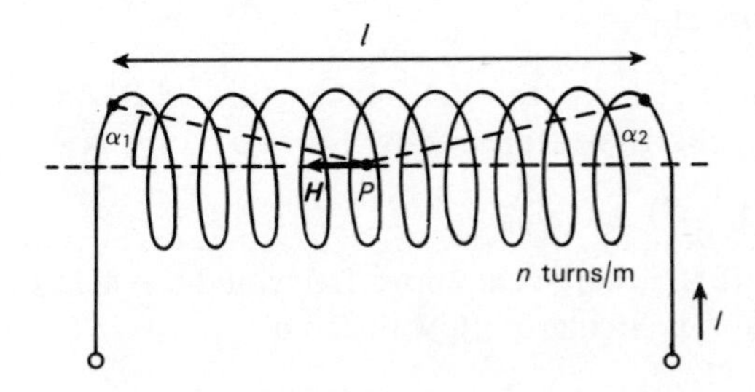

Fig. 3.8 The magnetic field on the axis of a solenoid

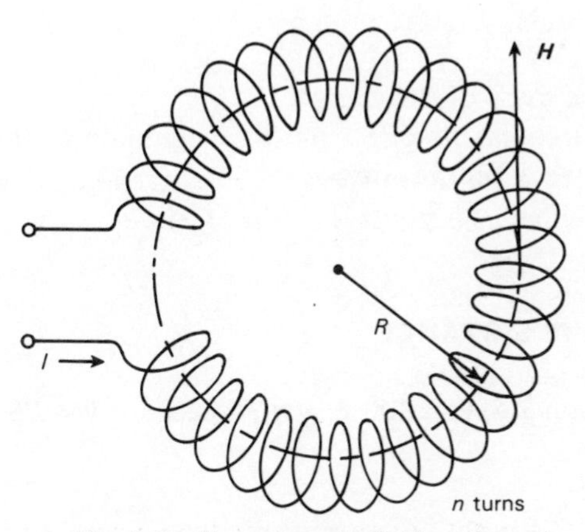

Fig. 3.9 The magnetic field on the axis of a toroid

3.7 FORCES IN MAGNETIC FIELDS

(a) A conductor carries a current I, its cross section is S, the electron concentration is N, the electron velocity $\boldsymbol{v}$. Then $I = NqvS$ (q = electron charge). The force exerted by a field $\boldsymbol{B}$ on a length l is expressed by

$$F = I(\boldsymbol{l} \times \boldsymbol{B})$$

(b) A rectangular loop of wire carries a current I. The loop area is S and the loop is placed in a uniform field $\boldsymbol{B}$ (Fig. 3.10). The forces on the vertical sides are equal and opposite ($F = IhB$) causing a torque:

$$T = Fl \sin \alpha = IBS \sin \alpha$$

The forces on the horizontal sides counteract. If the loop has n turns, the torque is

$$T = nIBS \sin \alpha$$

(c) A small charge Q moves through a magnetic field $\boldsymbol{B}$ with velocity $\boldsymbol{v}$. The force on the charge is expressed by Lorentz' force law:

$$F = Q(\boldsymbol{v} \times \boldsymbol{B})$$

(Fig. 3.11). If the charge can move freely and the field is uniform, the charge follows a circular path with radius

$$R = \frac{\mathrm{m}v}{\mathrm{QB}}$$

and angular velocity

$$\omega = \frac{QB}{m}$$

where m is the mass of the charge.

As the force has no effect on the magnitude of the velocity, the field does no work on the charge.

3.8 MAGNETIC RESISTANCE

According to Section 3.6, Fig. 3.8, $H = nI/l$. Because $\Phi = BS$ and $B = \mu H$ it follows that

$$\Phi = \frac{nI}{l/(\mu S)}$$

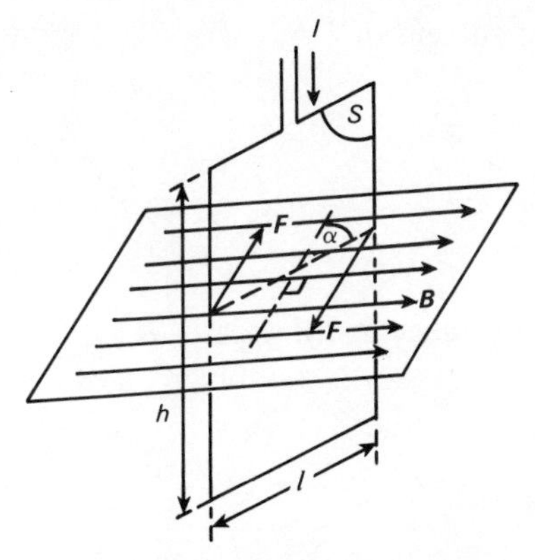

Fig. 3.10 Forces on a current carrying rectangular loop

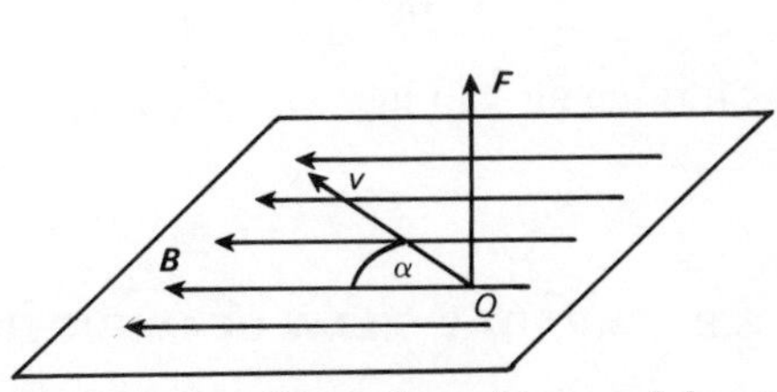

Fig. 3.11 Illustration of Lorentz' force law

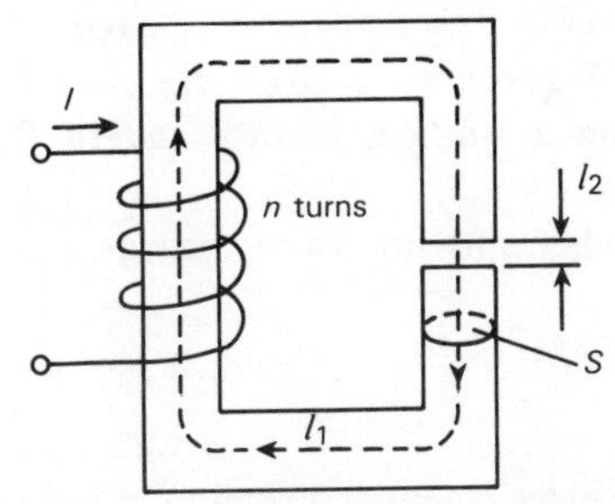

Fig. 3.12 A magnetic circuit with air gap

where

$nI = \Phi R_m$ (This is **Hopkinson's law** or Ohm's law for magnetic circuits)

$R_m = 1/\mu S$ is the **reluctance** or **magnetic resistance**

nI is the magnetomotive force

μ is the **specific magnetic conductivity** (permeability)

Example

The magnetic circuit, as in Fig. 3.12, has an air gap l_2. Assuming $\mu_r = 2000$, $l_1 = 0.5$ m, $l_2 = 2 \times 10^{-3}$ m, $S = 10^{-4}$ m^2, we find

$$
\begin{aligned}
R_m &= \frac{l_1}{\mu_1 S} + \frac{l_2}{\mu_v S} \\
&= 2.5 \times 10^6 + 20 \times 10^6 \\
&= 22.5 \times 10^6 \text{ A Wb}^{-1}
\end{aligned}
$$

If the circuit has 4000 turns and carries a current of 1 A,

$$\Phi = \frac{nI}{R_m} = \frac{4 \times 10^3}{22.5 \times 10^6} \simeq 0.2 \text{ mWb}$$

$$B = \frac{\Phi}{S} = \frac{0.2 \times 10^{-3}}{10^{-4}} = 2 \text{ Wb m}^{-2}$$

Thus the air gap increases the magnetic resistance and decreases B and Φ.

3.9 FARADAY'S LAW OF INDUCTION

In Fig. 3.13 a rectangular loop of wire moves into a uniform field $\boldsymbol{B}$ with velocity $\boldsymbol{v}$, perpendicular to the flux lines.

According to Lorentz' law a force is exerted on electrons in the wire. On a single electron this force is $\boldsymbol{F} = q(\boldsymbol{v} \times \boldsymbol{B})$. As q is negative, the electron moves in the direction as indicated and current flow is in the opposite direction by definition.

The work done when moving a unit charge around the loop is

$$W = \oint_s \boldsymbol{F} \cdot \mathrm{d}\boldsymbol{s} = \boldsymbol{F} \cdot \boldsymbol{l} = qvBl$$

As the electrons move in one direction a voltage between P and Q results. This is the **electromotive force** or **induced voltage** e. It is the work done per unit charge thus

$$e = \frac{W}{q} = vBl \text{ (volts)}$$

In a time $\mathrm{d}t$ the wire moves a distance $v\,\mathrm{d}t$ and the area which has entered the field is $\mathrm{d}S = vl\,\mathrm{d}t = \mathrm{d}\Phi/B$. The flux gained in $\mathrm{d}t$ is $\mathrm{d}\Phi = vBl\,\mathrm{d}t$. If the wire is closed a current will flow, the direction of which is such that it opposes the direction of the original field (Lenz' law). Thus the induced voltage can be written as

$$e = -\frac{\mathrm{d}\Phi}{\mathrm{d}t}$$

This is **Faraday's law of induction.**

In general, the induced voltage and the field $\boldsymbol{E}$ inside the wire are related by the expression

$$e = \oint_s \boldsymbol{E} \cdot \mathrm{d}\boldsymbol{s}$$

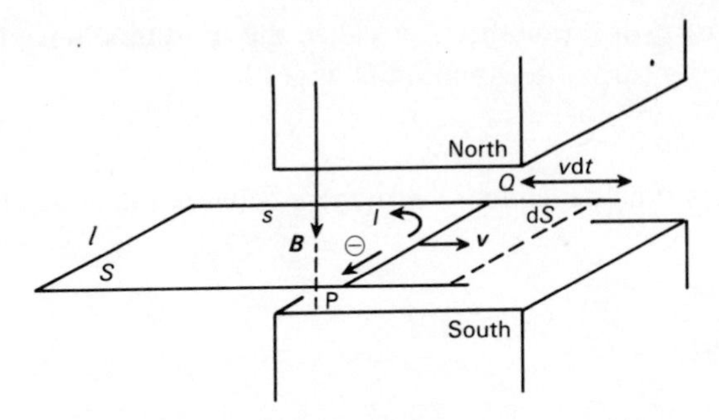

Fig. 3.13 Illustration of Faraday's law of induction

Appling Stokes' theorem:

$$\oint_s \boldsymbol{E} \cdot \mathrm{d}\boldsymbol{s} = \iint_S (\nabla \times \boldsymbol{E}) \cdot \mathrm{d}\boldsymbol{S}$$

so that

$$-\frac{\mathrm{d}\Phi}{\mathrm{d}t} = \iint_S (\nabla \times \boldsymbol{E}) \cdot \mathrm{d}\boldsymbol{s}$$

$$= -\iint_S \frac{\mathrm{d}\boldsymbol{B}}{\mathrm{d}t} \cdot \mathrm{d}\boldsymbol{S}$$

and finally

$$\nabla \times \boldsymbol{E} = -\frac{\mathrm{d}\boldsymbol{B}}{\mathrm{d}t}$$

If in Fig. 3.13 the wire were stationary and the field moving, nothing would change. This leads to the general conclusion: if the flux through a closed loop changes, a current will develop. This current causes a flux which opposes the original flux. Conversely, if a nonstationary current flows through a loop a varying flux results inducing a voltage between the ends of the loop.

3.10 SELF-INDUCTANCE

In an arbitrary loop carrying a stationary current I the flux is proportional to I (see for example Section 3.8):

$$\Phi = LI$$

The constant of proportionality L is called the **self-inductance** of the loop. If the loop has n turns the equation changes to

$$n\Phi = LI$$

If the current is time dependent a voltage e between the ends of the loop is induced:

$$e = -n\frac{d\Phi}{dt}$$

or

$$e = -L\frac{di}{dt}$$

This is the basic formula expressing the behavior of an inductive circuit (e.g. coil). It shows that in such a circuit a large induced voltage may develop when breaking the current abruptly. People have been killed by doing that!

In the example of Section 3.8 we found: $\Phi = nI/R_m$ so that $L = n^2/R_m \simeq 0.7$ henry (H). Thus, in an inductive circuit, L is proportional to the square of the number of turns.

3.11 MUTUAL INDUCTANCE

Figure 3.14 show two loops having n_1 and n_2 turns respectively. A current $i_1(t)$ flows through loop 1 causing a flux Φ_1. Part of this flux, $k\Phi_1$, is linked by loop 2 where it induces a voltage $e_2(t)$. Now

$$\Phi_1 = \frac{L_1 i_1}{n_1}, \quad \Phi_2 = k\Phi_1 = \frac{Mi_1}{n_2}$$

The constant M is the **coefficient of mutual inductance**. Thus

$$k\Phi_1 = \frac{kL_1 i_1}{n_1} = \frac{Mi_1}{n_2}$$

so that

$$M = kL_1\frac{n_2}{n_1}$$

It follows from Section 3.10 that $L_1/L_2 = n_1{}^2/n_2{}^2$ so that

$$M = k\sqrt{L_1 L_2}$$

The factor k is the **coupling factor**, depending upon the mutual position of the loops ($0 < k \leqslant 1$). The induced voltage is then

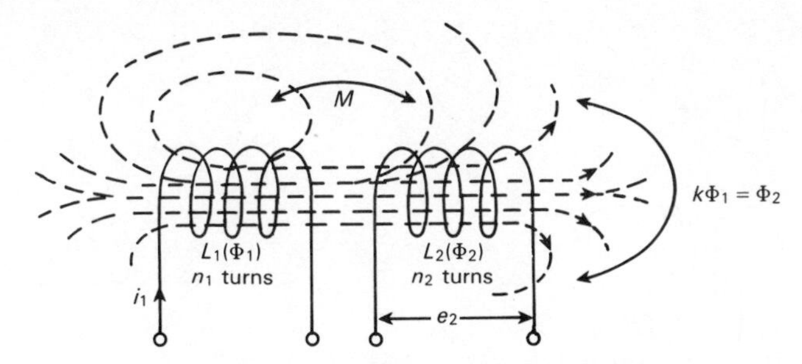

Fig. 3.14 Mutual inductance between two loops

$$e_2(t) = -n_2 \frac{d\Phi_2}{dt}$$

$$= -M \frac{di_1}{dt}$$

3.12 ENERGY

In an inductive circuit work is required to establish a magnetic field. At a time t after energizing the circuit the work done is

$$W = \int_0^t e(t)i(t)\, dt$$

Using $e = L(di/dt)$ this expression becomes

$$W = \int_0^I Li\, di = \tfrac{1}{2} LI^2$$

This is the energy required for establishing the magnetic field in a coil having a self inductance L.

In a coil of length l, cross section S and n turns we have

$$e = L \frac{di}{dt} = n \frac{d\Phi}{dt} = nS \frac{dB}{dt}$$

$$i = \frac{Hl}{n}$$

so that

$$ei = lSH \frac{dB}{dt}$$

and

$$W = \int ei\,\mathrm{d}t$$

$$= lS \int_0^B H\,\mathrm{d}B$$

$$= v \int_0^B H\,\mathrm{d}B$$

where v is the volume of the coil.

The energy density of the magnetic field is thus

$$w = \int_0^B H\,\mathrm{d}B$$

In a medium where μ is constant we find for the energy density:

$$w = \tfrac{1}{2}\mu H^2 = \tfrac{1}{2} BH$$

3.13 ANALOGY BETWEEN ELECTRIC AND MAGNETIC QUANTITIES

The theories of electricity and magnetism show a great many analogies. Therefore, a listing of corresponding quantities is presented in Table 7 for stationary fields.

3.14 THE MAXWELL EQUATIONS

All stationary and time-varying phenomena described in Chapters 2 and 3 can be derived from four fundamental equations: the Maxwell equations. They are listed below in vector form as well as in integral form.

A time-varying electric field or a current causes a magnetic field:

$$\nabla \times \boldsymbol{H} = \boldsymbol{J} + \frac{\partial \boldsymbol{D}}{\partial t} \qquad \oint_s \boldsymbol{H} \cdot \mathrm{d}\boldsymbol{s} = \sum_i I_i + \frac{\partial \psi}{\partial t} = \iint_S \left(\boldsymbol{J} + \frac{\partial \boldsymbol{D}}{\partial t}\right) \cdot \mathrm{d}\boldsymbol{S}$$

A time-varying magnetic field causes an electric field:

$$\nabla \times \boldsymbol{E} = -\frac{\partial \boldsymbol{B}}{\partial t} \qquad \oint_s \boldsymbol{E} \cdot \mathrm{d}\boldsymbol{s} = -\frac{\partial \Phi}{\partial t} = -\iint_S \frac{\partial \boldsymbol{B}}{\partial t} \cdot \mathrm{d}\boldsymbol{S}$$

Table 7

Current field	*Electrostatic field*	*Magnetostatic field*
$J = \gamma E$ (current density)	$D = \varepsilon E$ (displacement)	$B = \mu H$ (magnetic induction)
$I = JS$ (current)	$\psi = Q = DS$ (charge)	$\Phi = BS$ (flux)
$V = El$ (voltage)	$V = El$ (voltage)	$V = Hl$ (voltage)
$I = \dfrac{V}{R}$	$Q = CV$	$\Phi = \dfrac{LI}{n}$
$R = \dfrac{l}{\gamma S}$ (resistance)	$C = \dfrac{\varepsilon S}{l}$ (capacitance)	$L = \dfrac{\mu S}{l} n^2$ (self-inductance)
		$R_m = \dfrac{l}{\mu S} = \dfrac{n^2}{L}$ (magnetic resistance)
$W = I^2 Rt$ (energy)	$W = \frac{1}{2} CV^2$ (energy)	$W = \frac{1}{2} LI^2$ (energy)
$w = \frac{1}{2} JE = \frac{1}{2} \gamma E^2$ (energy density)	$w = \frac{1}{2} DE = \frac{1}{2} \varepsilon E^2$ (energy density)	$w = \frac{1}{2} BH = \frac{1}{2} \mu H^2$ (energy density)
	$\boldsymbol{P} = \boldsymbol{D} - \varepsilon_v \boldsymbol{E}$ (polarization)	$\boldsymbol{M} = \boldsymbol{B} - \mu_v \boldsymbol{H}$ (magnetization)
	$\varepsilon_r = \dfrac{\varepsilon}{\varepsilon_v}$ (permittivity)	$\mu_r = \dfrac{\mu}{\mu_v}$ (permeability)
	$\chi = \varepsilon_r - 1$ (electric susceptibility)	$\chi = \mu_r - 1$ (magnetic susceptibility)

The electric flux emanating from a closed surface equals the total charge enclosed by the surface:

$$\nabla \cdot \boldsymbol{D} = \rho \qquad \oiint_S \boldsymbol{D} \cdot \mathrm{d}\boldsymbol{S} = \sum_i Q_i = \iiint_v \rho \, \mathrm{d}v = \psi$$

The magnetic flux emanating from a closed surface is zero: or there are no free magnetic charges:

$$\nabla \cdot \boldsymbol{B} = 0 \qquad \oiint_S \boldsymbol{B} \cdot \mathrm{d}\boldsymbol{S} = 0$$

These equations are fundamental to the electromagnetic theory as applied in the theory of transmission lines, waveguides, etc.

Example 1

A parallel-plate capacitor has a distance d between the plates, the area of the plates is S and the dielectric constant of the medium is ε. The capacitor is connected to a time-varying voltage source V. The displacement current

$$I_D = SJ_D = S\frac{\partial D}{\partial t}, \quad D = \varepsilon E \quad \text{and} \quad E = \frac{V}{d}$$

Thus

$$I_D = \frac{\varepsilon S}{\mathrm{d}}\frac{\mathrm{d}V}{\mathrm{d}t} = C\frac{\mathrm{d}V}{\mathrm{d}t}$$

From $Q = CV$ we find for the conduction current

$$I_C = C\frac{\mathrm{d}V}{\mathrm{d}t}$$

This means that in a capacitor the conduction current is a displacement current.

Example 2

A copper wire has a diameter of 0.1 mm. Specific conductivity $\gamma = 6 \times 10^7\ \Omega^{-1}\,\mathrm{m}^{-1}$. At a frequency of 1 kHz a conduction current of 0.1 mA is flowing. The displacement current is $J_D = \partial D/\partial t = \varepsilon \partial E/\partial t$. If $E = E_m \cos \omega t$ it follows that $J_D = -\omega\varepsilon E_m \sin \omega t$. The ratio of displacement current to conduction current is

$$\frac{J_D}{J_C} = \frac{\omega\varepsilon}{\gamma}$$

$$= \frac{6.28 \times 10^3 \times 8.85 \times 10^{-12}}{6 \times 10^7}$$

$$\simeq 9.2 \times 10^{-16}\ \mathrm{A\,m^{-2}}$$

The conduction current density $J_C = 10^7\ \mathrm{A\,m^{-2}}$ so that

$$J_D = 9.2 \times 10^{-9}\ \mathrm{A\,m^{-2}}$$

4
Conduction and Basic Laws; Network Concepts

4.1 CONDUCTION IN SOLIDS

The atom consists of a number of protons forming the core or nucleus. According to quantum theory the electrons orbit around the nucleus in distinct layers or shells. An electron carries a negative charge (1.602×10^{-19} C), a proton an equal positive charge.

The position of a shell depends on the energy of the electrons in this shell. An electron further away from the nucleus exhibits a higher energy level than an electron in the proximity of the nucleus. Thus shells can be represented by energy levels (Fig. 4.1). The unit of energy in use is the **electron volt**. It is the potential energy lost by one electron when going through a potential difference of one volt ($1\ \text{eV} = 1.6 \times 10^{-19}$ J).

The gaps between the energy levels are the **forbidden energy gaps**. Here electrons cannot orbit.

Electrons in the outer shell are the most important since they determine the electrical, physical and chemical behavior of the material. This outer shell is the **valence shell** where the **valence electrons** orbit.

In most materials the atoms are spaced in a regular pattern where each valence electron forms an orbit around two atoms. This orbit-sharing binds the two atoms together and a **covalent bond** is said to exist.

An electron which is freed from the valence band becomes a conduction electron or **free electron** and it has entered the **conduction band**. It depends on the width of the energy gap whether a valence electron can be freed or not.

In an **insulator** the energy gap is too large (>4 eV) for an electron to cross the gap. The conduction band is empty, there are no free electrons and no current can be made to flow.

In a **conductor** the valence band and conduction band coincide which means that all valence electrons are free electrons. If there are sufficient numbers of open places in the conduction band the free electrons have 'room' to move and a current will flow when a voltage is applied across the conductor.

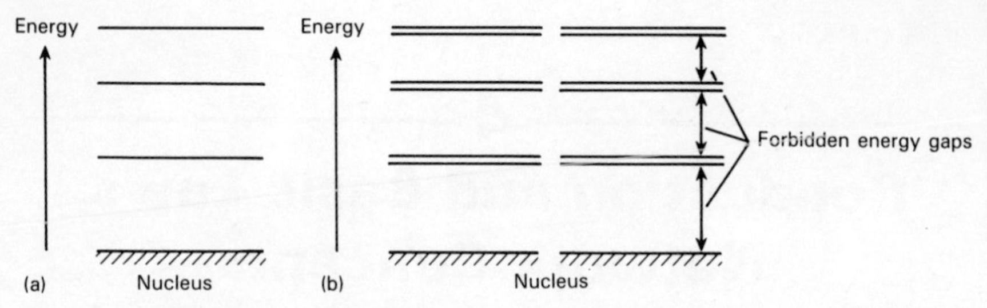

Fig. 4.1 (a) Energy levels of a single atom (b) Energy levels of two adjacent atoms

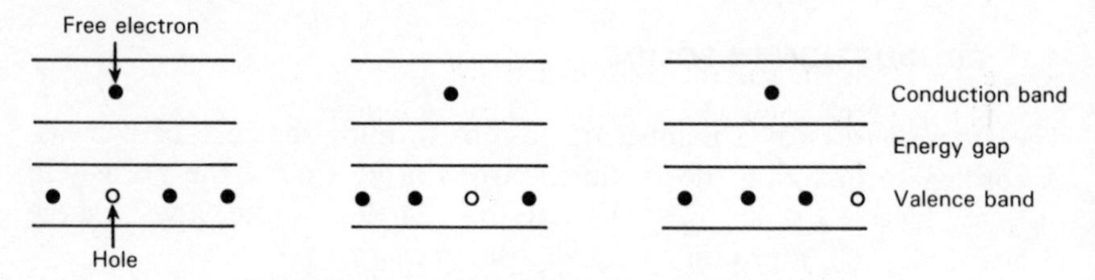

Fig. 4.2 Energy bands and hole conduction

In a **semiconductor** the energy gap is moderate (<3 eV) and electrons with sufficient energy can free themselves from the valence band and jump to the conduction band. When this happens a positive ion is left behind due to an electron deficiency in the valence band. This is called a **hole** and it behaves like a positive charge. Consequently, a valence electron in the proximity may fill this hole and neutralize the ion, thereby creating another hole. Thus a hole can move through the valence band (Fig. 4.2) as readily as an electron in the conduction band. This is a characteristic of semiconductors: free electrons in the conduction band and free holes in the valence band. Both charge carriers can participate in a current flow.

The process of an electron becoming free and leaving a hole is known as **generation**. Should a free electron meet a hole, both charges will cease to remain free charges. The free electron becomes a valence electron and the hole disappears. This reverse process is called **recombination**.

When raising the temperature of the semiconductor more electrons will become free electrons and the conductivity of the material increases. In other words, a semiconductor has a negative temperature coefficient.

In a conductor all electrons are free and raising the temperature has negligible effect on the number of free electrons. However, the more violent vibrations of the atoms in the crystal impair the motion of the free electrons through the material. Thus conduction decreases and a conductor thus has a positive temperature coefficient.

4.2 OHM'S LAW

A voltage V across a conductor of length l and cross section S results in a current I which is related to V by

$$I = \gamma \frac{S}{l} V$$

where γ is the specific conductivity ($\Omega^{-1}\ \text{m}^{-1}$).

Conversely, a current I flowing through a conductor results in a voltage V across its ends as expressed by

$$V = \rho \frac{l}{S} I$$

where $\rho = 1/\gamma$ is the specific resistance (Ω m).

This is Ohm's law which is commonly written as

$$V = IR$$

where $R = \rho l/S$ is the resistance of the conductor. The law is valid for time-dependent voltages and currents as well, thus

$$v(t) = i(t)R$$

The direction of current flow in a conductor or resistor is by definition opposite to the flow of electrons. Thus in a conductor current flows from the positive side to the negative side. Table 8 shows ρ for a number of materials.

Table 8

Material	ρ (Ω m)
Silver	16×10^{-9}
Copper	18×10^{-9}
Gold	22×10^{-9}
Aluminum	28×10^{-9}
Tungsten	55×10^{-9}
Zinc	60×10^{-9}
Iron	10×10^{-8}
Platinum	11×10^{-8}
Tantalum	15×10^{-8}
Lead	21×10^{-8}
Manganine (Cu + Mn + Ni)	44×10^{-8}
Constantan (Cu + Ni)	50×10^{-8}
Carbon	10^{-4}
Germanium	0.47
Silicon	2.3×10^{3}
Insulators	10^{8}–10^{22}

To express the **temperature coefficient** α of a conductor, a room temperature of 20 °C is usually chosen as the reference temperature. The resistance of the conductor at a temperature t can then be written as

$$R_t = R_{20}[1 + \alpha(t - 20)] \quad (\text{for copper } \alpha \simeq 0.004\,264\,^\circ\text{C}^{-1})$$

The **instantaneous power** which is dissipated in the conductor is

$$\begin{aligned} p(t) &= v(t)i(t) \\ &= i^2(t)R \\ &= \frac{v^2(t)}{R} \end{aligned}$$

With periodic voltages and currents, the **average power** is

$$\begin{aligned} P_{\text{av}} &= \frac{1}{T}\int_0^T i^2(t)R\,\mathrm{d}t \\ &= I^2R = \frac{V^2}{R} \end{aligned}$$

where

$$I = \sqrt{\frac{1}{T}\int_0^T i^2(t)\,\mathrm{d}t}$$

is the rms value of current $i(t)$. Likewise,

$$V = \sqrt{\frac{1}{T}\int_0^T v^2(t)\,\mathrm{d}t}$$

is the rms value of voltage $v(t)$. Power is usually expressed in watts, kilowatts, etc.

The amount of power consumed in a certain time is the **energy**. The formula is

$$W = \int_0^T p(t)\,\mathrm{d}t$$

With periodic quantities

$$W = P_{\text{av}}t = VIt = I^2Rt = \frac{V^2}{R}t$$

The units are joule (J), watt hour (W h) or kilowatt hour (kWh).

4.3 KIRCHHOFF'S LAWS

4.3.1 Current law

The algebraic sum of the currents flowing toward any point in a circuit is equal to the sum of the currents flowing away from that point (Fig. 4.3). Thus, in symbols:

$$\Sigma I = 0$$

or

$$\Sigma i(t) = 0$$

It is customary to count a current as positive when it is flowing toward the point and negative when it is flowing away from the point.

4.3.2 Voltage law

In any closed circuit (mesh) the algebraic sum of the source voltages and voltage drops is zero (Fig. 4.4):

$$\Sigma V = 0$$

or

$$\Sigma v(t) = 0$$

4.4 CAPACITOR

In Chapter 2 the basic equation for the capacitor was found as

$$Q = CV$$

Since

$$i = \frac{\mathrm{d}Q}{\mathrm{d}t}$$

it follows that

$$i = C\frac{\mathrm{d}V}{\mathrm{d}t}$$

and conversely

$$v = \frac{1}{C}\int_0^t i\,\mathrm{d}t$$

4.5 INDUCTOR

In Chapter 3 the inductance was defined by

$$\Phi = LI$$

Since

$$v = \frac{d\Phi}{dt}$$

the basic property of an inductor is expressed by the equation

$$v = L\frac{dI}{dt}$$

and conversely

$$i = \frac{1}{L}\int_0^t v\,dt$$

4.6 PHASOR REPRESENTATION OF VOLTAGES AND CURRENTS

When dealing with sinusoidal voltages and currents in a network or circuit it is often convenient to represent these signals as vectors or **phasors**.

A voltage $v = V_m \sin \omega t$ can be represented by choosing a coordinate system and drawing a circle around O with radius V_m (Fig. 4.5). At $t = 0$ the phasor is assumed to coincide with the positive x-axis. As time progresses the phasor rotates counterclockwise around O at an angular velocity ω. Thus at a time t_o the phasor makes the angle ωt_o with the positive x-axis, so that the instantaneous phasor position corresponds with the phase angle.

Since rotating phasors are difficult to draw, the position of the phasor is fixed at $t = 0$. The rotations can be visualized if necessary. Similarly, a voltage $v = V_m \cos \omega t$ is positioned at $t = 0$ along the positive y-axis.

A **resistor** obeys Ohm's law: $v = iR$. Apparently voltage and current phasors always coincide since

$$i = I_m \sin \omega t$$

$$= \frac{1}{R} V_m \sin \omega t$$

(Fig. 4.6).

If the voltage v is connected across a **capacitor** C the current is $\omega C V_m \cos \omega t$. Thus voltage and current have a 90° phase difference, the current leading the voltage (Fig. 4.7).

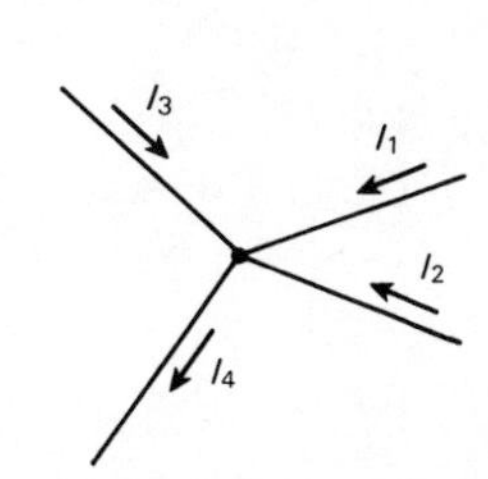

Fig. 4.3 Kirchhoff's current law

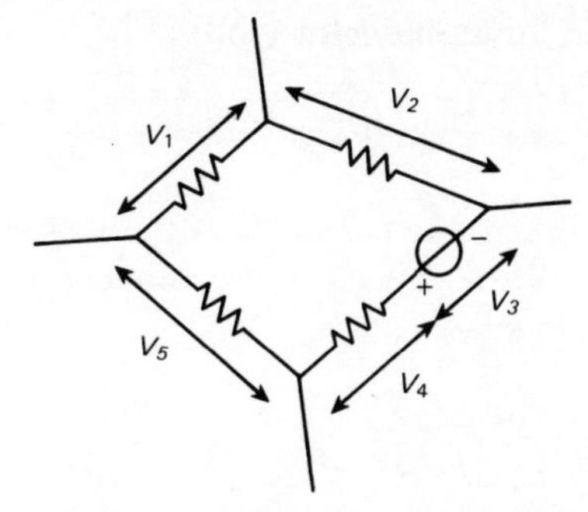

Fig. 4.4 Kirchhoff's voltage law

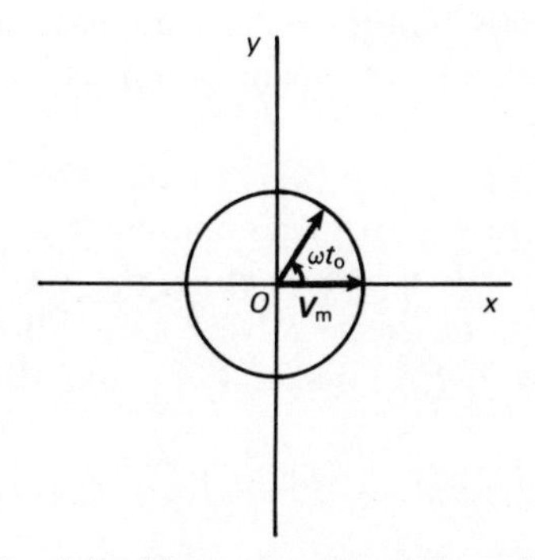

Fig. 4.5 Phasor representation of a sinusoidal voltage

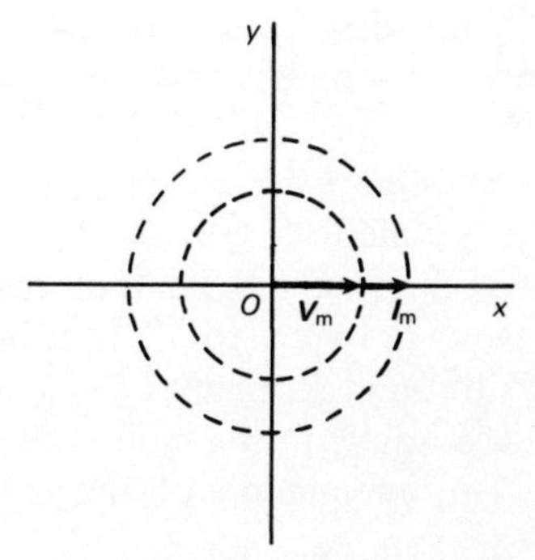

Fig. 4.6 Voltage and current phasor of a resistor

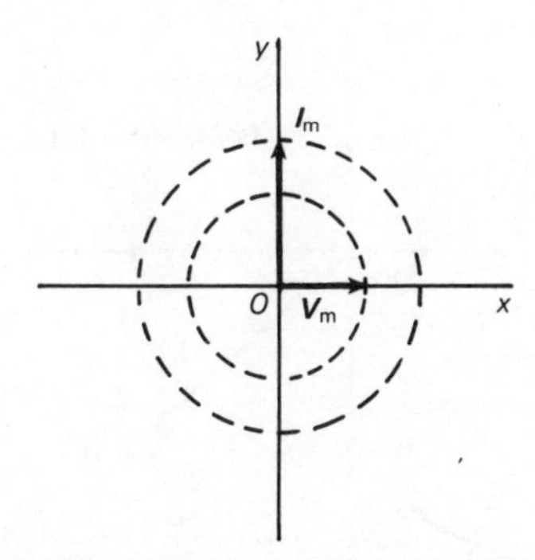

Fig. 4.7 Voltage and current phasor of a capacitor

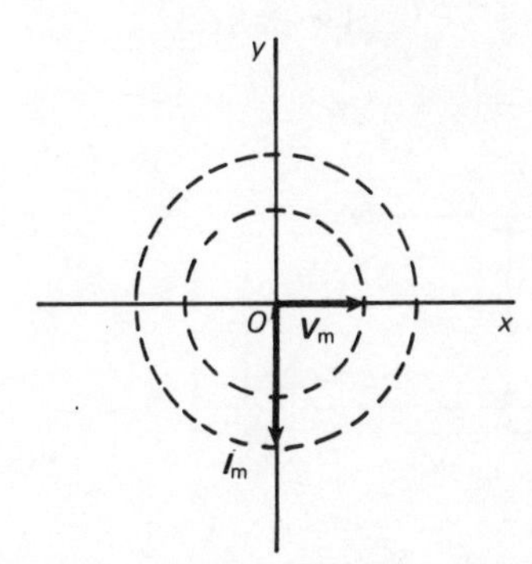

Fig. 4.8 Voltage and current phasor of an inductor

In an **inductor** (coil)

$$i = \frac{1}{L}\int_0^t V_m \sin \omega t \, dt = -\frac{V_m}{\omega L}\cos \omega t$$

and here the current lags the voltage by 90° (Fig. 4.8).

In a capacitor the **capacitive reactance** X_C is defined according to Ohm's law as

$$X_C = \frac{V_m}{I_m} = \frac{1}{\omega C}$$

and in an inductor the **inductive reactance** X_L is defined as

$$X_L = \frac{V_m}{I_m} = \omega L$$

With these basic forms of phasor representation we can increase the number of elements and represent voltages and currents in networks.

Example 1

In the simple RC network of Fig. 4.9 we draw all voltages and currents as shown. Starting at the output, current phasor I_m is chosen along the x-axis. The voltage $V_{o,m}$ across C is normal to I_m, along the negative y-axis. Since i flows through R, phasor $V_{R,m}$ coincides with I_m. Finally, $V_{R,m}$ and $V_{o,m}$ add vectorially to the input voltage $V_{s,m}$.

The attenuation of the network is $V_{o,m}/V_{s,m}$ while $V_{s,m}{}^2 = V_{o,m}{}^2 + V_{R,m}{}^2$ so that

$$\frac{V_{o,m}}{V_{s,m}} = \frac{I_m X_C}{\sqrt{I_m{}^2 X_C{}^2 + I_m{}^2 R^2}}$$

$$= \frac{1}{\sqrt{1 + (\omega RC)^2}}$$

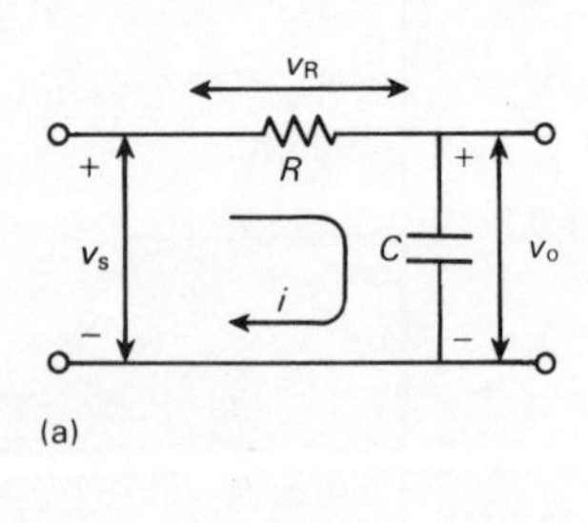

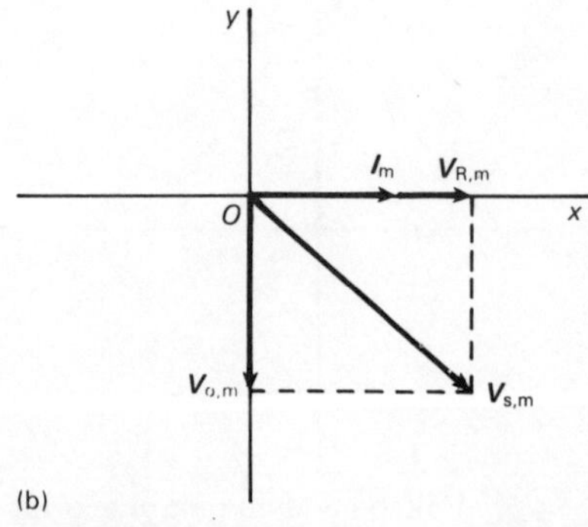

Fig. 4.9 (a) Voltages and currents in a series RC network (b) Phasor representation of the RC network

Example 2
In the network of Fig. 4.10 we take $V_{s,m}$ along the x-axis. Phasor $I_{2,m}$ is coincident with $V_{s,m}$. Phasor $I_{1,m}$ is normal to $V_{s,m}$ along the positive y-axis. Addition of $I_{1,m}$ and $I_{2,m}$ gives the total current I_m.
The impedance Z of the network is

$$\frac{V_{s,m}}{I_m} = \frac{V_{s,m}}{\sqrt{I_{1,m}^2 + I_{2,m}^2}}$$

$$= \frac{V_{s,m}}{\sqrt{\dfrac{V_{s,m}^2}{X_C^2} + \dfrac{V_{s,m}^2}{R^2}}}$$

$$= \frac{R}{\sqrt{1 + (\omega RC)^2}}$$

4.7 BODE PLOT

It is often desirable to graphically represent the behavior of a system as a function of ω, like $V_{o,m}/V_{s,m}$ of Example 1. The Bode plot is the customary way to do this.

A Bode plot is the logarithmic representation of the amplitude characteristic of a system. The **decibel** (dB) is the unit which is used for the modulus of the transfer ratio. In case the transfer ratio is a voltage ratio, this ratio is plotted as $20 \log(V_{o,m}/V_{s,m})$. The frequency scale is also logarithmic.

In Example 1 the voltage ratio is

$$\frac{V_{o,m}}{V_{s,m}} = \frac{1}{\sqrt{1 + (\omega RC)^2}}$$

In most cases a rectilinear approximation of the Bode plot provides sufficient information about the frequency behavior of the system. This requires the determination of the asymptotes of the Bode plot. Here we find

$$\text{for } \omega \to 0: \quad \frac{V_{o,m}}{V_{s,m}} \to 1 \quad (0\text{ dB})$$

$$\text{for } \omega \to \infty: \quad \frac{V_{o,m}}{V_{s,m}} \to \frac{1}{\omega RC}$$

Where these two asymptotes meet (Fig. 4.11), a characteristic frequency is defined, called the **cutoff frequency** ω_c. In this example ω_c follows from $1/\omega RC = 1$ so that $\omega_c = 1/RC$.

For $\omega > \omega_c$ the slope of the asymptote is -6 dB per octave. The deviation between Bode plot (dotted line) and rectilinear approximation is maximum at $\omega = \omega_c$, namely -3 dB ($\frac{1}{2}\sqrt{2}$). The phase difference between v_o and v_s at ω_c is 45°.

4.8 NETWORK THEOREMS

Many engineering problems can be translated into network problems. As these networks can be of considerable complexity, some guidelines in solving network problems would be very helpful. These guidelines are circuit principles which are called network theorems.

The theorems are valid under the assumption that the network is linear which means that the network elements have constant values, regardless of applied voltages and currents.

4.8.1 Superposition theorem

In a linear network the voltage across or the current through any circuit element equals the algebraic sum of the voltages or currents that exist in that element when each source is considered independently of the others.

To calculate the contribution of one source, all remaining sources must be turned off which means that all voltages and currents are made zero while the corresponding source resistances remain in the network.

Example

Find the voltage V_x in the network of Fig. 4.12.

(a) The voltage source is turned off (Fig. 4.13(a)). Then

$$V_{x_1} = 2 \text{ mA} \times \frac{2 \text{ k}\Omega \times 3 \text{ k}\Omega}{2 \text{ k}\Omega + 3 \text{ k}\Omega} = +2.4 \text{ volts}$$

(b) The current source is turned off (Fig. 4.13(b)).

$$V_{x_2} = \frac{2 \text{ k}\Omega}{5 \text{ k}\Omega} \times 10 \text{ V} = +4 \text{ volts}$$

(c) $V_x = V_{x_1} + V_{x_2} = +6.4$ volts

4.8.2 Thévenin's theorem

In so far as a load is concerned, any two-terminal linear network can be

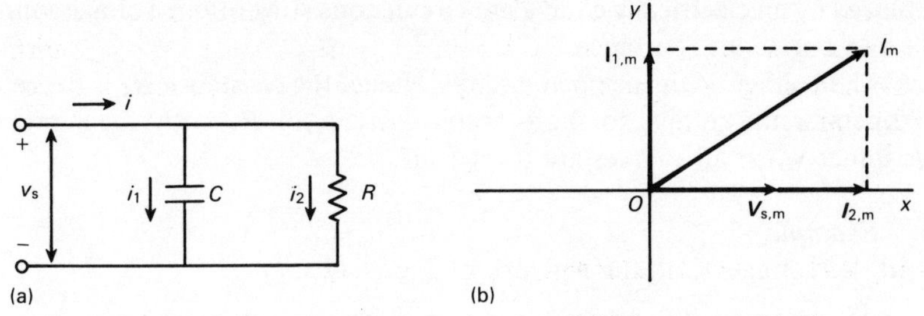

Fig. 4.10 (a) Voltages and currents in a parallel RC network (b) Phasor representation of the RC network

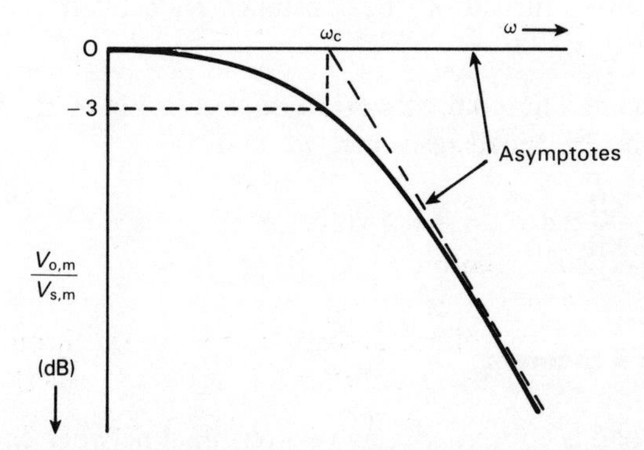

Fig. 4.11 Bode plot of the network of Fig. 4.9(a)

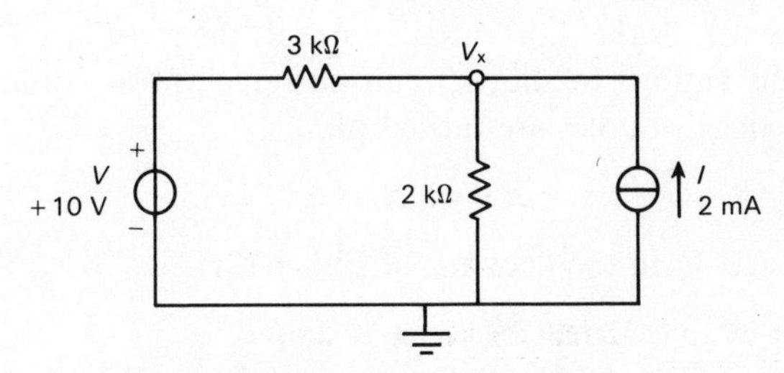

Fig. 4.12 Network example

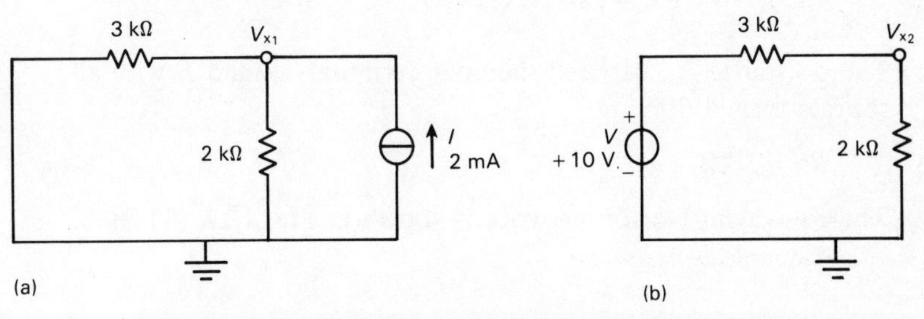

Fig. 4.13 Application of the superposition theorem

replaced by an electrically equivalent circuit consisting of one voltage source V_T and one series resistance R_T.

The voltage V_T is the open-circuit voltage, the resistance R_T is the ratio of open-circuit voltage to short-circuit current; or R_T is the open-circuit resistance when all sources are turned off.

Example

Find the voltage V_x in the network of Fig. 4.12.

(a) The load resistance (2 kΩ) is removed. The open-circuit voltage

$$V_T = +16 \text{ volts (Fig. 4.14(a))}.$$

(b) With all sources turned off the resistance R_T between the terminals (Fig. 4.14(b)) is 3 kΩ.

(c) The equivalent Thévenin network is shown in Fig. 4.15. When replacing the 2 kΩ load resistance we find

$$V_x = \frac{2 \text{ k}\Omega}{5 \text{ k}\Omega} \times 16 \text{ V} = +6.4 \text{ volts}$$

4.8.3 Norton's theorem

In so far as a load is concerned, any two-terminal network can be replaced by an electrically equivalent circuit consisting of one current source I_N and one parallel resistance R_N.

The current I_N is the short-circuit current, the resistance R_N is the ratio of open-circuit voltage to short-circuit current; or R_N is the open-circuit resistance when all sources are turned off.

Example

Find the voltage V_x in the network of Fig. 4.12.

(a) The short-circuit current between A and B is

$$I_N = 3\tfrac{1}{3} \text{ mA} + 2 \text{ mA} = 5\tfrac{1}{3} \text{ mA}$$

(Fig. 4.16).

(b) The resistance R_N between the open terminals A and B with all sources turned off is

$$R_N = 3 \text{ k}\Omega$$

(c) The equivalent Norton network is shown in Fig. 4.17. When

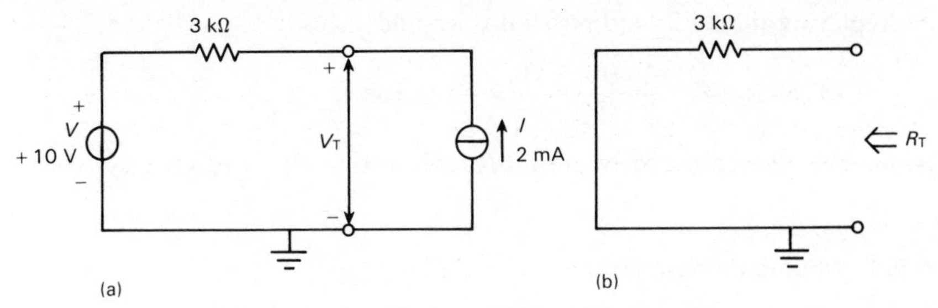

Fig. 4.14 Application of Thévenin's theorem

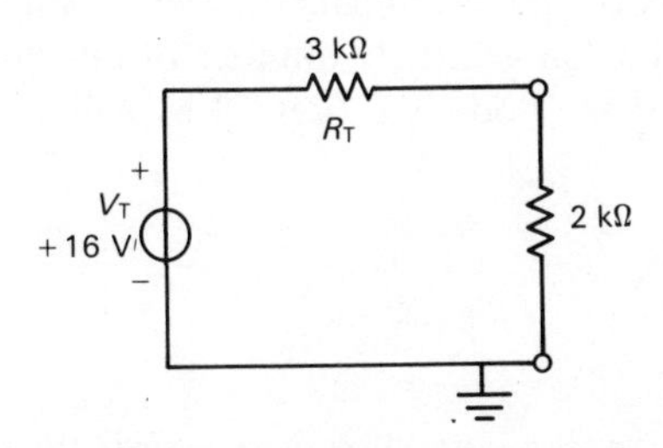

Fig. 4.15 Equivalent Thévenin network of the circuit of Fig. 4.12

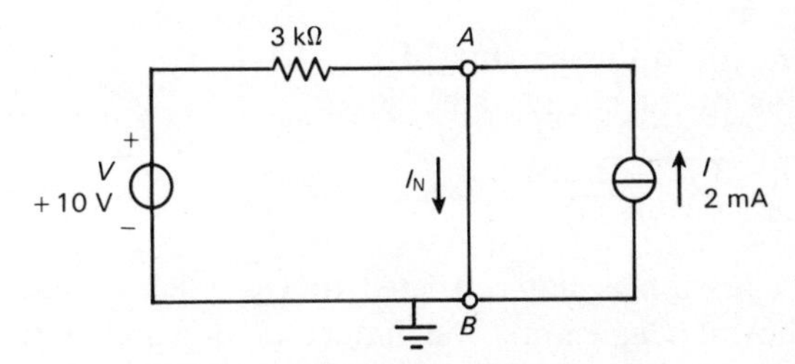

Fig. 4.16 Application of Norton's theorem

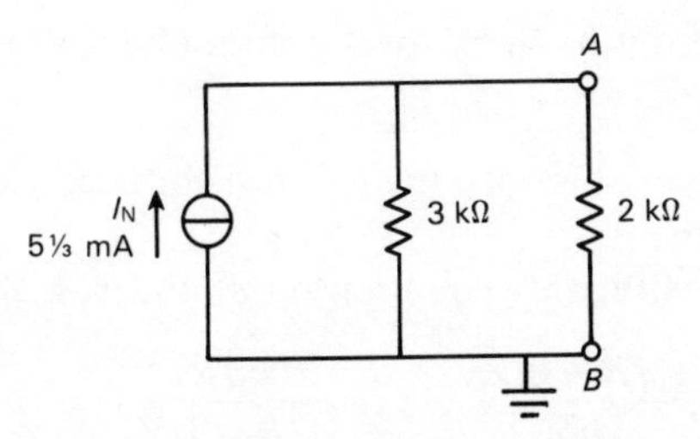

Fig. 4.17 Equivalent Norton network of the circuit of Fig. 4.12

replacing the 2 kΩ load resistance we find

$$V_x = 5\tfrac{1}{3}\ \text{mA} \times \frac{2\ \text{k}\Omega \times 3\ \text{k}\Omega}{2\ \text{k}\Omega + 3\ \text{k}\Omega} = +6.4\ \text{volts}$$

Note: The Thévenin and Norton theorems also apply to reactive systems.

4.8.4 Millman's equation

A very powerful tool in network analysis is Millman's node equation (Fig. 4.18). A node is the common junction where an arbitary number of resistors meet as well as an arbitrary number of current sources.

The voltage V_x at the node is expressed by Millman's equation:

$$V_x = \frac{\sum_{i=1}^{p} \frac{V_i}{R_i} + \sum_{k=1}^{q} I_k}{\sum_{i=1}^{p} \frac{1}{R_i}}$$

A current has a positive sign when flowing towards the node and a negative sign when flowing away from the node.

Example

Find the voltage V_x in the network of Fig. 4.12.

Substitution of the circuit values gives

$$V_x = \frac{+10\ \text{V}/3\ \text{k}\Omega + 2\ \text{mA}}{1/3\ \text{k}\Omega + 1/2\ \text{k}\Omega} = +6.4\ \text{volts}$$

The Millman equation is not restricted to DC voltages and currents or resistors as shown. Capacitors and inductors or combinations of these elements can replace the resistors and AC signals applied.

Example

Find the transfer ratio v_o/v_s of the two-section filter of Fig. 4.19 when v_s is a sinusoidal signal.

(a) Assume a voltage v_x between the two resistors and call Z the impedance of each capacitor.
(b) Application of Millman's equation on v_x and v_s gives

$$v_x = \frac{v_s/R + v_o/R + 0/Z}{2/R + 1/Z}, \quad v_o = \frac{v_x/R}{1/R + 1/Z}$$

Now $Z = 1/(j\omega C)$ and $R/Z = j\omega RC$. Substitute $j\omega RC = jx$. Then

$$v_x = \frac{v_s + v_o}{2 + jx}, \quad v_o = \frac{v_x}{1 + jx}$$

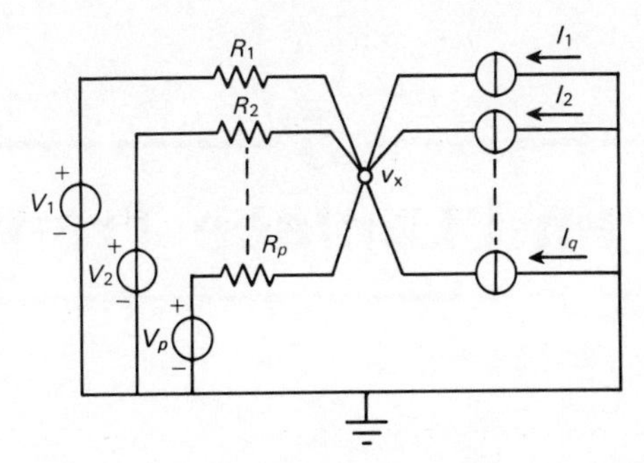

Fig. 4.18 Network node to illustrate Millman's equation

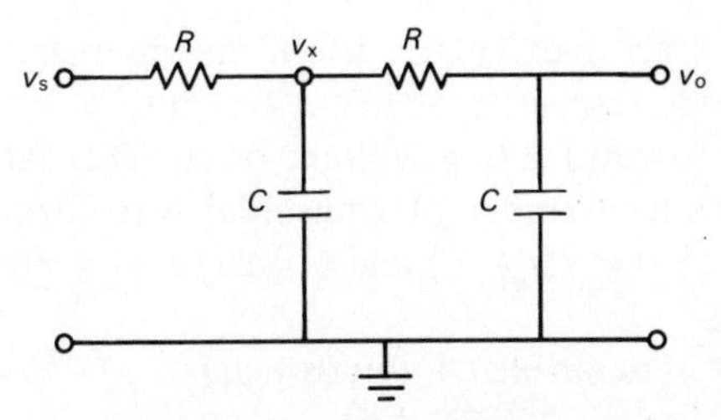

Fig. 4.19 Millman's equation applied to a two-section low-pass network

Elimination of v_x leads to

$$\frac{v_o}{v_s} = \frac{1}{1 - x^2 + 3jx}$$

$$= \frac{1}{1 - (\omega RC)^2 + 3j(\omega RC)}$$

$$\left|\frac{v_o}{v_s}\right| = \frac{1}{\sqrt{[1 - (\omega RC)^2]^2 + 9(\omega RC)^2}}$$

$$\tan \phi = \frac{-3(\omega RC)}{1 - (\omega RC)^2}$$

5 Fourier and Laplace Transforms

5.1 REAL FORM OF FOURIER SERIES

Fourier transforms are particularly useful in the analysis of signal waveforms and frequency responses of linear systems.

Any periodic waveform (e.g. voltage or current) can be represented as the sum of an infinite number of sinusoidal waveforms in addition to a constant. Thus if $f(t) = f(t + T)$ and $\omega = 2\pi/T$,

$$f(t) = a_o + \sum_{n=1}^{\infty} (a_n \sin n\omega t + b_n \cos n\omega t)$$

The coefficients are determined by

$$a_o = \frac{1}{T}\int_0^T f(t)\,\mathrm{d}t$$

This is the **average value** of $f(t)$.

$$a_n = \frac{2}{T}\int_0^T f(t)\sin n\omega t\,\mathrm{d}t$$

$$b_n = \frac{2}{T}\int_0^T f(t)\cos n\omega t\,\mathrm{d}t$$

Thus a periodic waveform can be expanded into a **discrete spectrum** or **line spectrum**.

Often a periodic function has special properties which simplify the Fourier series:

(a) Half-wave symmetry (Fig. 5.1):

$$f(t) = -f\left(t + \frac{T}{2}\right)$$

The negative portion of the wave is the mirror image of the positive portion displaced by a half period. The Fourier series contains only odd values of n (odd harmonics) and $a_o = 0$.

(b) $f(t)$ is an **odd** function (odd symmetry): $f(t) = -f(-t)$ (Fig. 5.2). The

Fourier series contains only sine terms and $a_o = 0$.

(c) $f(t)$ is an **even** function (even symmetry): $f(t) = -f(t)$ (Fig. 5.3). The Fourier series contains only cosine terms.

(d) Even quarter-wave symmetry or half-wave symmetry of an even function (Fig. 5.4). The Fourier series contains only odd values of n and contains only cosine terms.

(e) Odd quarter-wave symmetry or half-wave symmetry of an odd function (Fig. 5.5). The Fourier series contains only odd values of n and contains only sine terms.

Example

The waveform is the symmetrical square wave of Fig. 5.6.

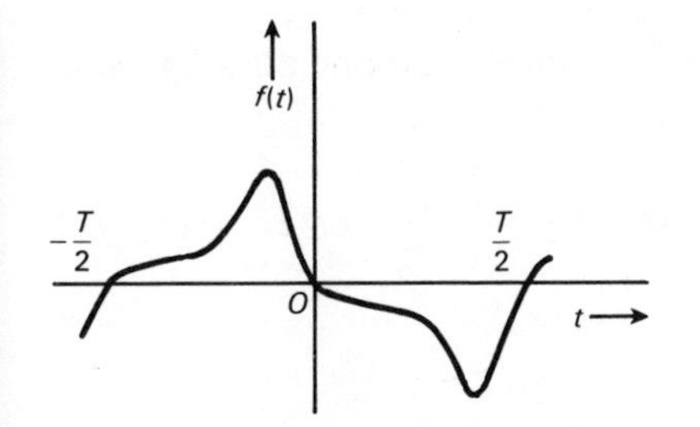

Fig. 5.1 Example of half-wave symmetry

Fig. 5.2 Example of odd symmetry

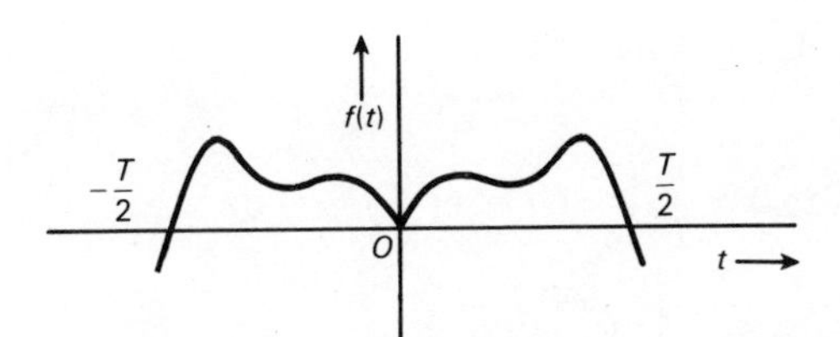

Fig. 5.3 Example of even symmetry

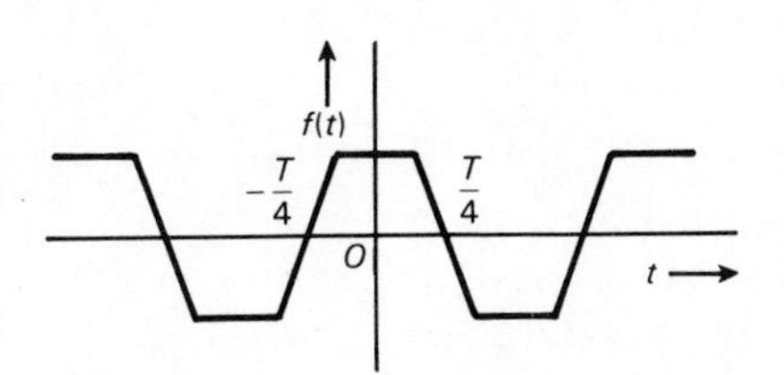

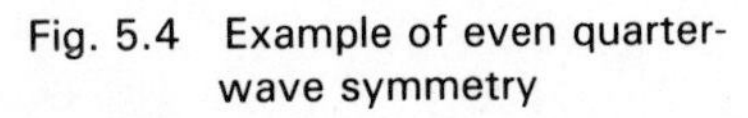

Fig. 5.4 Example of even quarter-wave symmetry

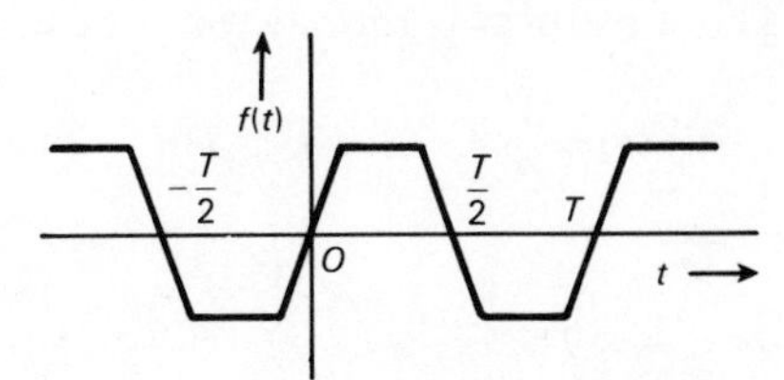

Fig. 5.5 Example of odd quarter-wave symmetry

$$a_0 = \frac{1}{T}\int_0^T f(t)\,dt = \frac{1}{T}(V\cdot\tfrac{1}{2}T) = \frac{V}{2}$$

$$a_n = \frac{2V}{T}\int_0^{T/2} \sin n\omega t\,dt = \frac{V}{n\pi}(1-\cos n\pi) = \frac{2V}{(2n-1)\pi}$$

$$b_n = 0$$

Hence the Fourier series is

$$f(t) = \frac{V}{2} + \frac{2V}{\pi}(\sin \omega t + \tfrac{1}{3}\sin 3\omega t + \tfrac{1}{5}\sin 5\omega t + \cdots)$$

5.2 COMPLEX FORM OF FOURIER SERIES

Fourier series can be expressed in terms of complex exponentials by using the identities

$$\sin x = \frac{1}{2j}[\exp(jx) - \exp(-jx)] \text{ and } \cos x = \tfrac{1}{2}[\exp(jx) + \exp(-jx)]$$

Substitution results in the complex Fourier series

$$f(t) = \sum_{n=-\infty}^{+\infty} c_n \exp(jn\omega t)$$

where

$$c_0 = a_0 = \frac{1}{T}\int_0^T f(t)\,dt$$

$$c_n = \tfrac{1}{2}(a_n - jb_n) = \frac{1}{T}\int_0^T f(t)\exp(-jn\omega t)\,dt$$

$$c_{-n} = \tfrac{1}{2}(a_n + jb_n) = \frac{1}{T}\int_0^T f(t)\exp(jn\omega t)\,dt$$

Example

The waveform is the sawtooth of Fig. 5.7:

$$f(t) = \frac{V}{T}t, \quad 0 < t < T$$

$$c_0 = \frac{V}{T^2}\int_0^T t\,dt = \frac{V}{2}$$

$$c_n = \frac{V}{T^2}\int_0^T t\exp(-jn\omega t)\,dt = \frac{V}{2\pi n}\exp\left(j\frac{\pi}{2}\right)$$

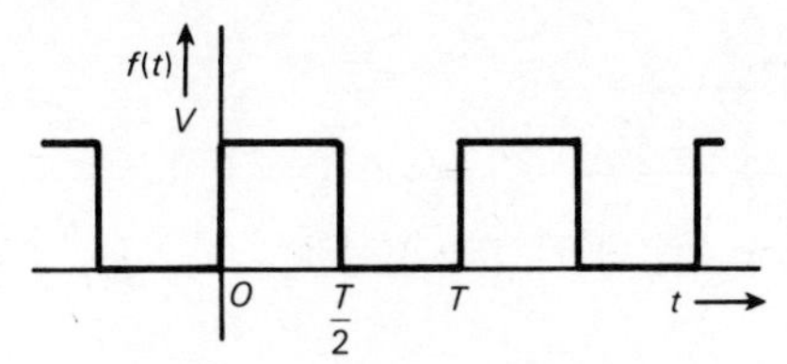

Fig. 5.6 Fourier analysis applied to a symmetrical square wave

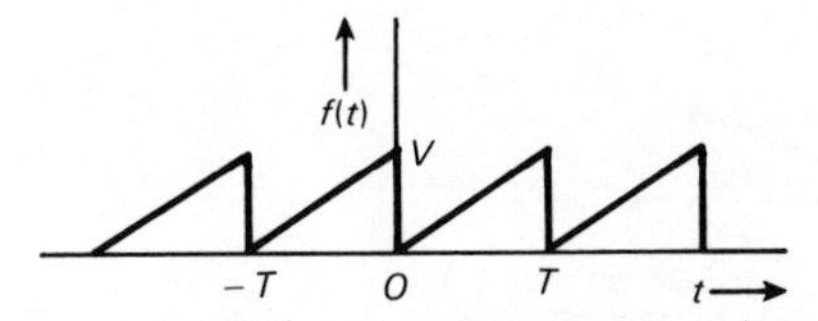

Fig. 5.7 Fourier analysis applied to a sawtooth waveform

Thus

$$f(t) = \frac{V}{2} + \frac{V}{2\pi} \sum_{-\infty}^{+\infty} \frac{1}{n} \exp\left[j\left(n\omega t + \frac{\pi}{2}\right)\right]$$

The **power content** of $f(t)$ is expressed by

$$\frac{1}{T} \int_0^T |f(t)|^2 \, dt = a_0{}^2 + \frac{1}{2} \sum_{n=1}^{\infty} (a_n{}^2 + b_n{}^2)$$

(see Section 5.3). A number of voltage waveforms and corresponding Fourier series are shown in Fig. 5.8.

5.3 FOURIER INTEGRAL

A nonperiodic waveform $f(t)$ can be represented by a Fourier integral provided the waveform is an integrable function. The transform is expressed by the integral

$$F(\omega) = \int_{-\infty}^{+\infty} f(t) \exp(-j\omega t) \, dt$$

(condition: $\int_{-\infty}^{+\infty} |f(t)| \, dt$ is finite). The inverse transform of $F(\omega)$ is

$$f(t) = \frac{1}{2\pi} \int_{-\infty}^{+\infty} F(\omega) \exp(j\omega t) \, d\omega$$

Generally, the function $F(\omega)$ is complex, thus $F(\omega) = R(\omega) + jX(\omega)$. $|F(\omega)|$ is the amplitude spectrum of $f(t)$ which is a **continuous spectrum** and $\tan \phi = X(\omega)/R(\omega)$ where $\phi(\omega)$ is the phase spectrum of $f(t)$. The

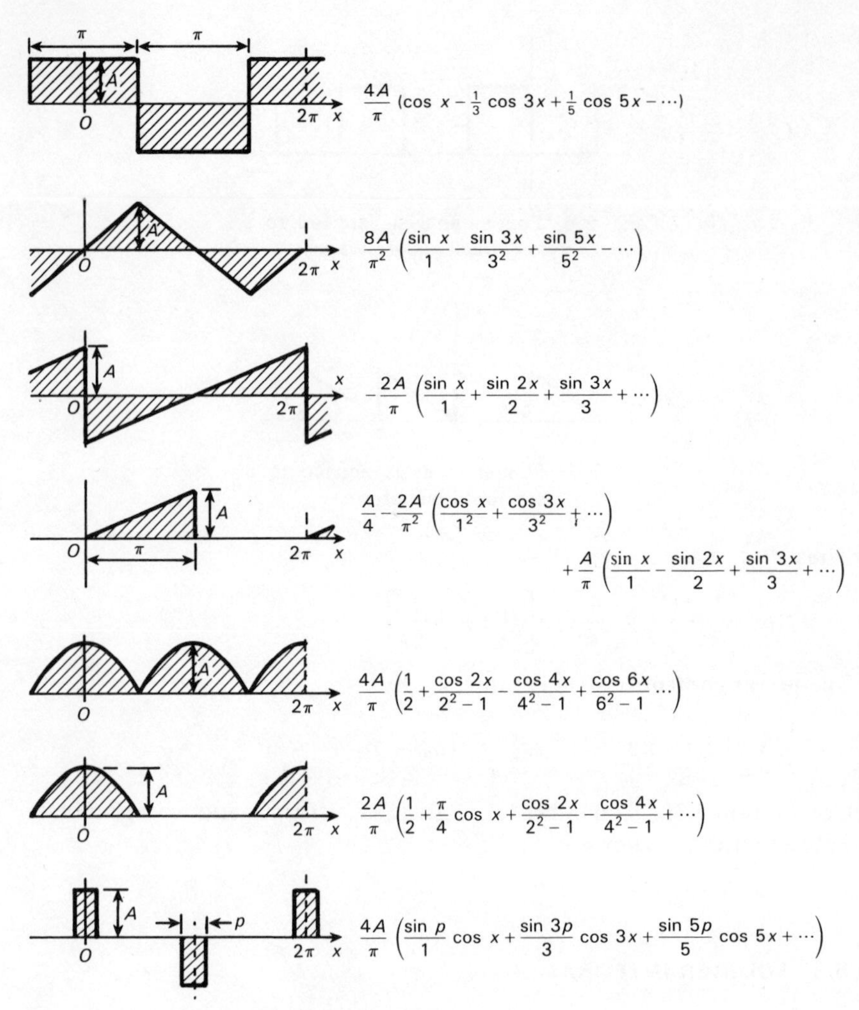

Fig. 5.8 Voltage waveforms and Fourier series

transform obeys a number of laws of which some are listed. If the transform is written in operational form as $F(\omega) = G[f(t)]$ we have

$$G[af_1(t) + bf_2(t)] = aF_1(\omega) + bF_2(\omega) \quad (a \text{ and } b \text{ are constants})$$

This is the **linearity property**.

$$G[f(t - t_o)] = F(\omega)\exp(-j\omega t_o) \quad \text{(time-shifting property)}$$

$$G[f(t)\exp(j\omega_o t)] = F(\omega - \omega_o) \quad \text{(frequency-shifting property)}$$

$$G[f(at)] = \frac{1}{|a|}F\left(\frac{\omega}{a}\right) \quad \text{(scaling property)}$$

$$\left.\begin{array}{l} G[F(t)] = 2\pi f(-\omega)] \\ G[F(-t)] = 2\pi f(\omega)] \end{array}\right\} \qquad \text{(reciprocity property)}$$

$$G\left[\frac{\mathrm{d}^n f(t)}{\mathrm{d}t^n}\right] = (j\omega)^n F(\omega)$$

$$G[(-jt)^n f(t)] = \frac{\mathrm{d}^n F(\omega)}{\mathrm{d}\omega^n}$$

$$G\left[\int_{-\infty}^{t} f(t)\,\mathrm{d}t\right] = \frac{1}{j\omega}\,F(\omega)$$

Parseval's Theorem

$$\int_{-\infty}^{+\infty} |f(t)|^2\,\mathrm{d}t = \frac{1}{2\pi}\int_{-\infty}^{+\infty} |F(\omega)|^2\,\mathrm{d}\omega$$

If $f(t)$ is the voltage across a 1 Ω resistor then $\int_{-\infty}^{+\infty} |f(t)|^2\,\mathrm{d}t$ is the total energy E delivered by the voltage source. Thus

$$E = \frac{1}{2\pi}\int_{-\infty}^{+\infty} |F(\omega)|^2\,\mathrm{d}\omega$$

$$= \int_{-\infty}^{+\infty} |F(\omega)|^2\,\mathrm{d}f$$

$|F(\omega)|^2$ is called the **energy spectrum** or **spectral density function** of $f(t)$.

Example

$f(t)$ is the cosine pulse of Fig. 5.9(a):

$$f(t) = V\cos\pi\frac{t}{t_o}, \quad f(t) = 0 \text{ when } t \leqslant -\tfrac{1}{2}t_o \text{ and } t \geqslant \tfrac{1}{2}t_o$$

$$|F(\omega)| = \frac{2}{\pi}\,Vt_o\,\frac{\cos\frac{1}{2}\omega t_o}{1-(\omega t_o/\pi)^2} \qquad \text{(Fig. 5.9b)}$$

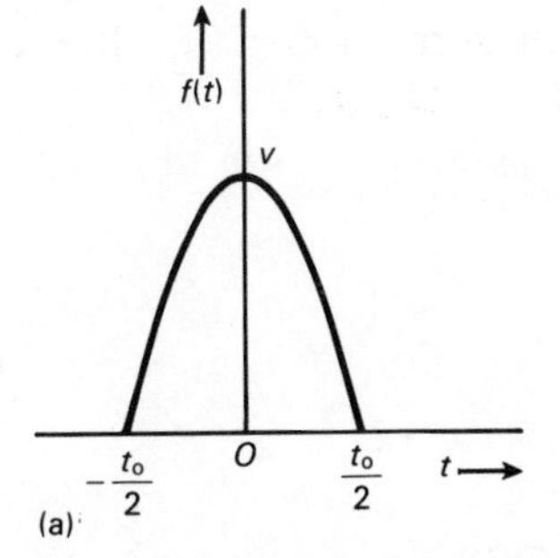

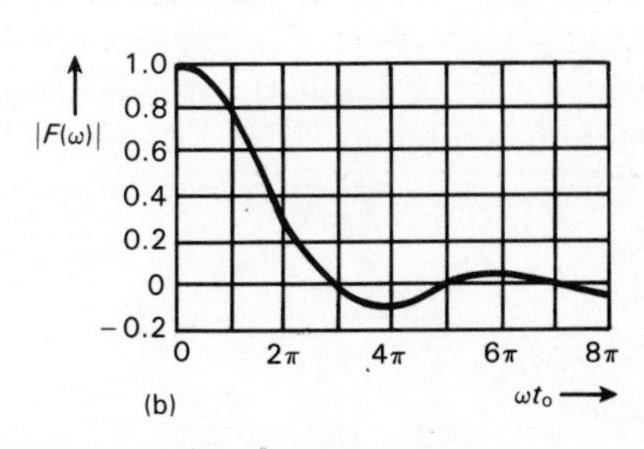

Fig. 5.9 (a) The cosine pulse (b) Spectrum of the cosine pulse

5.4 CONVOLUTION

When two functions $f_1(t)$ and $f_2(t)$ are given then the **convolution** of the two functions is defined as

$$f(t) = \int_{-\infty}^{+\infty} f_1(x) f_2(t-x)\, \mathrm{d}x$$

or symbolically

$$f(t) = f_1(t) * f_2(t)$$

A few properties of convolution are

$$f_1(t) * f_2(t) = f_2(t) * f_1(t) \qquad \text{(commutative law)}$$
$$[f_1(t) * f_2(t)] * f_3(t) = f_1(t) * [f_2(t) * f_3(t)] \qquad \text{(associative law)}$$

If $G[f_1(t)] = F_1(\omega)$ and $G[f_2(t)] = F_2(\omega)$ then

$$G[f_1(t) * f_2(t)] = F_1(\omega) F_2(\omega) \qquad \text{(time-convolution theorem)}$$

$$G[f_1(t) f_2(t)] = \frac{1}{2\pi} F_1(\omega) * F_2(\omega) \qquad \text{(frequency-convolution theorem)}$$

The convolution concept is important in the theory of correlation functions (Chapter 6).

5.5 LAPLACE TRANSFORMS

The Laplace transform is one of the algorithms in mathematics which lower the order of complexity of a problem. Specifically, the solution of differential equations is transformed into the solution of an algebraic equation.

The Laplace transform $F(s)$ of a function $f(t)$ is defined as

$$F(s) = \int_0^{\infty} f(t) \exp(-st)\, \mathrm{d}t \quad (f(t) = 0 \text{ for } t < 0)$$

where s is the Laplace variable $s = \sigma + j\omega$. In operational form the transform is written as

$$\mathscr{L}[f(t)] = F(s)$$

The inverse transform is

$$f(t) = \mathscr{L}^{-1}[F(s)]$$
$$= \frac{1}{2\pi j} \int_{\sigma - j\omega}^{\sigma + j\omega} F(s) \exp(st)\, \mathrm{d}s$$

Some properties of the Laplace transform are

$$\mathscr{L}[af_1(t) + bf_2(t)] = aF_1(s) + bF_2(s) \qquad \text{(linearity property)}$$

$$\mathscr{L}[f(t - t_0)] = F(s)\exp(-st_0) \qquad \text{(time-shifting property)}$$

$$\mathscr{L}[f(t)\exp(-at)] = F(s + a) \qquad \text{(shifting in the } s \text{ domain)}$$

$$\mathscr{L}[f(at)] = \frac{1}{|a|} F\left(\frac{s}{a}\right) \qquad \text{(scaling property)}$$

$$\mathscr{L}[f'(t)] = sF(s) - f(0) \qquad (f(0) = \text{initial condition})$$

$$\mathscr{L}[tf(t)] = -\frac{\mathrm{d}F(s)}{\mathrm{d}s}$$

$$\mathscr{L}\left[\int_0^t f(t)\,\mathrm{d}t\right] = \frac{F(s)}{s}$$

$$\mathscr{L}\left[\frac{1}{t} f(t)\right] = \int_s^\infty F(s)\,\mathrm{d}s$$

$$\mathscr{L}[f(t + T)] = f(t)$$

$$= (1 - \exp(-sT)^{-1} \int_0^T f(t)\exp(-st)\,\mathrm{d}t \qquad \text{(periodic function)}$$

$$\mathscr{L}[f_1(t) * f_2(t)] = F_1(s)F_2(s) \qquad \text{(convolution)}$$

$$\lim_{t\to\infty} f(t) = \lim_{s\to 0} sF(s) \qquad \text{(final value theorem)}$$

$$\lim_{t\to 0} f(t) = \lim_{s\to\infty} sF(s) \qquad \text{(initial value theorem)}$$

Example 1

In the network of Fig. 5.10 capacitor C is uncharged. Switch S is closed at $t = 0$. Find the current $i(t)$.

$$V = v_C(t) + v_R(t)$$

$$= \frac{1}{C} \int_0^t i(t)\,\mathrm{d}t + i(t)R$$

Transformation yields

$$\frac{V}{s} = \frac{1}{sC} i(s) + i(s)R$$

so that

$$i(s) = \frac{V}{R} \frac{1}{s + \dfrac{1}{RC}}$$

From the inverse transformation (see Table 5)

$$i(t) = \frac{V}{R} \exp\left(-\frac{t}{RC}\right)$$

Example 2

The unit-step function (Heaviside function) is defined as $v(t) = 0$ for $t < 0$ and $v(t) = 1$ for $t \geqslant 0$ (Fig. 5.11).

$$\mathscr{L}[v(t)] = \int_0^\infty v(t)\exp(-st)\,\mathrm{d}t$$

$$= \int_0^\infty \exp(-st)\,\mathrm{d}t = \frac{1}{s}$$

5.6 TRANSIENT RESPONSE OF NETWORKS

5.6.1 RC Network (Fig. 5.12(a))

Capacitor C is uncharged; at $t = 0$ switch S is closed. Find $v_o(t)$.

A capacitor is described by the equation

$$i(t) = C\frac{\mathrm{d}v_o(t)}{\mathrm{d}t}$$

or, after transformation, $i(s) = sCv_o(s)$. The impedance of the capacitor in Laplace notation can thus be written as $Z_C = 1/(sC)$. Now in Fig. 5.12(a) we have

$$\frac{v_o(s)}{V} = \frac{Z_C}{R + Z_C} \quad \text{or} \quad VZ_C = (R + Z_C)v_o(s)$$

so that

$$\frac{V}{sC} = v_o(s)R + \frac{v_o(s)}{sC}$$

Solving for $v_o(s)$ gives

$$v_o(s) = \frac{V}{1 + sRC}$$

$$v_o(t) = V\left[1 - \exp\left(\frac{-t}{RC}\right)\right]$$

(Fig. 5.12(b)).

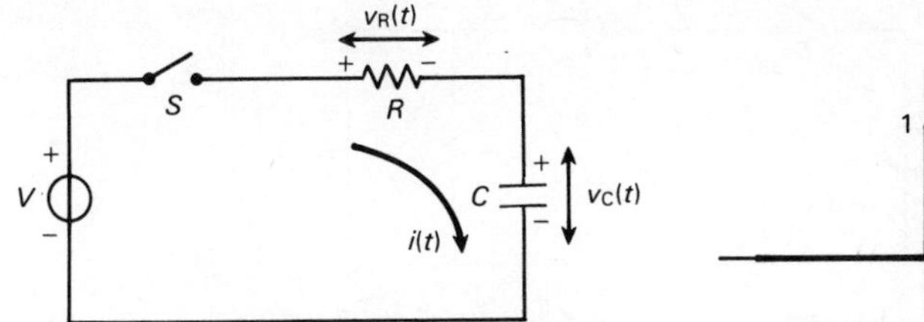

Fig. 5.10 Laplace transform applied to an RC network

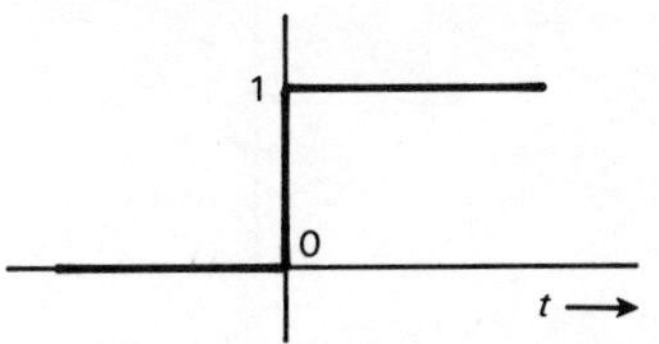

Fig. 5.11 The unit-step function

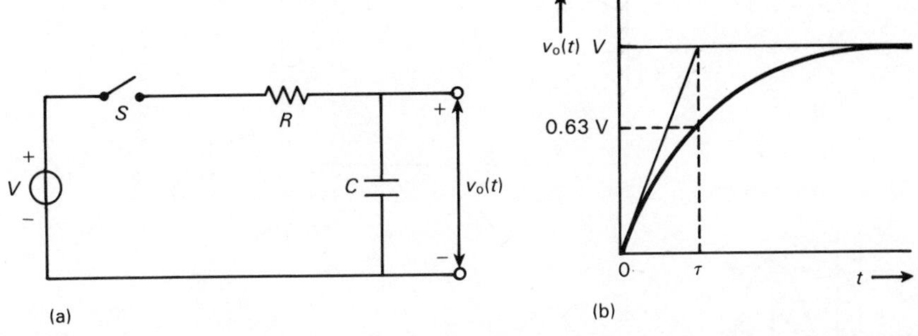

Fig. 5.12 (a) Laplace transform applied to an RC network (b) Transient response of the RC network

5.6.2 LR Network (Fig. 5.13)

The same solution as in Section 5.6.1 is obtained. Instead of RC the time constant is here L/R.

5.6.3 LCR Network (Fig. 5.14a)

L and C are uncharged; at $t = 0$ switch S is closed. The equation for an inductor is

$$v(t) = L\,\frac{\mathrm{d}i}{\mathrm{d}t}$$

and after transformation $v(s) = sLi(s)$. The impedance of the inductor is thus $Z_L = sL$. The current through the network is then

$$i(s) = V\,\frac{sC}{1 + sRC + s^2LC}$$

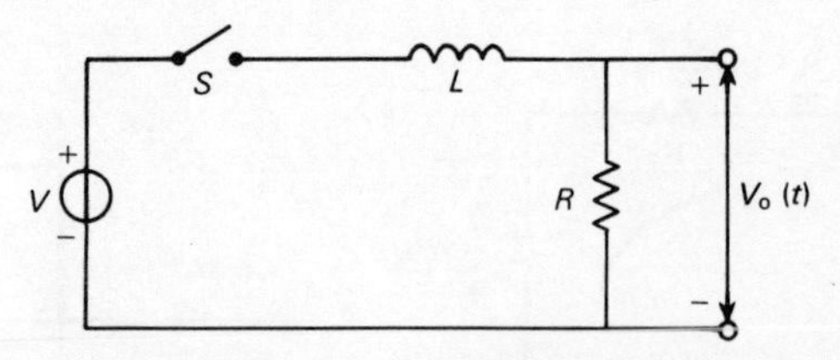

Fig. 5.13 Laplace transform applied to an LR network

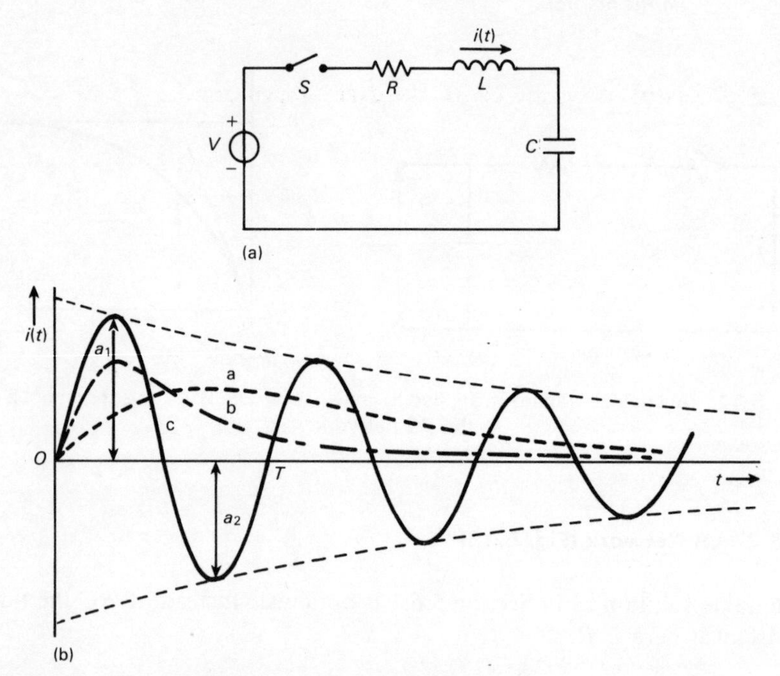

Fig. 5.14 (a) Laplace transform applied to an LCR network (b) Transient response of the LCR network

The roots of the equation $1 + sRC + s^2LC = 0$ are

$$s_1 = \frac{-1}{2\tau} + \sqrt{\left(\frac{R}{2L}\right)^2 - \frac{1}{LC}} = \frac{-1}{2\tau} + a$$

$$s_2 = \frac{-1}{2\tau} - \sqrt{\left(\frac{R}{2L}\right)^2 - \frac{1}{LC}} = \frac{-1}{2\tau} - a$$

where $\tau = L/R$. The general solution of $i(t)$ is

$$i(t) = K_1 \exp(s_1 t) + K_2 \exp(s_2 t)$$

where K_1 and K_2 are constants. The following three possibilities can be distinguished:

(a) $\left(\dfrac{R}{2L}\right)^2 - \dfrac{1}{LC} > 0$ or $R^2 > \dfrac{4L}{C}$

The solution is

$$i(t) = \exp\left(\frac{-t}{2\tau}\right)(K_1 \exp(at) + K_2 \exp(-at))$$

$$= \exp\left(\frac{-t}{2\tau}\right)(A \cosh at + B \sinh at)$$

(Fig. 5.14(b), curve a). This is the **overdamped** case.

(b) $R^2 = \dfrac{4L}{C}$

The solution is

$$i(t) = \exp\left(\frac{-t}{2}\right)(K_1 + K_2 t)$$

(Fig. 5.14(b), curve b). This is called **critical damping**.

(c) $R^2 < \dfrac{4L}{C}$

We see that a is imaginary so that we can write

$$a = j\sqrt{\frac{1}{LC} - \left(\frac{R}{2L}\right)^2} = j\omega$$

The solution is

$$i(t) = \exp\left(\frac{-t}{2\tau}\right)[K_1 \exp(j\omega t) + K_2 \exp(-j\omega t)]$$

$$= \exp\left(\frac{-t}{2\tau}\right)(A \cos \omega t + B \sin \omega t)$$

(Fig. 5.14(b), curve c). This is the **underdamped** case which results in a periodic waveform.

6
Error Calculations; Probability

When the value of a variable fluctuates in a random way, its behavior can be described by the laws of probability. These laws are laid down in formulas which are based on a number of definitions.

6.1 STATISTICAL AVERAGES AND ERRORS OF DISCRETE VARIABLES

Arithmetic mean or average of n data

$$\bar{x} = \frac{1}{n}(x_1 + x_2 + \cdots + x_n)$$

$$= \frac{1}{n}\sum_{i=1}^{n} x_i$$

Weighted arithmetic mean

$$\bar{x} = \frac{a_1x_1 + a_2x_2 + \cdots + a_nx_n}{a_1 + a_2 + \cdots + a_n}$$

$$\frac{\sum_{i=1}^{n} a_ix_i}{\sum_{i=1}^{n} a_i}$$

where $a_1, a_2, \ldots, a_n$ are weighting factors.

Absolute error of n data

$$f_x = \frac{1}{n}\sum_{i=1}^{n} |\bar{x} - x_i|$$

Relative error

$$f_{xr} = \frac{f_x}{\bar{x}}$$

Percentage error

$$f_{xp} = f_{xr} \times 100\%$$

Geometric mean of n data

$$x_g = \sqrt[n]{x_1 x_2 \ldots x_n}$$

$$= \sqrt[n]{\prod_{i=1}^{n} x_i}$$

Example

The population of a city increases at a rate proportional to the number of people in the city. Suppose in 1970 the population was 500 000 and in 1980 it was 980 000. Then in 1975 the population was

$$x_g = \sqrt{500\,000 \times 980\,000} = 700\,000$$

Harmonic mean of n data

$$x_h = \frac{n}{\dfrac{1}{x_1} + \dfrac{1}{x_2} + \cdots + \dfrac{1}{x_n}}$$

$$= \frac{n}{\sum_{i=1}^{n} \dfrac{1}{x_i}}$$

Example

A plane flies around a square whose side is 100 miles long. The first side is taken at 100 mph, the second at 200 mph, the third at 300 mph and the fourth at 400 mph. Then the average speed is $x_h = 192$ mph.

Root mean square (rms) or quadratic mean of n data

$$\sqrt{\overline{x^2}} = \sqrt{\frac{1}{n}({x_1}^2 + {x_2}^2 + \cdots + {x_n}^2)}$$

$$= \sqrt{\frac{1}{n} \sum_{i=1}^{n} {x_i}^2}$$

Weighted quadratic mean of n data

$$\sqrt{\overline{x^2}} = \sqrt{\frac{a_1 {x_1}^2 + a_2 {x_2}^2 + \cdots + a_n {x_n}^2}{a_1 + a_2 + \cdots + a_n}}$$

where $a_1, a_2, \ldots, a_n$ are weighting factors.

Standard deviation of n data

$$\sigma_x = \sqrt{\frac{\sum_{i=1}^{n} (x_i - \bar{x})^2}{n-1}} \quad \text{or} \quad \sigma_x = \sqrt{\frac{\sum_{i=1}^{n} (x_i - \bar{x})^2}{n}}$$

Variance (spread) of n data

$$\sigma_x{}^2 = \frac{\sum_{i=1}^{n} (x_i - \bar{x})^2}{n-1} \quad \text{or} \quad \sigma_x{}^2 = \frac{\sum_{i=1}^{n} (x_i - \bar{x})^2}{n}$$

Median of n data

The middle value or the arithmetic mean of the two middle values if the set of data is arranged in order of magnitude.

6.2 ERRORS IN FUNCTIONS OF MEASURED DATA

If we assume the relative errors in measured data to be sufficiently small ($f_{xr} \ll 1$), the following approximation applies:

$$(1 \pm f_{xr})^n \simeq 1 \pm n f_{xr}$$

If $n = -1$, it follows that

$$\frac{1}{1 - f_{xr}} \simeq 1 + f_{xr}, \quad \frac{1}{1 + f_{xr}} \simeq 1 - f_{xr}$$

If x and y are measured data, k a constant and p the resultant function, the expressions in Table 9 for the largest possible error (worst case) can be derived.

The following examples refer to Table 9.

Example 1(d)

Two resistors in series: $R_1 = 15\ \text{k}\Omega$ (2%), $R_2 = 10\ \text{k}\Omega$ (5%).

Then $x = 15\ \text{k}\Omega$, $f_x = 0.3\ \text{k}\Omega$, $y = 10\ \text{k}\Omega$, $f_y = 0.5\ \text{k}\Omega$ so that $p = 25\ \text{k}\Omega$, $f_p = 0.8\ \text{k}\Omega$, $f_{pr} = 0.032$ (3.2%).

Table 9

Function	*Absolute error*	*Relative error*
(a) $p = x + k$	$f_p = f_x$	$f_{pr} = \dfrac{f_x}{p} = \dfrac{x}{p} f_{xr}$
(b) $p = kx$	$f_p = kf_x$	$f_{pr} = f_{xr}$
(c) $p = x^k$	$f_p = kx^{k-1} f_x$	$f_{pr} = kf_{xr}$
(d) $p = x + y$	$f_p = f_x + f_y$	$f_{pr} = \dfrac{xf_{xr} + yf_{yr}}{p}$
(e) $p = x - y$	$f_p = f_x + f_y$	$f_{pr} = \dfrac{xf_{xr} + yf_{yr}}{p}$
(f) $p = xy$	$f_p = yf_x + xf_y$	$f_{pr} = f_{xr} + f_{yr}$
(g) $p = \dfrac{x}{y}$	$f_p = \dfrac{yf_x + xf_y}{y^2}$	$f_{pr} = f_{xr} + f_{yr}$
(h) $p = xe^{ky}$		$f_{pr} = \lvert f_{xr} + kf_y \rvert$
(i) $p = x \ln ky$		$f_{pr} = f_{xr} + \dfrac{x}{p} f_{yr}$
(j) $p = \dfrac{x}{\ln ky}$		$f_{pr} = \dfrac{f_x + pf_{yr}}{x - pf_{yr}}$

Example 2(e)

The difference of two currents: $I_1 = 100$ mA (4%), $I_2 = 60$ mA (5%).

Then $x = 100$ mA, $f_x = 4$ mA, $y = 60$ mA, $f_y = 3$ mA so that $p = 40$ mA, $f_p = 7$ mA, $f_{pr} = 0.175$ (17.5%).

Note that the relative error in the difference of two data can be very large.

Example 3(f)

Ohm's law: $R = 10$ kΩ (5%), $I = 5$ mA (3%).

Thus $x = 10$ kΩ, $f_x = 0.5$ kΩ, $f_{xr} = 0.05$, $y = 5$ mA, $f_y = 0.15$ mA, $f_{yr} = 0.03$.

Thus $p = V = IR = 50$ V, $f_p = 4$ V, $f_{pr} = 0.08$ (8%).

Example 4(g)

Ohm's law: $V = 50$ V (3%), $I = 5$ mA (2%).

Thus $x = 50$ V, $f_x = 1.5$ V, $f_{xr} = 0.03$, $y = 5$ mA, $f_y = 0.1$ mA, $f_{yr} = 0.02$.

Then $p = R = V/I = 10$ kΩ, $f_p = 0.5$ kΩ, $f_{pr} = 0.05$ (5%).

Example 5(h)

In the circuit of Fig. 6.1 it is assumed that C is charged to 10 V. The value of V_o is to be calculated 5 s after opening switch S. We have

$$V_o = V \exp\left(\frac{-t}{RC}\right)$$

so that

$$p = V_o = 10 \exp(-0.5) = 6.07 \text{ V}$$
$$x = V = 10 \text{ V}, \quad f_{xr} = 0.02, \quad k = -5$$

$$y = \frac{1}{RC} = 0.1 \text{ s}^{-1}, \quad f_{yr} = 0.07, \quad f_y = 0.007$$

Then $f_{pr} = |0.02 - 5(0.007)| = 0.015$ (1.5%) so that

$$V_o \simeq 6.07 \text{ V} \pm 0.09 \text{ V}$$

Example 6(j)

In the circuit of Fig. 6.2 the time constant RC has to be determined by measuring V_o 5s after closing switch S. The time measurement has an uncertainty of 0.2 s and $V_o = 4$ V (2%).

Since

$$V_o = V\left[1 - \exp\left(\frac{-t}{RC}\right)\right]$$

it follows that

$$RC = \frac{t}{\ln \dfrac{V}{V - V_o}}$$

Here

$$p = RC = \frac{5}{\ln \dfrac{10}{6}} = 9.79 \text{ s}$$

$$x = t = 5 \text{ s}, \quad f_x = 0.2 \text{ s}$$
$$k = 1$$

$$y = \frac{V}{V - V_o} = \frac{10}{6}, \quad f_{yr} = 0.05$$

so that

$$f_{pr} = \frac{0.2 + 9.79(0.05)}{5 - 9.79(0.05)} \simeq 0.15 \text{ (15\%)}$$

Thus $RC \simeq 9.79 \text{ s} \pm 1.5 \text{ s}$.

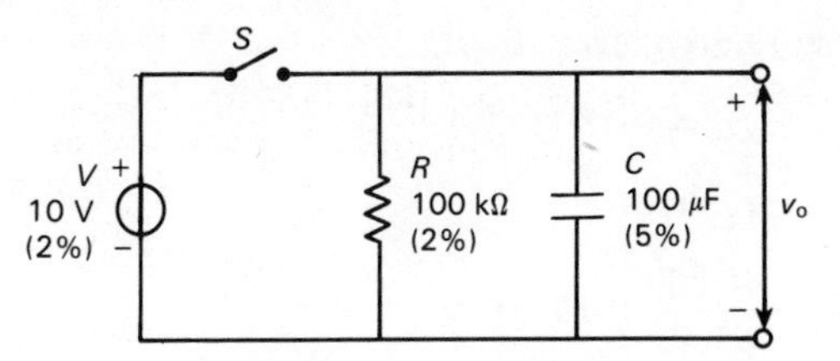

Fig. 6.1 Calculation of the error in V_o

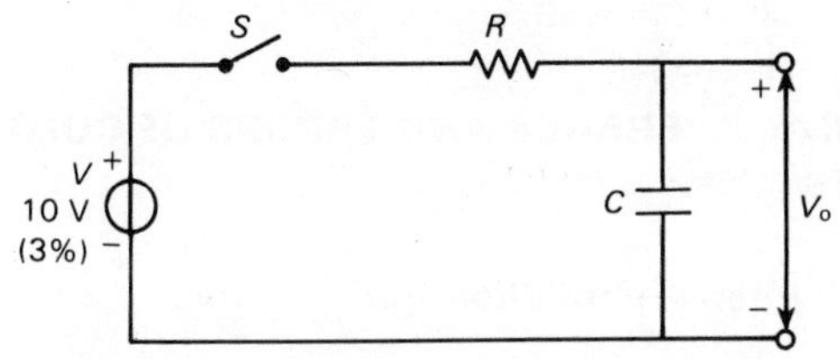

Fig. 6.2 Calculation of the error in RC

The following example is a combination of the functions (a) through (g) in Table 9.

Example 7

Four mutually independent measurements x, y, z and w are as follows:

$$x = 400\ (4\%),\quad y = 100\ (3\%),\quad z = 300\ (5\%),\quad w = 300\ (2\%)$$

Determine the relative and absolute error in the function

$$p = \frac{2x + y^2}{2z - w}$$

(a) Determine the absolute and relative error in the numerator $N = 2x + y^2$.

$$f_{2x} = 2f_x = 2xf_{xr} = 32$$

$$f_{y^2} = 2yf_y = 2y^2 f_{yr} = 600$$

$$f_N = f_{2x} + f_{y^2} = 632$$

$$f_{Nr} = \frac{f_N}{N} = \frac{632}{10\,800} = 0.058$$

(b) Determine the absolute and relative error in the denominator $D = 2z - w$.

$$f_{2z} = 2f_z = 2zf_{zr} = 30$$

$$f_w = wf_{wr} = 6$$

$$f_D = f_{2z} + f_w = 36$$

$$f_{Dr} = \frac{f_D}{D} = \frac{36}{300} = 0.120$$

(c) Determine the relative error in p.

$$f_{pr} = f_{Nr} + f_{Dr} \simeq 0.18 \ (18\%)$$

Then the absolute error is

$$f_p = pf_{pr} = \frac{10\,800}{300} 0.18 \simeq 6.5$$

6.3 STATISTICAL AVERAGES AND ERRORS OF CONTINUOUS VARIABLES

6.3.1 Probability density function *f(x)*

If $f(x)\,\mathrm{d}x$ is the probability that a continuous variable x lies in the interval $x - \frac{1}{2}\,\mathrm{d}x$ and $x + \frac{1}{2}\,\mathrm{d}x$, then $f(x)$ is called the probability density function of x.

In Fig. 6.3 the cross-hatched area represents the probability that x lies in the interval $\mathrm{d}x$ around x_o.

The probability of finding x somewhere in the interval $-\infty$, $+\infty$ is 1, so that

$$\int_{-\infty}^{+\infty} f(x)\,\mathrm{d}x = 1$$

6.3.2 Probability distribution function *F(x); definitions*

This is the probability that the variable will be less than or equal to some value x. In Fig. 6.4 $F(x_o)$ is the probability that $x \leqslant x_o$. In general $F(x)$ is expressed as

$$F(x) = \int_{-\infty}^{x} f(x)\,\mathrm{d}x$$

Arithmetic mean

$$\bar{x} = \int_{-\infty}^{+\infty} xf(x)\,\mathrm{d}x$$

Quadratic mean

$$\overline{x^2} = \int_{-\infty}^{+\infty} x^2 f(x)\,\mathrm{d}x$$

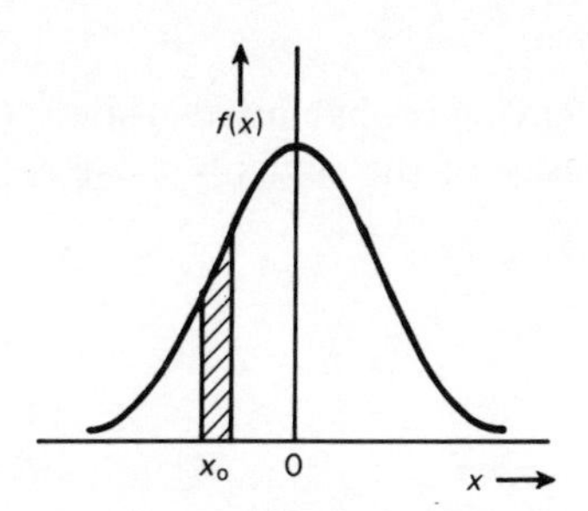

Fig. 6.3 The probability function $f(x)$

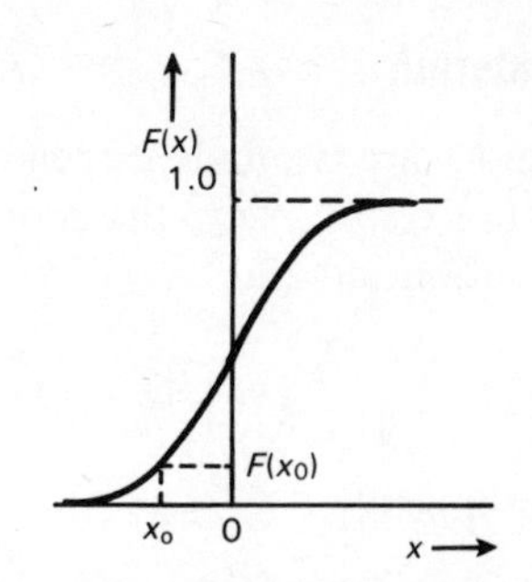

Fig. 6.4 The probability distribution function $F(x)$

Standard deviation

$$\sigma_x = \sqrt{\int_{-\infty}^{+\infty} (x-\bar{x})^2 f(x)\,\mathrm{d}x}$$

$$= \sqrt{\overline{x^2} - \bar{x}^2}$$

Variance

$$\sigma_x{}^2 = \int_{-\infty}^{+\infty} (x-\bar{x})^2 f(x)\,\mathrm{d}x$$

$$= \overline{x^2} - \bar{x}^2$$

Median

This is the value x_m of x for which the probabilities that $x < x_\mathrm{m}$ and $x > x_\mathrm{m}$ are equal. Thus

$$\int_{-\infty}^{x_\mathrm{m}} f(x)\,\mathrm{d}x = \int_{x_\mathrm{m}}^{+\infty} f(x)\,\mathrm{d}x$$

Covariance of two variables

$$\sigma_{xy} = \overline{(x-\bar{x})(y-\bar{y})}$$
$$= \overline{xy} - \bar{x}\bar{y}$$

Correlation coefficient of two variables

$$\rho = \frac{\sigma_{xy}}{\sigma_x \sigma_y}$$

Thus if two random variables x and y are uncorrelated:

$$\overline{xy} = \bar{x}\bar{y}$$

Convolution

If x and y are two independent variables, having probability densities $f(x)$ and $f(y)$ respectively, the probability density of the sum $z = x + y$ is the convolution integral

$$f(z) = \int_{-\infty}^{+\infty} f(x) f(z - x)\, \mathrm{d}x$$

or symbolically

$$f(z) = f(x) * f(y)$$

Normal or Gaussian distribution

This distribution is frequently encountered in statistics, physics and electronics when a large number of independent random causes act together on the quantity.

$$F(x) = \frac{1}{\sigma\sqrt{2\pi}} \int_{-\infty}^{x} \exp\left[-\frac{(x-\bar{x})^2}{2\sigma^2}\right] \mathrm{d}x$$

(Fig. 6.5(a))

$$f(x) = \frac{1}{\sigma\sqrt{2\pi}} \exp\left[-\frac{(x-\bar{x})^2}{2\sigma^2}\right]$$

(Fig. 6.5(b))

6.4 BANDWIDTH OF A PULSE

In Fig. 6.6(a) the area of the symmetrical pulse and that of the rectangle are equal so that

$$\tau f(0) = \int_{-\infty}^{+\infty} f(t)\, \mathrm{d}t$$

According to Chapter 5 the amplitude spectrum of $f(t)$ is

$$F(\omega) = \int_{-\infty}^{+\infty} f(t) \exp(-j\omega t)\, \mathrm{d}t$$

so that (Fig. 6.6(b))

$$F(0) = \int_{-\infty}^{+\infty} f(t)\, \mathrm{d}t$$

$$= \tau f(0)$$

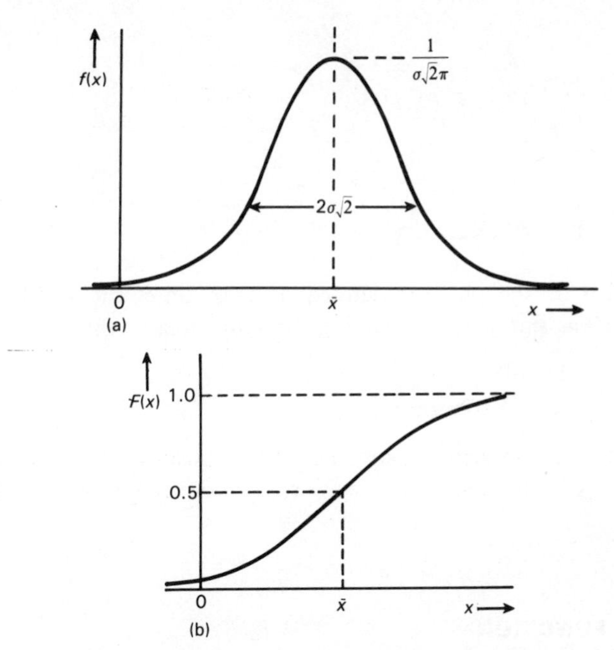

Fig. 6.5 (a) The Gaussian distribution function (b) The Gaussian density function

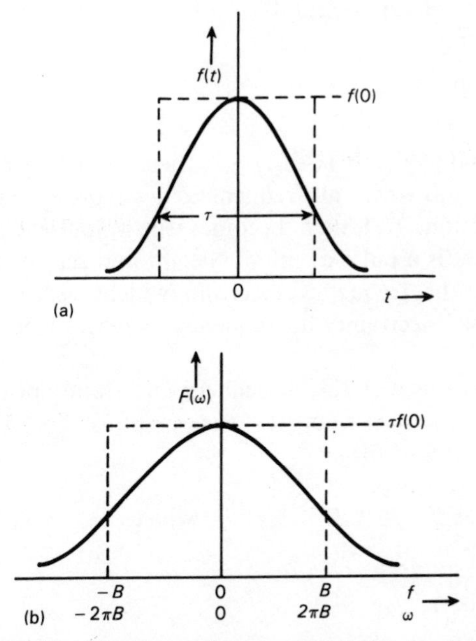

Fig. 6.6 (a) Bandwidth calculation of a symmetrical pulse (b) Spectrum of the symmetrical pulse of (a)

Also

$$f(t) = \frac{1}{2\pi} \int_{-\infty}^{+\infty} F(\omega)\exp(j\omega t)\, \mathrm{d}\omega$$

so that

$$f(0) = \frac{1}{2\pi} \int_{-\infty}^{+\infty} F(\omega)\, \mathrm{d}\omega$$

The bandwidth of the pulse is defined as $1/4\pi$ times the width of the equivalent rectangular spectrum having the same area so that

$$4\pi B\tau f(0) = 2\pi f(0)$$

and finally

$$B = \frac{1}{2\tau}$$

6.5 DELTA FUNCTION

The spectrum of the rectangular pulse of Fig. 6.7(a) is

$$F(\omega) = \int_{-\tau/2}^{+\tau/2} A \exp(-j\omega t)\, \mathrm{d}t$$

$$= A\tau \frac{\sin \frac{1}{2}\omega\tau}{\frac{1}{2}\omega\tau}$$

If, now, $\tau \to 0$ and $A \to \infty$ while $A\tau = 1$, then $F(\omega) \equiv 1$. A quantity which produces a spectrum where all frequencies are represented, is called **white noise**. The function $f(t)$ then becomes the impulse function or delta function $\delta(t)$, that is a pulse of infinite height and zero width (Fig. 6.7(b)).

Apparently, the frequency spectrum widens as the signal duration decreases; i.e. the uncertainty in frequency increases when the uncertainty in time decreases.

This is analogous with the Heisenberg uncertainty principle in modern physics.

Some properties of $\delta(t)$:

$$\delta(t) = \int_{-\infty}^{+\infty} \exp(-j\omega t)\, \mathrm{d}t, \quad \delta(t) = 0 \text{ when } t \neq 0, \delta(t) = \infty \text{ at } t = 0$$

$$\int_{-\infty}^{+\infty} \delta(t)\, \mathrm{d}t = 1$$

$$\int_{-\infty}^{+\infty} F(t)\delta(t)\, \mathrm{d}t = F(0) \text{ for any } F(t)$$

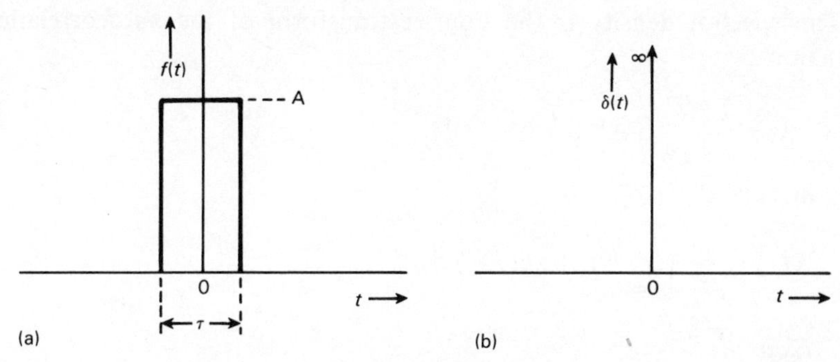

Fig. 6.7 (a) Rectangular pulse used to define the delta function (b) The delta function

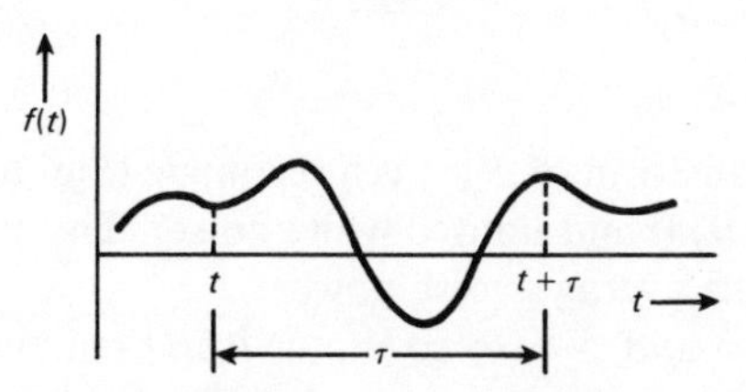

Fig. 6.8 Waveform for defining the autocorrelation function

6.6 AUTOCORRELATION FUNCTION

The autocorrelation function $R(t, \tau)$ describes the measure of dependence with time τ of a random variable $f(t)$ (Fig. 6.8)

$$R(t, \tau) = f(t)f(t + \tau)$$

If $\bar{f}$ is constant, then

$$R(\tau) = \overline{f(t)f(t + \tau)}$$

$$= \lim_{T \to \infty} \frac{1}{T} \int_0^T f(t)f(t + \tau)\, dt$$

$$R(\tau) = \overline{f^2} \quad \text{when } \tau = 0$$

$$R(\tau) = \bar{f}^2 \quad \text{when } \tau = \infty$$

6.7 SPECTRAL DENSITY, POWER DENSITY OR POWER SPECTRUM

In case $f(t)$ of Section 6.6 is a noise function $n(t)$, $\bar{n}$ is constant and $R(0) = \overline{n^2}$. This is the variance and also the average noise power in a 1 Ω resistor.

The spectral density is the Fourier transform of the autocorrelation function

$$N(\omega) = \int_{-\infty}^{+\infty} R(\tau)\exp(-j\omega\tau)\, d\tau$$

and inversely

$$R(\tau) = \frac{1}{2\pi}\int_{-\infty}^{+\infty} N(\omega)\exp(j\omega\tau)\, d\omega$$

Example

In Fig. 6.9(a)

$$R(\tau) = N\frac{\sin 2\pi B\tau}{2\pi B\tau}$$

The Fourier transform of $R(\tau)$ is a rectangle (Fig. 6.9(b)) having width $2B$ and height $N/(2B)$ (band-limited white noise). The area of this rectangle is $R(0) = N = \overline{n^2}$, the average noise power.

If, now, $B \to \infty$ and $N \to \infty$ so that $N/(2B)$ remains constant, then

$$R(\tau) \to \frac{N}{2B}\,\delta(\tau) = N_w\,\delta(\tau)$$

(Fig. 6.9(c)), where N_w is the spectral density of white noise (Fig. 6.9(d)). Thus in white noise $n(t)$ and $n(t+\tau)$ are uncorrelated for any value of τ.

6.8 NOISE BANDWIDTH

In Fig. 6.10 white noise of spectral density N_w is applied to the input of a linear system having a transfer function $H(j\omega)$. The average output noise power is

$$\overline{n_o^2(t)} = R(0)$$

$$= N_w \int_{-\infty}^{+\infty} |H(j\omega)|^2\, df$$

If in Fig. 6.10 the area of the rectangle is such that

$$H^2(0)B_n = \int_0^{\infty} |H(j\omega)|^2\, df$$

then the noise bandwidth is defined as

$$B_n = \frac{1}{H^2(0)}\int_0^{\infty} |H(j\omega)|^2\, df$$

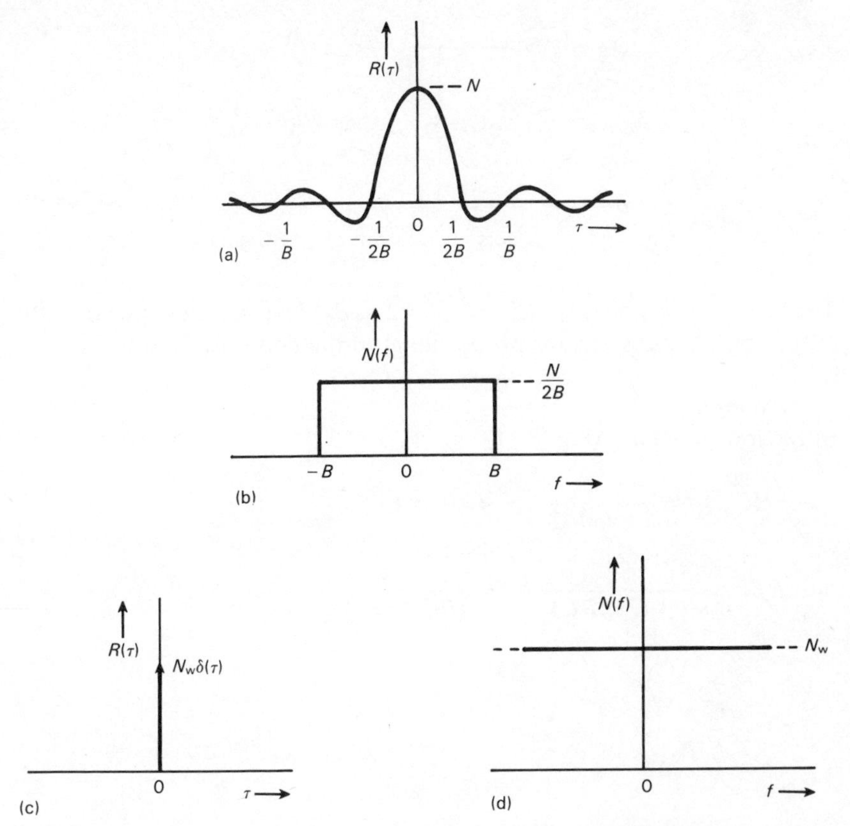

Fig. 6.9 (a) Example of autocorrelation function (b) Fourier transform of the autocorrelation function (c) The autocorrelation function when $B \to \infty$ and $N \to \infty$ (d) The Fourier transform of the delta function is white noise

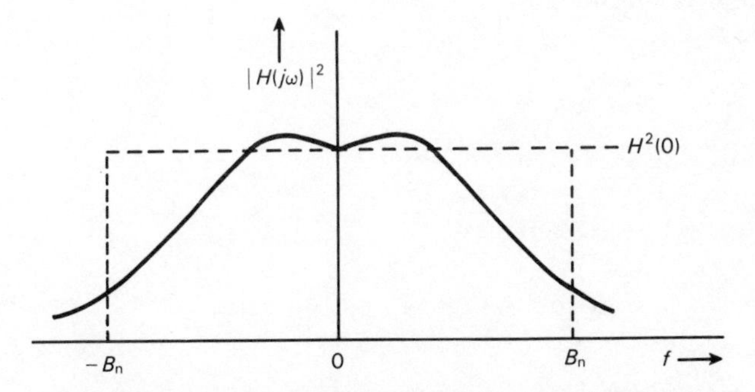

Fig. 6.10 A transfer function and the noise bandwidth

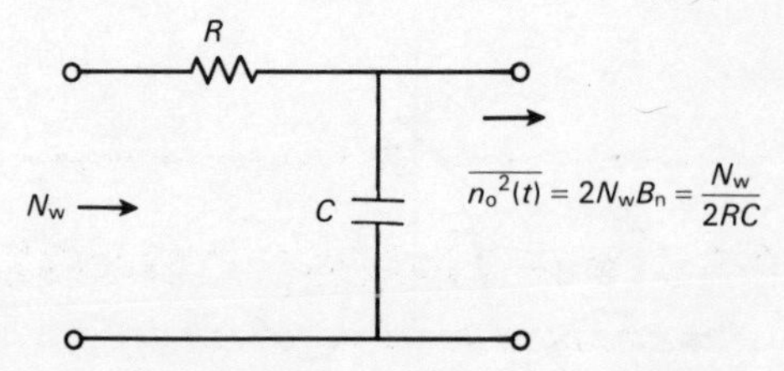

Fig. 6.11 The noise bandwidth of an RC circuit

The output noise power is then $\overline{n_o^2(t)} = 2N_wB_n$. Thus the noise power at the output of a linear system is proportional to the noise bandwidth.

Example

In the low-pass filter (Fig. 6.11)

$$|H(j\omega)|^2 = \frac{1}{1+(\omega RC)^2}, \quad H(0) = 1$$

$$B_n = \int_0^\infty \frac{1}{1+(2\pi fRC)^2}\,\mathrm{d}f = \frac{1}{4RC}$$

7
Passive Components

7.1 LINEAR RESISTORS

7.1.1 General

The characteristics of fixed resistors are usually indicated by a standard color code (IEC) as shown in Fig. 7.1(a). The resistance code consists of three or four bands followed by the tolerance band. If no tolerance band is shown, the tolerance is ± 20 percent.

Sometimes a temperature coefficient band is added to the right of the tolerance band. An example is shown in Fig. 7.1(b).

Resistance values and their tolerances are standardized according to the series E6, E12, E24, E48, E96, E192 (Fig. 7.1(c)).

7.1.2 Carbon resistors

Composition resistors

A mixture of powdered carbon and a binder is pressed together to form a rod. The resistor is then coated with a lacquer (Fig. 7.2(a)). The current noise ($1/f$-noise) of these resistors is high (see Chapter 20) due to intermittent contacts between connecting granules. They are used in applications where performance tolerances are rather loose. The temperature coefficient is negative and about 200–500 ppm $^\circ C^{-1}$.

Film resistors

A film of pure carbon is deposited on a ceramic body at a high temperature. A helical groove is cut in the resistive layer which trims the resistance to the required value. Finally, the resistor is coated with lacquer (Fig. 7.2(b)). The stability ($\Delta R/R$) is very good; the (negative) temperature coefficient is less than 200 ppm $^\circ C^{-1}$.

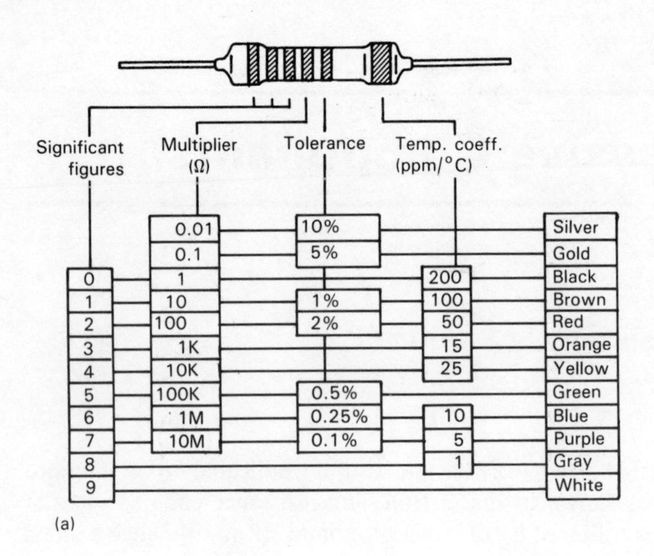

(a)

Orange Yellow Gray Red Brown

34.8 kΩ ± 1%

(b)

series	E6	E12	E24
tolerance	±20%	±10%	±5%
resistance values	10	10	10
			11
		12	12
			13
	15	15	15
			16
		18	18
			20
	22	22	22
			24
		27	27
			30
	33	33	33
			36
		39	39
			43
	47	47	47
			51
		56	56
			62
	68	68	68
			75
		82	82
			91
	100	100	100

(c)

Fig. 7.1 (a) Illustration of the color code for resistors (b) Example of the color code of a 1 percent resistor (c) E series of fixed resistors

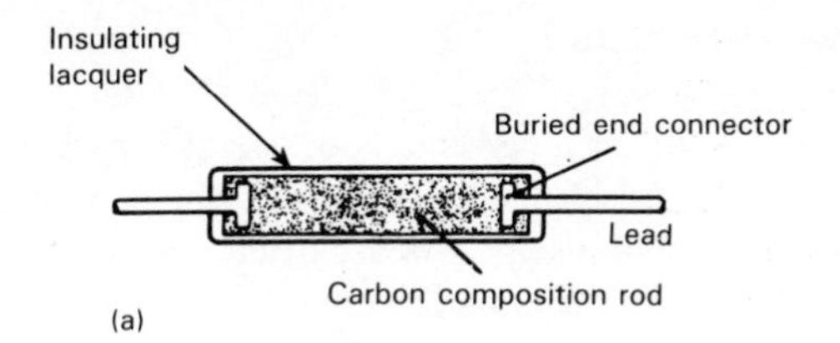

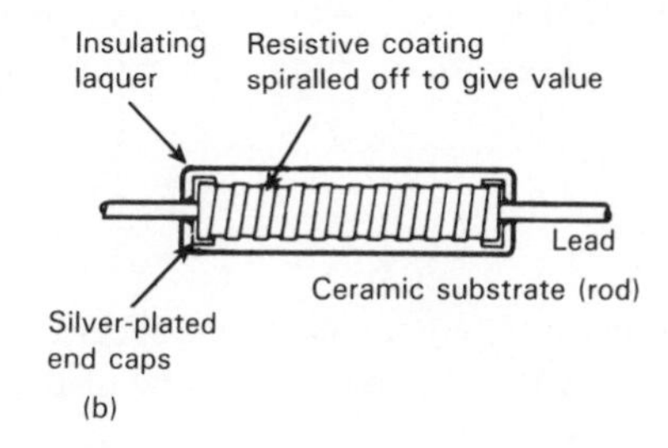

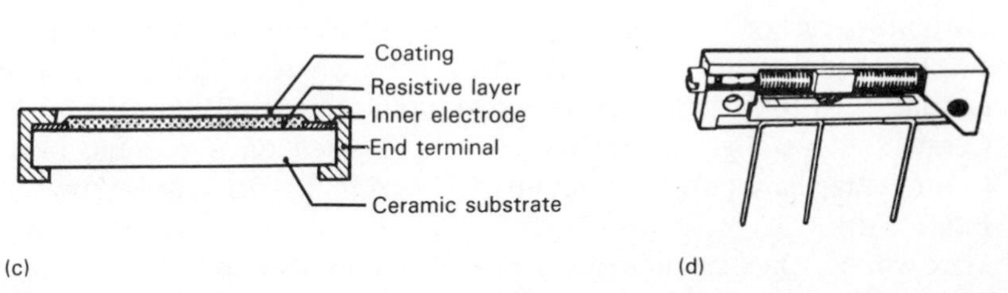

Fig. 7.2 (a) Structure of carbon composition resistors (b) Structure of carbon film resistors (c) Structure of chip resistor (d) Multiturn cermet resistor

7.1.3 Metal film resistors

A film of metal alloys (Ni, Au, Al) is deposited on a ceramic body and coated with a lacquer. The current noise is low; the temperature coefficient is about 50–100 ppm $^{\circ}C^{-1}$ or less and can be positive or negative. They are used in precision applications, e.g. active filters.

7.1.4 Chip resistors

A metal-glaze layer is screened on a ceramic body (Fig. 7.2(c)). The composition of the metal glaze determines the resistance value. A protective coating covers the body.

Chip resistors are used in communications equipment, radio, TV, etc. The temperature coefficient is less than 200 ppm $^{\circ}C^{-1}$.

7.1.5 Wire wound resistors

Resistive wire is wound on a ceramic rod. Metal caps are pressed over the ends of the rod. The resistor is coated with cement or enamel. Resistance value, tolerance and rated dissipation are usually printed on the body. They are used in high precision applications.

Some alloys and temperature coefficients are listed below:

constantan (55% Cu, 45% Ni) $\alpha \simeq 2$ ppm $^\circ C^{-1}$
nichrome (65% Ni, 12% Cr, 23% Fe) $\alpha \simeq 1.7$ ppm $^\circ C^{-1}$
manganin (84% Cu, 12% Mn, 4% Ni) $\alpha \simeq 0.2$ ppm $^\circ C^{-1}$

7.1.6 Variable resistors

Variable resistors are used in all areas of engineering, either as preset resistors or as control elements. Three types can be distinguished according to the materials used:

- Carbon: A carbon track is fixed on a layer of resin-bonded paper and a slider moves along the track. These are made in values of 220 Ω to 4.7 MΩ and intended for general purposes (e.g. volume control).
- Cermet: A metal-glaze resistive element is fixed on a ceramic base. Cermets may have power ratings up to 5 W. A multi-turn type is shown in Fig. 7.2(d).
- Wire wound: On a ceramic ring a wire or ribbon is wound. The windings are coated with a layer of cement to prevent them from shifting. These are made in values of 0.5 kΩ to 10 kΩ and power ratings up to 100 W.

Single turn and multiturn types are available and the law of resistance variation can be linear or logarithmic.

7.2 NONLINEAR RESISTORS

7.2.1 Voltage dependent resistors (VDR)

VDR resistors are also called **varistors**. The materials used are SiC, ZnO or TiO_2. The crystals of these materials are pressed together with a ceramic binder in the shape of a disk. The contact of particles results in a large number of series and parallel rectifying actions (Fig. 7.3(a)).

A voltage–current characteristic is shown in Fig. 7.3(b) (linear) and 7.3(c) (log–log). The characteristic equation is $V = CI^K$ where C and K are constants which depend on the composition of the material and the

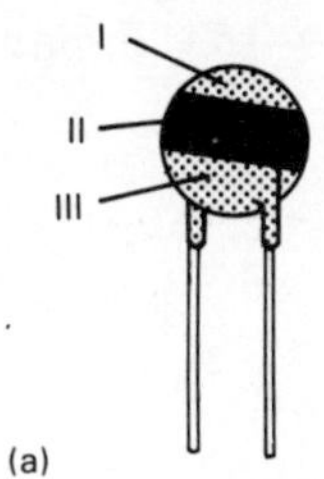

Band I	Band II	Band III				
		Black	Red	Yellow	Blue	Gray
Brown = 100 mA	Brown	—	—	—	8	10
Red = 10 mA	Red	12	15	18	22	27
Orange = 1 mA	Orange	33	39	47	56	68
	Yellow	82	100	120	150	180
	Green	220	270	330	—	—

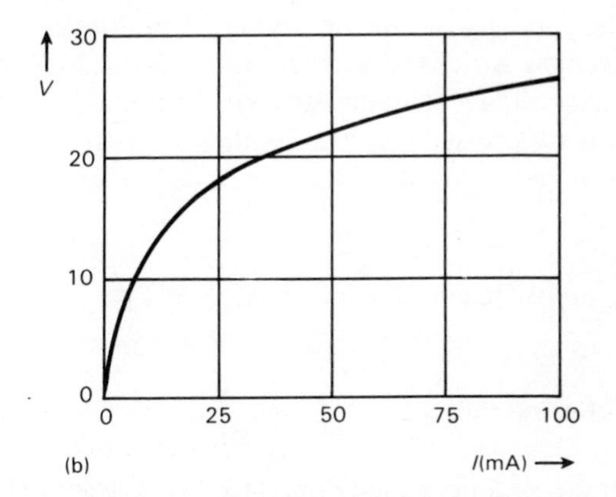

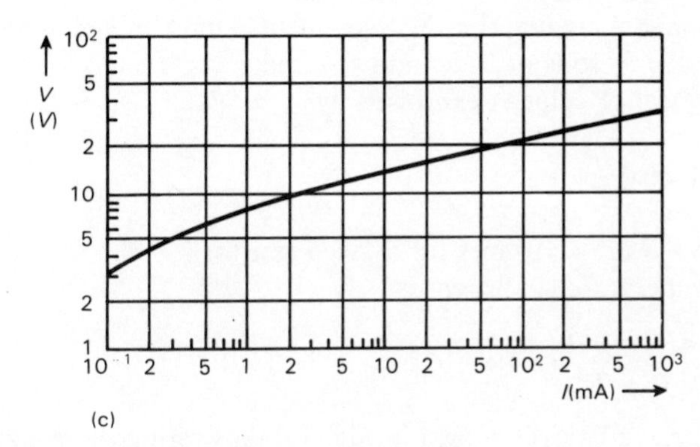

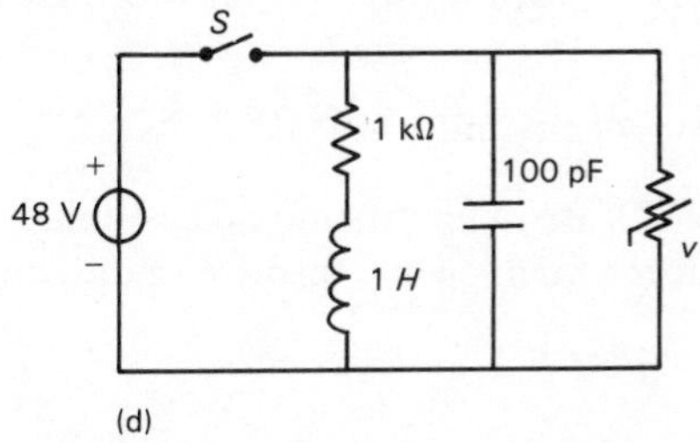

Fig. 7.3 (a) Structure and color code of VDRs (b) Typical voltage–current characteristic of a VDR (linear scales) (c) Typical voltage–current characteristic of a VDR (log–log scales) (d) A protection circuit using a VDR

manufacturing process. C is the **form factor**; its value is 40–1600 Ω. Some K-values are

	K
SiC	0.2–0.4
TiO_2	0.15
ZnO	0.035

A characteristic with $C = 500$, $K = 0.4$ is shown in Fig. 7.3(b).

VDR resistors are used for voltage stabilization purposes and in protection circuits. An example is shown in Fig. 7.3(d). When switch S is opened the high voltage transient between the contacts (due to the energy of the inductor) is suppressed since this voltage decreases the resistance of the VDR. The color code consists of three bands and is illustrated in Fig. 7.3(a).

Example

Orange, red, yellow means $I_{nom} = 1$ mA, V at I_{nom} is 18 volts.

7.2.2 NTC thermistors

These resistors have a negative temperature coefficient. They are manufactured from metal oxides (Fe_2O_3, NiO, CoO, etc.) mixed with a plastic binder, pressed in rods or disks and sintered.

The resistance value is expressed by

$$R_T = A \exp\left(\frac{B}{T}\right)$$

where A and B are constants for a given resistor.

The temperature coefficient is

$$\alpha = \frac{1}{R}\frac{dR}{dT} = \frac{-B}{T^2} \quad (-5 \times 10^{-2} < \alpha < -3 \times 10^{-2})$$

A set of characteristics is shown in Fig. 7.4 illustrating the equation

$$\frac{R_{25}}{R_t} = \exp\left[B\left(\frac{1}{298} - \frac{1}{T}\right)\right] \quad (t \text{ in } ^\circ\text{C}, \ T \text{ in } K)$$

Applications of NTCs are in temperature measurements, compensation of unwanted temperature coefficients, protection of circuits, stabilizing purposes, etc.

7.2.3 PTC thermistors

PTC resistors are manufactured from materials such as $BaTiO_3$, $SrTiO_3$, doped with La, Bi, Sb or Nb, mixed with a binder and sintered. The temperature coefficient is only positive within a certain temperature range.

A typical resistance characteristic is shown in Fig. 7.5. Typical values of the positive temperature coefficient are $7 \times 10^{-2} < \alpha < 60 \times 10^{-2}$.

Applications of PTCs are in temperature stabilization, current limiting, protection of circuits.

7.2.4 Light dependent resistors (LDR)

LDRs are made from pure cadmium sulphide which is mixed with a binder, disk-shaped and sintered (Fig. 7.6(a)). This semiconductor has a high resistance in the dark. When illuminated, generation of electrons and holes takes place increasing the conductivity. The resistance is expressed by the equation

$$R = BL^{-K}$$

where B and K are constants ($K \simeq 0.7$–0.9), depending on the manufacturing process. L is the illumination level in lux.

A typical characteristic is shown in Fig. 7.6(b) on a log–log scale for $B \simeq 10^5$ and $K \simeq 1$.

The resistors are (slightly) temperature dependent and they have their maximum sensitivity at wavelengths of about 650 nm. They also exhibit a recovery time when a certain illumination level is suddenly reduced to darkness due to a time delay in the recombination process. The recovery rate is typically $200\ \text{k}\Omega\,\text{s}^{-1}$.

NTC thermistors, PTC thermistors and LDRs are actually sensors since changes in resistance values are caused by nonelectrical quantities. They are members of the large family of transducers which are described in Chapter 8.

7.3 CAPACITORS

7.3.1 General

A capacitor is basically a system of two parallel plates separated by a dielectric. The capacitance value is expressed as

$$C = \varepsilon \frac{S}{d} = \varepsilon_r \varepsilon_v \frac{S}{d}$$

where

C is the capacitance in farads

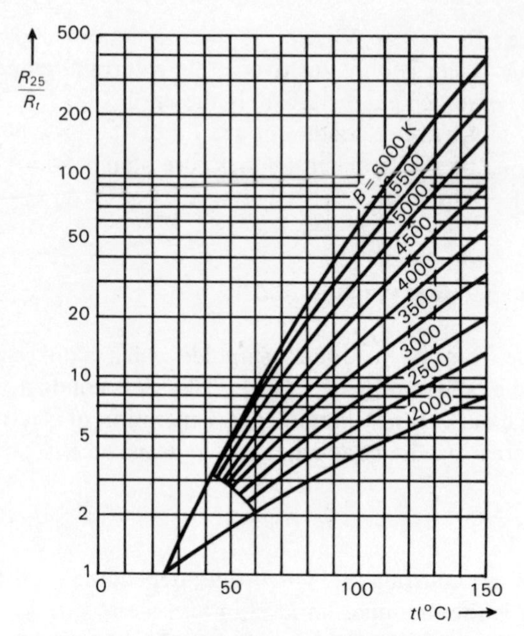

Fig. 7.4 Typical characteristics of an NTC resistor

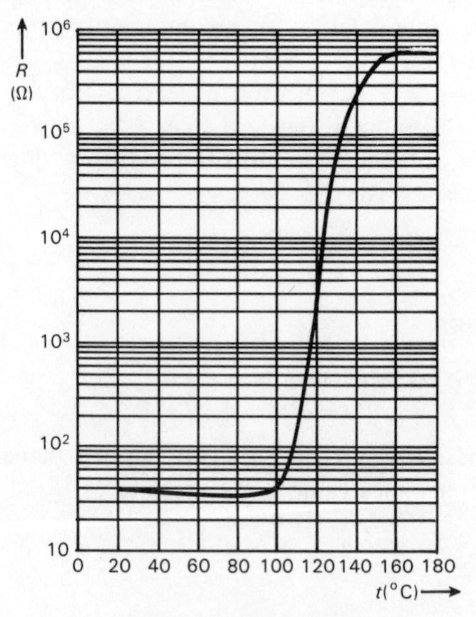

Fig. 7.5 Typical characteristics of a PTC resistor

(a)

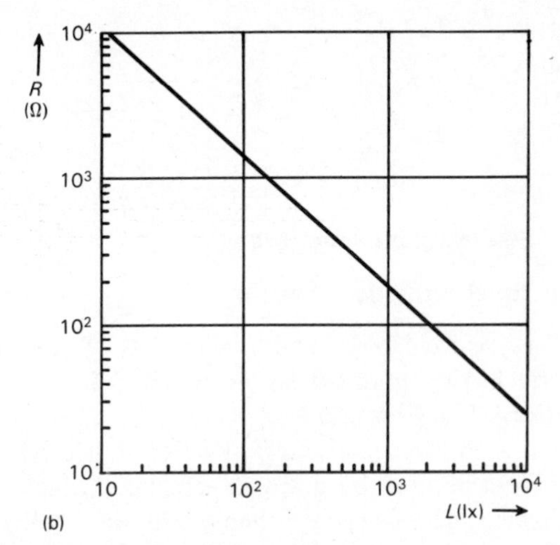

Fig. 7.6 (a) An LDR (b) Typical characteristic of an LDR

ε is the dielectric constant or permittivity
ε_r is the relative dielectric constant (in air $\varepsilon_r \simeq 1$),
$\varepsilon_v = 8.85 \times 10^{-12}\ \mathrm{F\,m^{-1}}$ is the dielectric constant of vacuum
d is distance of the plates in m
S is area of the plates in m^2

The dielectric has a resistance

$$R = \frac{d}{S}\rho$$

where

R is the resistance between the electrodes in Ω
ρ is the specific resistance of the dielectric in Ω m

Thus

$$C = \varepsilon \frac{\rho}{R}$$

A voltage V across the capacitor causes a leakage current $I_l = V/R$ so that

$$C = \frac{\varepsilon \rho I_l}{V}$$

This shows that the leakage current of a capacitor is proportional to its capacitance value. Some values of ε_r are listed below.

vacuum	1
air	1.0006
paper	2

glass	5–7
mica	7
Al_2O_3	8
Ta_2O_5	26

7.3.2 Electrolytic capacitors

Aluminum electrolytic capacitor

Al-foil is oxidized on one side (Fig. 7.7). The oxide layer (Al_2O_3) is the dielectric having a thickness of about 0.1 μm and a high electric field strength ($\approx 7 \times 10^5$ V mm^{-1}).

A second layer (cathode), made of etched Al-foil is inserted. The two layers are separated by a spacer of paper which is impregnated with an organic acid. The layers are then wound in a roll and mounted.

The electrolytic is polarized: the anode should be positive with respect to the cathode. If not, hydrogen gas will form which damages the dielectric layer causing a high leakage current or blow-up.

The equivalent circuit of the electrolytic is shown in Fig. 7.8 as well as the (approximate) vector diagram.

The losses of the capacitor are expressed by the **dissipation factor** which is defined as

$$\tan \delta = \omega R_s C$$

Al-electrolytics have large tolerances (often −10% to +50%). Values up to 10 000 μF can be manufactured. They are used in noncritical applications like coupling, bypass, filtering, etc.

Solid aluminum electrolytic

The cathode is the semiconductor MnO_2 (Fig. 7.9) which is mixed with glass fiber to prevent contact between anode and aluminum contact foil. The electrolytic is nonpolarized; it is reliable and stable. The losses are smaller than in the aluminum electrolytic.

Tantalum electrolytic

The anode consists of sintered Ta-powder (Fig. 7.10). The dielectric is Ta_2O_5 which has a high value of ε_r. Surrounding the dielectric is a semiconducting layer of MnO_2. Then a graphite cathode is deposited around the MnO_2 and finally the capacitor is sealed.

These capacitors have a large temperature range (−85 °C–+100 °C) and they are stable and reliable. A construction, the corresponding color code, the temperature coefficient α and the voltage rating are shown in Fig. 7.11.

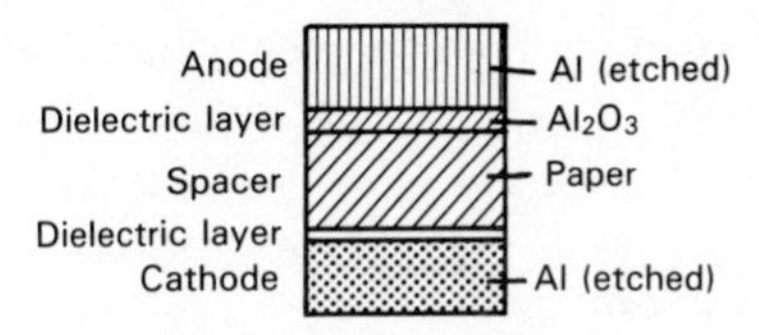

Fig. 7.7 Construction of an Al electrolytic capacitor

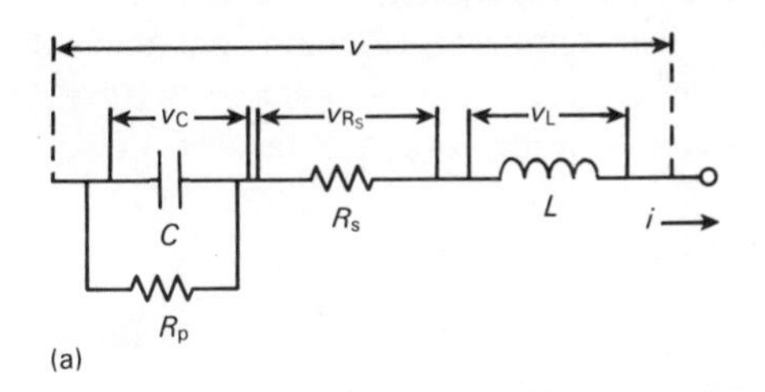

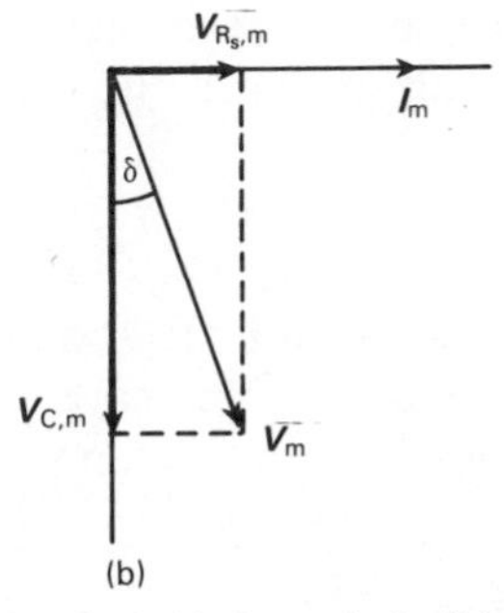

Fig. 7.8 (a) Equivalent circuit of an Al electrolytic (b) Vector representation of an Al electrolytic

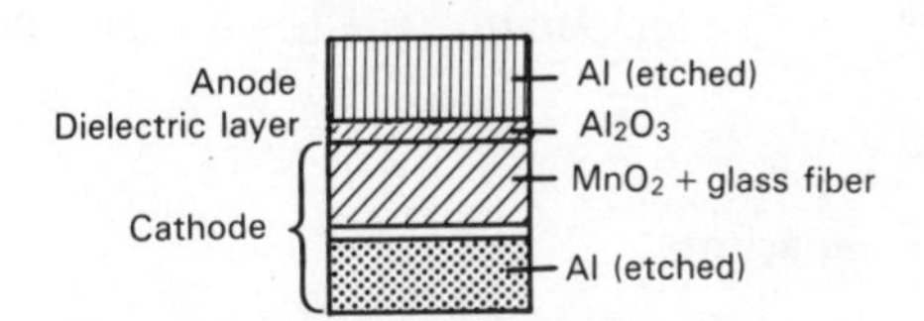

Fig. 7.9 Structure of solid Al electrolytic

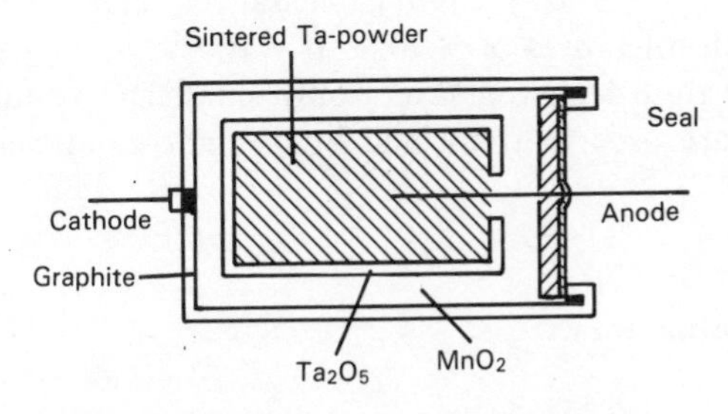

Fig. 7.10 Structure of Ta electrolytic

They are often used in high-frequency applications since the self-inductance is very small.

7.3.3 Ceramic capacitors

The dielectric is ceramic material with deposited metals. They are usually rod or disk shaped. The temperature coefficient depends on the ceramic material used and may vary between -1500 and $+100$ ppm $°C^{-1}$. They are reliable, have a small volume and are largely used in high-frequency applications.

Low *K* ceramic

The dielectric is $TiO_2 + MgO + SiO_2$. The temperature coefficient can be positive or negative, 0–30 ppm $°C^{-1}$, the tolerance is usually 5–10%. They are used in high-frequency filters, tuned circuits, etc.

High *K* ceramic

The dielectric is $BaTiO_3$ mixed with $PbTiO_3$ or $PbZrO_3$ resulting in values of ε_r between 250 and 10 000. They have rather poor stability and considerable losses. They are used in coupling and bypass applications.

Miniature ceramic capacitors

These are used in critical high-frequency applications (filters, tuned circuits, etc.) where the temperature coefficient is important. They are made in the range 0.25 pF to 1 nF. The temperature coefficient α is indicated by a color code (Fig. 7.12).

Dielectric ceramic capacitors

The material is a semiconducting ceramic (Fig. 7.13) with deposited metal on both sides. This results in two depletion layers which make up the very thin dielectric. In this way high capacitance values can be obtained (22–100 nF) with tolerances of -20% to $+100\%$.

Due to the thin depletion layers only small DC voltages ($\simeq$6V) are allowed. They are used in small and lightweight equipment like hearing aids.

7.3.4 Paper capacitors

The dielectric consists of paraffin impregnated paper. The electrodes are

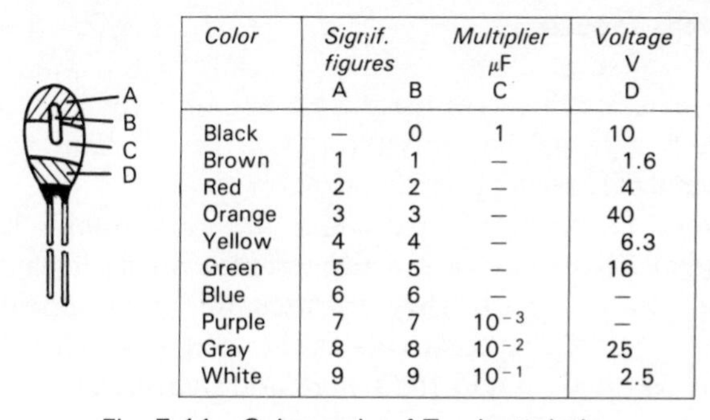

Color	Signif. figures A	B	Multiplier μF C	Voltage V D
Black	—	0	1	10
Brown	1	1	—	1.6
Red	2	2	—	4
Orange	3	3	—	40
Yellow	4	4	—	6.3
Green	5	5	—	16
Blue	6	6	—	—
Purple	7	7	10^{-3}	—
Gray	8	8	10^{-2}	25
White	9	9	10^{-1}	2.5

Fig. 7.11 Color code of Ta electrolytics

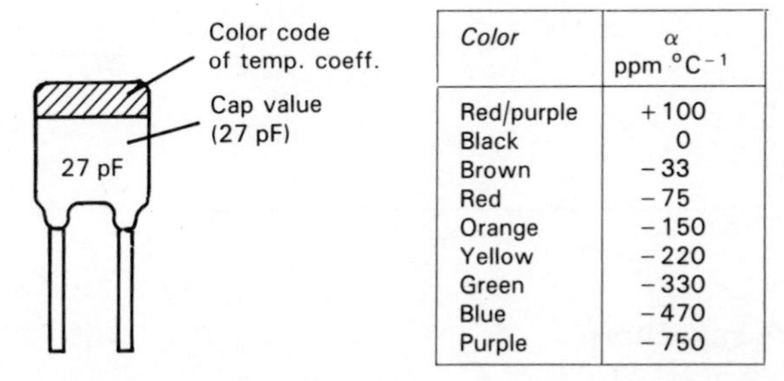

Color	α ppm $^\circ C^{-1}$
Red/purple	+100
Black	0
Brown	−33
Red	−75
Orange	−150
Yellow	−220
Green	−330
Blue	−470
Purple	−750

Fig. 7.12 Miniature ceramic capacitors and color code

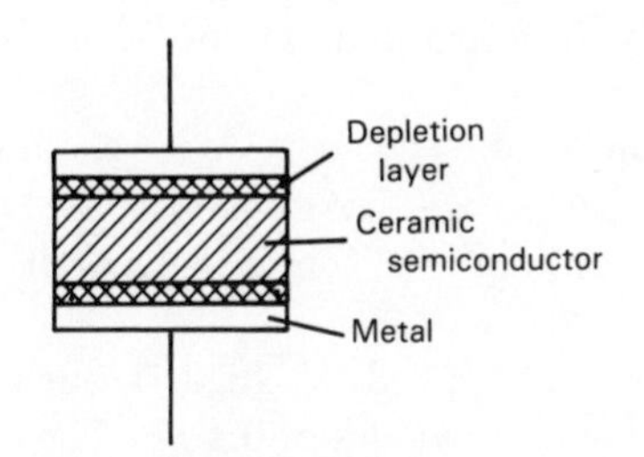

Fig. 7.13 Structure of dielectric ceramic capacitor

made of Sn- or Al-foil (thickness about 4 μm) or of metallized paper (Fig. 7.14(a)). The losses are considerable and frequency dependent: $\tan\,\delta \simeq 10^{-2}$. The self-inductance is often not negligible. They have self-healing properties. Applications are in the low-frequency range and high-voltage equipment (250–1000 V). The color code is shown in Fig. 7.14(b).

When impregnated with Si-oil they can withstand voltages of 20–300 kV. These types are used (for example) in X-ray equipment.

7.3.5 Polymer capacitors

Various polymers are in use as the dielectric: polycarbonate, polystyrene, polyethylene, polypropylene, polystyrol (styroflex), teflon, etc. The construction is similar to that of paper capacitors.

Polystyrene capacitors in particular are very stable, are virtually frequency independent, have a low temperature coefficient and the losses are small ($\tan\delta \simeq 5 \times 10^{-4}$). They are used for tuning purposes and as capacitance standards. The color code is that of Fig. 4.14(b).

Some values of ε_r, ρ and the temperature coefficient α are listed in Table 10.

Table 10

Material	ε_r	ρ (Ω cm)	α (ppm $^\circ C^{-1}$)
polystyrene	2.6	10^{18}	−120
polycarbonate	3.0	10^{17}	−75
polypropylene	2.2	10^{17}	−200

7.3.6 Mica capacitors

Thin layers of mica ($\geqslant$0.003 mm) are stapled with Cu-foil or coated with a layer of deposited silver (Fig. 7.15(a)). They are then vacuum impregnated and coated with epoxy. The temperature coefficient is very small

$$\alpha = \frac{1}{C}\frac{dC}{dT} \simeq 1 \text{ ppm } ^\circ C^{-1}$$

The field strength is about 10^5 V mm^{-1}, $\rho \simeq 10^6$–10^{15} Ω m, $\tan\delta \simeq 10^{-4}$–$10^{-3}$.

These capacitors are applied in high-frequency and high-voltage circuits and are available up to values of 0.1 μF. The color code is shown in Fig. 7.15(b).

7.3.7 Variable capacitors

Variable capacitors usually have air as the dielectric and consist of two assemblies of spaced plates positioned together by insulating members. One set of plates can be rotated (Fig. 7.16(a)).

The losses of these capacitors are very low. To minimize these even more, the insulating members are made from steatite which has a very small value of $\tan\delta$.

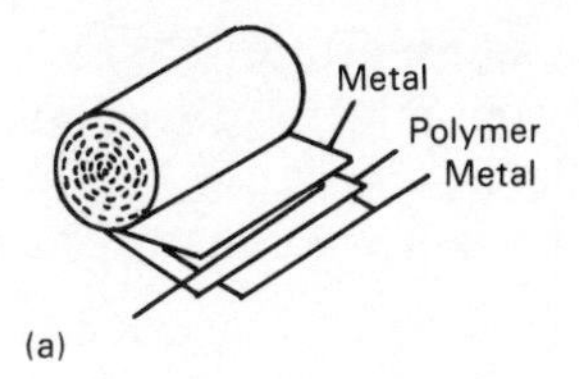

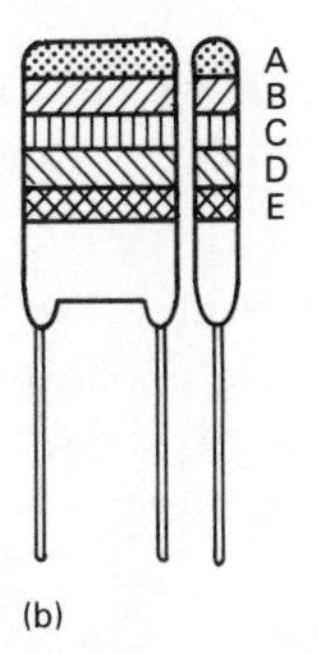

Color	*Signif. figures* A	B	*Multiplier* pF C	*Tolerance* % D	*Voltage* V E
Black	—	0	1	±20	125
Brown	1	1	10		160
Red	2	2	10^2		250
Orange	3	3	10^3		—
Yellow	4	4	10^4		400
Green	5	5	10^5		—
Blue	6	6	10^6		630
Purple	7	7	10^7		—
Gray	8	8	—		—
White	9	9	—	±10	1000

Fig. 7.14 (a) Construction of paper capacitors (b) Color code of paper capacitors

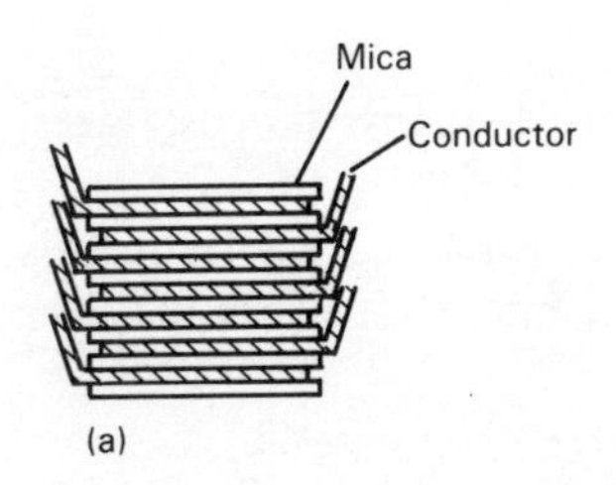

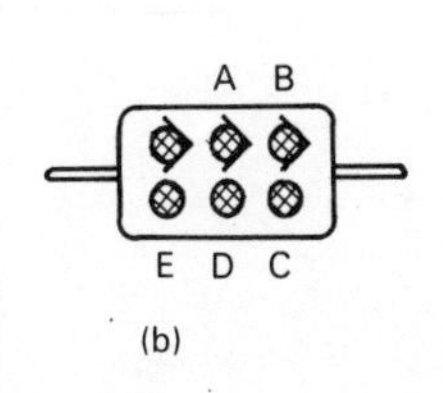

Color	*Signif. figures* A	B	*Multiplier* pF C	*Tolerance* % D	α ppm $^{\circ}C^{-1}$ E
Black	0	0	1	20	±1000
Brown	1	1	10	1	±500
Red	2	2	10^2	2	±200
Orange	3	3	10^3	3	±100
Yellow	4	4	10^4	4	−20...+100
Green	5	5	10^5	5	0...+70
Blue	6	6	10^6	6	—
Purple	7	7	10^7	12.5	—
Gray	8	8	10^{-2}	30	−50...+150
White	9	9	10^{-1}	10	−50...+100
Gold	—	—	10^{-1}	5	—
Silver	—	—	10^{-2}	10	—
No color	—	—	—	20	—

Fig. 7.15 (a) Construction of mica capacitors (b) Color code of mica capacitors

Their main use is in adjustment of the resonance frequency of tuned circuits in receivers and transmitters. Capacitance values of 5–500 pF in different ranges are available.

Various laws of capacitance variation can be made by shaping the plates:

- Linear capacitance: The capacitance value is a linear function of the angle of rotation (Fig. 7.16(b)).
- Logarithmic capacitance: The logarithm of the capacitance value (and also of the wavelength) is a linear function of the angle of rotation (Fig. 7.16(c)).

Air trimmers (Fig. 7.16(d)) are based on the same principles. They are used for adjustment of tuned circuits, filters, etc. in IF and HF applications.

Mica trimmers are available in values of 1.5 pF to 750 pF divided in a number of ranges as well as trimmers with various polymers as the dielectric.

Varicap diodes and monolithic capacitors are covered in Chapters 9 and 13 respectively. A number of capacitor types are shown in Fig. 7.17 in relation to their usable frequency ranges.

7.4 INDUCTORS

Inductors or coils exist in many shapes, sizes, winding patterns and inductance values.

The single layer coil with air core is frequently used in tuning heads of receivers (Fig. 7.18). The inductance formula is approximately

$$L \simeq 0.2n^2 l \frac{(d/l)^2}{9(d/l)+20} \ \mu\text{H} \quad \left(\frac{d}{l} < 3\right)$$

where

n is the number of turns
l is the winding length in cm
d is the diameter in cm

The majority of coils have ferrite cores

- To increase the inductance so that less turns are required.
- To adjust the inductance to the proper value.

Ferrites have a very high resistivity (up to $10^7\ \Omega\,\text{m}$) so that eddy current losses are negligible, even at high frequencies. They also have a high permeability (10^3–10^4) thus concentrating the electromagnetic field lines. They are made by pressing and sintering a mixture of metallic oxides into a

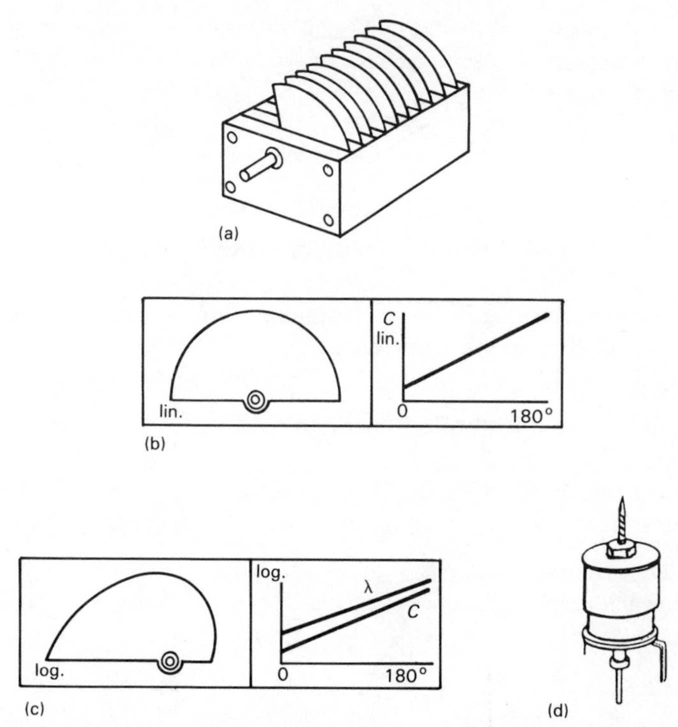

Fig. 7.16 (a) Example of a variable capacitor construction (b) Linear law of capacitance variation (c) Logarithmic law of capacitance variation (d) Example of air trimmer construction

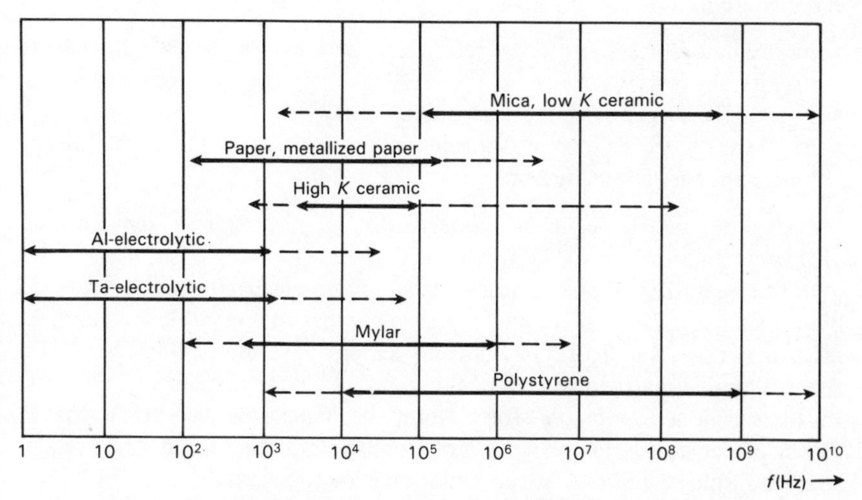

Fig. 7.17 Operating frequencies of a number of capacitor types

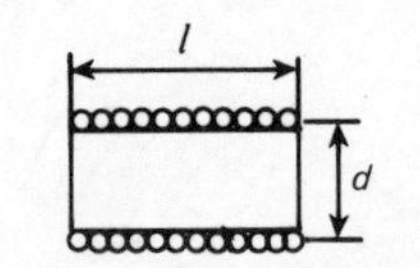

Fig. 7.18 Single layer coil

composition $MO.Fe_2O_3$ where M stands for a divalent metal such as Zn, Ni, Co, Cd, Cu, Mn, etc.

During sintering the mixture crystallizes into a very hard material. Applications of ferrites are numerous:

- Deflection yokes in television picture tubes.
- Cores of flyback transformers in television.
- Antenna cores.
- Cores for tuned inductors and IF transformers.
- Potcores for low-loss filters, magnetic recording heads, resonant cavities, etc.

In microwave applications rare earths are used such as yttrium and gadolinium.

7.5 CRYSTALS

The very sensitive frequency characteristic of a crystal makes it an extremely useful component, in particular in the field of communications. Here crystals are used:

- In oscillators, as frequency references and in frequency-synthesizing circuits.
- To construct very selective and stable filters.
- As transducers to convert mechanical energy into electrical energy (microphones, sound systems).

Crystal material is usually quartz (SiO_2), although in acoustic and ultrasonic applications $BaTiO_3$ is used.

The operating principle of a crystal is **piezo-electricity** which is caused by unbalancing the positive and negative charges in a crystal. Unlike most other materials, a quartz crystal has no center of symmetry (Fig. 7.19). If a stress is applied in the direction of the x-axis, a charge displacement occurs, thus creating a dipole. A stress along the y-axis also causes a charge displacement along the x-axis but of opposite polarity. On a larger scale, stresses result in voltages at the boundaries of the crystal.

When applying the stress periodically, the crystal will resonate at a

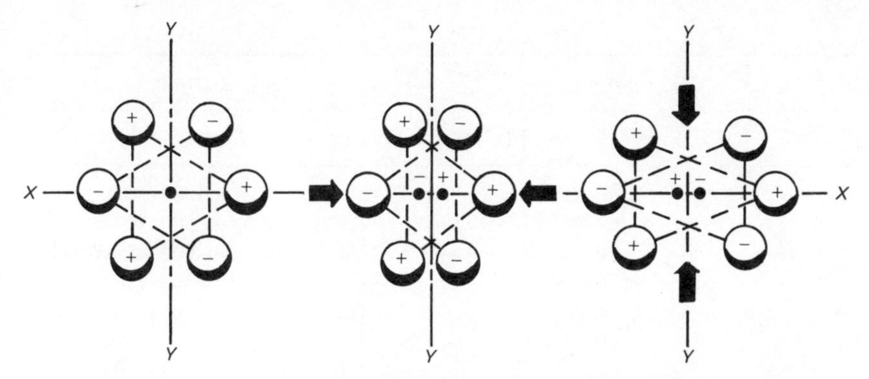

Fig. 7.19 Crystal structure and charge displacements due to stress

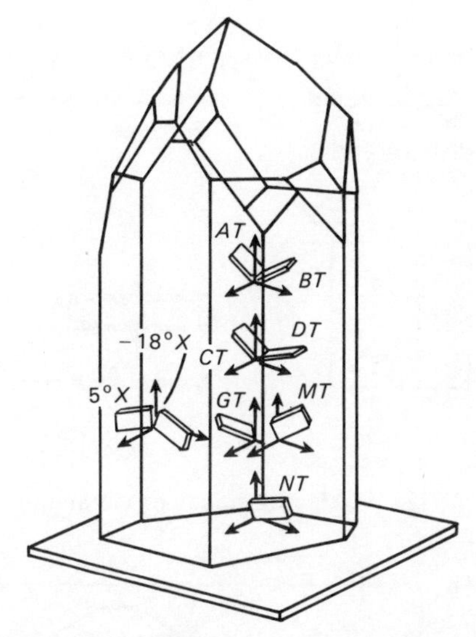

Fig. 7.20 Different cuts of a crystal

certain frequency. Here the required energy is minimum. The resonant frequency is determined by the size of the crystal; the majority of frequencies lie between 50 kHz and 50 MHz. Above 50 MHz resonance is possible at higher harmonics (overtone mode).

An important characteristic is the temperature coefficient

$$\alpha = \frac{1}{f}\frac{\mathrm{d}f}{\mathrm{d}t}$$

which is largely determined by the way the crystal is cut (Fig. 7.20). Two main groups can be distinguished: the *X*-group and the *Y*-group. The relevant frequency ranges are shown in Table 11.

Table 11

X-group		*Y-group*	
Cut	*Frequency range* (kHz)	*Cut*	*Frequency range* (kHz)
X	40–20 000	Y	1000–20 000
$5°X$	0.9–500	AT	500–100 000
$-18°X$	60–350	BT	1000–75 000
MT	50–100	CT	300–1100
NT	4–50	DT	60–500
V	60–20 000	ET	600–1800
		FT	150–1500
		GT	100–550

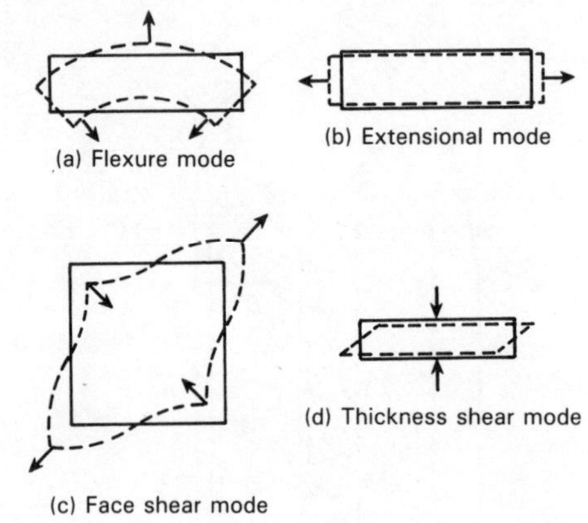

Fig. 7.21 Different modes of vibration

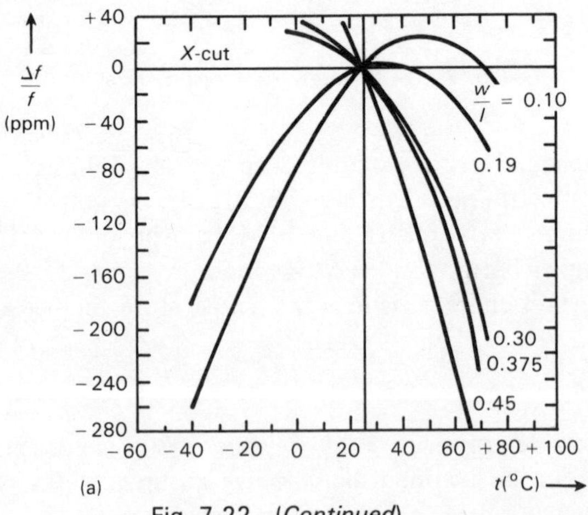

Fig. 7.22 (*Continued*)

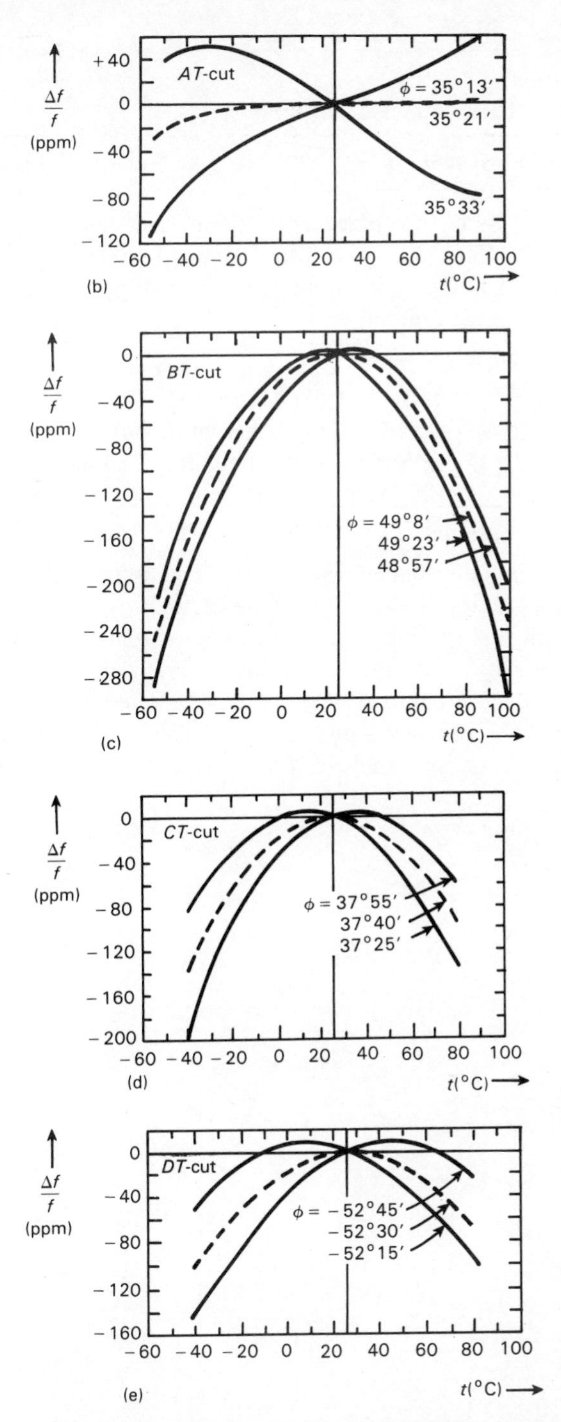

Fig. 7.22 (a) Temperature coefficient of X-cut crystals (b) Temperature coefficients of AT-cut crystals (c) Temperature coefficient of BT-cut crystals (d) Temperature coefficient of CT-cut crystals (e) Temperature coefficient of DT-cut crystals

There are three basic modes of vibration:

(a) Flexure mode (bending or bowing, Fig. 7.21(a)). Cuts: 5°*X*, *NT*. Frequency ≃ 100 kHz.
(b) Extensional mode (displacement along the length of the plate, Fig. 7.21(b)). Cuts: *MT*, *GT*, Frequency: 40–200 kHz.
(c) Shear mode (sliding two parallel planes in opposite directions). This mode is subdivided into:

- Face shear (Fig. 7.21(c)). Cuts: *CT*, *DT*. Frequency: 100–600 kHz.
- Thickness shear (Fig. 7.21(d)). Cuts: *AT*, *BT*.
  Frequency: 0.6–15 MHz (fundamental mode)
  15–60 MHz (overtone mode, third harmonic)
  60–100 MHz (overtone mode, fifth harmonic)
  100–150 (overtone mode, seventh harmonic)

The temperature coefficient of a number of cuts is shown in Fig. 7.22.

The quality factor Q of a crystal is very high (up to 10^6) because crystal losses are much smaller than those of resonant circuits and are mainly due to internal dissipation and mechanical mounting.

The equivalent circuit is shown in Fig. 7.23 where typical values of the elements are: $C = 0.001–0.1$ pF; $C_o = 1–30$ pF, $R = 10\ \Omega–100\ \text{k}\Omega$. The magnitude of L is indicated in Fig. 7.24 as a function of resonant frequency.

If $C_o \gg C$, the crystal impedance is

$$Z = \frac{1}{j\omega}\frac{C}{C_o}\frac{R + j\gamma_s}{R + j\gamma_p\sqrt{1 + \dfrac{C}{C_o}}}$$

where

$$\gamma_s = \frac{\omega}{\omega_s} - \frac{\omega_s}{\omega}$$

$$\gamma_p = \frac{\omega}{\omega_p} - \frac{\omega_p}{\omega}$$

$$\omega_s = \frac{1}{\sqrt{LC}}$$

$$\omega_p = \frac{1}{\sqrt{L\dfrac{CC_o}{C + C_o}}}$$

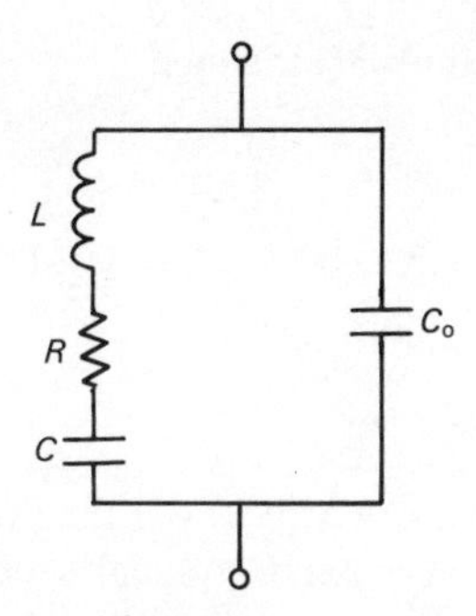

Fig. 7.23 Equivalent circuit of a crystal

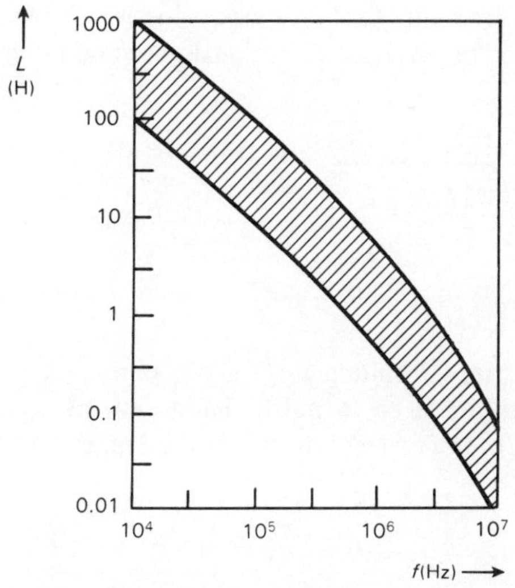

Fig. 7.24 Range of inductance magnitudes as a function of frequency

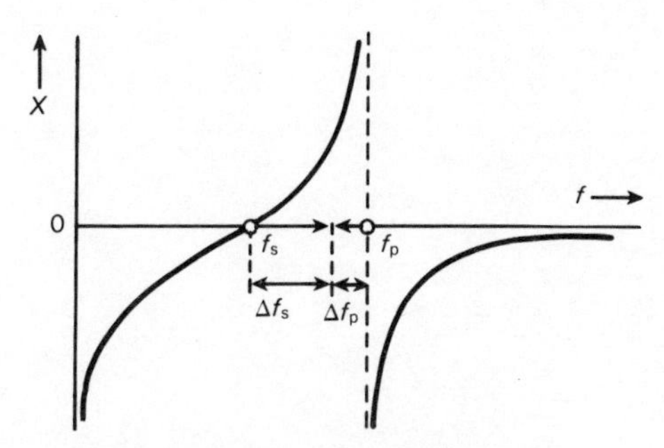

Fig. 7.25 Reactance characteristic of a crystal

Thus two resonant frequencies are defined:

series resonant frequency $f_s = \dfrac{\omega_s}{2\pi}$

parallel resonant frequency $f_p = \dfrac{\omega_p}{2\pi}$

The reactance X is shown in Fig. 7.25 as a function of frequency (where R is ignored; this hardly affects the results).

Offsetting (pulling) of crystal frequencies can (to a small extent) be accomplished by connecting a capacitor in series or in parallel with the crystal.

A series capacitor C^* pulls f_s to a higher value (Fig. 7.25). The same capacitor in parallel pulls f_p to a lower value which is the same value as previously obtained by series pulling. The degree of pulling can be expressed by the equations

$$\frac{\Delta f_p}{f_p} \simeq \frac{C^*}{2\left(1 + \dfrac{C_o}{C}\right)(C_o + C^*)}$$

$$\frac{\Delta f_s}{f_s} \simeq \frac{C}{2(C + C^*)}$$

Generally, the degree of pulling $\Delta f/f \leqslant 500$ ppm.

Crystal filters are used in narrowband systems, e.g. single-sideband filters and IF filters in receivers or comb filters in radar.

8
Transducers

One of the main functions of electronics is the transmission of information. Often this information is of a nonelectrical nature. A **transducer** is required to convert it to an electrical signal. Information in electrical form is easy to manipulate due to the nature of the electron allowing fast transmission and handling of very small to very large signals (current, voltage, power).

Transducers can be grouped into two main categories: passive transducers which need a current or voltage to operate, and self-generating transducers which generate an electrical signal directly.

Of the many nonelectrical quantities only the most frequently occurring will be dealt with.

8.1 TEMPERATURE

8.1.1 Thermistor (see Chapter 7)

Both NTC and PTC thermistors are used, in particular to control, protect or stabilize circuits. Typical resistance versus temperature characteristics are shown in Fig. 8.1 on a semi-log scale. A few typical applications are illustrated in Fig. 8.2.

8.1.2 Semiconductor

For measurement purposes the characteristic of a *pn* junction can be used. A forward biased Si junction is described by the equation

$$I \simeq I_o \exp\left(5800 \frac{V}{T}\right)$$

where V is the voltage (V) across the junction and T the absolute temperature (K). An IC differential circuit is shown in Fig. 8.3. The asymmetry causes a differential voltage ΔV which is a highly linear function of T. Sensitivities of 1 μA $^{\circ}$C^{-1} or 10 mV $^{\circ}$C^{-1} can be obtained.

8.1.3 Crystal

The dimensions of a crystal (often quartz) change with temperature changes. When the crystal is part of an oscillator circuit the varying output frequency is a measure of temperature.

8.1.4 Thermocouple

When two dissimilar conductors are bonded together at one end, a voltage develops across the open ends, the value of which depends upon the temperature. This is caused by the difference in energy levels in both materials: electrons from a higher energy level move to a lower level (Fig. 8.4).

A few frequently used materials and their temperature ranges are shown below.

copper–constantan	-200 to $+400$ °C
iron–constantan	0 to $+750$ °C
chromel–alumel	0 to $+1100$ °C
platinum–platinum rhodium	0 to $+1300$ °C

The common set-up for temperature measurements is shown in Fig. 8.5. One of the junctions is the reference junction, the other the sensing junction. In accurate measurements the reference junction is placed in melting ice or in a calibrated oven.

For greater sensitivity a number of thermocouples may be connected in series to form a **thermopile**. Such a device is used as a radiation pyrometer to measure the temperature of heat radiating objects.

Thermocouples are also used to measure RMS values of voltages and currents. The **Seebeck effect** is based on the same principle. The materials are two different types of semiconductors (Fig. 8.6). When heating the junctions a voltage is produced which can be converted to a current. The sensitivity is about 0.1 mV $°C^{-1}$.

The **Peltier effect** is the reverse of the Seebeck effect (Fig. 8.7): a voltage source connected between the two semiconductors results in a current flow which heats the cold junction and cools the hot junction. Constructions called **frigistors** are based on this effect. The semiconducting material used is Bi_2Te_3 (*n* and *p* material) joined by a copper strip. At the junction, generation takes place. The required energy is withdrawn from the crystal materials thus cooling the copper strip. At the other ends recombination takes place which heats the copper contacts.

Peltier elements are used for cooling purposes on a small scale in chemistry and biology.

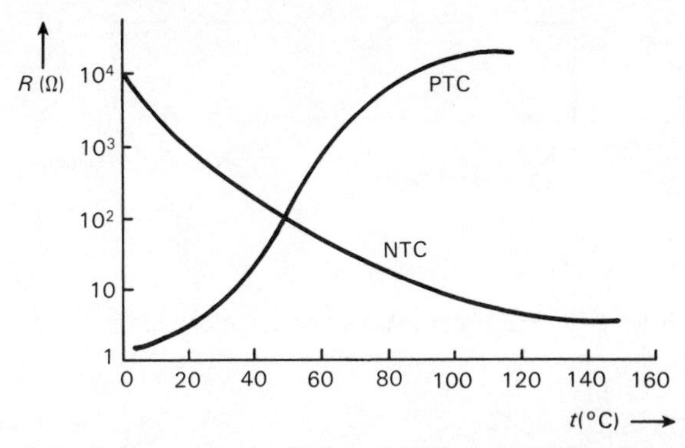

Fig. 8.1 Typical characteristics of NTC and PTC thermistors

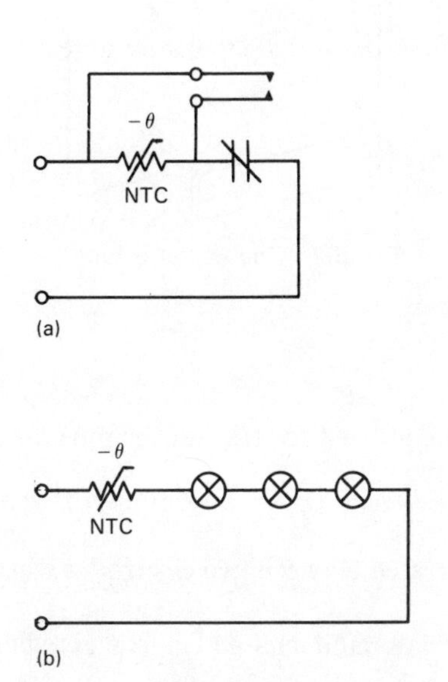

Fig. 8.2 (a) Delaying the turn-on of a relay (b) Delayed turn-on of incandescent lamps to suppress current transient

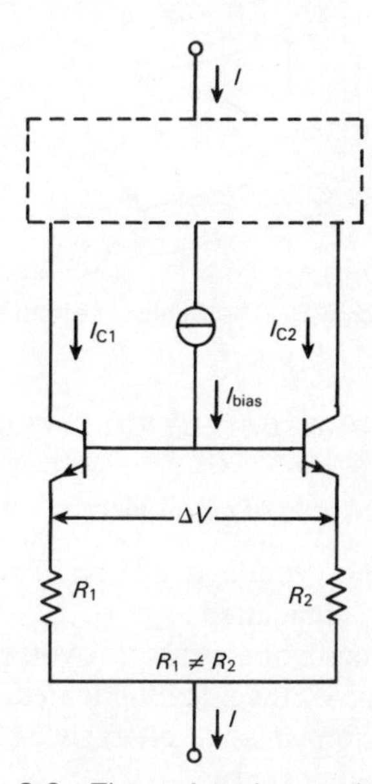

Fig. 8.3 The *pn* junction used for temperature measurements

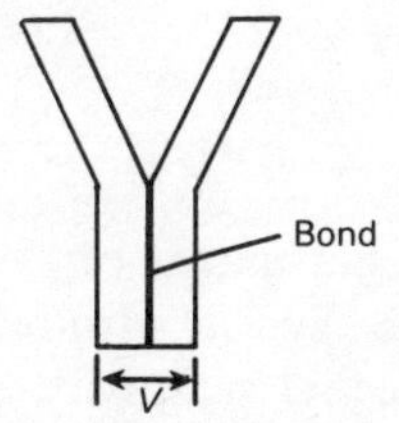

Fig. 8.4 Principle of a thermocouple

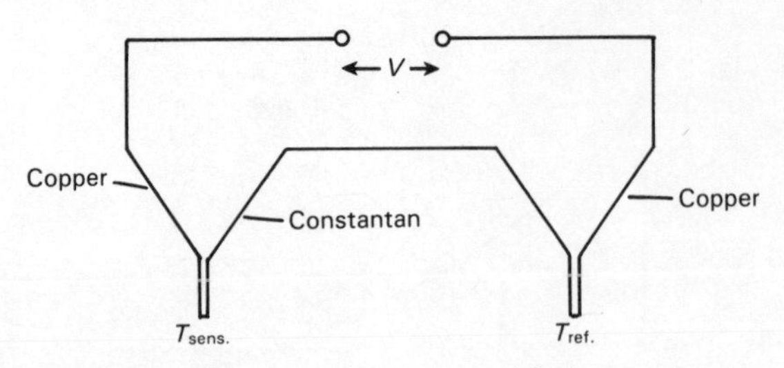

Fig. 8.5 Temperature measurements with thermocouples

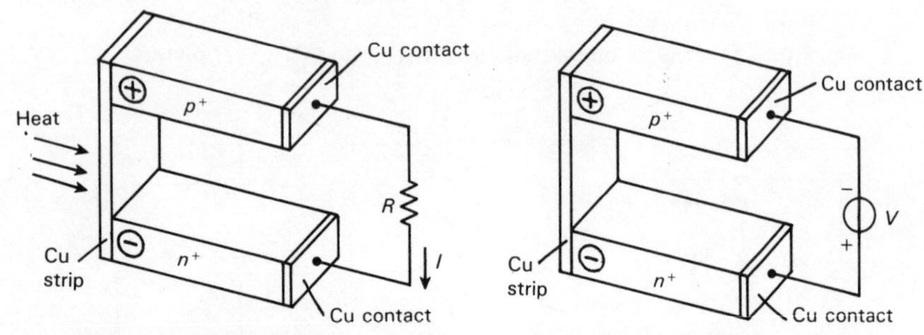

Fig. 8.6 The Seebeck effect

Fig. 8.7 The Peltier effect

8.2 PHOTOELECTRIC EFFECT

Three kinds of photoelectric effect can be utilized for transducer purposes:

(a) Photoconductive effect: The conductivity of a material changes when it is illuminated.
(b) Photovoltaic effect: A voltage is generated between two electrodes when one of them is illuminated.
(c) Photoemissive effect: Certain materials emit charged particles when illuminated.

Devices based on the first two effects are termed **photocells**.

8.2.1 Photoconductive effect

LDR

As stated in Chapter 7 the disadvantage of LDRs is their slow response to changes in illumination levels. Typical turn-on and turn-off characteristics are shown in Fig. 8.8(a). Spectral sensitivity curves of CdS and CdSe are shown in Fig. 8.8(b). Two simple circuits illustrating the use of LDRs are given in Fig. 8.9.

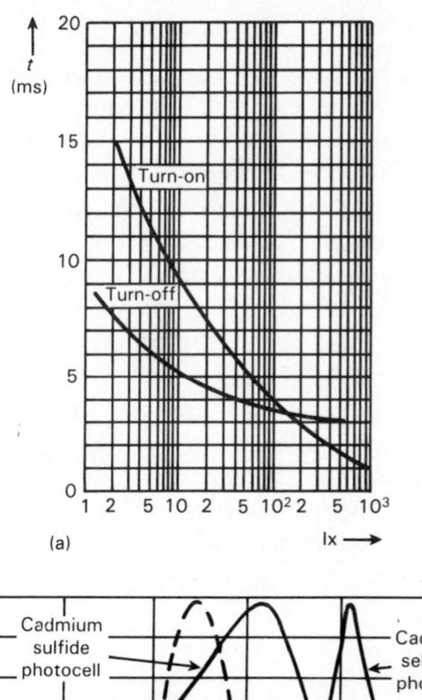

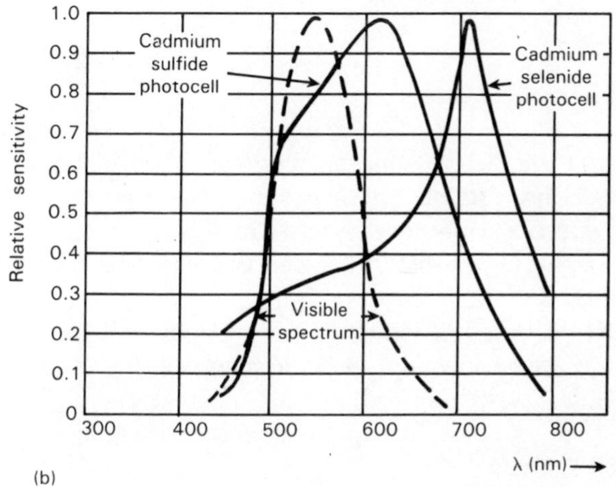

Fig. 8.8 (a) Turn-on and turn-off characteristics of LDR (b) Spectral response of Cd photocells

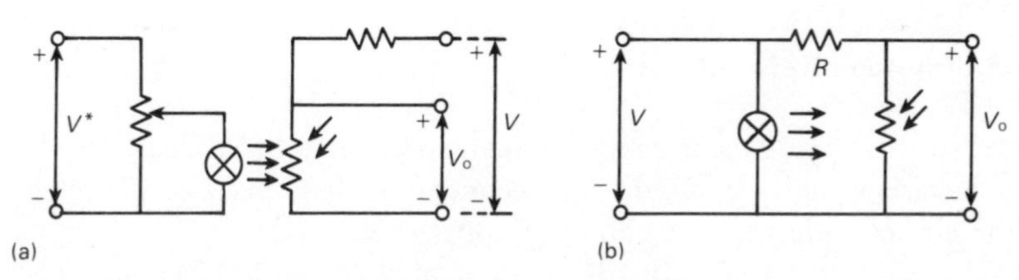

Fig. 8.9 (a) Noise-free potentiometer control using an LDR (b) Voltage stabilizing circuit using an LDR

Photodiode

When illuminating a *pn* junction (Fig. 8.10(a)) electrons in the *p* region and holes in the *n* region will be generated. The contact potential of the junction causes electrons to move to the *n* region and holes to the *p* region. A **photocurrent** will flow when biasing the diode. The magnitude of this current is a function of the illumination level (Fig. 8.10(b)).

The spectral sensitivity of a Si photodiode is maximum at about 800 nm. A circuit for light measurements is shown in Fig. 8.11 where the diode is connected across the inputs of an operational amplifier. Any current through the diode flows through R_f thus producing an output voltage V_o.

Reverse biased *pn* junctions are used in radiation measurements (Fig. 8.12). A single α-particle can produce millions of electron/hole pairs and several volts across the junction.

A special photodiode is the PIN photodiode (see Chapter 9). When reverse biased the *i* layer is depleted and has a very high resistivity. The 'dark current' of such a diode is of the order of 0.5 nA.

8.2.2 Photovoltaic effect

It follows from Fig. 8.10(b) that an illuminated *pn* junction generates a potential when no external voltage source is connected. This is the photovoltaic effect. The **photocell** is based on this principle. The structure is essentially that of Fig. 8.10(a), the semiconductor material CdS for the visible range, PbS or InSb for the infrared.

Photocells are used in industry, photography and for light measurements. Another well-known device is the **solar cell** (Fig. 8.13). The heavily doped *n* region is very thin ($\approx$500 nm); to prevent a high value of the internal resistance, a comb structure is used for the contact. The efficiency of a solar cell is about 15%.

A **phototransistor** is merely a combination of a photodiode and an amplifier (see Chapter 10). Some phototransistors are equipped with a lens making them direction sensitive.

8.2.3 Photoemissive effect

If electrons with sufficient energy strike certain materials, electrons from these materials will be emitted. This **secondary emission** is used in the most sensitive photodetector, the photomultiplier (Fig. 8.14).

The semitransparent cathode is illuminated and emits electrons. These are accelerated to the first **dynode** which is at a positive potential.

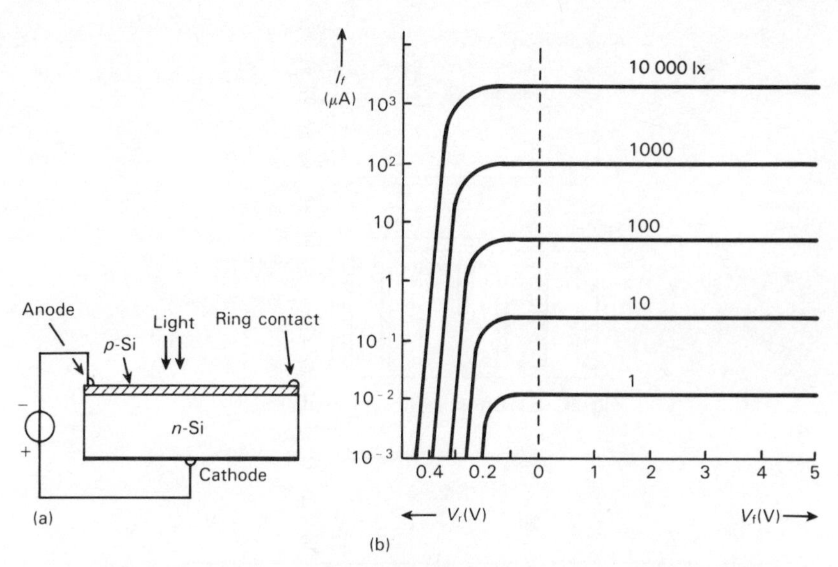

Fig. 8.10 (a) Basic structure of photodiode (b) Typical characteristics of a photodiode

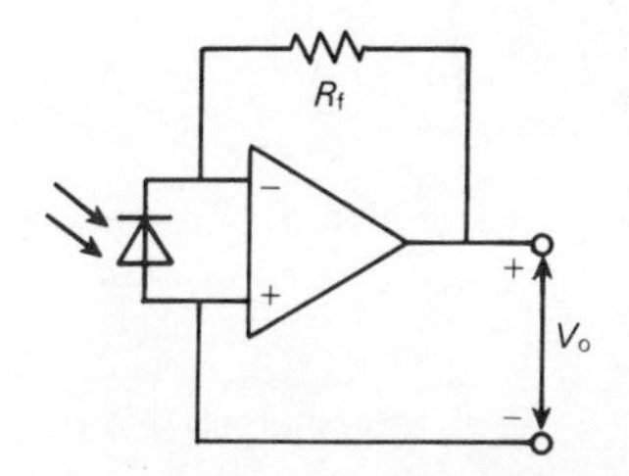

Fig. 8.11 Application of photodiode for light measurements

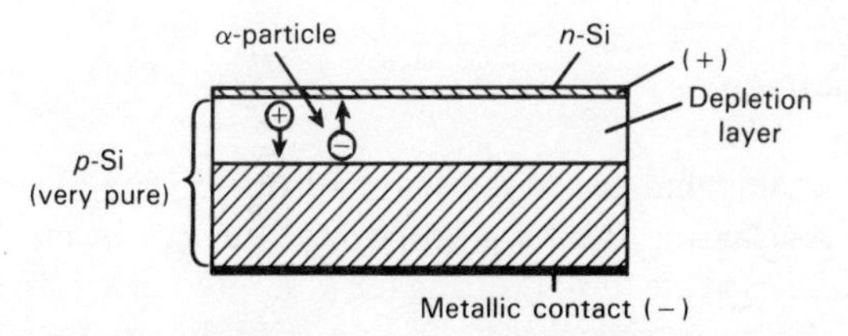

Fig. 8.12 Photodiode used for radiation measurements

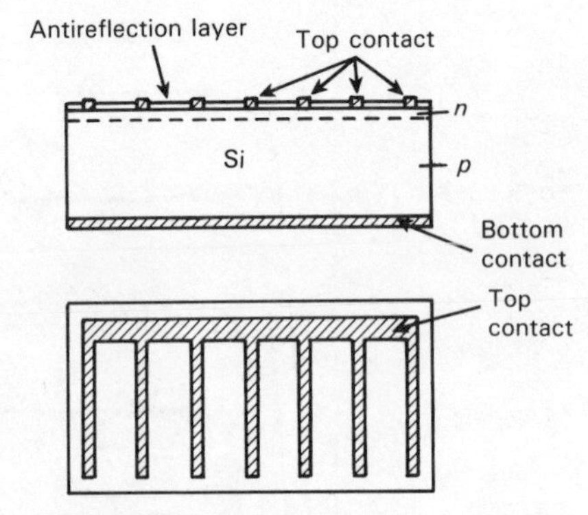

Fig. 8.13 Construction of a solar cell

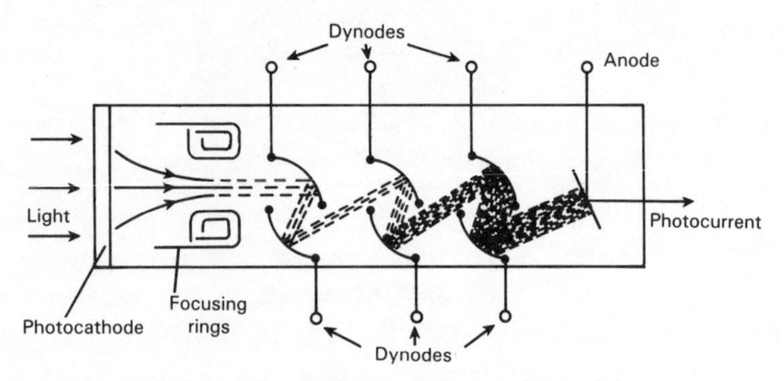

Fig. 8.14 Construction of photomultiplier tube

Secondary electrons (3–6 per primary electron) from the first dynode travel to the second which has a higher potential than the first. The final electrode is the anode which collects all electrons. Amplifications of up to 10^9 are possible. Even a single photon can be detected.

8.3 MECHANICAL TRANSDUCERS

8.3.1 Strain gage

Static and dynamic changes in the shape of objects are measured by strain gages which are among the most important devices in mechanical engineering. They are capable of measuring tension, pressure, force, moment and acceleration. A strain gage consists of resistive material which changes its resistance when the shape changes. A change in length causes a change in

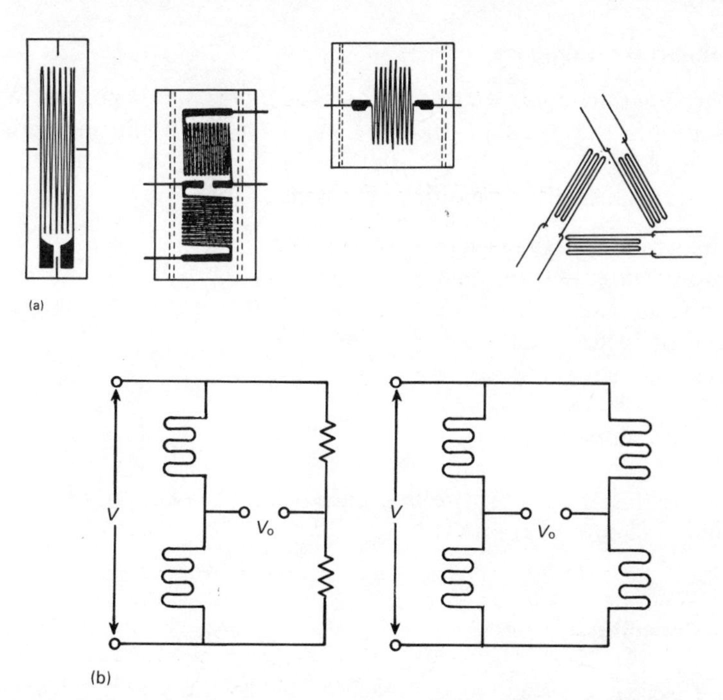

Fig. 8.15 (a) A few patterns of wire strain gauges (b) Strain gauges used in bridge circuits

resistance according to

$$\frac{\Delta R}{R} = K \frac{\Delta l}{l}$$

where K is a sensitivity constant depending upon the material used. Three types of strain gages are used.

Wire strain gage

A wire strain gage consists of resistive wire wound on a carrier. Wire strain gages are made in a large assortment of patterns (Fig. 8.15(a)) often in two or more sections for reasons of temperature compensation. They are usually connected in bridge circuits (Fig. 8.15(b)). A typical resistance value is 120 Ω.

Foil strain gage

A resistive metal is deposited on a carrier and etched in a pattern to form a resistive wire (Fig. 8.16).

Semiconductor strain gage

On a carrier a thin layer ($\simeq 0.02$ mm) of *n*- or *p*-type Si is deposited. When force is applied the crystalline structure alters resulting in the **piezoresistive effect** (Fig. 8.17).

Typical values of the sensitivity constant K are

wire strain gage (manganine)	$K \simeq 0.5$
wire strain gage (constantan)	$K \simeq 2$
foil strain gage	$K \simeq 2$
semiconductor strain gage	$K \simeq 130$

8.3.2 Potentiometer

The potentiometer or slide-wire transducer (Fig. 8.18) converts a displacement into a change of resistance.

8.3.3 Capacitor

- The dielectric constant of a capacitor can be the sensing element to measure displacement (Fig. 8.19). This type is often used to detect the level of liquids.
- Rotating one set of plates with respect to the other changes the value of the capacitance (Fig. 7.16(a)).

8.3.4 Piezoelectric crystal

The stress applied to such a crystal may be compression, shear or bending. Thus mechanical quantities such as pressure, force and acceleration (vibrations and shocks) can be measured.

The high output impedance of crystals requires a connection to a high input impedance amplifier.

Barium titanate and lead zirconate are ceramics which can be formed into complex shapes. Their response is considerably greater than that of quartz.

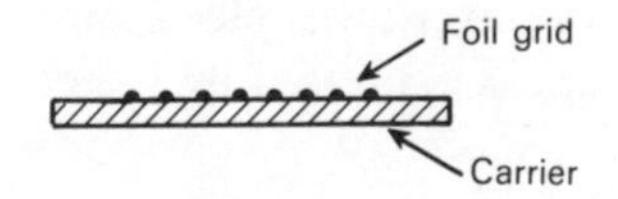

Fig. 8.16 Construction of foil strain gage

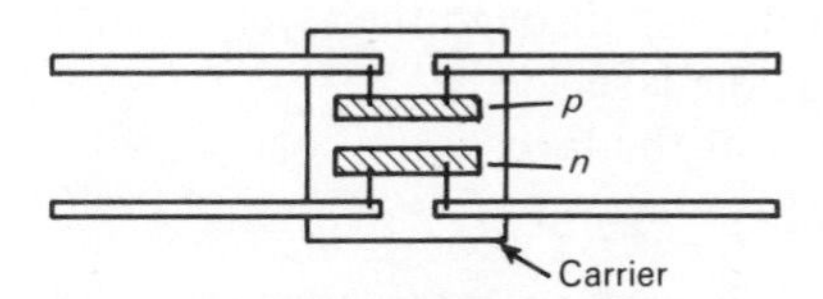

Fig. 8.17 Construction of semiconductor strain gage

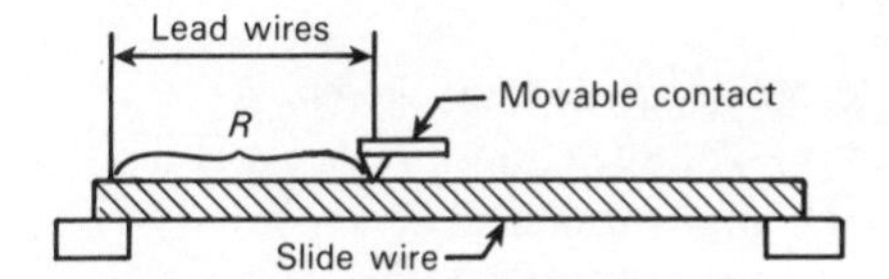

Fig. 8.18 Potentiometer transducer

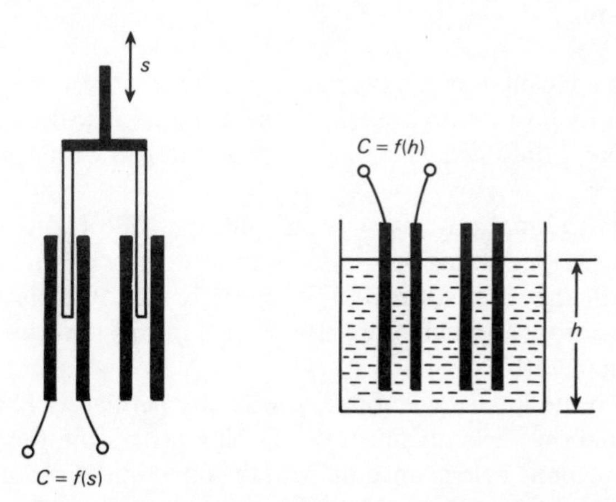

Fig. 8.19 Capacitance transducer to measure displacements

8.4 MAGNETIC TRANSDUCERS

8.4.1 Hall element

When a thin layer of semiconducting material (InSb or InAs) is placed in a magnetic field (Fig. 8.20), moving charges caused by a current I will deviate from a rectilinear path due to the Lorentz force. Thus a charge concentration on one side of the layer takes place causing a voltage across the ends of the layer. The voltage is expressed by the equation

$$V = bM_H(\boldsymbol{J} \times \boldsymbol{B})$$

where

M_H is the Hall constant
J is the current density through the layer
b is the width of the layer
d is the thickness of the layer
B is the magnetic induction

Hall elements are often used to measure magnetic field strengths. Since $I = bd/J$ the magnitude of B is expressed by

$$B = \frac{Vd}{M_H I \sin \alpha}$$

where α is the angle between field and layer.

It follows from the voltage equation that a Hall element can be used in DC as well as in AC applications. The sensitivity is about 10^{-1}–10^{-3} T (tesla), the frequency range about 1 MHz.

8.4.2 Fluxistor

When InSb is deposited on a substrate of ferrite the resistance of the layer will vary with its position in a magnetic field (Fig. 8.21(a)). An increase in resistance value is obtained when the material is doped with a small amount of NiSb.

By proper thermal processing techniques the NiSb forms a needle-like structure, the needles acting as short circuits for the electrons. When they contact a needle they are returned to the opposite side of the layer. Thus the path of the electrons is greatly increased which means that the resistance is increased (Fig. 8.21(b)).

Related to the fluxistor is a device whereby permalloy (20% Fe, 80% Ni) is deposited on a silicon substrate. During processing the direction of the internal magnetic field is lined up with the direction of current flow. The presence of an external magnetic field alters the internal field direction

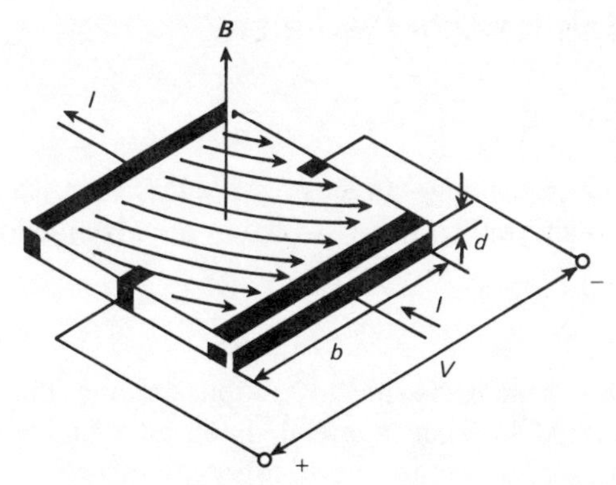

Fig. 8.20 Hall element in a magnetic field

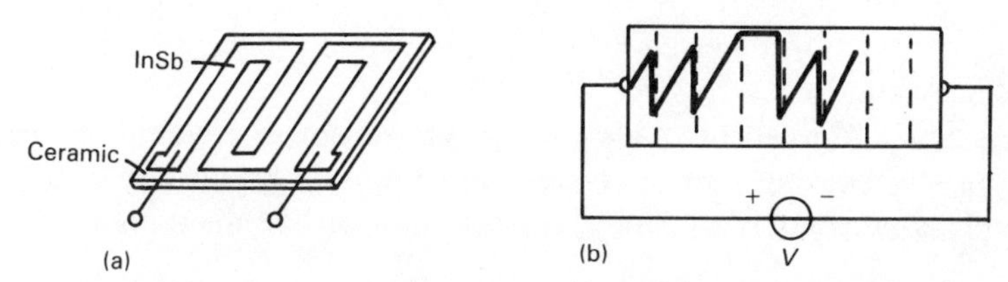

Fig. 8.21 (a) Fluxistor in a magnetic field (b) Path of electrons in a fluxistor

causing a change in resistance. The sensitivity of these devices is very high (10^{-3}–10^{-6} T).

Aside from magnetic field measurements, Hall elements and fluxistors are used to measure RPM values, shaft positions and angles of rotation of engines.

8.5 INDUCTIVE TRANSDUCERS

8.5.1 Moving coil meter

A force acts on a current-carrying conductor when it is placed in a stationary magnetic field (Section 3.7).

8.5.2 AC generator

A coil, rotating in a stationary magnetic field, generates an AC voltage (Fig. 8.22).

8.6 ELECTROACOUSTIC TRANSDUCERS

8.6.1 Microphone

A microphone converts changes in air pressure into electrical signals. A few principles used to accomplish this conversion are given below.

Resistance variations

An example of this principle is the carbon microphone (Fig. 8.23). Carbon granules are enclosed between a metal diaphragm and a metal case. A movable contact is attached to the diaphragm in order to transmit the movements of the diaphragm to the layer of granules. When the granules are compressed by a sound wave, the resistance is lowered. These resistance variations are converted to current variations by a battery and a load resistor R_L thus forming a closed circuit. Hence resistance variations result in voltage variations across R_L.

As resistance can be made to vary proportional to variations in air pressure, current and output voltage vary inversely proportional to resistance. This property is a cause of distortion in carbon microphones.

Inductance

A coil is wound on a cylindrical body which is attached to a diaphragm (Fig. 8.24). The coil is mounted in the air gap of a permanent magnet. Sound waves cause the coil to move in the magnetic field. These movements induce an electromotive force in the coil.

The response of this type is very linear and no bias voltage is needed. The source impedance is usually very low (about 25 Ω).

Capacitance variations

A capacitor microphone uses a metal diaphragm separated and insulated from a metal plate (Fig. 8.25). A high voltage is applied between diaphragm and plate.

Sound waves change the value of the capacitance thereby changing the voltage across the capacitor. This voltage is linearly proportional to pressure variations.

Capacitor values are in the range 10–20 pF; bias voltages are a few hundred volts.

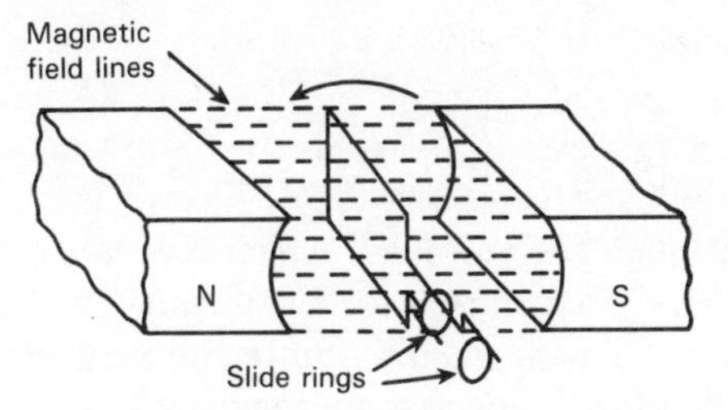

Fig. 8.22 Principle of AC generator

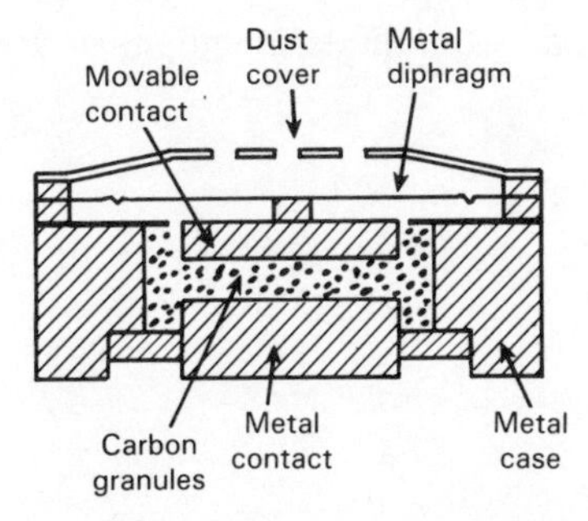

Fig. 8.23 The carbon microphone

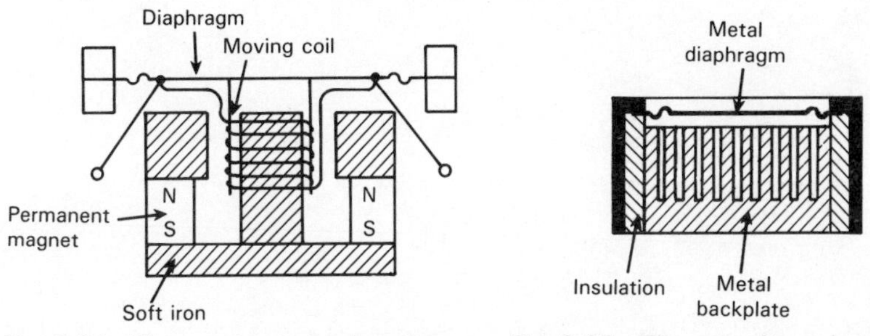

Fig. 8.24 The moving coil microphone

Fig. 8.25 The capacitor microphone

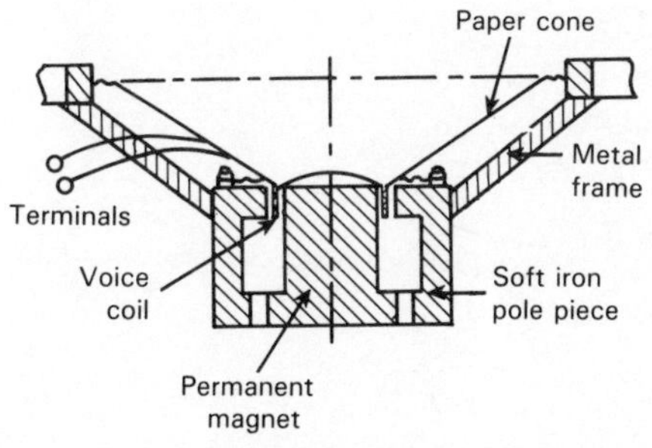

Fig. 8.26 The electrodynamic loud-speaker

8.6.2 Loudspeaker

A loudspeaker converts electrical signals to sound waves. Most frequently used are moving coil speakers (Fig. 8.26). Electrical currents representing the sound pass through the voice coil which is located in the air gap of a permanent magnet. When currents flow through the coil, forces result between coil and permanent magnet. Since the magnet cannot move, the coil moves axially along the permanent magnet.

A large cone of soft material (paper) is attached to the coil. This cone moves the air in front of the speaker.

Typical coil impedances in modern equipment are 4, 8 and 16 Ω.

9
Semiconductor Materials; The *pn* Junction; Diodes

9.1 SEMICONDUCTOR MATERIALS; DOPING

The most common semiconductor materials in use are

Ge	InAs
Si	InP
InSb	GaAs
	PbS

The conductivity of these materials varies greatly as illustrated in the specific resistivity graph of Fig. 9.1.

The two most important materials are Si and Ge. A Si atom has a total of 14 electrons, a Ge atom has 32; both have four valence electrons.

The crystal structure or lattice forms a tetrahedral pattern, a cross section of which is shown in Fig. 9.2.

Each atom is bonded to four other atoms. Two atoms are bonded by two valence electrons (**covalent bond**). In Chapter 4 a brief outline was given of the energy levels, generation and recombination.

The **intrinsic** or pure materials have insufficient conductivity for practical purposes: in Ge one electron out of 10^9 is in the conduction band, in Si it is one out of 10^{12}.

To enhance the conductivity, foreign materials are added to the pure Si and Ge. This is called **doping** and the material is said to be **extrinsic.** Two types of doping are used: pentavalent elements (N, P, As, Sb) and trivalent elements (B, Al, Ga, In).

Figure 9.3(a) shows the crystal structure of Si (or Ge) doped with pentavalent atoms. The covalent bond is complete with four valence electrons of the foreign atom. Thus the energy of the fifth electron is higher, and at room temperature the thermal energy is already sufficient to make it a free electron. Thus the foreign atom is left behind as a positive ion. Such an atom is called a **donor**. The material now has a greater conductivity and is termed ***n*-type material**. The electrons are the **majority carriers**, the holes the **minority carriers**. The doping level of material is often indicated by '*n*' which is about $1:10^7$ ($\rho \simeq 5\ \Omega$ cm) or 'n^+' which amounts to $1:10^4$ ($\rho \simeq 0.03\ \Omega$ cm).

When the material is doped with trivalent atoms (Fig. 9.3(b)) the covalent bond is incomplete. The open place or hole can be filled easily by a neighboring valence electron. When this happens the foreign atom becomes a negative ion; the hole wanders freely and is a free hole. The trivalent atom is called an **acceptor**, the material **p-type material,** the holes are majority carriers and the electrons minority carriers. The doping levels are indicated by 'p' meaning $1:10^6$ ($\rho \simeq 2\ \Omega$ cm) or 'p^+' meaning $1:10^4$ ($\rho \simeq 0.005\ \Omega$ cm).

In a uniform electric field E the charges are accelerated. Superimposed on the random motions by collisions with the atoms, there is a velocity component in the direction of the field. This average velocity is the **drift velocity** which is proportional to the field strength E, according to

$$v_d = \mu E$$

The constant μ is called the **mobility.** The mobilities in Ge and Si are

	μ_n	μ_p	
Ge	0.390	0.190	($m^2\,V^{-1}s^{-1}$)
Si	0.135	0.048	($m^2\,V^{-1}s^{-1}$)

The motion of electrons and holes due to a field E constitutes a current density

$$J = (n\mu_n + p\mu_p)qE = \sigma E$$

where

q is the electron charge
n is the concentration of electrons
p is the concentration of holes

Thus the conductivity σ is

$$\sigma = q(n\mu_n + p\mu_p)$$

In a pure semiconductor $n = p = n_i$ is the intrinsic concentration.

The **thermal generation rate** Φ_g is the number of generated electron/hole pairs per second per m^3. The energy needed for generation of a free electron is E_g where E_g is the band gap of the material (in Si : $E_g \simeq 1.1$ eV, in Ge : $E_g \simeq 0.7$ eV). It follows from statistical analysis that the probability of a valence electron becoming free can be expressed by the **Boltzmann relation** $\exp(-E_g/kT)$.

The **recombination rate** Φ_r is the number of recombining electron/hole pairs per second per m^3. Its value depends on the concentration of both

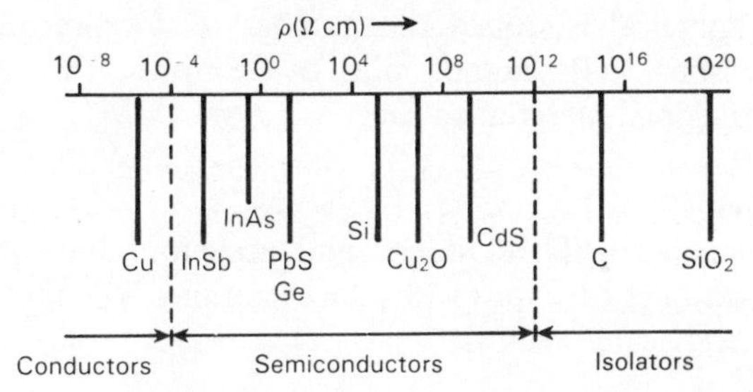

Fig. 9.1 Specific resistance of a number of materials

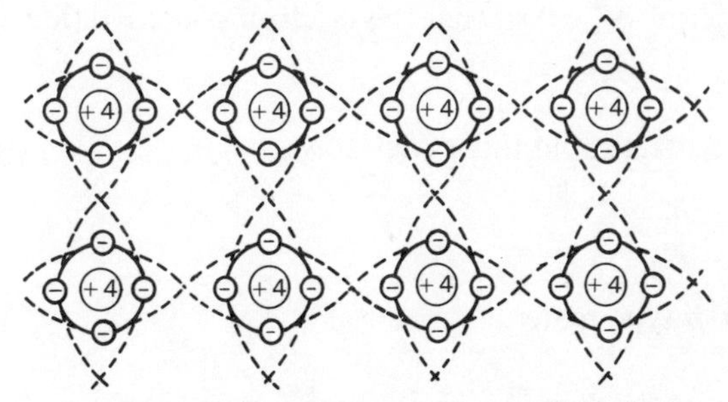

Fig. 9.2 Crystal structure of pure Si or Ge

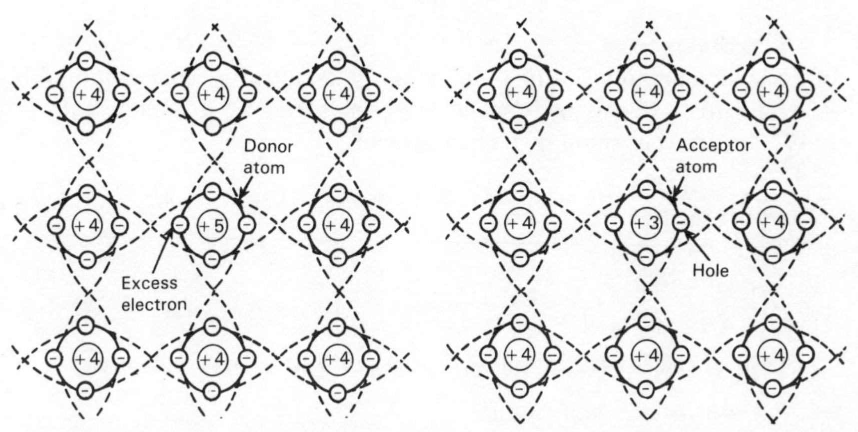

Fig. 9.3 (a) Crystal structure of *n*-doped material (b) Crystal structure of *p*-doped material

electrons and holes, thus

$$\Phi_r = rnp$$

where r is a proportionality constant, depending on the material.

At equilibrium, generation and recombination rates are equal so

$$\Phi_g = \Phi_r = rnp$$

In intrinsic semiconductors $\Phi_g = \Phi_r = rn_i^2$.

When the material is doped the number of foreign atoms is still negligible with respect to the number of Si or Ge atoms. Thus the equation holds as well for doped material so that

$$np = n_i^2$$

In doped material virtually all donor and acceptor atoms are ionized, leaving positive and negative ions with concentrations N_d and N_a. But since the material is electrically neutral

$$p + N_d = n + N_a$$

In n-type material $N_a = 0$ so that the electron concentration is

$$n_n = N_d + p_n \simeq N_d$$

The hole concentration in this material is

$$p_n = \frac{n_i^2}{n_n} \simeq \frac{n_i^2}{N_d}$$

Similarly, for p-type material

$$p_p \simeq N_a \quad \text{and} \quad n_p \simeq \frac{n_i^2}{N_a}$$

Example

Intrinsic Si contains 5×10^{28} atoms m^{-3}. Its intrinsic concentration is $n_i = 1.5 \times 10^{16}$. If this Si is made n-type material by a doping level of $1:10^8$, the concentration of free electrons is

$$n_n = 5 \times 10^{28} \times 10^{-8} = 5 \times 10^{20}$$

Thus

$$p_n = \frac{n_i^2}{N_d} = \frac{2.25 \times 10^{32}}{5 \times 10^{20}} = 0.45 \times 10^{12}$$

The ratio of the concentrations is thus $n_n/p_n \simeq 10^9$. The conductivity is

$$\sigma = (n_n\mu_n + p_n\mu_p)q \simeq N_d\mu_n q$$
$$= 5 \times 10^{20} \times 0.135 \times 1.6 \times 10^{-19} = 11.5 \text{ S m}^{-1}.$$

The intrinsic conductivity is

$$\sigma_i = n_i(\mu_n + \mu_p)q$$
$$= 1.5 \times 10^{16} \times 0.183 \times 1.6 \times 10^{-19} = 0.000\,45 \text{ S m}^{-1}.$$

This shows that the doping has increased the conductivity by a factor of about 30 000.

9.2 THE *pn* JUNCTION

When forming a junction of p and n material (Fig. 9.4) electrons from the n region move into the p region due to the excess concentration of electrons in the n region. The same is true for the holes that move in the opposite direction. This motion of charge carriers is called **diffusion**.

An electron moving into the p region will soon recombine and fill a hole, thereby creating a negative ion. Likewise, a moving hole creates a positive ion in the n region; so two adjoining layers of ions are formed. The combined layers are called **space charge layer** or **depletion layer.** The space charge forms an internal electric field which counteracts the flow of more diffusion carriers.

The potential which has developed across the junction is called **contact potential.** (Ge : $\simeq 0.4$ V, Si : $\simeq 0.8$ V). It causes a second flow of carriers: electrons from the p to the n region and holes from the n to the p region. This is a flow of minority carriers which is called **drift**. A dynamic equilibrium of diffusion and drift develops which keeps the depletion layer balanced at a certain width. When w_p represents the width of the depletion layer in the p region, S the area of the junction and p_p the doping level, the total charge is $Q_p = w_p S p_p$. Similarly, in the n region: $Q_n = w_n S n_n$. Electric neutrality demands that $Q_p = Q_n$ from which it follows that $w_p / w_n = n_n / p_p$. Thus the width of a depletion layer is inversely proportional to the doping level in that region.

When connecting an external voltage source so that the p region is positive with repect to the n region the potential barrier for the diffusion carriers is lowered, causing an increase in diffusion current. The drift current remains the same. The junction is said to be **forward biased.**

When reversing the polarity of the voltage source the potential barrier is raised and the diffusion current is practically reduced to zero. The drift

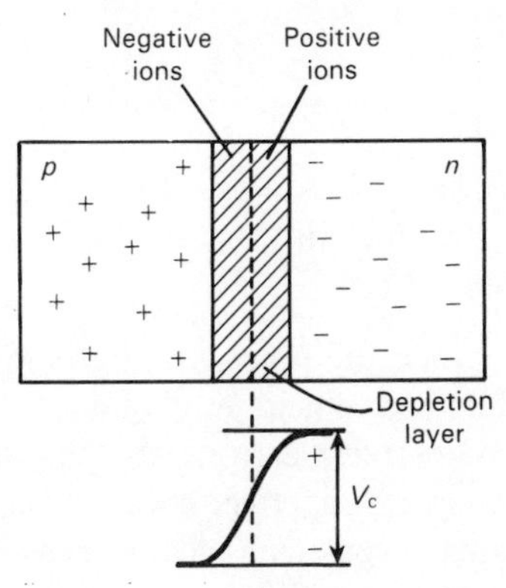

Fig. 9.4 The *pn* junction and its contact potential

current, which only depends on the rate of thermal generation, reaches its maximum value I_s at a small reverse bias voltage. It is called **saturation current** or **leakage current**. The junction is now **reverse biased.**

The equation relating the current flow and the externally applied voltage in a *pn* junction is usually called the **diode equation**:

$$I_D = I_s\left[\exp\left(\frac{qV_D}{\eta kT}\right) - 1\right]$$

where

I_D is the current flowing through the junction
V_D is the external voltage across the junction
k is Boltzmann's constant (1.38×10^{-23} J K^{-1})
T is the absolute temperature (K),
η is the junction ideality factor (for Ge $\eta = 1$, for Si $\eta = 2$)
I_s is the reverse saturation current

The reverse saturation current can be expressed as

$$I_s = I_{so} \exp[\alpha(T - 293)]$$

where $\alpha \simeq 0.08$ for Ge and $\alpha \simeq 0.12$ for Si and I_{so} is the saturation current at $T = 293$ K. The diode equation is illustrated in Fig. 9.5.

At a constant forward bias current the voltage temperature coefficient is

$$\frac{dV_D}{dT} \simeq -2.3 \text{ mV °C}^{-1}$$

The **dynamic resistance** r_d is the slope of the tangent of the forward biased junction at the bias point (Fig. 9.5). When ignoring the ohmic resistance of the semiconductor material it can be expressed at room temperature as

$$r_d \simeq \frac{25 \text{ mV}}{I_D}$$

where

r_d is the dynamic resistance in Ω
I_D is the bias current in mA

Breakdown occurs when the junction is reverse biased and the bias is increased sufficiently. The strong field pulls valence electrons directly out of covalent bonds to become free electrons so that the number of carriers increases by a large amount. This effect is called the **Zener effect.**

At a somewhat higher reverse voltage another avalanche effect occurs: the internal field strength is so high that carriers gain sufficient kinetic energy to knock electrons from covalent bonds (ionization) and a large current begins to flow abruptly.

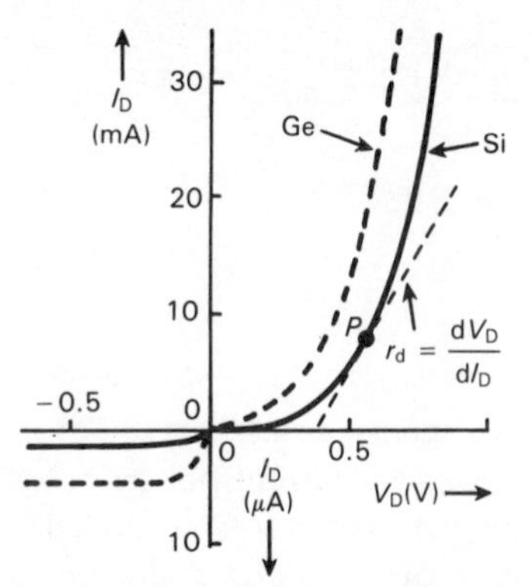

Fig. 9.5 Characteristics of Si and Ge diodes showing the dynamic resistance

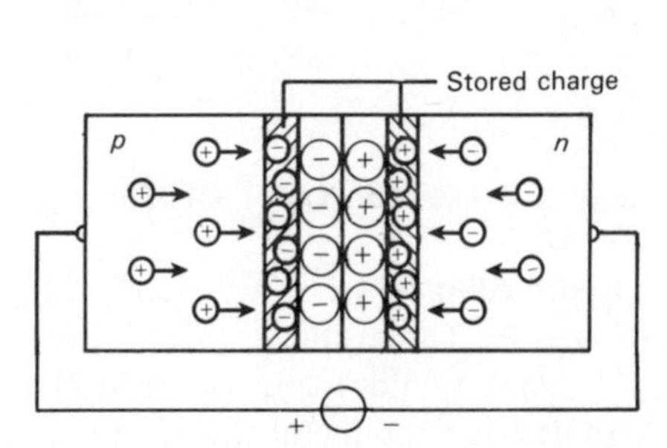

Fig. 9.6 The diffusion capacitance of a diode

A reverse biased *pn* junction resembles a parallel-plate **capacitor**, the depletion layer acting as the dielectric enclosed between the (semi)conducting *p* and *n* regions. The width of the depletion layer increases as the reverse voltage increases thus decreasing the capacitance value.

The junction thus has a **barrier capacitance** which can be written as

$$C_b \simeq \frac{C_o}{\sqrt{1 + \frac{V_D}{V_{Do}}}}$$

where

V_D is the reverse voltage
$V_{Do} \simeq 0.6$ V(Si), 0.3 V (Ge)
C_o is the capacitance value at $V_D = 0$ V

The forward biased junction also has a capacitance (Fig. 9.6). The concentration of diffusion carriers which have just crossed the depletion layer is still high, becoming smaller by recombination when these carriers penetrate further into the semiconducting material. When suddenly changing the current flow (e.g. by reverse biasing the junction), the excess concentration or stored charge must first be changed before the static condition is achieved. Therefore, a diffusion current flows during a short time. This charge–current relation implies a capacitive reactance which in a model can be represented by a capacitance. This capacitance is called **diffusion capacitance**. Its value is directly proportional to the forward bias current.

The description of the *pn* junction is in essence the description of the most widely used diode, the **junction diode.** Many other diode types have been developed.

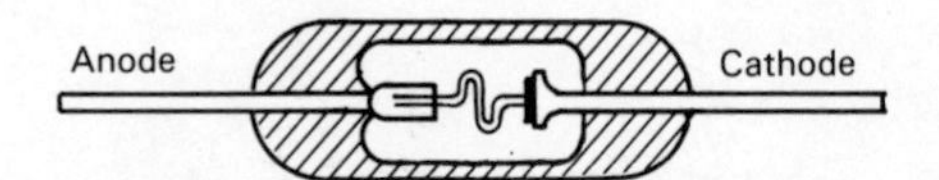

Fig. 9.7 Construction of a point contact diode

9.3 POINT-CONTACT DIODE

A pointed metal wire (tungsten, gold) presses against a small crystal of *n*-Ge (Fig. 9.7). The contact is briefly heated by passing a current pulse through it (like spot welding). Due to the local melting process the *n*-Ge is doped with atoms from the metal thus forming a region of *p*-Ge and a *pn* junction. Only small values of current and reverse voltage are allowed.

Point-contact diodes are suitable for detecting high-frequency signals since the barrier capacitance is small.

9.4 ZENER DIODE

By controlling the doping levels, Zener diodes can be manufactured with breakdown voltages from a few volts to several hundred. At low breakdown voltages (<5 V) the Zener effect is dominant; higher breakdown voltages (>6 V) are due to the avalanche (ionization) effect. Zener diodes are employed as voltage reference sources and for voltage stabilizing and protection purposes.

The constancy of the stabilized voltage depends upon the dynamic resistance and the temperature coefficient of the diode. Figure 9.8(a) shows a plot of dynamic resistance versus Zener voltage. Apparently, minimum r_d values are obtained when using Zener diodes of about 7 V.

Figure 9.8(b) shows the temperature coefficient versus Zener voltage. An approximate equation for the temperature coefficient is

$$\frac{\mathrm{d}V_z}{\mathrm{d}T} \simeq V_z - 6 \ (\mathrm{mV\ ^\circ C^{-1}})$$

when V_z is expressed in volts.

9.5 STABISTOR

Stabistors are used to stabilize small voltages ($\leqslant 3$ V). They are operated in the foward biased direction. Two typical stabistor characteristics are shown in Fig. 9.9.

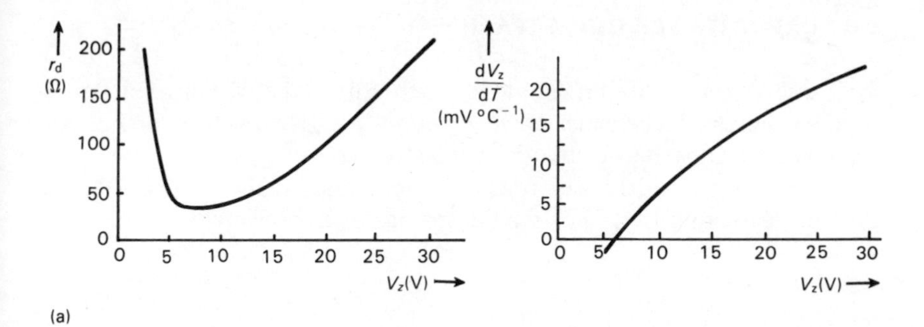

Fig. 9.8 (a) The dynamic resistance of Zener diodes (b) The temperature coefficient of Zener diodes

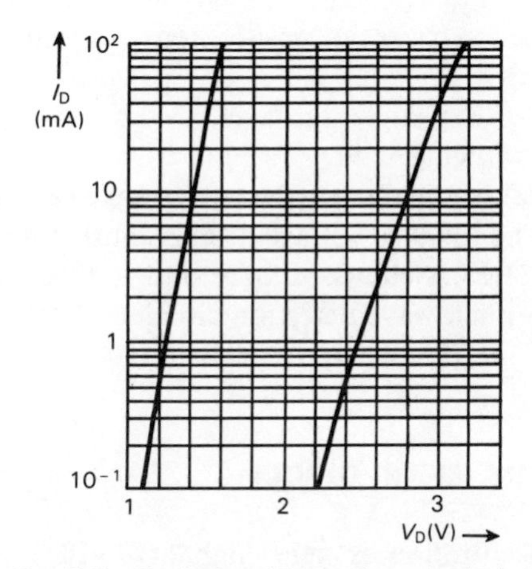

Fig. 9.9 Typical stabistor characteristics

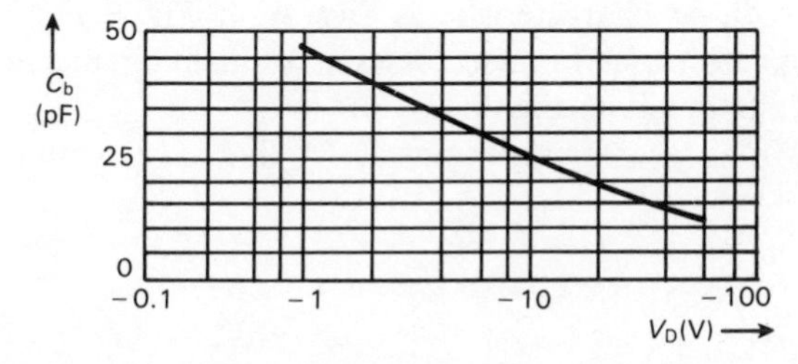

Fig. 9.10 Typical characteristic of a capacitance diode

9.6 CAPACITANCE DIODE (VARICAP)

The varicap is a reverse biased diode which utilizes the barrier capacitance of a *pn* junction. The capacitance equation was given in Section 9.2. An example of a capacitance characteristic is shown in Fig. 9.10.

Varicaps are widely used for tuning resonant circuits in AM, FM, TV and for generating frequency-modulated signals in oscillators.

9.7 SCHOTTKY DIODE

The Schottky diode or **hot carrier diode** is a metal-semiconductor diode (Fig. 9.11(a)). A contact potential develops across the junction. When forward biased, free electrons in the *n* region having sufficient energy to cross the contact potential enter the metal, a negative charge builds up and the *n*-Si develops a space charge of positive ions. Current flow is a majority carrier (electron) flow.

Due to the very thin depletion layer ($\simeq 0.1\ \mu m$) and the absence of minority carriers, switching is very fast since virtually no charge storage takes place. A typical storage time is 50 ns. Leakage current and noise have very small values. The forward voltage drop is considerably smaller than in a *pn* diode (Fig. 9.11(b)). Schottky diodes find applications in digital ICs, UHF mixers and in microwave detection circuits.

9.8 TUNNEL DIODE (ESAKI DIODE)

If the doping concentration is very high ($\simeq 1:10^3$) the width of the depletion layer decreases to about 10^{-5} mm resulting in field strengths of about 1 MV cm^{-1}. Electrons may tunnel through the thin layer without possessing sufficient energy to climb over the potential barrier. Tunneling is a wave phenomenon, i.e. electron transfer occurs at the speed of light. This makes the tunnel diode applicable at very high frequencies.

A tunnel diode characteristic is shown in Fig. 9.12. The negative resistance property is used to offset losses in resonant circuits making simple oscillator circuits possible (Fig. 9.13).

Breakdown at reverse bias begins at 0 V due to the very thin depletion layer.

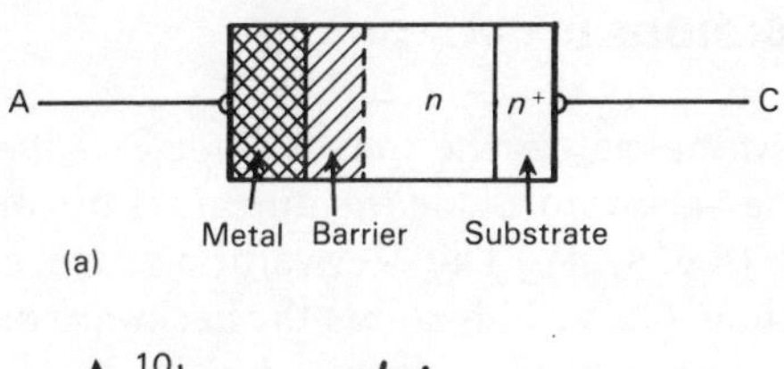

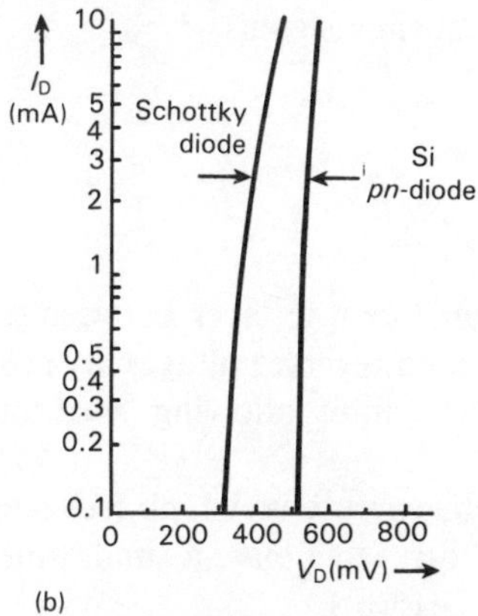

Fig. 9.11 (a) Structure of a Schottky diode (b) Comparison of Schottky and Si-diode characteristics

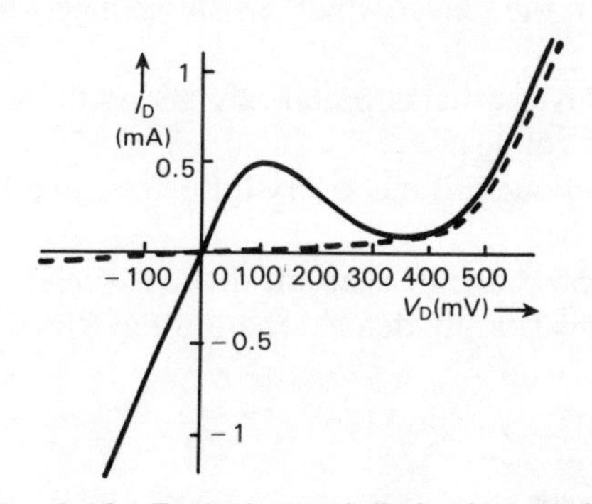

Fig. 9.12 Typical tunnel diode characteristic

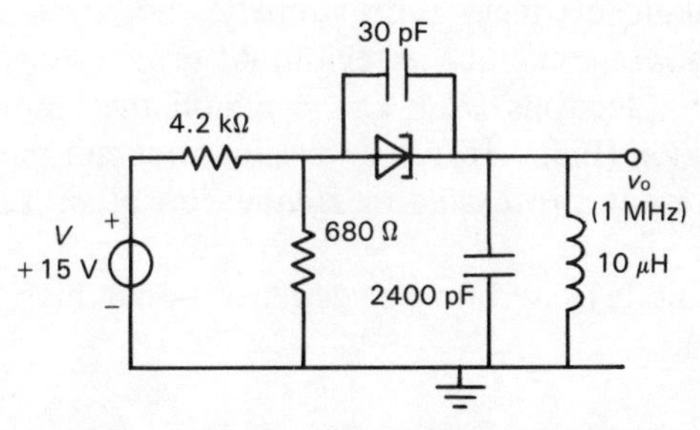

Fig. 9.13 1 MHz tunnel diode oscillator circuit

9.9 BACKWARD DIODE

The backward diode resembles the tunnel diode except that the tunneling effect is less. In the backward diode the directions of forward and reverse bias are reversed (Fig. 9.14). The 'forward' voltage is very small, the 'reverse' voltage about 0.5 V. This allows the backward diode to be used for detection of small high-frequency signals.

9.10 PIN DIODE

The PIN diode contains an intrinsic layer between a *p* region and an *n* region (Fig. 9.15(a)). Very high reverse voltages are possible (a few kV); the barrier capacitance is very small allowing applications at microwave frequencies.

PIN diodes are used to generate very high frequencies; when the diode is switched from forward to reverse bias, a small time delay occurs which has very steep slopes (Fig. 9.15(b)).

The structure of a **PIN photodiode** is shown in Fig. 9.15(c). A very thin (transparent) *p* layer is accessible for incident photons. These generate electron/hole pairs in the *i* layer which are immediately swept apart by the reverse field.

The depletion layer extends practically across the whole *i* layer, even with a small reverse voltage.

The PIN photodiode produces very little noise and has very short rise and fall times ($\simeq$1 ns).

The **step-recovery diode** or **snap-off** diode is similar to the PIN diode. The junctions are less sharply defined than in the PIN diode.

9.11 IMPATT DIODE

The impact avalanche transit time (Impatt) diode has a lightly doped (almost intrinsic) *n* region and a *p* region. At reverse bias, breakdown will eventually occur. Electrons from the *p* region then move to the anode through the *n* region (Fig. 9.16). In an oscillator circuit the transit time can be made to correspond to oscillator frequencies of up to a few hundred GHz.

The Impatt diode is used for power generation at high frequencies (e.g. 10 W at 10 GHz).

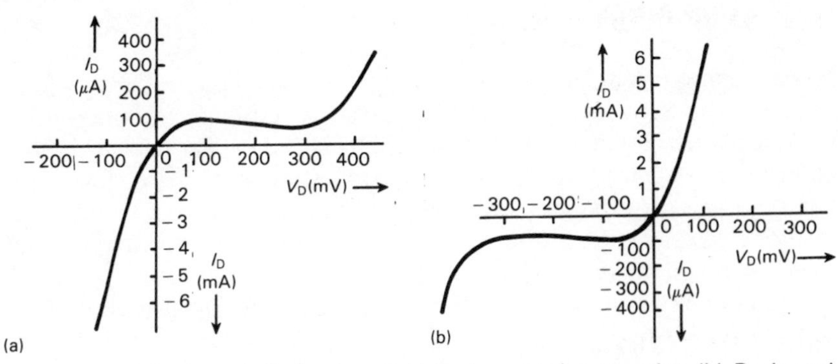

Fig. 9.14 (a) Backward diode characteristic in normal connection (b) Backward diode characteristic in backward connection

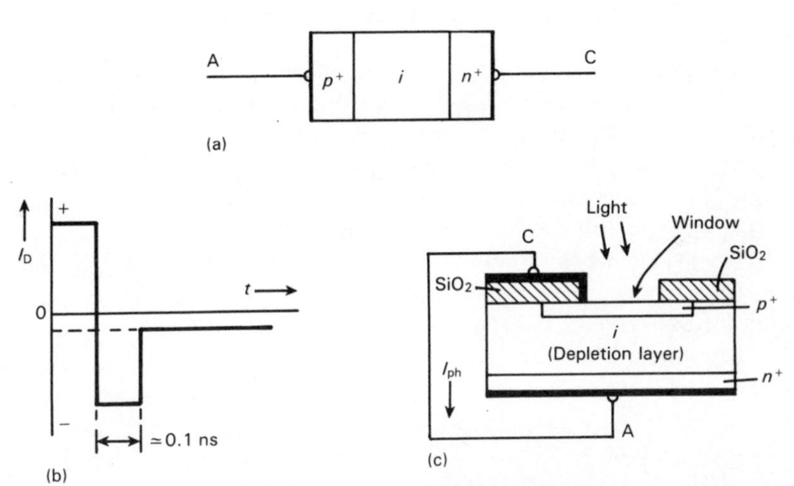

Fig. 9.15 (a) Structure of PIN diode (b) Switching characteristic of PIN diode (c) Structure of PIN photodiode

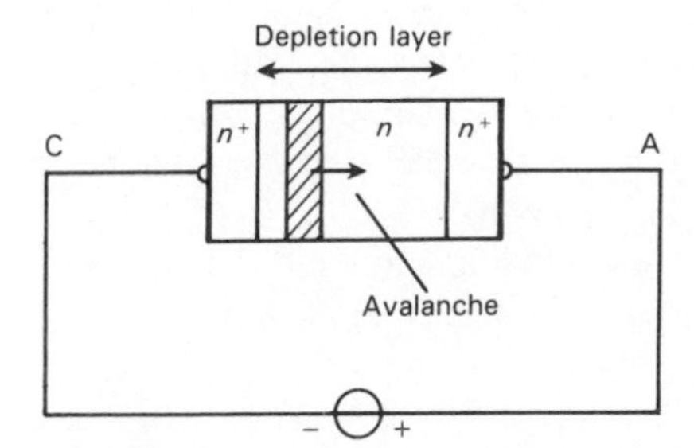

Fig. 9.16 Structure of Impatt diode

9.12 GUNN DIODE

When connecting a DC voltage of 10 V or more, the current–voltage characteristic of a Gunn diode shows a negative resistance region resulting in oscillations at microwave frequencies (Fig. 9.17). The reason is that electrons reach energy levels where the mobility (current) decreases. This has the effect of domains with high field strengths moving through the material. The thickness of the crystal can be chosen so that the frequency of these moving domains is of the order of 10 GHz.

Gunn diodes are used as microwave oscillators in radar (police radar, security systems).

9.13 AVALANCHE DIODE

Two lightly doped regions of *p* and *n* material are reverse biased (Fig. 9.18). At a sufficiently high reverse bias ($\simeq$ 100 V) the wide depletion layer breaks down and avalanche ionization occurs. Atoms near the boundary between the *p* region and the depletion layer lose electrons which move to the *n* region. The frequency of the recurrent avalanches can be varied by varying the bias voltage. Applications are in the field of microwaves.

The **avalanche photodiode** is a kind of Zener diode having a breakdown voltage between 100 V and 200 V. The photocurrent amplification factor is about 150. The noise level is very low and applications up to 50 GHz are possible.

9.14 JOSEPHSON JUNCTION

The Josephson junction, discovered in 1962, is at present the fastest semiconducting switching device. The structure is that of a sandwich of two electrodes of lead based alloy separated by a very thin oxide layer of about 5 nm (Fig. 9.19(a)). The junction is required to have the temperature of liquid helium (4.2 K) which makes the lead superconducting so that the resistance of the material vanishes completely.

The oxide layer forms a tunnel barrier where electrons can pass with zero voltage drop (Fig. 9.19(b)).

Tunneling is a wave phenomenon and the electromagnetic field of this wave causes the barrier to change from insulator to superconductor and vice versa. The switching speed is of the order of 50 ps.

Applications are in microwave mixers, detectors and in the digital field. In digital applications switching speeds are comparable to the velocity of

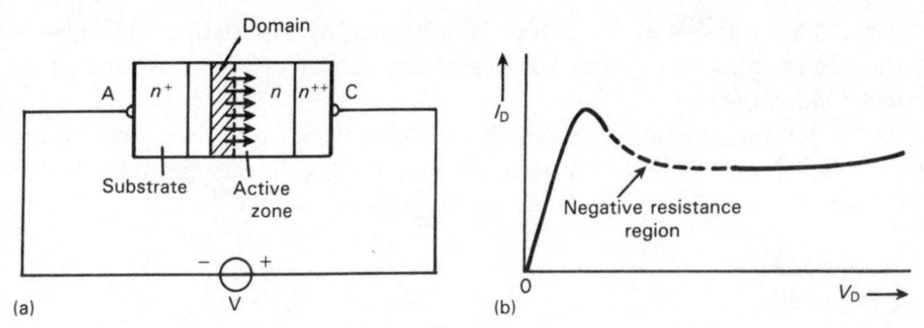

Fig. 9.17 (a) Structure of Gunn diode (b) Typical characteristic of Gunn diode

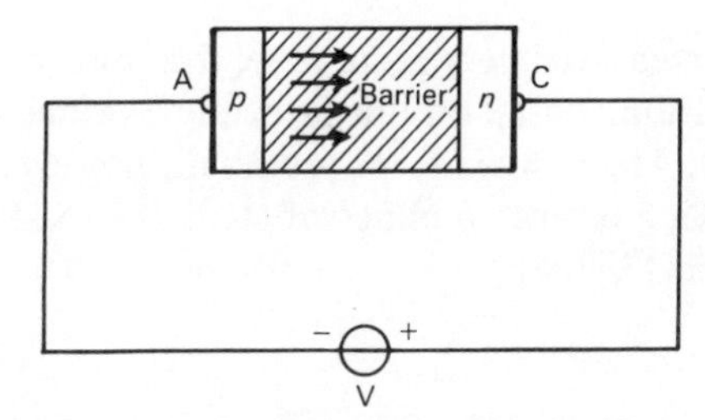

Fig. 9.18 Structure of avalanche diode

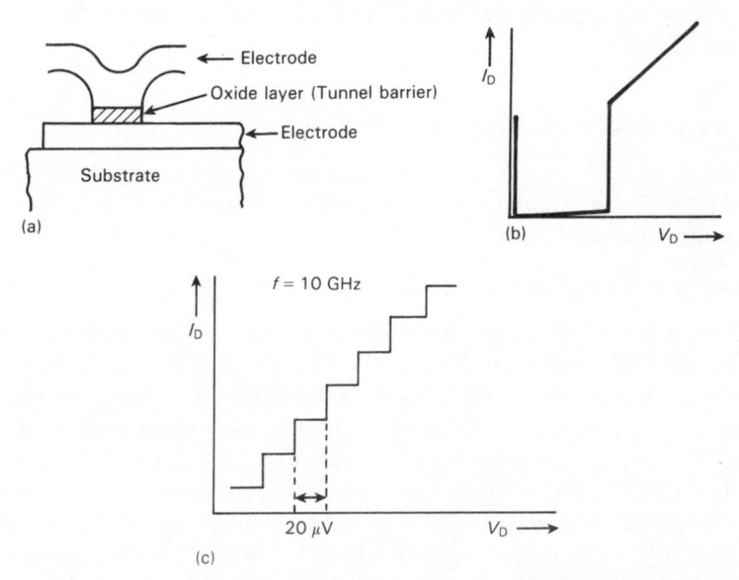

Fig. 9.19 (a) Structure of Josephson junction (b) Typical characteristic of Josephson junction (c) Characteristic of the Josephson junction in an electromagnetic field

light. The storage of data bits is achieved by circulating currents in superconducting loops. The absence of any resistance allows storing times to be indefinite.

When the junction is placed in a high-frequency electromagnetic field, the characteristic is that of a staircase (Fig. 9.19(c)). The magnitude of each voltage step is

$$v = \frac{fh}{2q}$$

where

f is the frequency of the electromagnetic field
h is Planck's constant

Since f can be measured very accurately, the voltage v is known with a high degree of precision. Based on this principle a primary standard of the volt can be realized. The voltage steps are small, however, about 20 μV at $f = 10$ GHz. By taking a large number of steps the total voltage accumulates to more practical values.

9.15 VARACTOR

The varactor is a capacitance diode (Si or GaAs) which is used in microwave applications.

For parametric amplification (Fig. 9.20(a))

A signal v_s with frequency f_s is connected to varactor C together with an internally generated signal v_i with frequency f_i. If the two signals have the same amplitude, their sum is:

$$v_s + v_i = A \cos \tfrac{1}{2}(\omega_s + \omega_i)t \sin \tfrac{1}{2}(\omega_s - \omega_i)t$$

A large amplitude square wave voltage v_p with frequency f_p, the **pump voltage**, is also connected across C. The pump voltage changes the capacitance value periodically (Fig. 9.20(b)). At the moments when the capacitance of C decreases, the pump voltage supplies energy into C. At the moments when the capacitance of C increases, the pump voltage takes energy out of C. The moments of capacitance increase are synchronized with the moments when $v_s + v_i = 0$ (no charge on C) so that no energy is taken out of C. This procedure results in a flow of energy ΔE_i from v_p to C each time the capacitance decreases. This energy is transferred to a resonant circuit of which C is a part. The signal v_i automatically forms as C is pumped. The input impedance seen by v_s can be shown to have a negative

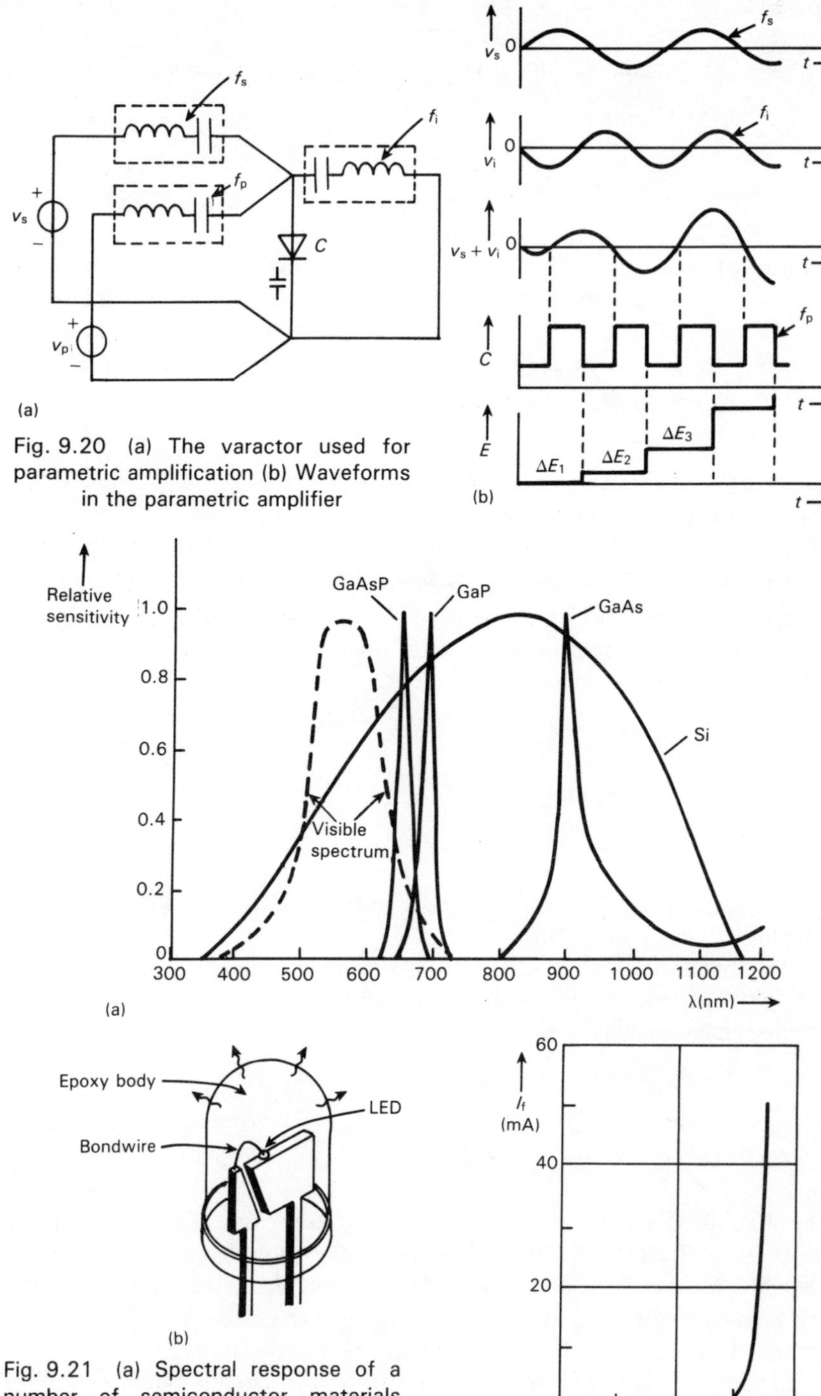

Fig. 9.20 (a) The varactor used for parametric amplification (b) Waveforms in the parametric amplifier

Fig. 9.21 (a) Spectral response of a number of semiconductor materials (b) Construction of a LED (c) Typical LED characteristic

real part which means that more energy comes out of C at frequency f_s than comes in. The circuit therefore amplifies.

For harmonic power generation

A large signal voltage with frequency f_s is applied so that the varactor swings from near breakdown to forward conduction. Due to the nonlinear capacitance characteristic, harmonics of f_s are generated. If the varactor is part of a resonance circuit tuned to f_s, the nth harmonic is thus obtained. Usually n is 2–3 but cascading multipliers results in a high multiplication ratio.

9.16 LIGHT EMITTING DIODE (LED)

Recombination of the excess carriers in a forward biased diode releases energy (photons). The wavelengths of these photons can be chosen in the visible part of the spectrum by a proper choice of materials (Fig. 9.21(a)). A number of materials and wavelengths are given below:

GaP	550–690 nm
GaAsP	585–670 nm
GaPZnO	690 nm
GaAs	905 nm
GaAsSi	940 nm

The forward bias voltage of an LED is rather high: 1.5–3 V, the allowed reverse voltage is about 3 V.

A construction of an LED and a typical characteristic are shown in Fig. 9.21(b). One of the well-known applications of the LED is the seven-segment display.

9.17 DIAC

The diac (diode AC switch) consists of two p regions separated by an n region (Fig. 9.22(a)). At low voltages of either polarity (Fig. 9.22(b)) the diac blocks the flow of current. When increasing the voltage to the **breakover voltage** V_{BO}, avalanche occurs, a large current will flow and the voltage drops sharply. Thus the diac acts as a two-way switch.

Diacs are used for overvoltage protection in circuits but foremost as trigger devices for triacs (Section 9.20).

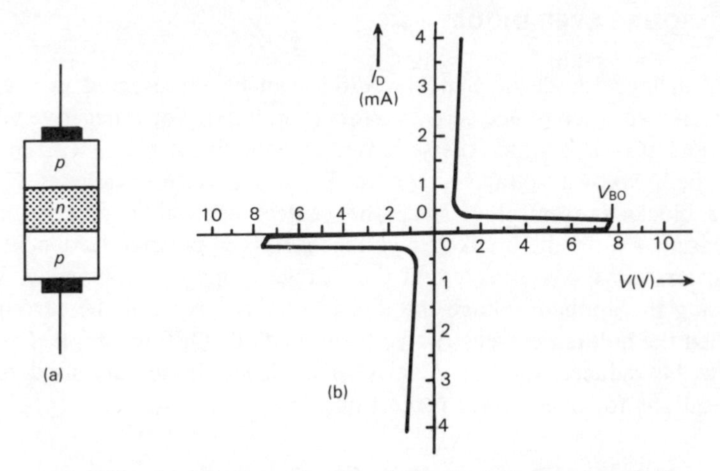

Fig. 9.22 (a) Structure of a diac (b) Typical diac characteristic

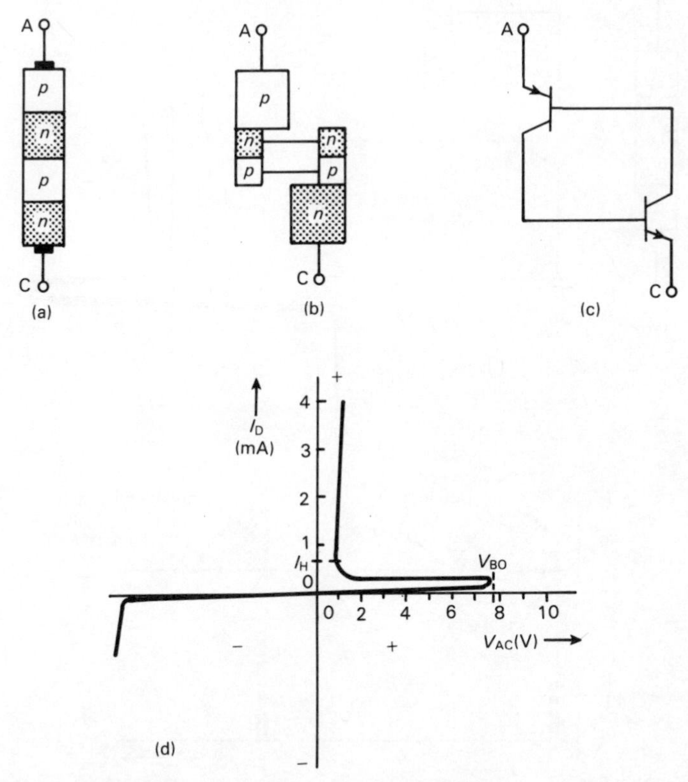

Fig. 9.23 (a) Structure of four-layer diode (b) The four-layer diode shown to consist of two transistors (c) Equivalent circuit of the four-layer diode (d) Typical characteristic of the four-layer diode

9.18 FOUR-LAYER DIODE

The four-layer diode or Shockley diode can be represented in a circuit model as two interconnected transistors (Fig. 9.23). For a negative voltage V_{AC} both transistors are reverse biased and the diode blocks current flow until the breakover voltage is reached. For a positive low value of V_{AC} the diode blocks current flow since the center junction is reverse biased. Increasing V_{AC} to the breakover voltage results in positive feedback: both transistors abruptly saturate and the voltage drops to a low value. When reducing the applied voltage the diode remains ON until the current has reached the **holding current** I_H. To turn the diode OFF the applied voltage has to be reduced so that $I_D < I_H$. Four-layer diodes are used in DC applications for overvoltage protection.

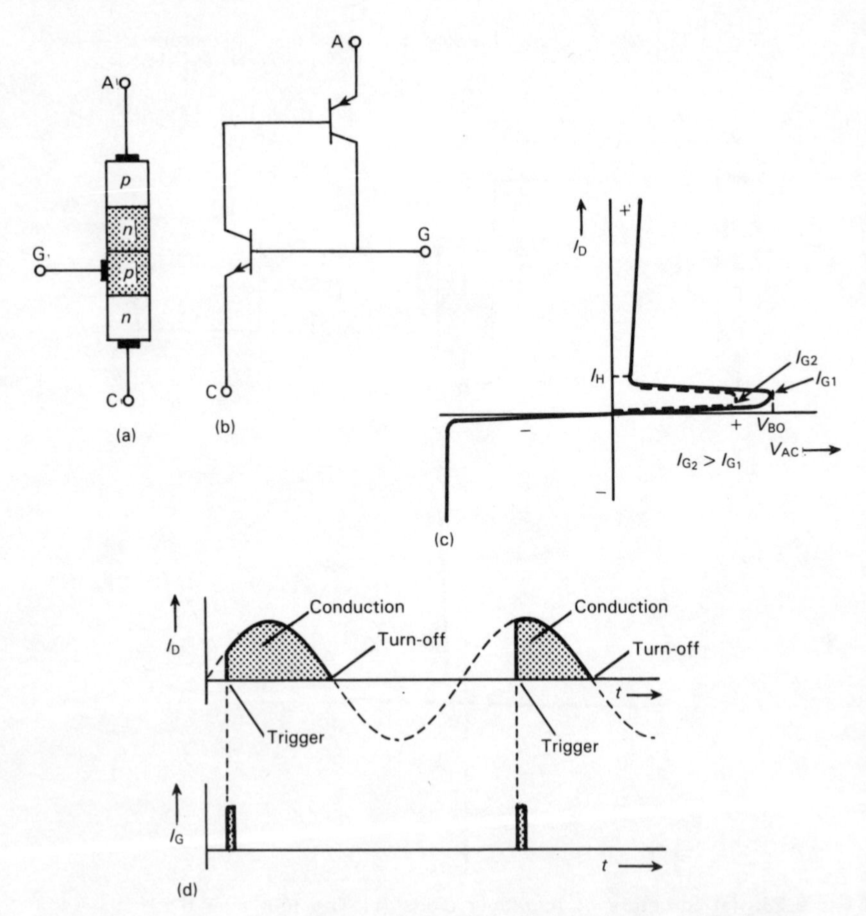

Fig. 9.24 (a) Structure of the thyristor (b) Equivalent circuit of the thyristor (c) Typical characteristic of the thyristor (d) Example of thyristor switching

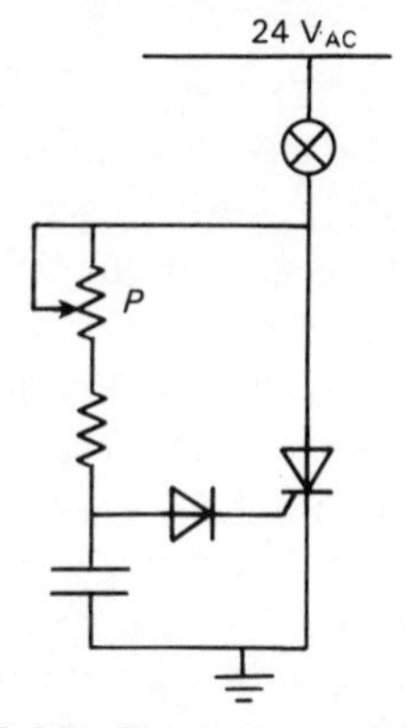

Fig. 9.25 Thyristor used in lamp control circuit

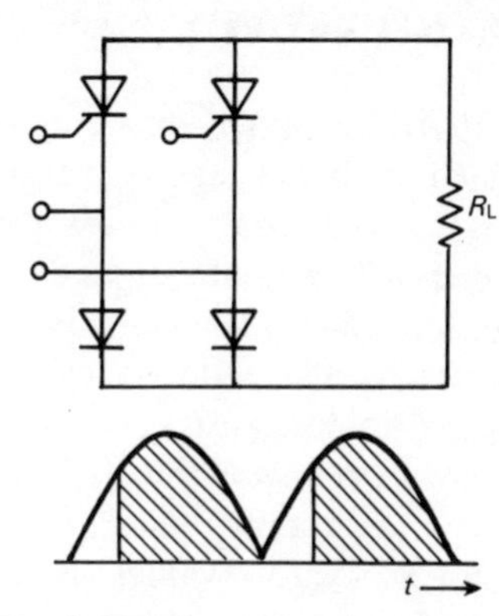

Fig. 9.26 Two thyristors used in bridge circuit

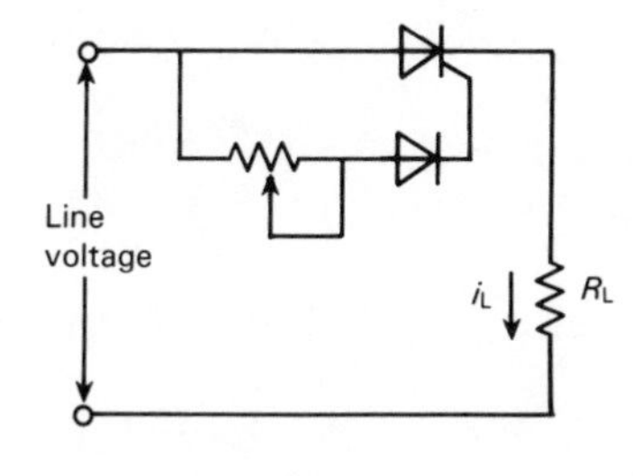

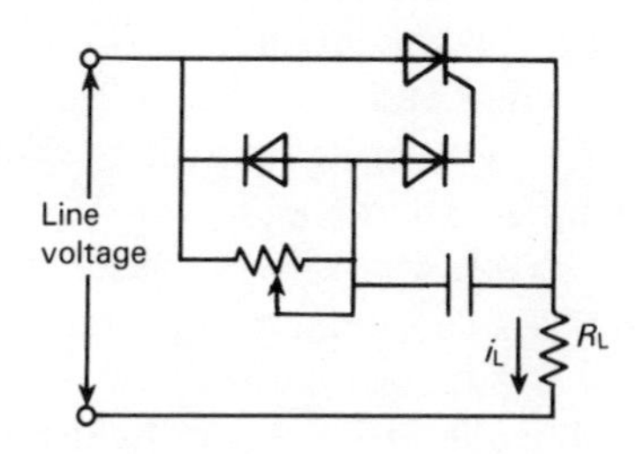

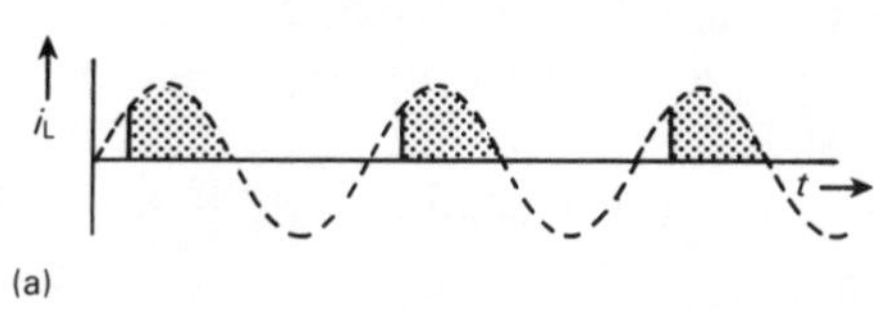

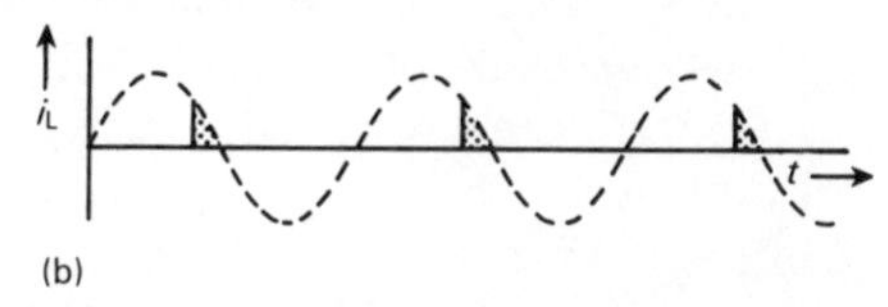

Fig. 9.27 (a) Control of conduction angle up to 90° (b) Control of conduction angle up to 180°

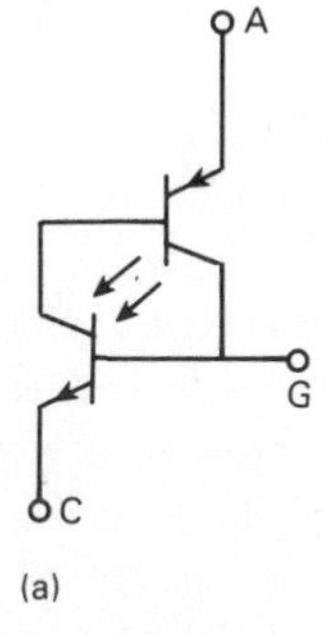

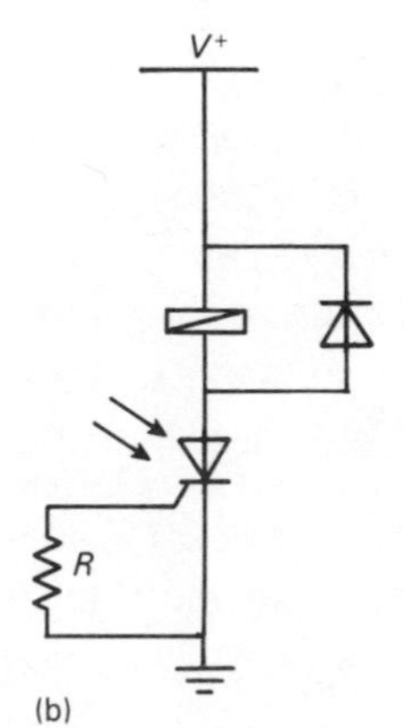

Fig. 9.28 (a) Basic photothyristor circuit (b) Application of thyristor in relay switching circuit

9.19 THYRISTOR

The thyristor or silicon controlled rectifier (SCR) is a four-layer diode supplied with a trigger electrode (Fig. 9.24(a)). A positive voltage V_{AC} which is smaller than the breakover voltage V_{BO} keeps the diode OFF unless a trigger or firing pulse at gate G of a certain voltage and current is applied. Then the SCR turns ON. This occurs also when V_{AC} reaches the breakover voltage V_{BO}. The voltage then drops to a low value ($\simeq 1.5$ V). After turn-on the gate no longer has control of the diode and turn-off can only be achieved by reducing the diode current below the holding current I_H (e.g. by reducing the supply voltage to zero, Fig. 9.24(d)).

With negative values of V_{AC} the SCR blocks conduction until the reverse breakdown voltage is reached. SCRs thus require a minimum of control energy to control DC and AC voltages.

SCRs are used for rectification, in timing circuits, ignition circuits, etc. A lamp control circuit is shown in Fig. 9.25. Potentiometer *P* controls the gate current and the firing angle.

An SCR bridge circuit is shown in Fig. 9.26 as well as the voltage waveform. Figure 9.27 shows two circuits for control of the conduction angle of 90° and 180° respectively.

SCRs can handle large signals and power levels (2000 V, 600 A). The turn-on time is about 1 μs.

The **photothyristor** or light activated SCR (LASCR) consists of a *pnp* transistor and a photo *npn* transistor (Fig. 9.28). Firing of the SCR is achieved by light pulses or a positive gate voltage. In a simple application a relay is triggered when light strikes the SCR. Optical sensitivity is controlled by the value of *R*. Photothyristors have their maximum sensitivity at infrared wavelengths ($\simeq$900 nm).

9.20 TRIAC

The triac (triode AC switch) is constructed with two SCRs in reverse parallel (Fig. 9.29). The connections are first main terminal (A_1), second main terminal (A_2) and gate (G).

The triac can conduct in both directions and can be fired by positive as well as negative pulses. Turn-off occurs when the diode current becomes less than the holding current I_H (e.g. when $V_{A_1A_2}$ goes through zero).

The triac is often triggered by a diac as shown in the light dimming circuit of Fig. 9.30(a). Operation is based on control of the firing angle (Fig. 9.30(b)). Like the thyristor the triac can handle large currents and voltages. One of their main uses is in AC motor speed control.

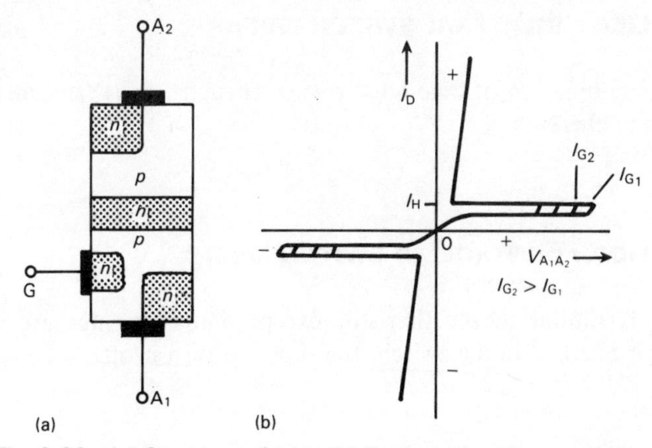

Fig. 9.29 (a) Structure of triac (b) Typical triac characteristics

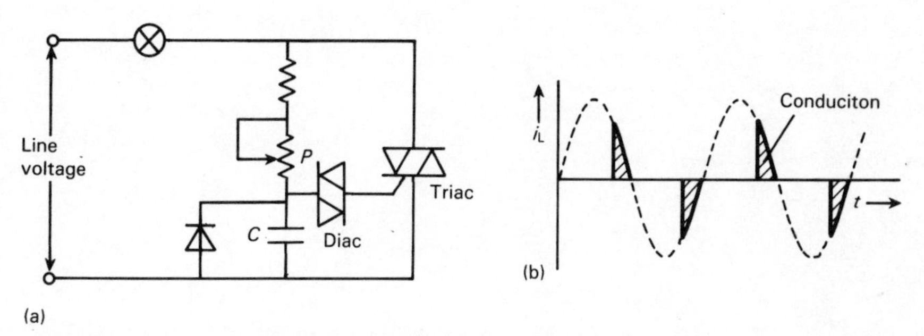

Fig. 9.30 (a) Diac and triac used in light dimming circuit (b) Waveform and switching characteristic of light dimming circuit

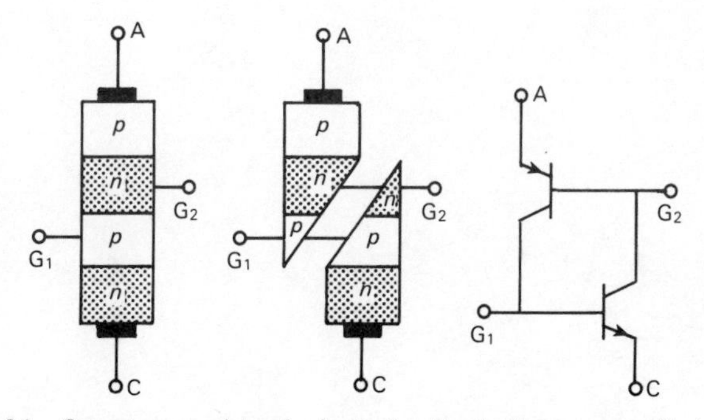

Fig. 9.31 Structure and equivalent circuit of silicon controlled switch

9.21 SILICON BILATERAL SWITCH (SBS)

This is a combination of diac plus triac constructed as IC and intended to switch small currents.

9.22 SILICON CONTROLLED SWITCH (SCS)

The SCS is similar to the thyristor except that two gates are available (Fig. 9.31): one to fire the switch, the other to turn it off.

10
Bipolar Transistors

10.1 BASIC OPERATION

J. Bardeen and W. H. Brattain invented the transistor in 1948. It was the beginning of an electronic revolution and the downfall of the vacuum tube as an amplifying device.

A transistor is constructed by sandwiching a region of semiconductor material (e.g. *p*-type) between two regions of *n*-type material thus forming two *pn* junctions. The connections to the outside world are made by metallic contacts. The three semiconductor regions are **collector, base** and **emitter** (Fig. 10.1(a)) and the construction is an *npn* transistor. The circuit symbol is that of Fig. 10.1(b). When connected to external voltage sources the collector–base junction is reverse biased, the base–emitter junction is forward biased (Fig. 10.1(c)).

When reversing the types of semiconductor materials, the *pnp* transistor results and its circuit symbol is that of Fig. 10.1(d).

If, in Fig. 10.1(c), only V_{CB} is connected, the reverse bias implies that only minority carriers can pass the junction: holes move from collector to base and electrons from base to collector. This is the drift current. If in addition V_{BE} is connected, a diffusion current of majority carriers enters the base region. These electrons, as well as intrinsic electrons in the base region move to the collector region. Thus the emitter injects electrons into the base and collector regions.

Conversely, the base returns holes to the emitter region. The current picture is shown in Fig. 10.2. The upper current is the injected current of electrons which travel from emitter to collector. The lower current is the drift current called **leakage current** and designated as I_{CBO}. Like all leakage currents, it increases exponentially with temperature (Chapter 9).

Some of the electrons of the emitter current are lost in the base region due to recombination. Holes in the base region thus disappear but new holes are created near the base contact where electrons are 'pulled' out of the base region due to V_{BE}. Recombination in the base means that part of the emitter current is lost. To minimize recombination as much as possible the base doping level is made very light.

In addition, the thickness of the base is made less than the **diffusion length**, that is, the average distance an electron penetrates into the base region before recombining. By these controllable mechanisms most of the emitter electrons are swept across to the collector region. Here the electrons are removed by the voltage V_{CB} leaving a supply of holes in the collector region available for recombination.

As shown in Fig. 10.2 the total emitter current is the sum of an electron current and a hole current. The hole current, however, is not a part of the collector current. The ratio of electron current to total emitter current is the **emitter efficiency** γ. To increase this efficiency the emitter doping level is made orders of magnitude larger than that of the base.

The final result is a collector current which is nearly equal to the emitter current, the difference being the base current.

The amplifying property of a transistor can now be shown as follows (Fig. 10.3). If V_{BE} increases by an amount ΔV_{BE} the emitter current increases (approximately) by an amount $\Delta I_E = \Delta V_{BE}/R_e$. This current change occurs also in the collector region so that $\Delta I_C \simeq \Delta V_{BE}/R_e$. As the collector current flows through the collector resistor R_c, $\Delta V_{CE} = \Delta I_C R_c = \Delta V_{BE}(R_c/R_e)$. If $R_c > R_e$ the transistor amplifies the input voltage change ΔV_{BE}.

In a *pnp* transistor the same mechanism of current flow exists except that all polarities are reversed and electrons and holes interchange their roles.

10.2 DC BEHAVIOR

It is obvious from Section 10.1 that $I_E = I_B + I_C$. The ratio $I_C/I_E = \alpha$ is the **emitter-to-collector current amplification factor**. The transistor construction has resulted in α-values very close to 1. Thus the following equations apply:

$$I_C = \alpha I_E + I_{CBO}$$

$$I_B = (1 - \alpha) I_E + I_{CBO}$$

When relating I_C and I_B directly

$$I_C = \frac{\alpha}{1-\alpha} I_B + \frac{1}{1-\alpha} I_{CBO}$$

The factor $\alpha/(1 - \alpha) = \beta$ is the **base-to-collector current amplification factor** and since α is close to 1, β values can be very high. Because

$$\frac{\alpha}{1-\alpha} \simeq \frac{1}{1-\alpha} \simeq \beta$$

$$I_C \simeq \beta I_B + \beta I_{CBO}$$

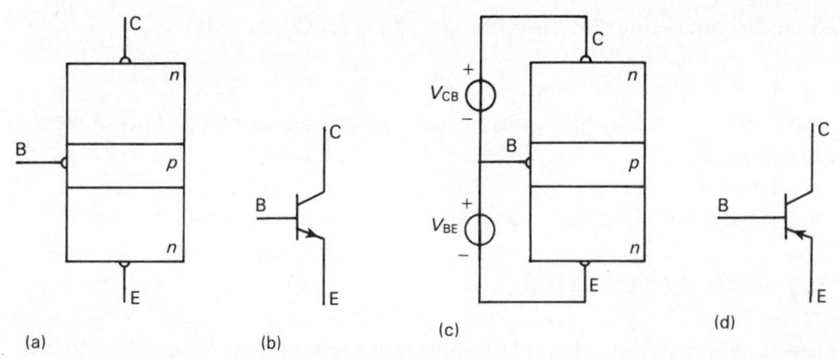

Fig. 10.1 (a) Basic structure of *npn* transistor (b) Symbol of *npn* transistor (c) Voltage sources connected to *npn* transistor (d) Symbol of *pnp* transistor

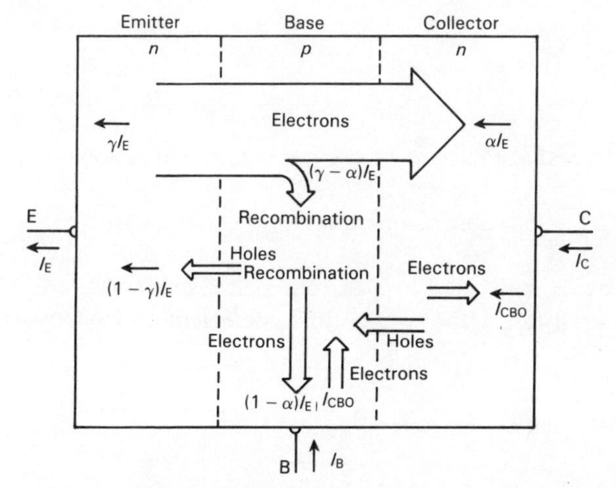

Fig. 10.2 Current flow in an *npn* transistor

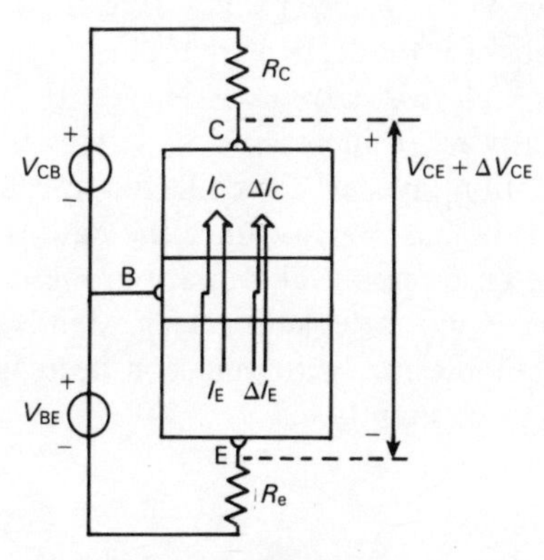

Fig. 10.3 Amplification in an *npn* transistor

When disconnecting the base (making $I_B = 0$) we see that

$$I_C = \beta I_{CBO}$$

Thus the effect of an open base is a large leakage current. This current is designated as I_{CEO}.

10.3 CHARACTERISTICS

Three characteristics adequately illustrate the behavior of a transistor:

Input characteristic. I_B versus V_{BE} (Fig. 10.4(a)). Since the base-emitter junction is forward biased

$$I_B \simeq \frac{I_{ES}}{\beta}\left[\exp\left(\frac{q}{kT}V_{BE}\right) - 1\right]$$

where

I_{ES} is the leakage current of the base-emitter junction

$\dfrac{q}{kT} = 40\ V^{-1}$ at room temperature

When the base is connected to a current source or a voltage source with a high internal resistance, the temperature coefficient of V_{BE} equals that of a *pn* diode:

$$\frac{dV_{BE}}{dT} \simeq -2.3\ mV\ ^\circ C^{-1}$$

Transfer characteristic. I_C versus I_B (Fig. 10.4(b)). This relation is basically the current amplification factor β.

Output characteristics. I_C versus V_{CE} (Fig. 10.4(c)). Here I_B is used as parameter. The transistor model of Fig. 10.2 implies that changes in V_{CB} have no effect on the collector current. However, the characteristics show that I_C increases slightly when increasing V_{CB}. This is the **Early effect**. In Fig. 10.4(d) the depeletion layer of the collector-to-base junction is shown to extend mainly into the base region since the base is only lightly doped. Recombination in the base region takes place in the remaining part of the base. When increasing V_{CB}, the depletion layer extends further into the base so that the region available for recombination decreases. As a result, the collector current increases slightly.

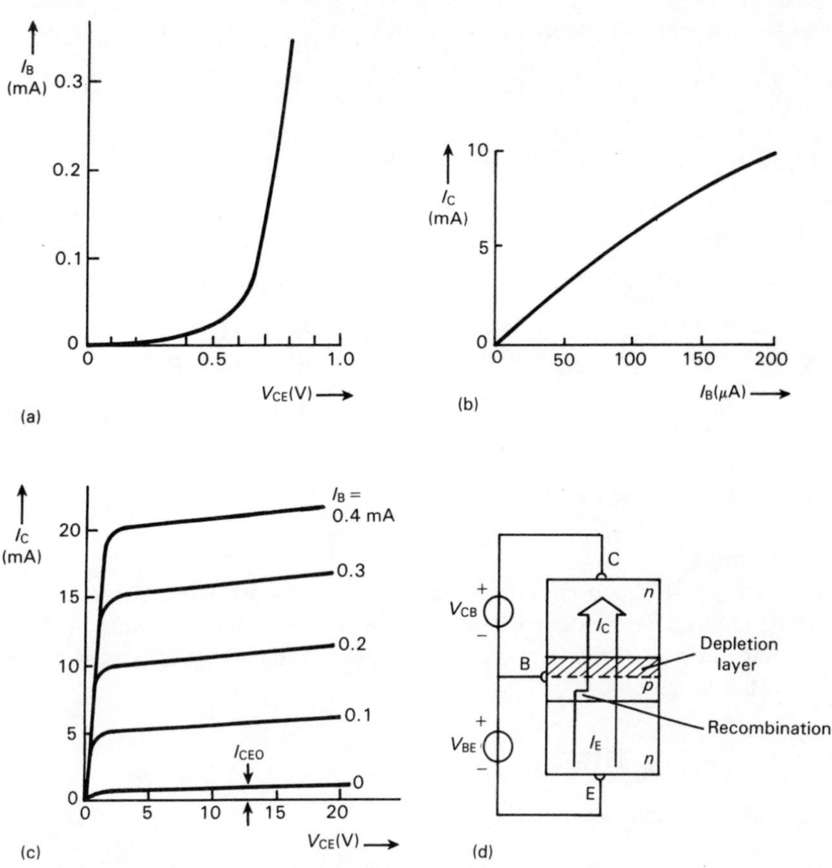

Fig. 10.4 (a) Typical input characteristic of *npn* transistor (b) Typical transfer characteristic of *npn* transistor (c) Typical output characteristics of *npn* transistor (d) The Early effect in an *npn* transistor

10.4 BIASING

The commonly used biasing arrangements of a transistor are shown in Fig. 10.5. When solving the voltage and current equations for the circuit of Fig. 10.5(a) the bias current is found as

$$I_C \simeq \left[\frac{V^+}{R_1} - \frac{V_{BE}}{R_s} + I_{CBO}\left(1 + \frac{R_e}{R_s}\right)\right] \Big/ \left(\frac{1}{\beta} + \frac{R_e}{R_s}\right)$$

where

$$\frac{1}{R_s} = \frac{1}{R_1} + \frac{1}{R_2}$$

The **current stability factor** S_i is defined as

$$S_i = \frac{\partial I_C}{\partial I_{CBO}} = \left(1 + \frac{R_e}{R_s}\right) \Big/ \left(\frac{1}{\beta} + \frac{R_e}{R_s}\right)$$

The effect of I_{CBO} on I_C thus becomes less if

- R_e is increased,
- R_s is decreased.

Similarly, in the circuit of Fig. 10.5(b) the bias current is

$$I_C \simeq \left[\frac{V^+}{R_1} - V_{BE}\left(\frac{1}{R_s} + \frac{R_c}{R_1 R_2}\right) + I_{CBO}\left(1 + \frac{R_c}{R_1}\right)\right] \Big/ \left(\frac{1}{\beta} + \frac{R_c}{R_1}\right)$$

$$S_i = \left(1 + \frac{R_c}{R_1}\right) \Big/ \left(\frac{1}{\beta} + \frac{R_c}{R_1}\right)$$

Here S_i decreases when R_c/R_1 increases.

Example

In the circuit of Fig. 10.5(c) the Si transistor has $\beta = 50$. The leakage current can be ignored. A bias current of 1 mA is desired. Find the value of R_1.

Substitution of the known circuit values in the bias equation, assuming $V_{BE} = 0.6$ V, gives

$$1 = \left(\frac{12}{R_1} - \frac{0.6}{R_s}\right) \Big/ \left(\frac{1}{50} + \frac{1}{R_s}\right)$$

from which it follows that $R_1 \simeq 100$ kΩ.

10.5 THE EBERS–MOLL EQUATIONS

When considering all currents flowing in a transistor, not only the injected emitter carriers which reach the collector with current amplification factor α_f should be taken into account but also the injected collector carriers which reach the emitter with current amplification factor α_r.

In Fig. 10.6 the junction currents are represented by diodes and the current amplification by current sources. From this model the following equations can be derived:

$$\begin{aligned} I_E &= I_f - \alpha_r I_r \\ &= I_{ES}[\exp(\gamma V_{BE}) - 1] - \alpha_r I_{CS}[\exp(\gamma V_{BC}) - 1] \\ I_C &= -I_r + \alpha_f I_f \\ &= -I_{CS}[\exp(\gamma V_{BC}) - 1] + \alpha_f I_{ES}[\exp(\gamma V_{BE}) - 1] \end{aligned}$$

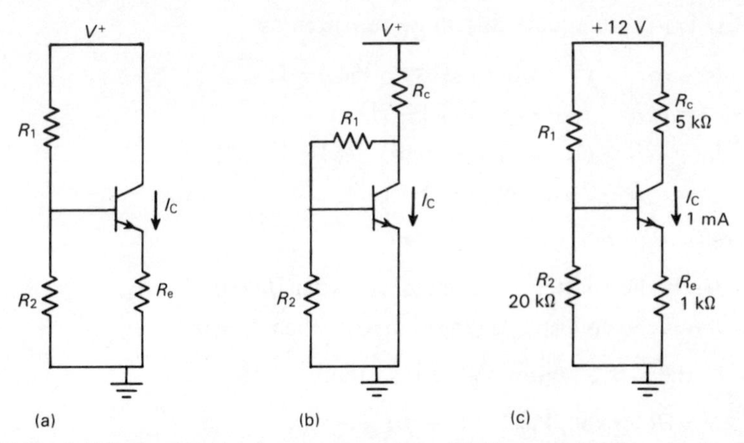

Fig. 10.5 (a) Biasing circuit of an *npn* transistor (b) Alternate biasing circuit of an *npn* transistor (c) Example of bias current calculation

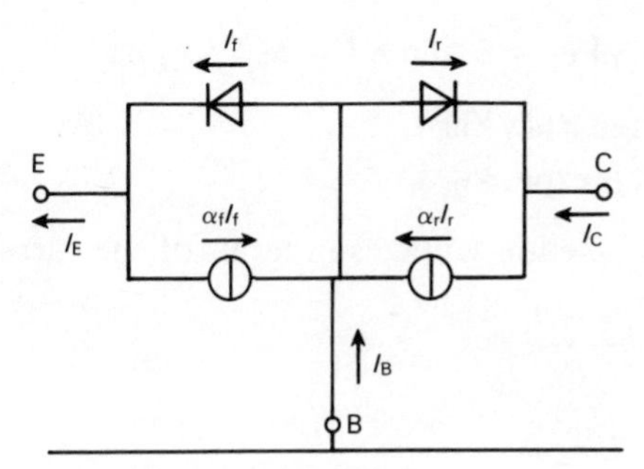

Fig. 10.6 Equivalent circuit of a transistor under forward and reverse bias

These are the **Ebers–Moll equations** where

$$\gamma = \frac{q}{kT}$$

I_{ES} is the leakage current of the emitter–base junction with the collector-base junction shorted

I_{CS} is the leakage current of the collector–base junction with the emitter–base junction shorted

It can be shown that the following relationship between these leakage currents holds:

$$\frac{I_{ES}}{I_{CS}} = \frac{\alpha_r}{\alpha_f}$$

The Ebers–Moll equations can be rewritten as

$$\begin{aligned} I_E &= \alpha_r I_C + (1 - \alpha_r \alpha_f) I_{ES} [\exp(\gamma V_{BE}) - 1] \\ &= \alpha_r I_C + I_{EO} [\exp(\gamma V_{BE}) - 1] \\ I_C &= \alpha_f I_E - (1 - \alpha_r \alpha_f) I_{CS} [\exp(\gamma V_{BC}) - 1] \\ &= \alpha_f I_E - I_{CO} [\exp(\gamma V_{BC}) - 1] \end{aligned}$$

where

I_{EO} is the emitter leakage current when $I_C = 0$

I_{CO} is the collector leakage current when $I_E = 0$

In the **active region** $V_{BC} \ll 0$ so that

$$\begin{aligned} I_E &= I_{ES} [\exp(\gamma V_{BE}) - 1] + \alpha_r I_{CS} \\ I_C &= \alpha_f I_E + I_{CO} \end{aligned}$$

The last equation shows that $I_{CO} = I_{CBO}$ is the leakage current with open emitter.

In **saturation** $\gamma V_{BE} \gg 1$ and $\gamma V_{BC} \gg 1$ so that

$$\begin{aligned} I_E &= \alpha_r I_C + I_{EO} \exp(\gamma V_{BE}) \\ I_C &= \alpha_f I_E - I_{CO} \exp(\gamma V_{BC}) \end{aligned}$$

When solving the junction voltages in terms of the currents we find

$$V_{BE} = \frac{1}{\gamma} \ln \frac{I_E - \alpha_r I_C}{I_{EO}}$$

$$V_{BC} = \frac{1}{\gamma} \ln \frac{\alpha_f I_E - I_C}{I_{CO}}$$

Since $V_{CE} = V_{BE} - V_{BC}$ we find

$$\begin{aligned} V_{CE} &= \frac{1}{\gamma} \ln \frac{I_E - \alpha_r I_C}{\alpha_f I_E - I_C} \frac{I_{CO}}{I_{EO}} \\ &= \frac{1}{\gamma} \ln \frac{I_B + (1 - \alpha_r) I_C}{\alpha_f I_B - (1 - \alpha_f) I_C} \frac{I_{CS}}{I_{ES}} \end{aligned}$$

When saturating the transistor by making I_B large, the saturation voltage is

$$V_{CE,s} = \frac{1}{\gamma} \ln \frac{1}{\alpha_r}$$

When operating the transistor in the inverse mode by interchanging the roles of emitter and collector ($I_C \leftrightarrow I_E$, $\alpha_r \leftrightarrow \alpha_f$, $I_{CS} \leftrightarrow I_{ES}$) we find for the saturation voltage

$$V_{EC,s} = \frac{1}{\gamma} \ln \frac{1}{\alpha_f}$$

The numerical values of these two saturation voltages differ considerably since $\alpha_r \ll \alpha_f$ because

- the emitter junction area is much smaller than the collector junction,
- the collector doping is relatively light.

Values of α_r are in the range of 0.01–0.2 while α_f is close to unity. If for example $\alpha_f = 0.98$ and $\alpha_r = 0.05$, $V_{CE,s} \simeq 75$ mV, $V_{EC,s} \simeq 0.5$ mV. For this reason inverted transistors are often employed in chopper circuits since in this mode the transistor resembles much more a closed switch.

10.6 SMALL-SIGNAL LOW-FREQUENCY BEHAVIOR; EQUIVALENT CIRCUITS

If the transistor is considered as an unknown object in a black box (Fig. 10.7) two currents and two voltages are available as measurable information. Several sets of equations are possible according to which variables are dependent and which are independent. If linear relationships between the variables are required (otherwise the equations become unmanageable) only small signals are allowed because of the nonlinear current and voltage characteristics.

Three transistor configurations are possible:

- Common base (CB) configuration (Fig. 10.8(a)). As shown, the base is common to input and output signal.
- Common emitter (CE) configuration (Fig. 10.8(b)).
- Common collector (CC) configuration (Fig. 10.8(c)). This configuration is normally called **emitter follower**.

The two systems of equations in common use are (for the CE configuration)

$$i_b = y_{ie} v_{be} + y_{re} v_{ce}$$

$$i_c = y_{fe} v_{be} + y_{oe} v_{ce}$$

The coefficients y are all **admittance parameters** which are defined as follows:

$$y_{ie} = \left.\frac{i_b}{v_{be}}\right]_{v_{ce}=0}$$ is the input admittance with shorted output

$$y_{re} = \left.\frac{i_b}{v_{ce}}\right]_{v_{be}=0}$$ is the reverse transfer admittance with shorted output

$$y_{fe} = \left.\frac{i_c}{v_{be}}\right]_{v_{ce}=0}$$ is the forward transfer admittance with shorted output

$$y_{oe} = \left.\frac{i_c}{v_{ce}}\right]_{v_{be}=0}$$ is the output admittance with shorted input

The y parameters are also called short-circuit parameters. Because the input impedance of a transistor is low, a short-circuit input measurement at low frequencies is difficult. Therefore, y parameters are mostly used at high frequencies where a short circuit is easily accomplished by placing a small capacitor across the terminals. The y parameters then become complex quantities. At lower frequencies the following equations are favored:

$$v_{be} = h_{ie} i_b + h_{re} v_{ce}$$

$$i_c = h_{fe} i_b + h_{oe} v_{ce}$$

The h parameters have mixed dimensions since

$$h_{ie} = \left.\frac{v_{be}}{i_b}\right]_{v_{ce}=0}$$ is the input impedance with shorted output

$$h_{re} = \left.\frac{v_{be}}{v_{ce}}\right]_{i_b=0}$$ is the reverse voltage transfer ratio with open input

$$h_{fe} = \left.\frac{i_c}{i_b}\right]_{v_{ce}=0}$$ is the forward current transfer ratio with shorted output

$$h_{oe} = \left.\frac{i_c}{v_{ce}}\right]_{i_b=0}$$ is the output admittance with open input

With this system open input and shorted output measurements are easy to realize.

The y and h equations can be used to construct equivalent circuits of the transistor. Figures 10.9(a) and 10.9(b) show the y-equivalent circuit and h-equivalent circuit respectively, both for the CE configuration.

The relationships between y and h parameters can be easily derived:

$$h_{ie} = \frac{1}{y_{ie}}$$

$$h_{re} = -\frac{y_{re}}{y_{ie}}$$

$$h_{fe} = \frac{y_{fe}}{y_{ie}}$$

$$h_{oe} = \frac{y_{oe} y_{ie} - y_{re} y_{fe}}{y_{ie}}$$

Very often the T-equivalent circuit is used. The relevant CB and CE circuits are shown in Figs. 10.10(a) and 10.10(b). The resistances in these circuits

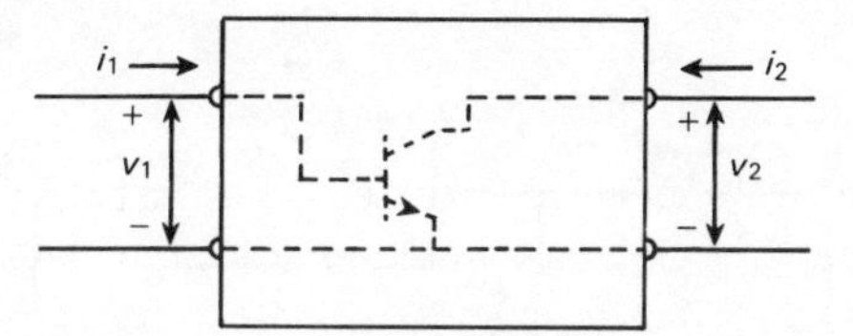

Fig. 10.7 Black box representation of *npn* transistor in CE configuration

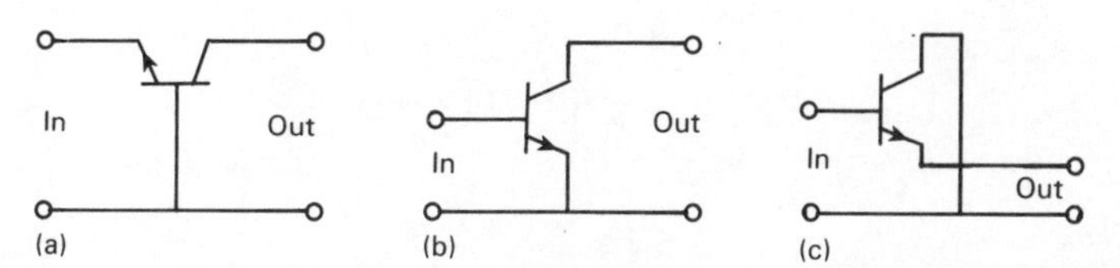

Fig. 10.8 (a) The CB configuration (b) The CE configuration (c) The CC configuration

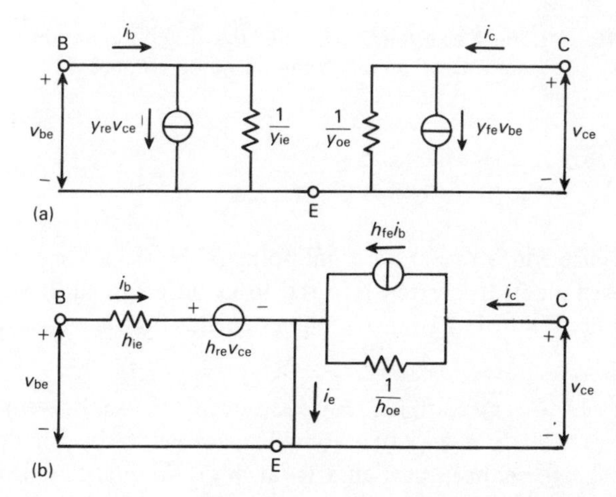

Fig. 10.9 (a) The *y*-equivalent circuit (b) The *h*-equivalent circuit

are as follows:

- The dynamic emitter resistance r_e which is the same as the dynamic resistance of a forward biased diode:

$$r_e \simeq \frac{25\text{ mV}}{I_E} \quad (I_E \text{ in mA, } r_e \text{ in } \Omega)$$

- The dynamic collector resistance r_c which is due to the Early effect. Its value is inversely proportional to the bias current I_E. The order of magnitude is several MΩ s at 1 mA bias current.
- The base resistance r_b which is considered as the sum of two other resistances:

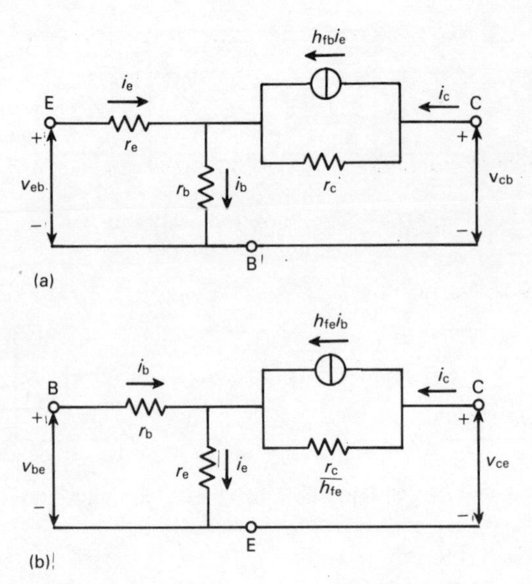

Fig. 10.10 (a) The *T*-equivalent circuit of the CB-configuration (b) The *T*-equivalent circuit of the CE-configuration

$r_{bb'}$ from base contact to an internal point B' between the two junctions (lateral resistance). It is often referred to as base spreading resistance. $r_{b'}$ from junction to junction. Its value depends on the transistor current and V_{CB}.

Although r_b is a very complicated resistance, it is usually simply written as $r_b = r_{b'} + r_{bb'}$. It is an objectionable parameter since it deteriorates transistor performance, particularly at high frequencies. An average value of r_b is 50–200 Ω.

In circuit design, instead of using r_e, r_b and r_c, more practical parameters, the characteristic parameters, are commonly employed:

- Transconductance

$$g_m = \left.\frac{i_c}{v_{be}}\right|_{v_{ce}=0}$$

$$= \left.\frac{h_{fe}}{h_{ie}}\right|_{v_{ce}=0}$$

$$= \frac{1}{r_e\left(1+\frac{1}{\beta}\right)+\frac{r_b}{\beta}} \simeq \frac{1}{r_e+\frac{r_b}{\beta}}$$

Very often $r_b/\beta \ll r_e$ which allows g_m to be expressed in the bias current:

$$g_m \simeq \frac{1}{r_e} \simeq 40 I_E \simeq 40 I_C \text{ (mA V}^{-1}\text{)}$$

- Base-to-collector current amplification factor

$$\beta \simeq \left.\frac{i_c}{i_b}\right]_{v_{ce}=0} = h_{fe}$$

- Voltage amplification factor

$$\mu = -\left.\frac{v_{ce}}{v_{be}}\right]_{i_c=0} \simeq \frac{r_c}{r_e + r_b} \simeq \frac{1}{(h_{oe}h_{ie}/h_{fe}) - h_{re}}$$

- Voltage transfer ratio

$$\mu' = \left.\frac{v_{ce}}{v_{be}}\right]_{i_b=0} \simeq \frac{r_c}{\beta r_e} \simeq \frac{1}{h_{re}}$$

The last two parameters are caused by the Early effect and are often ignored. Their values are very high.

10.7 SMALL-SIGNAL LOW-FREQUENCY AMPLIFIERS

In Fig. 10.11 voltage gain A_v, input resistance r_i and output resistance r_o for the three configurations are summarized. The formulas are based on the assumptions:

- $r_b = 0$
- $h_{oe} = h_{re} = 0$ (no Early effect)

Example 1

A single stage amplifier (Fig. 10.12) with $h_{ie} = 1$ kΩ, $h_{fe} = 100$ (thus $g_m = h_{fe}/h_{ie} = 100$ mA V^{-1}).

$$A_v = \frac{-R_c}{\frac{1}{g_m} + R_e + \frac{R_b}{\beta}}$$

$$= \frac{-5}{0.01 + 0.1 + 0.002} \simeq -44.6$$

$$r_i = R_b + \beta\left(\frac{1}{g_m} + R_e\right)$$

$$= 0.2 + 100(0.01 + 0.1) = 11.2 \text{ k}\Omega$$

$$r_o = R_c = 5 \text{ k}\Omega$$

Example 2

A two-stage amplifier (Fig. 10.13). Both transistors have $h_{ie} = 2\ \mathrm{k\Omega}$, $h_{fe} = 50$. Reactances of coupling and bypass capacitors are negligible. In Fig. 10.13 an auxiliary voltage v_a is assumed and the voltage gain is calculated as

$$A_v = \frac{v_o}{v_s}$$

$$= \frac{v_o}{v_a}\frac{v_a}{v_s}$$

$$\frac{v_o}{v_a} = -g_m R_{c2}$$

$$= -25 \times 5 = -125$$

$$\frac{v_a}{v_s} = \left(\frac{R_{c1} r_{i,2}}{R_{c1} + r_{i,2}}\right) \Big/ \left(\frac{1}{g_m} + R_{e1} + \frac{R_{b1}}{\beta}\right)$$

$$= \frac{1}{0.04 + 0.1 + 0} \simeq 7.1$$

$$A_v = -125 \times 7.1 \simeq -888$$

$$r_i = \frac{1}{g_m} + R_e + \frac{R_b}{\beta}$$

$$= 0.04 + 0.1 = 0.14\ \mathrm{k\Omega}$$

$$r_o = R_{c2} = 5\ \mathrm{k\Omega}$$

10.8 HIGH-FREQUENCY BEHAVIOR

In high-frequency applications parasitic capacitances have to be accounted for. Figure 10.14(a) shows the CE-configuration including these capacitances. Based on this circuit is the **hybrid-π equivalent circuit** (Fig. 10.14(b)) the advantage of which is its applicability over a wide frequency range. The elements are as follows:

- $C_{cb'}$ is the Miller capacitance. Its value depends on V_{CB}.
- $C_{b'e} = C_{b'} + C_d$ = barrier + diffusion capacitance of the emitter–base junction. As mentioned before, C_d is proportional to I_E.
- C_{ce} is the collector-emitter capacitance, which can often be ignored.

From this circuit the frequency dependence of a number of intrinsic

	CB-configuration	CE-configuration	CC-configuration
$A_v \approx$	$\dfrac{R_c}{\dfrac{1}{g_m} + R_e + \dfrac{R_b}{\beta}}$	$\dfrac{-R_c}{\dfrac{1}{g_m} + R_e + \dfrac{R_b}{\beta}}$	$\dfrac{R_e}{\dfrac{1}{g_m} + R_e + \dfrac{R_b}{\beta}}$
$r_i \approx$	$\dfrac{1}{g_m} + R_e + \dfrac{R_b}{\beta}$	$\beta\left(\dfrac{1}{g_m} + R_e\right) + R_b$	$\beta\left(\dfrac{1}{g_m} + R_e\right) + R_b$
$\dfrac{1}{r_o} \approx$	$\dfrac{1}{R_c}$	$\dfrac{1}{R_c}$	$\dfrac{1}{R_e} + \dfrac{1}{\left(\dfrac{1}{g_m} + \dfrac{R_b}{\beta}\right)}$

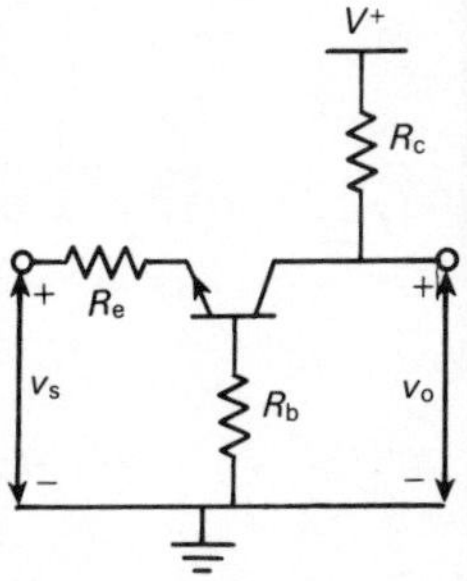

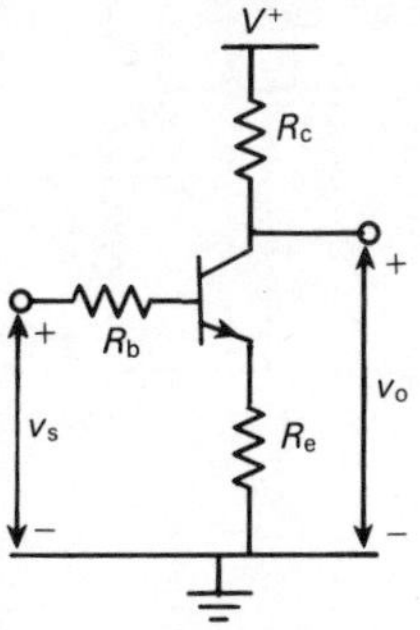

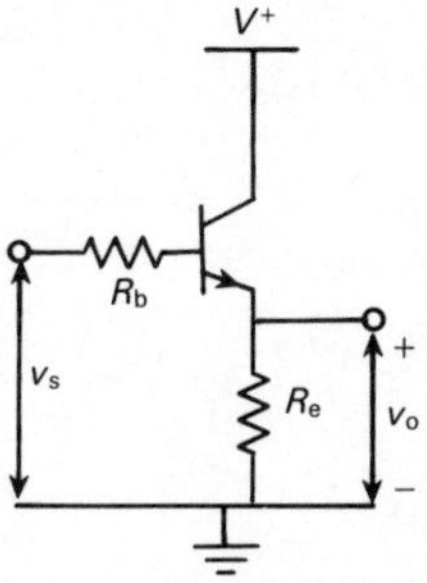

Fig. 10.11 The three configurations and relevant formulas

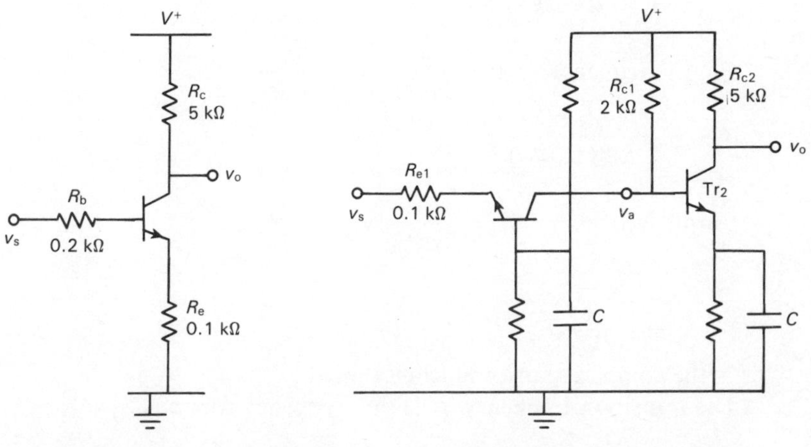

Fig. 10.12 Example of single-stage amplifier

Fig. 10.13 Example of two-stage amplifier

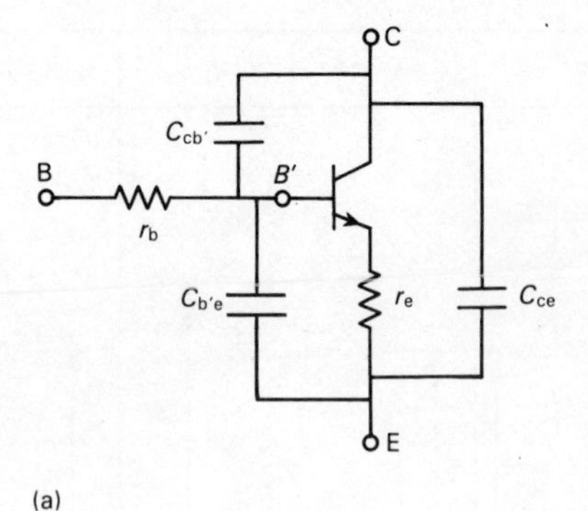

Fig. 10.14 (a) Parastic capacitances in a transistor at high frequencies (b) The hybrid-π equivalent circuit

parameters can be derived:

$$\alpha_\omega = \frac{\alpha}{1 + j\omega\alpha(C_d/g_m)}$$

$$\text{Cutoff frequency: } f_{c,\,\alpha} = \frac{g_m}{2\pi\alpha C_d} \simeq \frac{1}{2\pi\alpha C_d r_e}$$

$$\beta_\omega = \frac{\beta}{1 + j\omega\beta(C_d/g_m)}$$

$$\text{Cutoff frequency: } f_{c,\,\beta} = \frac{g_m}{2\pi\beta C_d} \simeq \frac{1}{2\pi\beta C_d r_e}$$

$$g_{m,\,\omega} = \frac{g_m}{1 + j\omega g_m C_d r_b r_e} \simeq \frac{g_m}{1 + j\omega C_d r_b}$$

This is the complex transconductance y_{fe}.

The **transition frequency** f_T is the frequency where $|\beta_\omega| = 1$:

$$f_T \simeq \frac{g_m}{2\pi C_d}$$

The gain of a voltage amplifier consisting of one or more stages will begin to

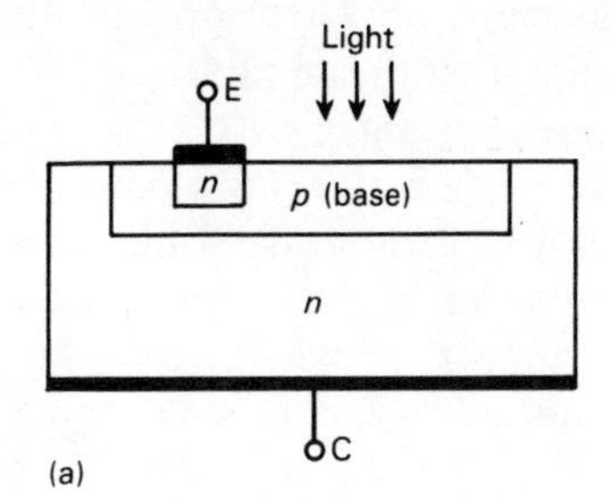

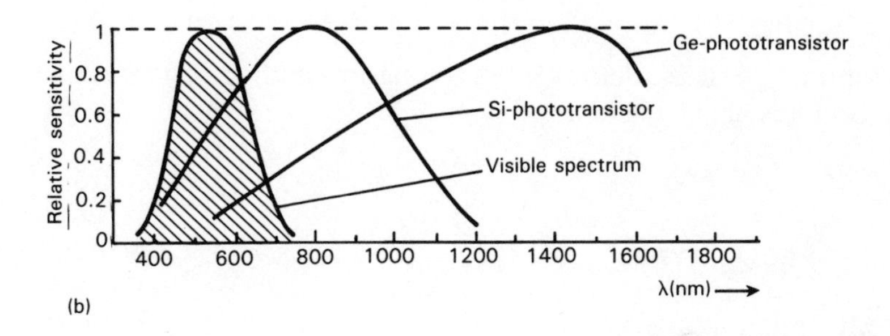

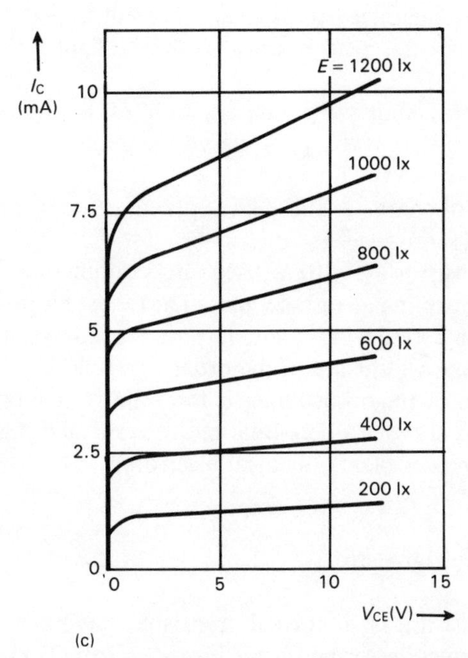

Fig. 10.15 (a) Basic structure of phototransistor (b) Spectral response of phototransistors (c) Typical characteristics of a Si planar epitaxial phototransistor

roll off at some frequency f_c. The slope of this roll off is usually 6 dB octave^{-1} which means that the amplifier can be represented by a single low-pass circuit with time constant $\tau = RC$.

Often this frequency f_c is the approximate bandwidth B of the amplifier. A very useful waveform to determine B is the **step voltage**. The response of a low-pass network to a step voltage is an exponential rise (Section 5.6.1) the rise time t_r of which (between 10 and 90 percentage points) equals $2.2RC$. Since $B \simeq f_c = 1/2\pi RC$ and $t_r = 2.2RC$ we have the relation

$$t_r \simeq \frac{0.35}{B}$$

If, for example, a rise time of 3.5 μs is measured at the amplifier output, its bandwidth is about 100 kHz.

10.9 SPECIAL TYPES

10.9.1 Phototransistor

The phototransistor utilizes the leakage current I_{CEO} of an open base transistor. This current is generated when incident light strikes the collector-base junction.

Figure 10.15(a) shows the construction of a phototransistor, Fig. 10.15(b) the spectral sensitivity. Apparently, phototransistors have their maximum sensitivity in the infrared.

A typical set of characteristics of a Si planar epitaxial phototransistor is shown in Fig. 10.15(c).

A phototransistor is often used in conjunction with an LED (Fig. 10.16) thus forming an **optocoupler**. The main feature of optocouplers is the DC isolation between input and output. Breakdown voltages can be as high as 50 kV. Bandwidths are in the order of 250 kHz.

An extension of the optocoupler is the **solid state relay** (Fig. 10.17).

Optocouplers are used when situations are critical or hazardous such as explosive areas, power plants, medical electronics.

10.9.2 Schottky transistor

A Schottky transistor is a normal transistor having a Schottky diode connected between collector and base (Fig. 10.18(a)). They are mostly used in digital ICs and designated by the symbol of Fig. 10.18(b).

When the transistor is in the active region, the diode is reverse biased.

Forward bias of the diode occurs when the base-collector voltage is about 0.4 V. The diode prevents the collector voltage from going lower than 0.4 V below the base voltage.

In saturation $V_{BE,s} \simeq 0.75$ V, $V_{CE,s} \simeq 0.2$ V so that $V_{CB} \simeq 0.55$ V. Thus the collector-base junction is prevented from becoming forward biased and saturation will not occur. Hence no charge storage in the transistor takes place.

The absence of charge storage in the Schottky diode allows very fast switching of the transistor from maximum conduction to cutoff.

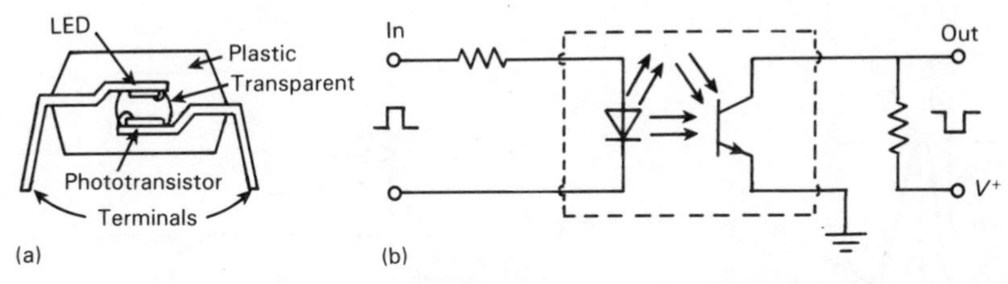

Fig. 10.16 (a) Construction of optocoupler circuit (b) Schematic of optocoupler circuit

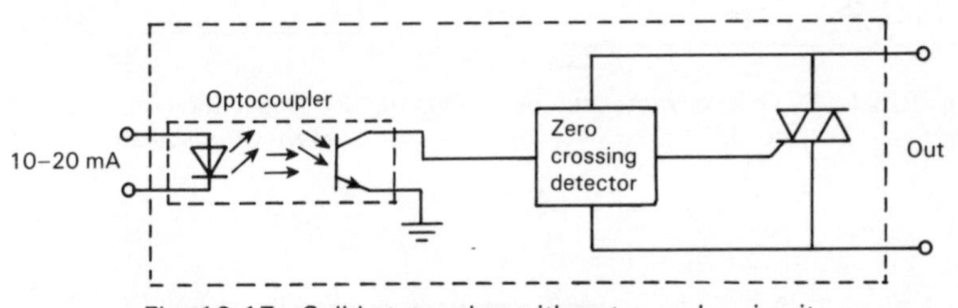

Fig. 10.17 Solid state relay with optocoupler circuit

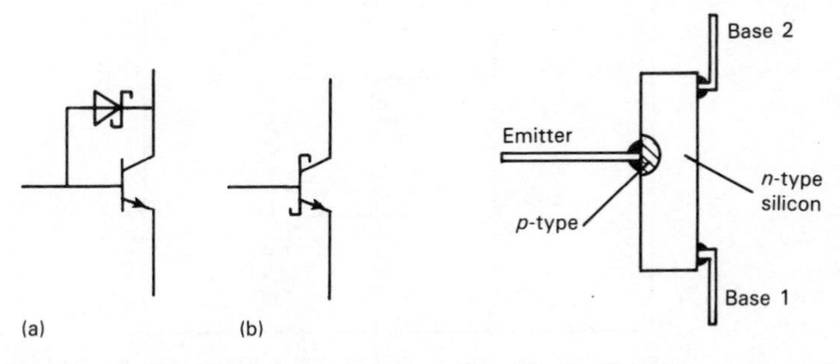

Fig. 10.18 (a) The Schottky transistor (b) Symbol of the Schottky transistor

Fig. 10.19 Structure of unijunction transistor

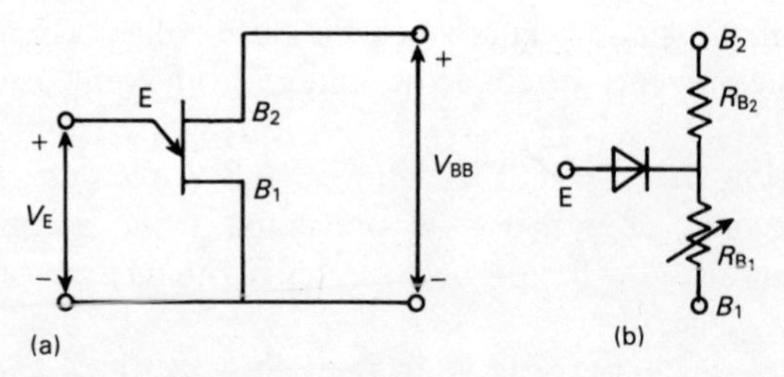

Fig. 10.20 (a) Symbol of the unijunction transistor (b) Equivalent circuit of the unijunction transistor

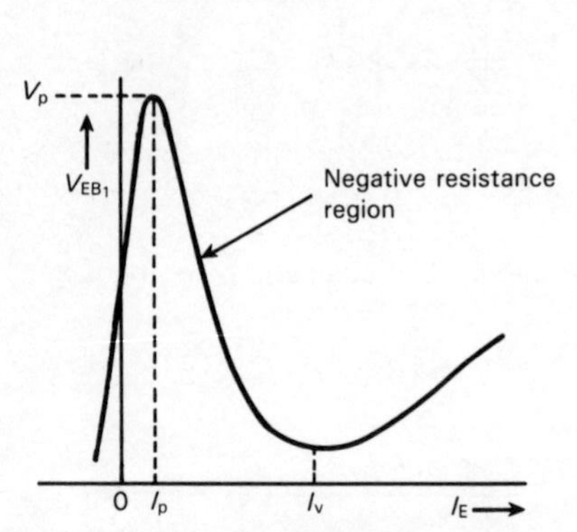

Fig. 10.21 Typical characteristic of the UJT

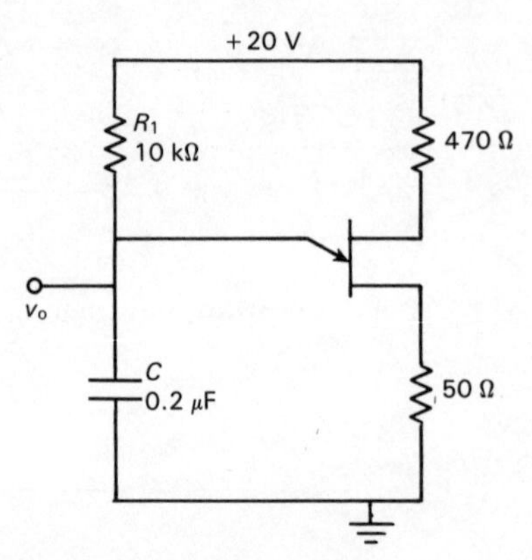

Fig. 10.22 Application of a UJT in a sawtooth oscillator

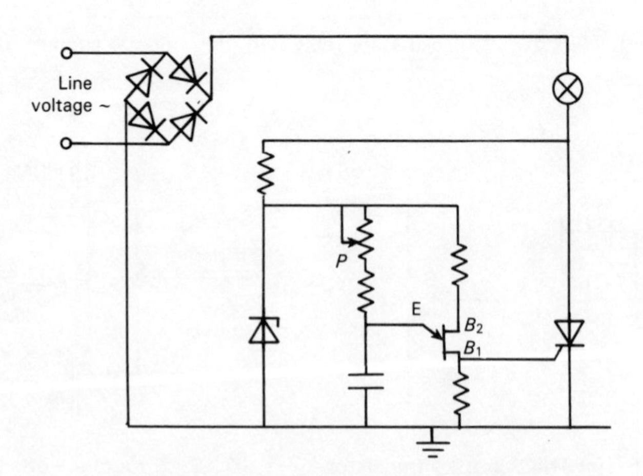

Fig. 10.23 Application of a UJT in thyristor triggering

10.9.3 Unijunction transistor

The operation of a unijunction transistor (UJT) is quite different from that of a conventional transistor.

The UJT is constructed of a bar of lightly doped Si (Fig. 10.19) with ohmic contacts at each end, designated **base 1** and **base 2**. Somewhere near the center of the bar a small *p*-type region is diffused. This contact is the **emitter**. B_2 is made positive with respect to B_1. This voltage (V_{BB}) is the **interbase voltage** (Fig. 10.20) which causes a gradual voltage drop along the bar. If V_E is less positive than the voltage of the Si opposite the emitter then the emitter junction is reverse biased and the bar resistance is then 5–10 kΩ. The voltage across R_{B_1} is called ηV_{BB} where η is the **intrinsic stand-off ratio** ($\eta = 0.5$–0.8).

If V_E is increased a value will be reached where the emitter junction is almost forward biased. This voltage is the **peak voltage** V_p corresponding with a small **peak current** I_p. When V_E is further increased, holes from the *p*-Si move to B_1 thus lowering the resistance R_{B_1}. Since R_{B_2} remains almost unchanged, the voltage across R_{B_1} falls, thereby increasing the forward bias across the junction. This further decreases the resistance of R_{B_1}; as a result R_{B_1} returns rapidly to a low value.

The UJT characteristic is shown in Fig. 10.21 where the negative resistance region is obvious.

The equation expressing the triggering voltage V_E is

$$V_E \simeq 0.7 + + \eta V_{BB} \text{ (V)}$$

UJTs offer a simple solution to the construction of sawtooth oscillators, trigger circuits, timing circuits, etc.

The circuit of Fig. 10.22 is a sawtooth oscillator having a frequency

$$f \simeq \frac{1}{R_1 C \ln \dfrac{1}{1-\eta}}$$

In Fig. 10.23 the UJT is used to trigger the thyristor. The firing moment is adjusted by potentiometer P thus controlling the light intensity.

11 Field Effect Transistors

11.1 BASIC OPERATION

The effect of an electric field on semiconductor resistivity was discovered in 1928 but the prototype of the present field effect transistor (FET) was realized by Shockley in 1952.

The principle of FET operation is that the resistance of a semiconductor path is controlled by an electric field perpendicular to the current flow. The electric field is obtained by reverse biasing a *pn* junction.

The two basic FET-types are the junction FET (JFET) and the MOSFET.

11.1.1 JFET

The structure of a JFET is shown in Fig. 11.1(a): a bar of semiconducting material, the **channel** is lightly doped (e.g. *n*-type Si). A heavily doped *p*-type diffusion is made inside the channel. This *p* region is called the **gate**. Ohmic contacts are made at the gate and at both ends of the channel, called **drain** and **source.** This is the *n*-channel JFET the circuit symbol of which is shown in Fig. 11.1(b). By reversing the types of semiconductor material, the *p*-channel JFET results with the circuit symbol of Fig. 11.1(c).

The construction of an *n*-channel JFET is shown in Fig. 11.1(d): on a *p*-type substrate the *n* channel is deposited by epitaxy. In the channel the p^+-type gate is diffused. The small n^+ regions in the channel improve the electric connection between channel and Al-contacts.

In Fig. 11.2 the voltage sources V_{DG} and V_{GS} are connected. The voltage V_{GS} reverse biases the junction. The resulting depletion layer extends principally into the channel due to the low doping level of the channel. This layer is wider near the drain since here the reverse voltage is larger than near the source, giving the layer a wedge shape.

The current flow is a flow of majority carriers (here electrons) from source to drain. Therefore, an FET is often referred to as a **unipolar** transistor.

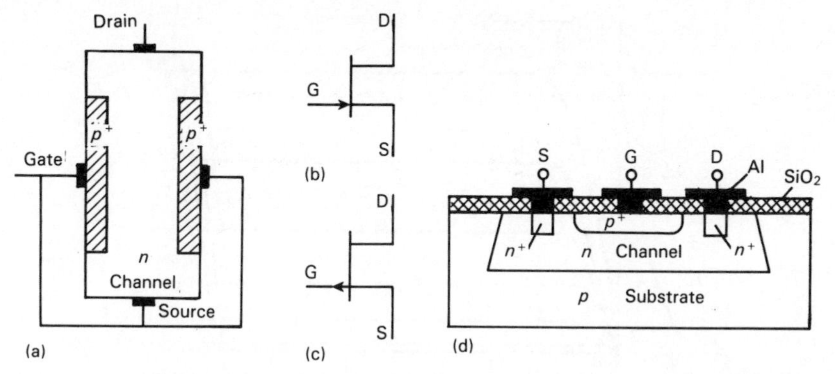

Fig. 11.1 (a) Basic structure of *n*-channel JFET (b) Symbol of *n*-channel JFET (c) Symbol of *p*-channel JFET (d) Structure of *n*-channel JFET

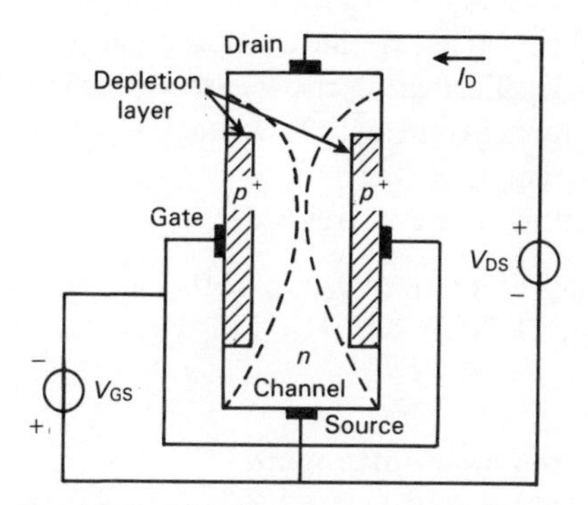

Fig. 11.2 Voltage sources connected to *n*-channel JFET

The bias voltage V_{GS} controls the width of the depletion layer which determines the magnitude of the channel current I_D.

Since the gate is reverse biased the only gate current is the flow of minority carriers. This current is usually less than 100 pA and often negligible. Therefore, the input resistance of a JFET is very high, about $10^{12}\ \Omega$. The characteristics of an *n*-channel JFET are shown in Fig. 11.3.

The **pinch-off voltage** V_P is that value of V_{GS} which makes $I_D = 0$.

If $|V_{DS}| < |V_{GS} - V_P|$ the FET operates in the **triode region**. Between drain and source the FET behaves as a dynamic resistance, the value of which is controlled by V_{GS} (see Section 11.3).

If $|V_{DS}| > |V_{GS} - V_P|$ the electrons have reached their maximum mobility when moving through the channel and a further increase of V_{DS} has little effect on the current I_D. The FET is biased in the **pinch-off region** where it can be used for linear amplification.

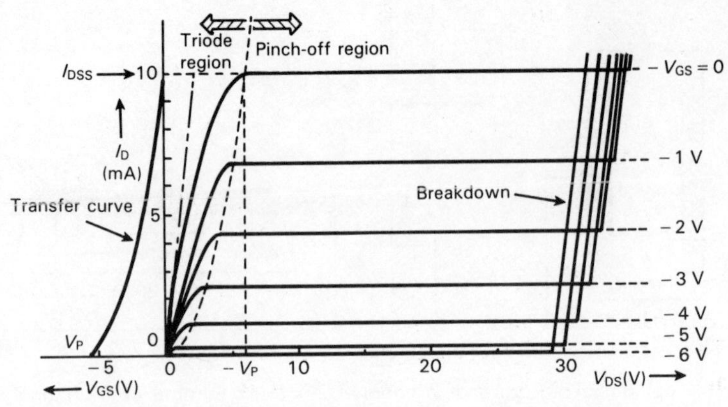

Fig. 11.3 Typical characteristics of *n*-channel JFET

When $|V_{DS}|$ exceeds a certain value, breakdown between gate and drain will occur. The curve in the second quadrant of Fig. 11.3 is the transfer characteristic, illustrating the **transconductance** of the FET.

Maximum channel current will flow when $V_{GS} = 0$. This current is the **saturation drain current** I_{DSS}.

The signs of voltages and currents are as follows:

for an *n*-channel FET: $I_D > 0, V_{GS} < 0, V_P < 0$
for a *p*-channel FET: $I_D < 0, V_{GS} > 0, V_P > 0$

11.1.2 Biasing in the pinch-off region

In the pinch-off region the relationship between drain current I_D and gate-source voltage V_{GS} is expressed by the square-law equation:

$$I_D \simeq I_{DSS}\left(1 - \frac{V_{GS}}{V_P}\right)^2$$

The biasing method of a JFET which is commonly employed is the **self-bias** method of Fig. 11.4(a). The bias voltage is produced by the FET's own drain current. When solving the voltage and current equations of the circuit, the relative drain current can be expressed as

$$\frac{I_D}{I_{DSS}} \simeq \frac{1 - 2\left(\frac{I_{DSS}R_s}{V_P}\right)\left(1 - \frac{V_G}{V_P}\right)\sqrt{1 - 4\left(\frac{I_{DSS}R_s}{V_P}\right)\left(1 - \frac{V_G}{V_P}\right)}}{2\left(\frac{I_{DSS}R_s}{V_P}\right)^2}$$

Example

Assume $I_{DSS} = 10$ mA, $V_P = -5$ V, $V_G = 0$ V and $R_s = 1$ kΩ, then

$$\frac{I_D}{I_{DSS}} \simeq \frac{1-2(-2)(+1)-\sqrt{1-4(-2)(+1)}}{2(-2)^2} = \frac{1}{4}$$

Hence $I_D \simeq 2.5$ mA.

In the bias circuit of Fig. 11.4(b) the loading effect of gate resistor R_g is decreased. Here

$$\frac{I_D}{I_{DSS}} \simeq \left[1-2\left(\frac{I_{DSS}R_{s1}}{V_P}\right)\sqrt{1-4\left(\frac{I_{DSS}R_{s1}}{V_P}\right)}\right] \Big/ \left[2\left(\frac{I_{DSS}R_{s1}}{V_P}\right)^2\right]$$

In general, the bias current is affected by temperature changes as shown in the transfer characteristics of Fig. 11.4(c).

The pinch-off voltage changes due to the variation of the contact potential. An increase in temperature causes a decrease of the barrier width so that I_D increases. Opposing this effect is the positive temperature coefficient of the channel resistance causing a decrease in I_D. In Fig. 11.4(c) the two effects are shown to cancel when

$$|V_{GS} - V_P| = 0.63 \text{ V}$$

and

$$\frac{I_D}{I_{DSS}} = \left(\frac{0.63}{V_P}\right)^2$$

For practical reasons this bias current is often not desirable and a larger value of I_D is chosen. In that case the bias is thermally stable since I_D tends to decrease with increasing temperature.

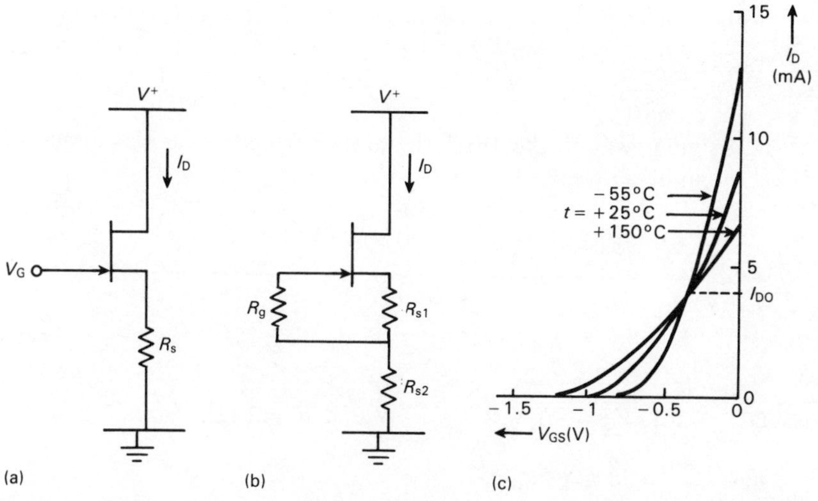

Fig. 11.4 (a) Self-bias of n-channel JFET (b) Alternate self-bias circuit (c) Drain current as a function of temperature

11.1.3 MOSFET (IGFET)

A thin layer of SiO_2 ($\simeq 100$ nm) is deposited between gate and channel. This allows an n-channel MOSFET (MOS means metal oxide semiconductor, IG means insulated gate), to be used with positive values of V_{GS} with no danger of gate current. The gate input resistance is of the order of $10^{14}\ \Omega$, the gate current often less than 1 pA.

Control of the channel resistance is achieved by the field of a parallel-plate capacitor with SiO_2 as the dielectric.

MOSFETs are extensively used in digital logic circuitry. One of the advantages is that a MOSFET requires about 15 percent of the chip area of a bipolar transistor. There are two MOSFET-types:

Depletion FET ('normally on' type). The structure and characteristics of an n-channel type are shown in Figs. 11.5(a) and 11.5(b) respectively. The channel is intentionally diffused into the structure.

Enhancement FET ('normally off' type). When V_{GS} exceeds a certain **threshold voltage** V_T (positive in an n-channel FET), electrons are attracted under the gate oxide, thus forming a channel and allowing drain current to flow. The structure and characteristics of an n-channel type are shown in Figs. 11.6(a) and 11.6(b) respectively.

A small static charge across the very thin SiO_2 layer may cause such a high field strength that breakdown of the oxide layer occurs. Therefore, most MOSFETs are incorporated with gate oxide protection diodes (Fig. 11.6(c)) which enter into avalanche breakdown when the input voltages exceed about 10 V.

In the pinch-off region I_D and V_{GS} for both types are related by

$$I_D \simeq \gamma(V_{GS} - V_T)^2$$

where γ is a constant.

In a similar way to the JFET the drain current in the bias circuit of Fig. 11.7 can be written as

$$\frac{I_D}{\gamma} \simeq \frac{1 - 2(\gamma V_T R_s)\left(1 - \dfrac{V_G}{V_T}\right) - \sqrt{1 - 4(\gamma V_T R_s)\left(1 - \dfrac{V_G}{V_T}\right)}}{2(\gamma R_s)^2}$$

Example

Assume $\gamma = 0.1$, $V_G = -1$ V, $V_T = -4$ V and $R_s = 1$ kΩ, then

$$I_D \simeq \frac{1.6 - \sqrt{2.2}}{0.2} = 0.58 \text{ mA}$$

An alternate and very simple biasing method for the enhancement FET is

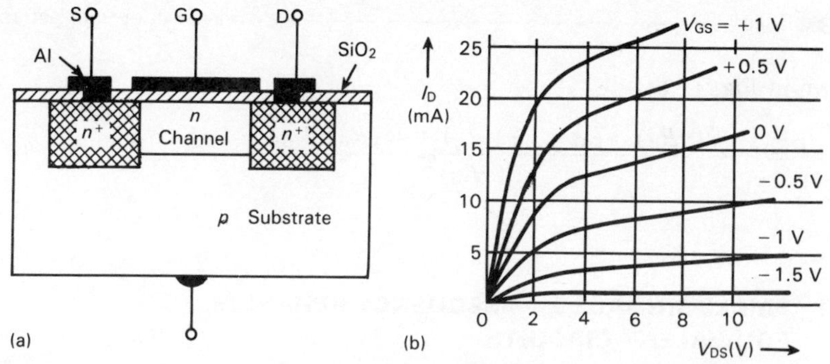

Fig. 11.5 (a) Structure of n-channel depletion FET (b) Typical characteristics of n-channel depletion FET

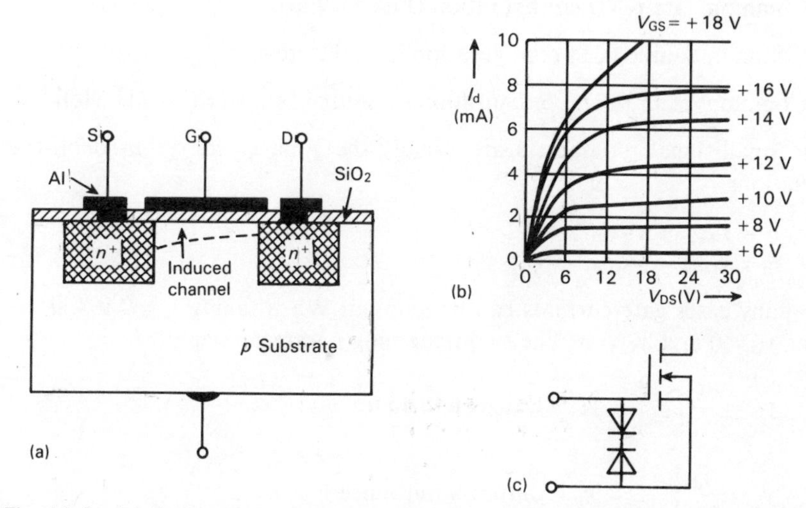

Fig. 11.6 (a) Structure of n-channel enhancement FET (b) Typical characteristics of n-channel enhancement FET (c) Gate oxide protection diodes in MOSFET

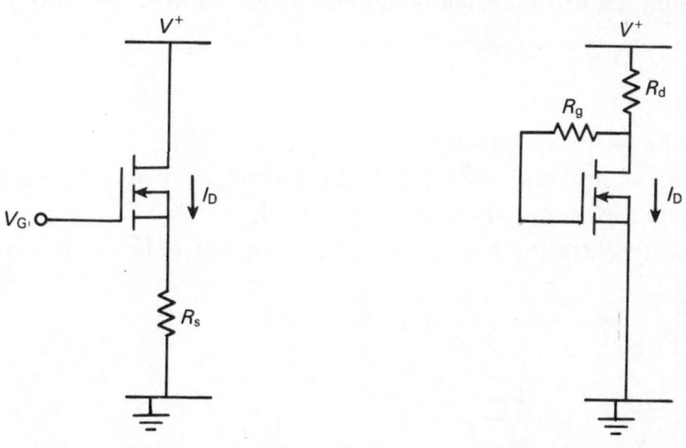

Fig. 11.7 Bias circuit of n-channel enhancement FET

Fig. 11.8 Alternate bias circuit of n-channel enhancement FET

shown in Fig. 11.8:

$$\frac{I_D}{\gamma} \simeq \frac{1 + 2(\gamma R_d)(V_{DD} - V_T) - \sqrt{1 + 4(\gamma R_d)(V_{DD} - V_T)}}{2(\gamma R_d)^2}$$

11.2 SMALL-SIGNAL LOW-FREQUENCY BEHAVIOR; EQUIVALENT CIRCUITS

Similar to the bipolar transistor there are three possible configurations:

- Common gate (CG) configuration (Fig. 11.9(a)).
- Common source (CS) configuration (Fig. 11.9(b)).
- Common drain (CD) configuration or **source follower** (Fig. 11.9(c)).

The small-signal parameters are usually the y parameters, based on the equations

$$i_g = y_{is} v_{gs} + y_{rs} v_{ds}$$
$$i_d = y_{fs} v_{gs} + y_{os} v_{ds}$$

In many cases gate currents can be ignored. When taking $i_g = 0$ it follows that $y_{is} = 0$ and $y_{rs} = 0$. The two remaining parameters are:

$$y_{fs} = \left.\frac{i_d}{v_{gs}}\right|_{v_{ds}=0} = g_m = \textbf{transconductance}$$

$$y_{os} = \left.\frac{i_d}{v_{ds}}\right|_{v_{gs}=0} = g_o = \textbf{output admittance}$$

Thus $1/g_o = r_o$ is the **output resistance.**

The amplification factor μ is defined as the ratio of y_{fs} and y_{os}:

$$\mu = \frac{y_{fs}}{y_{os}} = -\left.\frac{v_{ds}}{v_{gs}}\right|_{i_d=0} = g_m r_o$$

This is **Barkhausen's equation.**

The equivalent circuit of a JFET in CS configuration is based upon the parameter equations and shown in Fig. 11.10.

When differentiating the $I_D - V_{GS}$ equation for a JFET we obtain

$$g_m = \frac{\partial I_D}{\partial V_{GS}}$$

$$= \frac{-2I_{DSS}}{V_P}\left(1 - \frac{V_{GS}}{V_P}\right)$$

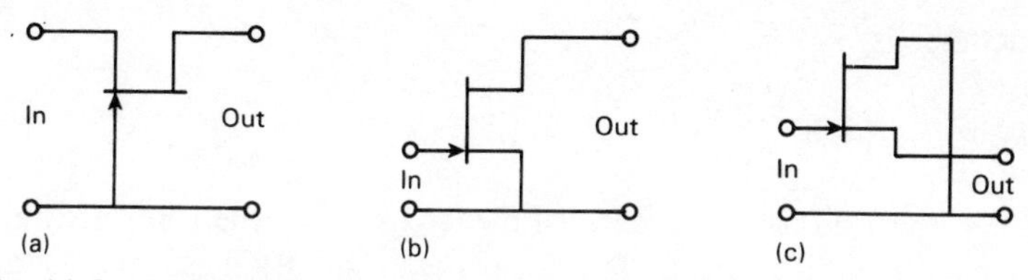

Fig. 11.9 (a) The CG configuration (b) The CS configuration (c) The CD configuration

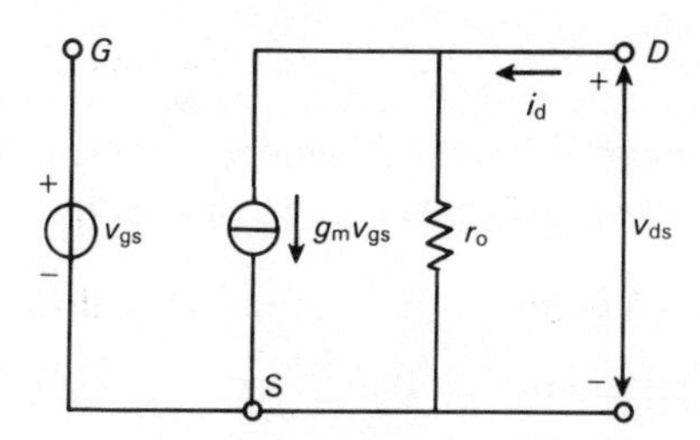

Fig. 11.10 Equivalent circuit of JFET in CS configuration

$$= g_{max}\left(1 - \frac{V_{GS}}{V_P}\right)$$

$$= g_{max}\sqrt{\frac{I_D}{I_{DSS}}}$$

where $g_{max} = -2I_{DSS}/V_P$ is the maximum transconductance. Similarly for a MOSFET we have

$$g_m = 2\gamma(V_{GS} - V_T)$$
$$= 2\sqrt{\gamma I_D}$$

11.3 THE FET AS A VOLTAGE CONTROLLED RESISTOR

In the triode region the transconductance of a JFET is

$$g_m = g_{max}\left(1 - \frac{V_{GS} - V_{DS}}{V_P}\right)$$

where $1/g_{max} = r_{DS,\,on}$ is the minimum channel resistance. This is the slope of the zero bias output curve.

With zero bias current ($I_D = 0$, $V_{DS} = 0$), the small-signal transconductance is

$$g_m = g_{max}\left(1 + \frac{V_{GS}}{V_P}\right)$$

Basically, the FET is a symmetrical device in which the roles of drain and source may be interchanged. Thus near the origin drain–source voltages of either polarity are permitted. The range of V_{DS} is then about ± 0.2 V. Figure 11.11(a) shows the expanded characteristics near the origin.

Example

Voltage controlled voltage divider (AGC circuit, Fig. 11.11(b)).

The transconductance becomes independent of V_{GS} (linearization) when two equal resistors are connected as shown in Fig. 11.11(c). In this circuit

$$g_m = g_{max}\left(1 - \frac{V_S}{2V_P}\right)$$

This method allows attenuation of larger signal voltages without distortion.

Example

In the circuit of Fig. 11.11(b) the data are $I_{DSS} = 10$ mA, $V_P = -4$ V, $V_{GS} = -2$ V.

Then $g_{max} = 5\ \text{k}\Omega^{-1}$, $g_m = 5\left(1 - \frac{-2}{-4}\right) = 2.5\ \text{k}\Omega^{-1}$ and $r_{DS} = 400\ \Omega$.

The same value of r_{DS} in the circuit of Fig. 11.11(c) requires $V_{GS} = -4$ V.

If the drain current of a MOSFET is written in the form

$$I_D = \gamma(V_{GS} - V_T)^2$$

then

$$g_m = 2\gamma(V_{GS} - V_T - V_{DS}) \qquad \text{(without linearization)}$$

$$g_m = \gamma(V_{GS} - 2V_T) \qquad \text{(with linearization)}$$

11.4 SMALL-SIGNAL LOW-FREQUENCY AMPLIFIERS

Figure 11.12 summarizes voltage gain A_v, input resistance r_i and output resistance r_o for the three configurations.

Example 1

JFET amplifier (Fig. 11.13) with $I_{DSS} = 12$ mA, $V_P = -3$ V, $\mu = 100$.

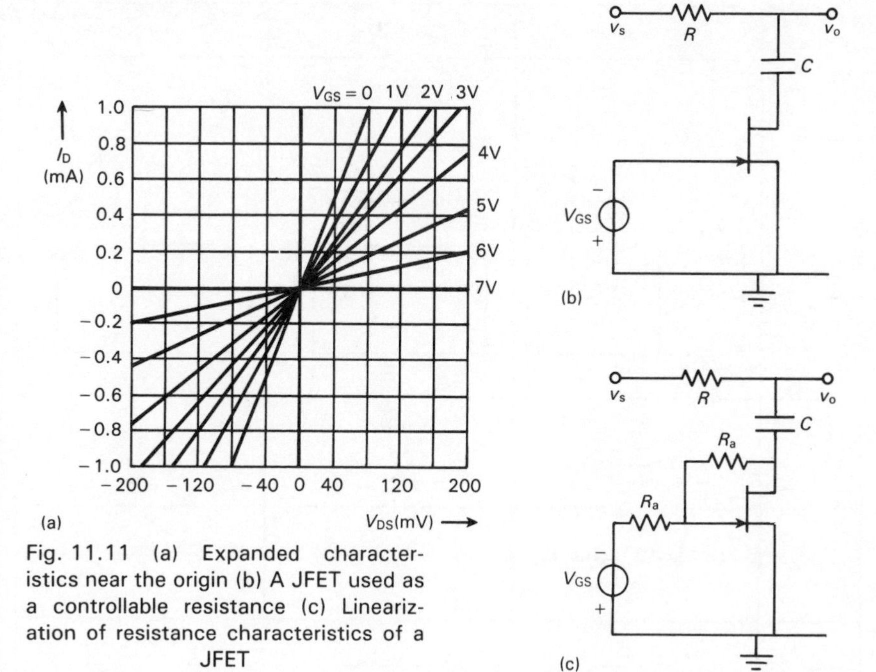

Fig. 11.11 (a) Expanded characteristics near the origin (b) A JFET used as a controllable resistance (c) Linearization of resistance characteristics of a JFET

From Section 11.1.2 we find $I_D = 3$ mA and from Section 11.2, $g_{max} = 8$ mA V^{-1}, $g_m = 4$ mA V^{-1}. Then from Fig. 11.12

$$A_v = \frac{-5}{0.25 + 0.5 + 0.05} = -6.25$$

Example 2

MOSFET amplifier (Fig. 11.14) with $I_D = 4(V_{GS} - 1)^2$.

It follows from Section 11.1.3 that $I_D = 4$ mA, so that $V_{GS} = +2$ V.

$$g_m = \frac{\partial I_D}{\partial V_{GS}} = 8(V_{GS} - 1) = 8 \text{ mA V}^{-1}$$

From Fig. 11.12

$$A_v = \frac{-5}{0.125 + 0.5} = -8$$

11.5 CASCODE AMPLIFIER

In Figs. 11.15(a) and 11.15(b) the cascode structure can be considered as a

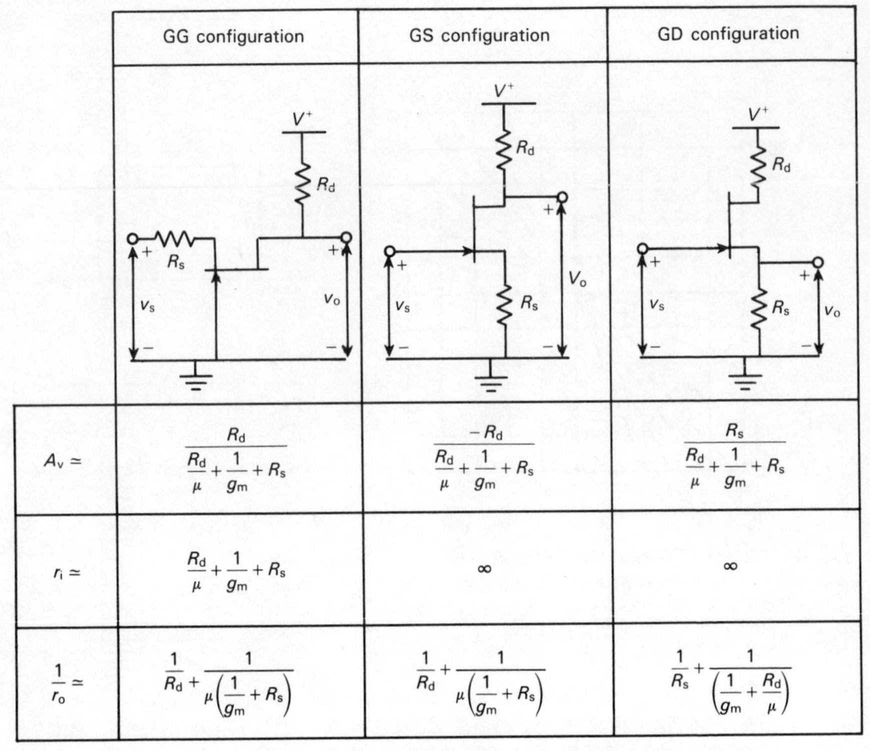

	GG configuration	GS configuration	GD configuration
$A_v \simeq$	$\dfrac{R_d}{\dfrac{R_d}{\mu}+\dfrac{1}{g_m}+R_s}$	$\dfrac{-R_d}{\dfrac{R_d}{\mu}+\dfrac{1}{g_m}+R_s}$	$\dfrac{R_s}{\dfrac{R_d}{\mu}+\dfrac{1}{g_m}+R_s}$
$r_i \simeq$	$\dfrac{R_d}{\mu}+\dfrac{1}{g_m}+R_s$	∞	∞
$\dfrac{1}{r_o} \simeq$	$\dfrac{1}{R_d}+\dfrac{1}{\mu\left(\dfrac{1}{g_m}+R_s\right)}$	$\dfrac{1}{R_d}+\dfrac{1}{\mu\left(\dfrac{1}{g_m}+R_s\right)}$	$\dfrac{1}{R_s}+\dfrac{1}{\left(\dfrac{1}{g_m}+\dfrac{R_d}{\mu}\right)}$

Fig. 11.12 The three configurations and relevant formulas

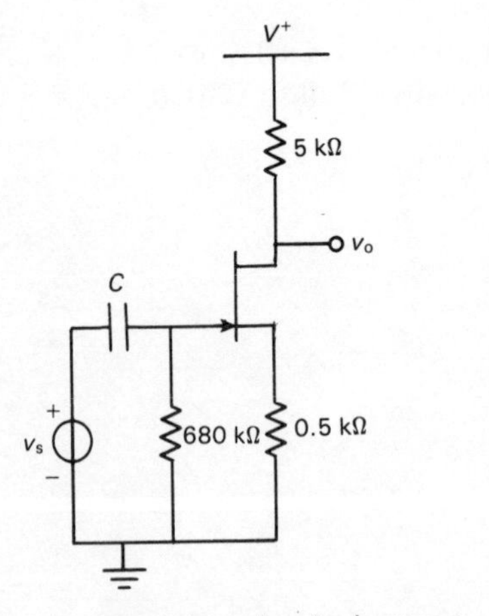

Fig. 11.13 Example of single-stage JFET amplifier

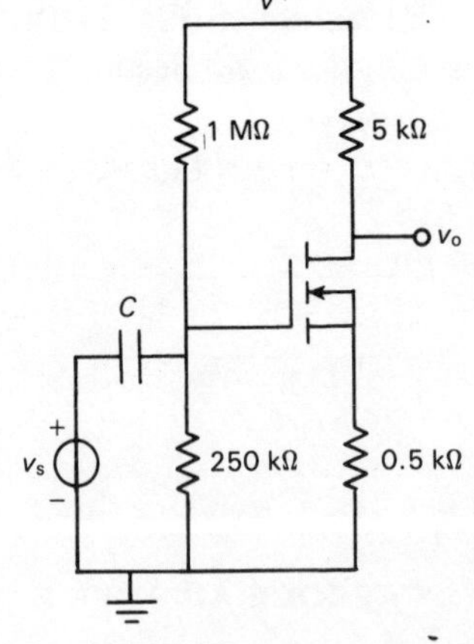

Fig. 11.14 Example of single-stage MOSFET amplifier

single FET having the following parameters:

$$\mu_{casc} = \mu_1(1 + \mu_2) \simeq \mu_1\mu_2 \quad \text{so that } \mu_{casc} \gg \mu_1, \mu_2$$

$$r_{o,casc} = r_{o,2} + r_{o,1}(1 + \mu_2) \simeq r_{o,1}\mu_2 \gg r_{o,1} \quad \text{(useful for a current source)}$$

$$g_{m,casc} \simeq \frac{g_{m,1}}{1 + \dfrac{1}{r_{o,1}g_{m,2}}} \simeq g_{m,1}$$

$$C_i = C_{DG}\frac{2}{1 + g_{m,1}R_s} \ll C_{DG}$$

The last expression shows a drastic decrease of the effect of C_{DG}. In other words, the cascode amplifier shows negligible Miller effect and is therefore useful in high-frequency applications.

11.6 HIGH-FREQUENCY BEHAVIOR

The high-frequency equivalent circuit of a JFET in CS configuration is shown in Fig. 11.16. The input capacitance can be shown to be

$$C_i \simeq C_{GS} + C_{DG}(1 - A_v)$$

Typical values are $C_{GS} \simeq 10$ pF, $C_{DG} \simeq 0.7$ pF.

Figure 11.17 shows the equivalent circuit of a MOSFET in CS configuration. The input capacitance equation is the same as for the JFET. Typical values here are $C_{GS} \simeq 1$ pF, $C_{DG} \simeq 0.5$ pF.

11.7 SPECIAL TYPES

11.7.1 Dual-gate MOSFET

Between drain and gate G_1 a second gate (G_2) is grown (Fig. 11.18) in order to reduce the effect of the drain–gate capacitance C_{DG}. In this way the value of C_{DG1} is reduced to about 0.02 pF allowing operation at high frequencies.

G_2 is usually biased at a positive voltage so reducing the resistance of channel 2. Figure 11.19 shows a dual-gate MOSFET operating as a high-frequency amplifier. The control voltage at G_2 is used to control the amplifier gain (AGC).

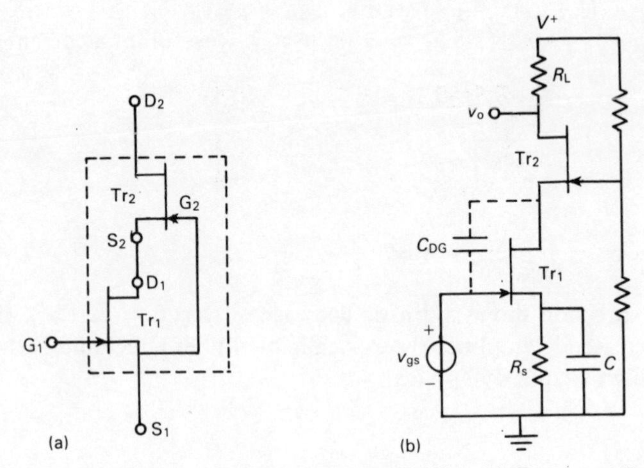

Fig. 11.15 (a) Equivalent circuit of cascode amplifier (b) Circuit configuration of cascode amplifier

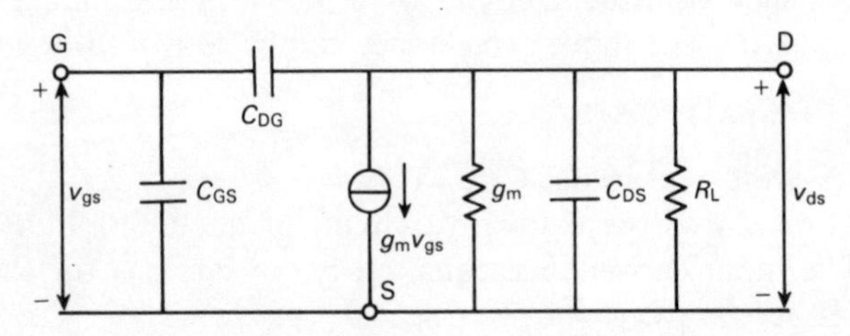

Fig. 11.16 High-frequency equivalent circuit of JFET in CS configuration

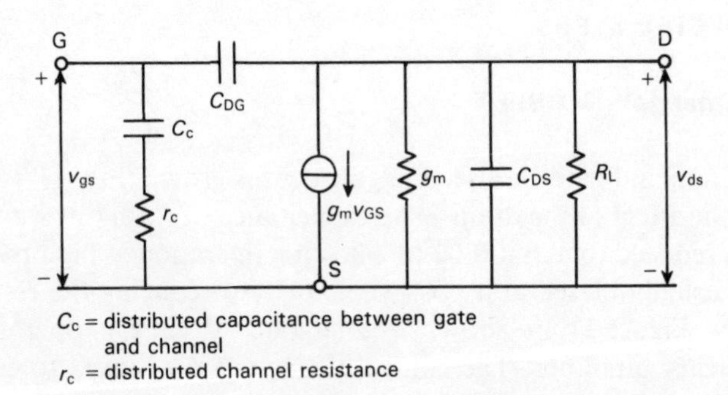

Fig. 11.17 High-frequency equivalent circuit of MOSFET in CS configuration

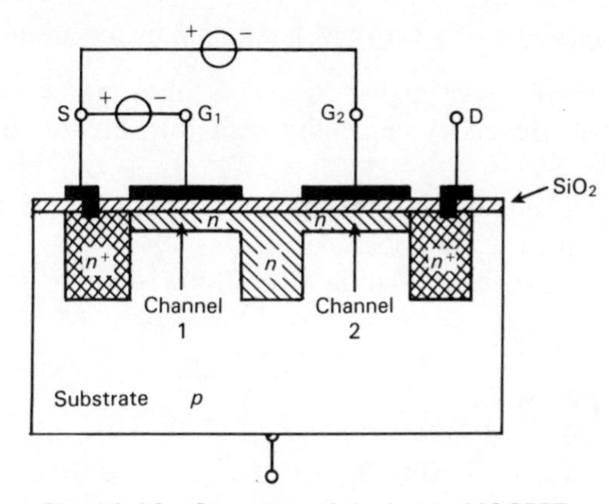

Fig. 11.18 Structure of dual-gate MOSFET

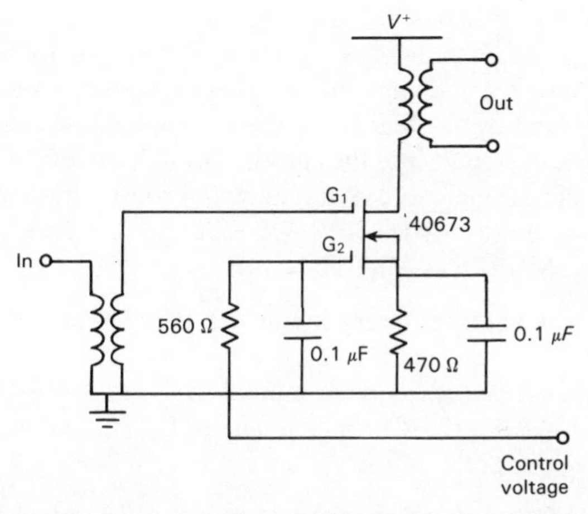

Fig. 11.19 Application of dual-gate MOSFET in RF amplifier circuit

11.7.2 MESFET

The structure of the MESFET (MES means metal semiconductor) is shown in Fig. 11.20. The semiconductor material is usually GaAs. The channel region is very thin ($\simeq 200$ nm) and of epitaxial *n*-type GaAs which is grown on a semi-insulating chromium-doped GaAs substrate. The gate is a Schottky-barrier metal contact having a length of about 1000 nm. The characteristics of a MESFET are similar to those of a JFET. A dual-gate MESFET has already been developed.

The advantages of using GaAs as a semiconductor are as follows:

- Electron mobility is much higher than in Si allowing operation at high frequencies. The frequency capability of a MESFET is approximately 15 GHz.
- As GaAs can withstand higher temperatures than Si, power figures of 1 W at high frequencies are possible.
- Leakage currents are lower than in Si resulting in lower noise figures.

11.7.3 VMOS power FET

On a heavily doped Si wafer (Fig. 11.21(a)) a lightly doped epitaxial *n* layer is grown. This layer increases the drain–source breakdown voltage and reduces the drain-gate capacitance.

Two diffusions, *p* and n^+ are then made into the epitaxial layer, followed by a V-groove which is etched into the material. Finally, SiO_2-isolation and metal contacts are made.

In operation, drain and gate are positive with respect to the source. The gate voltage inverts the *p* region into an *n* region thereby forming a channel between source and drain. Increasing the gate voltage widens the channel and increases the current. Since the current through the FET flows laterally, the channel width is small ($\simeq 1.5\ \mu$m) allowing continuous currents of tens of amperes. The characteristics of VMOS FETs differ somewhat from those of conventional MOSFETs (Fig. 11.21(b)):

- The output conductance is very low which is apparent from the almost flat curves.
- The transconductance g_m is proportional to V_{GS}^2 at low current levels. Above about 0.5 A, g_m is proportional to V_{GS} rather than quadratic. Typical g_m values are 0.5–2 A V^{-1}.

VMOS power FETs can be operated at power levels of about 100 W at frequencies up to 200 MHz.

11.7.4 HEXFET

One of the many different constructions of power FETs is the HEXFET. The hexagonal structure (Fig. 11.22(a)) has resulted in values of on-resistance which are comparable with those of bipolar transistors. They can handle currents up to 100 A, drain–source voltages of 500 V, power levels of a few kW and they have r_{DS} values of about 50 mΩ. The required gate currents are only a few nA.

The structure of a (double-diffused) HEXFET is shown in

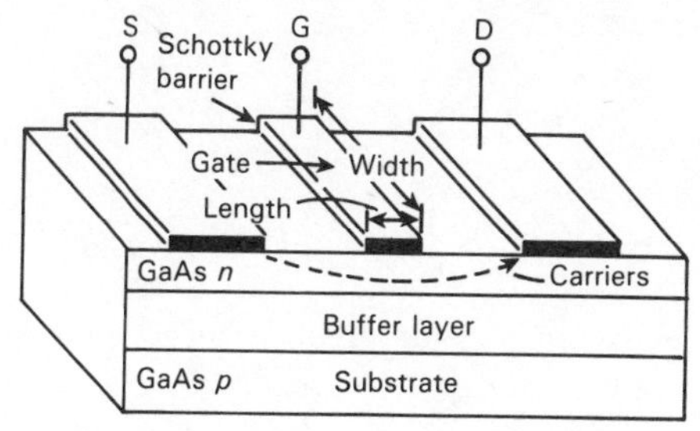

Fig. 11.20 Basic MESFET structure

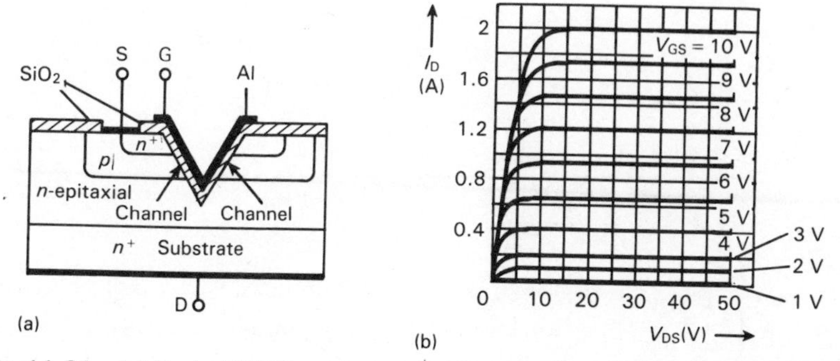

Fig. 11.21 (a) Basic VMOS structure (b) Typical characteristics of VMOS power FET

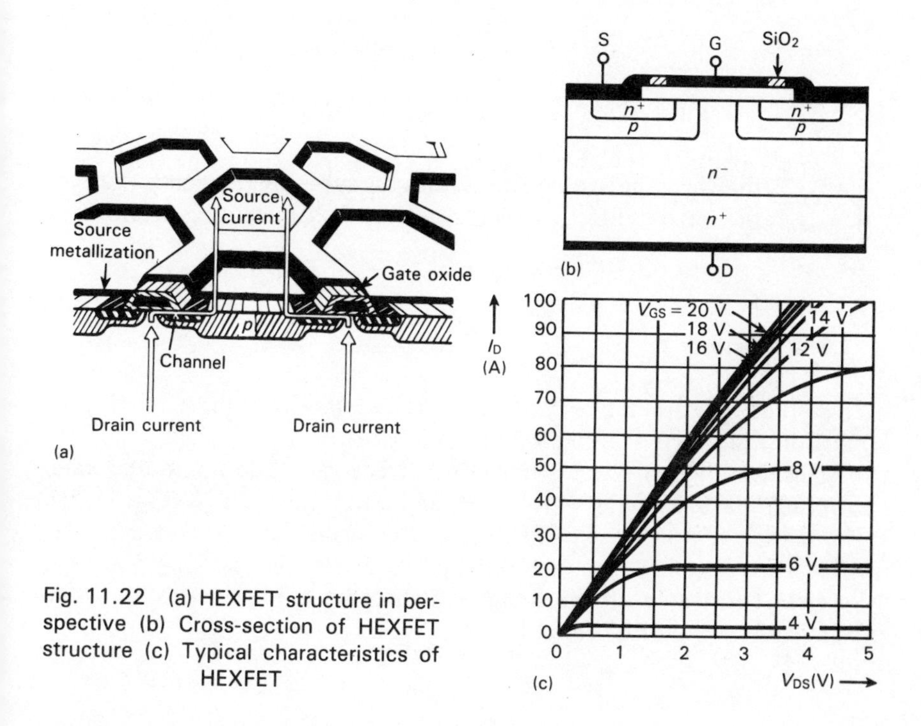

Fig. 11.22 (a) HEXFET structure in perspective (b) Cross-section of HEXFET structure (c) Typical characteristics of HEXFET

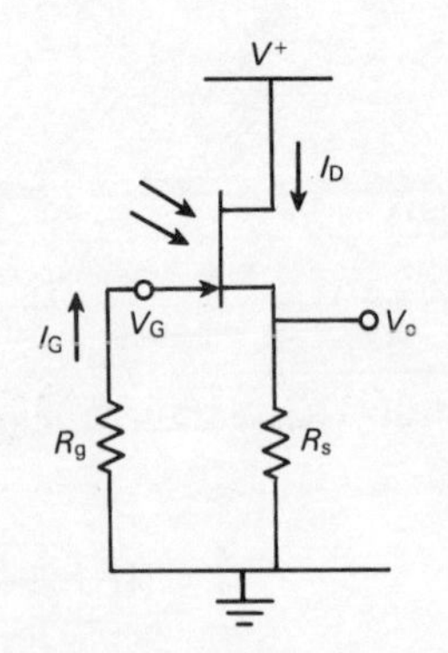

Fig. 11.23 PhotoFET circuit

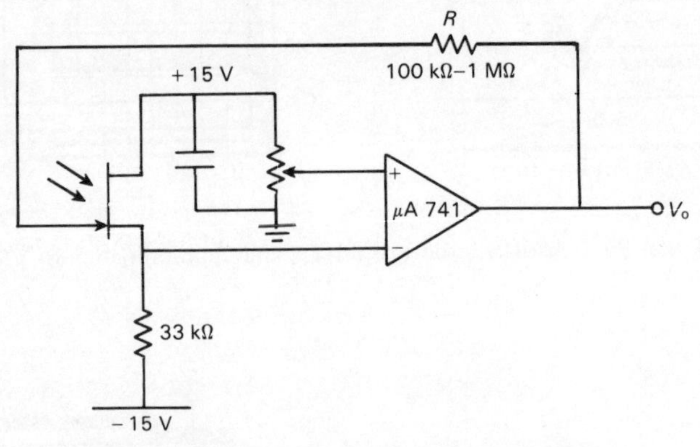

Fig. 11.24 Example of photoFET amplifier

Fig. 11.22(b) where the two *p* regions form the substrate. Evidently, current flow is in the vertical direction. A typical set of characteristics is shown in Fig. 11.22(c).

11.7.5 PhotoFET

This is basically a JFET with a transparent window over the gate–channel junction which acts as a photodiode.

Incident light causes a junction current I_G to flow into the gate, resulting in a bias voltage $V_G = I_G R_g$ and a drain current I_D (Fig. 11.23). The sensitivity is increased when using larger values of R_G. However, this reduces the response time due to the parasitic capacitances C_{DG} and C_{GS}. Spectral sensitivity is similar to that of a photodiode. Figure 11.24 shows an example of a photoFET amplifier.

12 Semiconductor and IC Technology

12.1 PURIFICATION OF SEMICONDUCTOR MATERIALS

Semiconductor materials require a very high degree of purity and often a single crystal structure in order to allow their use in high-tech electronics.

The initial purification of GeO_2 involves a reaction with NaCl whereby $GeCl_4$ is formed. After fractionating, water is added producing relatively pure GeO_2. At about 600 °C reduction with H_2 is performed giving very pure Ge powder which is melted into a bar at about 960 °C.

Further purification is achieved by **zone refining**. This method is based on the fact that impurities are more soluble when the metal is in the liquid state. Zone refining of Ge is sketched in Fig. 12.1(a). By local induction heating, a narrow molten zone traverses the length of the bar whereby the impurities remain in solution. The Ge bar is held in position by graphite or quartz and the process takes place in an inert atmosphere.

Impure SiO_2 is first reduced with carbon: $SiO_2 + 2C \rightarrow Si + 2CO$. This results in a purity of about 99.9 percent. Then the Si is combined with HCl giving $SiHCl_3$. Further purification is achieved by fractionating processes. Finally the $SiHCl_3$ is reduced with hydrogen according to $SiHCl_3 + H_2 \rightarrow Si + 3HCl$. The impurity content is further reduced by zone refining. As molten Si is chemically very active, zone refining is accomplished by keeping the Si bar in a vertical position (Fig. 12.1(b)). By repeated zone refining the impurity content is down to about $1:10^{10}$.

After purification it is necessary to grow the crystal into **single-crystal material**. This is done by the Czochralski method (Fig. 12.1(c)). A crucible maintains the molten metal a few degrees above its melting point in an inert atmosphere. A small piece of single crystal, the **seed**, is lowered into the metal. By slowly revolving and withdrawing the seed all the metal grows into a crystal of the same molecular pattern.

The single-crystal bars are then sawn into wafers for further processing.

12.2 JUNCTION FORMATION

12.2.1 Alloying

This process was important in the germanium era and is still used for power transistor fabrication.

On both sides of a wafer of *n*-type Ge a small amount of In (*p* type) is deposited (Fig. 12.2). The wafer is then heated to about 500 °C in an inert atmosphere. The In melts at 155 °C and a thin film of doped Ge results, thus forming the two junctions for the *pnp* transistor.

12.2.2 Diffusion

An inert carrier gas, usually N or A, is mixed with the gaseous impurity (e.g. PH_3 or AsH_3) and the impurity diffuses into the wafer.

The diffusion process is critical with respect to temperature but as diffusion takes several hours, good process control is possible. For example, it takes about two days to obtain a 10 μm phosphorus diffusion with a doping level of 10^{15} atoms cm^{-3}. A typical diffusion transistor is shown in Fig. 12.3.

12.2.3 Epitaxy

Epitaxy is an extension of the diffusion process. In addition to the impurity gas, $SiCl_4$ in hydrogen is added at a temperature of 1150 °C.

Si atoms now diffuse into the substrate so that the atoms are aligned in a continuation of the original single-crystal structure. The substrate has to be very clean otherwise polycrystallinity results.

12.2.4 Ion implantation

Impurity ions are electromagnetically focused and accelerated in vacuum to a sufficiently high energy (100–200 keV) and shot into the substrate. The ions penetrate the substrate to a fraction of a μm (e.g. boron at 100 keV penetrates about 0.4 μm, phosphorus about 0.1 μm).

The disturbed crystal structure is then annealed (either by heating at 950 °C or by laser techniques) whereby the foreign atoms further diffuse into the substrate and the original crystal structure is restored.

Ion implantation finds more and more use today since it is accurate and gives a good uniformity. It is, however, expensive due to the requirements of a high vacuum, high voltages, strong magnets, etc.

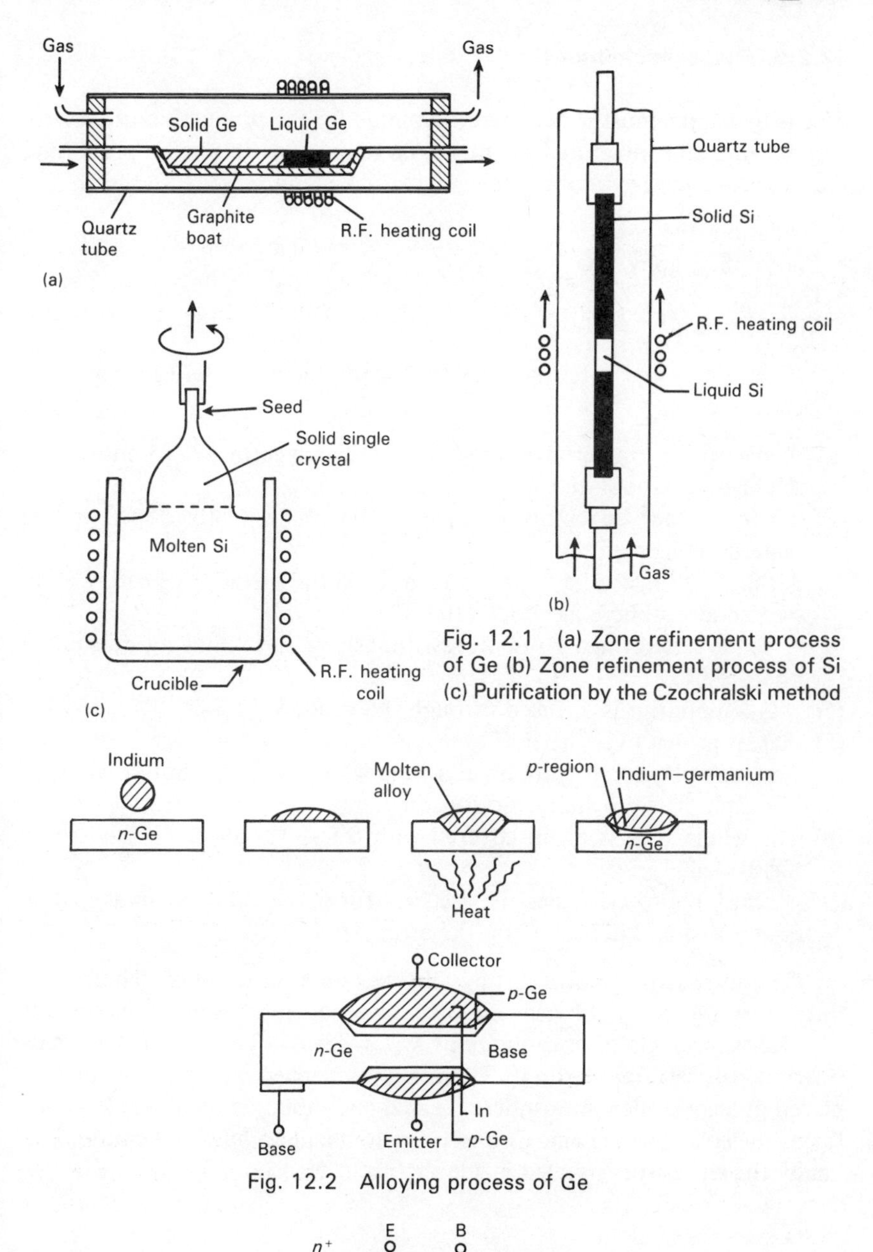

Fig. 12.1 (a) Zone refinement process of Ge (b) Zone refinement process of Si (c) Purification by the Czochralski method

Fig. 12.2 Alloying process of Ge

Fig. 12.3 Structure of diffusion transistor

12.2.5 Planar technology

This is by far the most widely used technique for the production of discrete components and integrated circuits. The main steps in planar processing are:

- Oxidation.
- Photolithography.
- Doping.
- Metallization.

Figure 12.4 illustrates the steps for the production of an *npn* transistor in detail.

(a) A slice of n^+ material is used with a diameter of 50–75 mm and a thickness of about 300 μm.
(b) On its surface an epitaxial layer ($\simeq 10$ μm) of lightly doped *n*-type material is formed.
(c) A thin SiO_2 layer (0.5–1 μm) is formed on the surface by heating in an oxygen atmosphere at about 1100 °C.
(d) A window is formed by photoresist masking and removing unexposed photoresist.
(e) *p*-type impurity is diffused through the window.
(f) SiO_2 is grown over the slice.
(g) A second photoresist process and etching produces the emitter window. The emitter is grown by *n* diffusion.
(h) The whole slice is again covered with SiO_2. The device is said to be **passivated**.
(i) A third photoresist step is used to form the contact areas. Al is evaporated and deposited on the areas.

The final steps involve testing all transistors on the slice. The passed transistors are removed by sawing the slice along grooves and breaking off the transistors. Gold or aluminum wires are bonded to the Al pads (thermocompression bonding). Then they are etched to clean the surface, placed in a controlled atmosphere, heated and mounted on a header which forms the collector contact. Connections are made to base and emitter and finally the transistor is sealed inside the package.

12.3 IC FABRICATION PROCESSES

12.3.1 General

The monolithic chip or IC was developed around 1958 whereby active and passive components were formed by diffusion and deposition including the

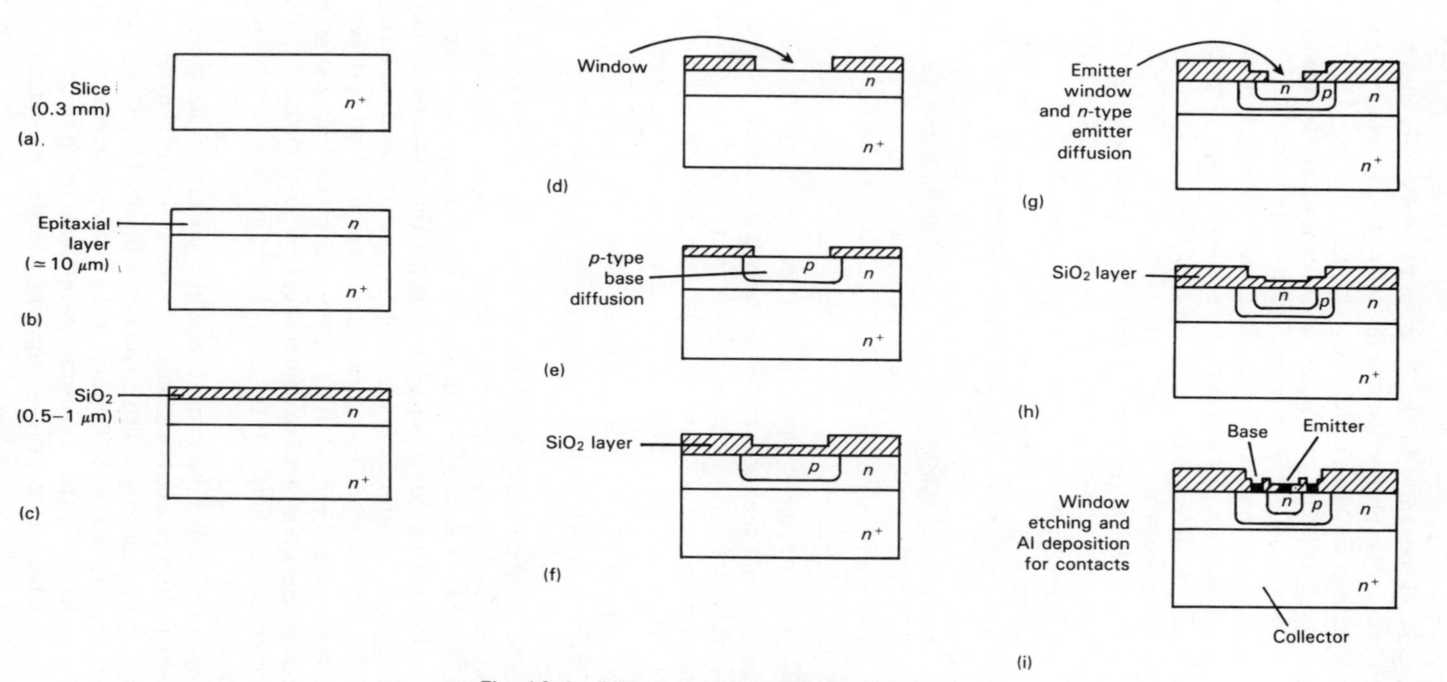

Fig. 12.4 Diffusion steps in planar process

interconnections. In the early 1970s, accurate control of dimensions and doping levels led to mass production of a diversity of IC types of very small size, few connections and high reliability.

Some of the processing techniques are very critical and require an extremely clean environment with controlled humidity and temperature. Materials and chemicals used must often be up to 99.999 percent pure.

Passive components in ICs (capacitors, resistors) occupy excessive chip area. If possible, passive components are therefore replaced by their semiconductor counterparts.

The basis for the IC fabrication is the planar technology as described in Section 12.2.5. The 'ground plane' is a thin wafer ($\simeq 0.5$ mm) of p-type Si, 50–100 mm in diameter which forms the substrate. An n-type layer is deposited by epitaxy. The wafer is oxidized by heating it in an O_2 atmosphere, coated with photoresist and covered with a mask of the desired geometry (Fig. 12.5).

After exposure to ultraviolet light and etching the SiO_2 away, a pattern of windows is left through which the diffusion takes place.

Components in a circuit need to be isolated from each other. This can be achieved in different ways:

Diode isolation. Islands in the n layer are formed by diffusing p^+ layers all the way into the substrate. When eventually reverse biasing the substrate by connecting it to the most negative voltage in the circuit, each component is surrounded by a reverse biased pn junction. In this way, isolation resistances between 10 and 100 MΩ are obtained (Fig. 12.5).

Local oxide isolation. Each component is surrounded by a layer of SiO_2. This is a costly method and only used for special circuits.

In the discrete planar transistor of Section 12.2.5 a low resistance path between collector and emitter exists since the n^+ substrate is heavily doped and accessible from the underside. In an integrated transistor, all connections must be brought out to the top of the surface. Due to the lightly doped n-epitaxial region, the saturation voltage $V_{CE,s}$ and the internal collector resistance are relatively high. To reduce this resistance, a heavily doped n^+ region is diffused into the p substrate before the epitaxial layer is grown. This is called the **buried layer**. The collector current first flows down to the buried layer, continues laterally through the layer and then vertically to the emitter contact (Fig. 12.6).

Window forming on a very small scale may pose problems due to the etching process which proceeds not only downwards but also laterally. This can be avoided by a process called **sputtering** whereby high-energy argon ions are used to bombard the SiO_2 locally and give a vertical etching profile.

After completion of the processing, each circuit is electrically tested and cut up into individual dies each comprising a single circuit. Each circuit

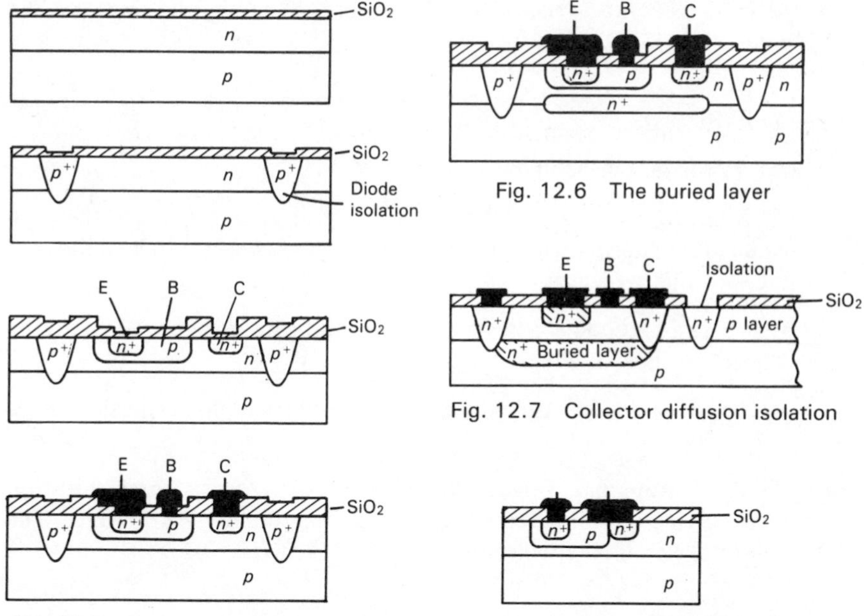

Fig. 12.5 Steps in IC planar process

Fig. 12.6 The buried layer

Fig. 12.7 Collector diffusion isolation

Fig. 12.8 IC diode structure

is then bonded on a header and wires are bonded to the Al pads and header terminals. Finally, the encapsulation is sealed.

ICs are available in three standard packages:

- Multipin circular type.
- Flat pack.
- Dual-in-line (DIL) package.

12.3.2 Component formation

The *npn* transistor

The buried layer in the *npn* transistor can be used as the collector in a construction, called **collector diffusion isolation** (CDI, Fig. 12.7). On top of the buried layer a *p* region is epitaxially grown to form the base. The n^+ diffusions now provide isolation and allow the collector leads to be attached. In a final step the emitter is diffused.

Diode

A diode is obtained by interconnecting base and collector (Fig. 12.8). The n^+ region is placed adjacent to the base region, both regions being covered by the same Al contact. Base and collector are the anode, the emitter is the cathode.

The *pnp* transistor

A *pnp* transistor can be made in two ways:

(a) As a lateral *pnp* (Fig. 12.9(a)). The n^+ and n regions form the base while individual p regions serve as emitter and collector. The disadvantages of this type are:

- A parasitic diode exists between base and emitter since the emitter is located under the base.
- The relative wide base reduces β and f_T substantially; typical β-values lie between 2 and 10.

(b) As a substrate *pnp* (Fig. 12.9(b)). This type requires three diffusion steps, giving an improvement of β to about 50 and $f_T \simeq 50$ MHz. The substrate here is the collector.

FETs

The structures of a depletion and an enhancement FET are shown in Fig. 12.10. Processing is along the same lines as described for the Si-planar transistor. Current flow is controlled by the field of a capacitor, one plate being the Al layer of the gate on top of the oxide, the other plate being the p^+ substrate. This construction requires few process steps and permits the interconnection of FETs by the Al conductors.

MOS capacitor

The structure of a MOS capacitor is very simple (Fig. 12.11(a)): the electrodes are an Al plate and the substrate (n or p type).

A p substrate type requires a negative bias voltage, an n type a positive bias. A capacitor requires considerable chip area: a 100 pF capacitor requires an area of about 3×10^5 $(\mu m)^2$.

Junction capacitor

A p region and an n^+ region are diffused into the n substrate (Fig. 12.11(b)). A *pn* junction thus exists between the two electrodes and a reverse bias produces the barrier capacitance.

Integrated capacitors require much chip area ($\simeq 15$ pF $(\mu m)^{-2}$) and are therefore avoided as much as possible.

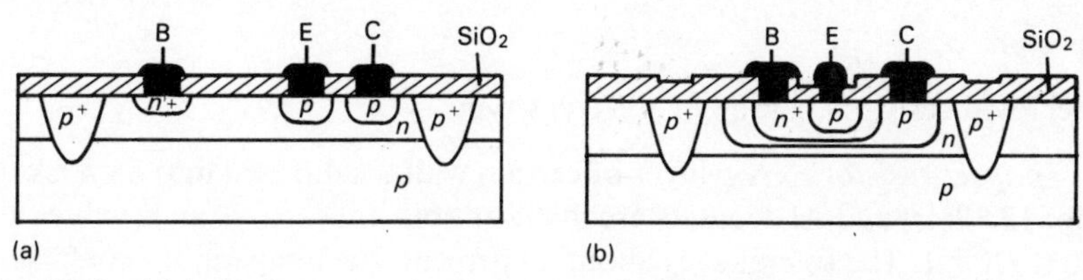

Fig. 12.9 (a) Lateral *pnp* transistor structure (b) Substrate *pnp* transistor structure

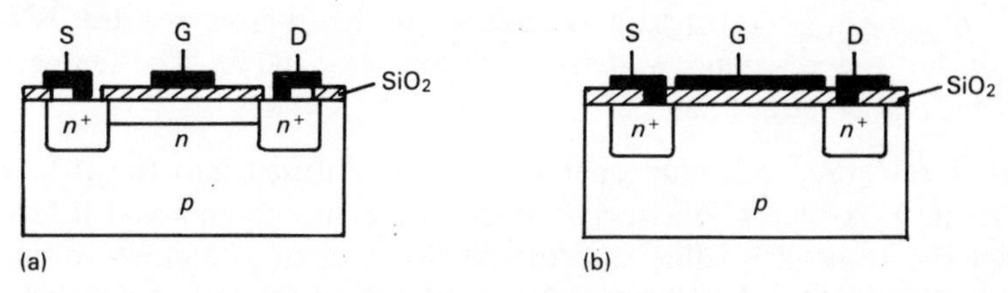

Fig. 12.10 (a) Structure of depletion FET (b) Structure of enhancement FET

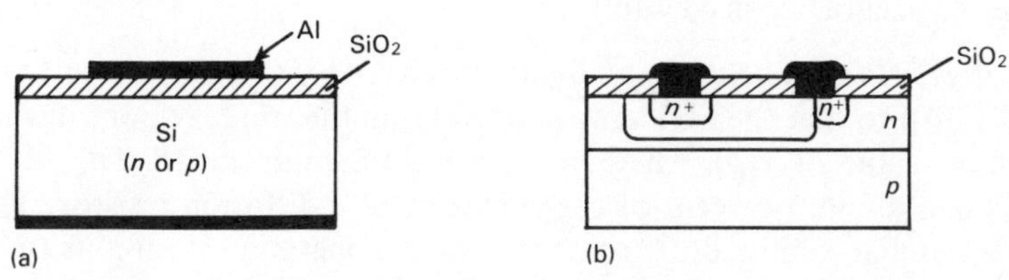

Fig. 12.11 (a) Structure of MOS capacitor (b) Structure of junction capacitor

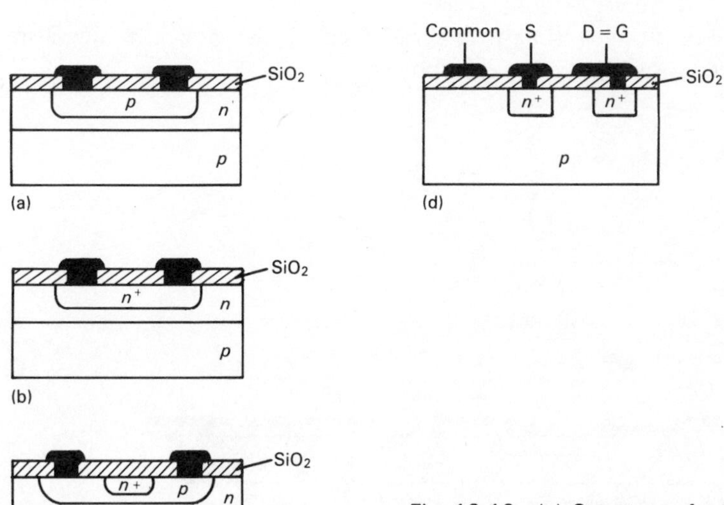

Fig. 12.12 (a) Structure of *p*-layer resistor (b) Structure of n^+ resistor (c) Structure of pinch resistor (d) Structure of MOS resistor

Resistor

Integrated resistors can be of different types:

p-layer resistor. A p layer of certain width is diffused into an n island (Fig. 12.12(a)) and Al contacts are made at both ends. Resistance values are 100 Ω–20 kΩ. The tolerance is about 20 percent, the temperature coefficient 0.25% $°C^{-1}$.

n^+ resistor. The structure is similar to the p-layer resistor (Fig. 12.12(b)), but low resistance values are obtained (< 100 Ω). The temperature coefficient is about 0.2% $°C^{-1}$.

Pinch resistor. An additional n^+ layer is diffused into the p layer (Fig. 12.12(c)). A reverse bias between the p region and epitaxial n layer increases the resistance value due to the decrease of the effective area between n^+ and p regions. This way large resistance values can be obtained but accurate values are difficult to achieve. Ion implantation techniques improve the accuracy considerably.

MOS resistor. The drain and gate of a MOSFET are interconnected (Fig. 12.12(d)) so that the FET always operates in the triode region, having a resistance value of $1/g_m$ where g_m is the transconductance. The MOS resistor takes about 5 percent of the chip area of a diffusion resistor. The inaccuracy and instability of IC resistors requires operation of circuits to be based on resistance ratios rather than on absolute values. These ratios are very constant since the resistors on a chip are manufactured in the same process and are very close together.

In Fig. 12.13 a simple integrated circuit is shown: a single-stage amplifier consisting of three components.

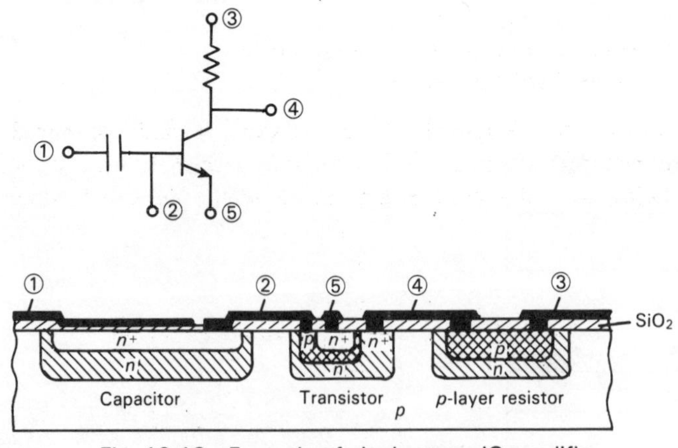

Fig. 12.13 Example of single-stage IC amplifier

13 Feedback

13.1 GENERAL MODEL

In analog electronics, feedback is probably the most important principle. It was described by H. S. Black in 1927 and analyzed by H. Nyquist, both at Bell Telephone Laboratories.

A general model of a feedback amplifier is shown in Fig. 13.1. The feedback network has a transfer k from output to input. A small amount of input signal passes directly to the output. This direct transfer is indicated by ρ. The transfer function or **transactance** of the amplifier (here voltage gain) is

$$A_f = \frac{v_o}{v_s} = \frac{A + \rho}{1 - Ak} = \frac{A + \rho}{1 - A_r}$$

$$= \frac{1}{k}\frac{A_r}{1 - A_r} + \frac{\rho}{1 - A_r}$$

where

A is the open loop gain
$Ak = A_r$ is the loop gain
A_f is the closed loop gain or transactance
$1 - A_r$ is the feedback factor

In practical amplifier designs the direct transfer ρ is small compared with k. Therefore the term including ρ will be omitted in the following sections. For the time being, A and k are assumed to be independent of frequency.

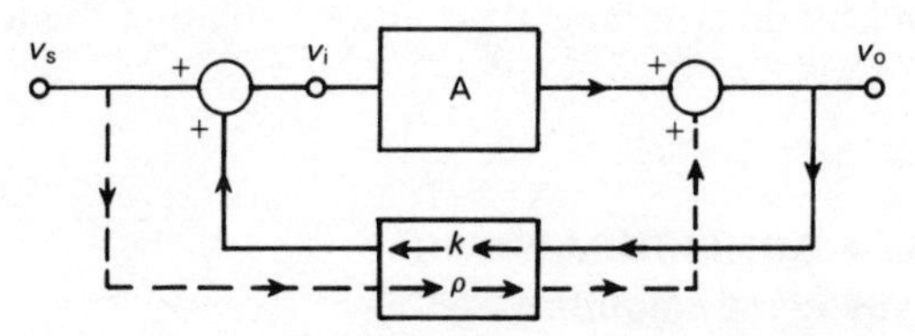

Fig. 13.1 General model of feedback amplifier

13.2 CLASSIFICATION OF FEEDBACK

Feedback can be classified on the basis of sign and magnitude of loop gain A_r.

$A_r < 0$, then $|A_f| < |A|$. The feedback is called **negative feedback**.

$0 < A_r < +1$, then $|A_f| > |A|$. The feedback is called **positive feedback.**

$A_r \geqslant +1$, then $|A_f| = \infty$. The amplifier is unstable and **oscillatory**.

Of these, negative feedback is almost exclusively used in linear amplifiers. The principal effects of negative feedback are:

- Biasing is more stable.
- Gain is more constant (Section 13.8.1).
- Distortion and noise decrease (Section 13.8.2).
- Bandwidth increases (Section 13.9.1).
- Input and output impedances change drastically (Sections 13.6 and 13.7).

13.3 IDEAL AMPLIFIER

As a reference model we introduce the **ideal amplifier** which has an infinite open loop gain $A = \infty$. Then the transfer function (asymptotic gain) is defined as

$$A_{f,\infty} = \lim_{A \to \infty} A_f = -\frac{1}{k}$$

Thus increasing values of open loop gain ultimately lead to a transfer function which is only dependent on the feedback network transfer k. This network usually consists of passive components.

The transfer function of the nonideal amplifier is now described by

$$A_f = A_{f,\infty} \frac{-A_r}{1 - A_r}$$

Thus A_f is found by determining $A_{f,\infty}$ and calculating the loop gain A_r.

13.4 FEEDBACK CONFIGURATIONS (frequency independent amplifier)

The four most commonly used methods of feedback are shown in Fig. 13.2. Also listed are the expressions of $A_{f,\infty}$, input and output resistances.

Name	*Transadmittance amplifier*	*Current amplifier*	*Voltage amplifier*	*Transimpedance amplifier (current source input)*	*Transimpedance amplifier (voltage source input)*
Configuration	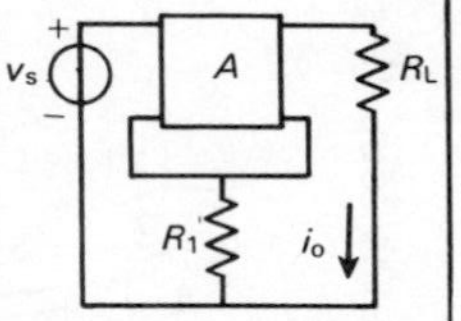 v_s, +, −, A, R_L, R_1, i_o	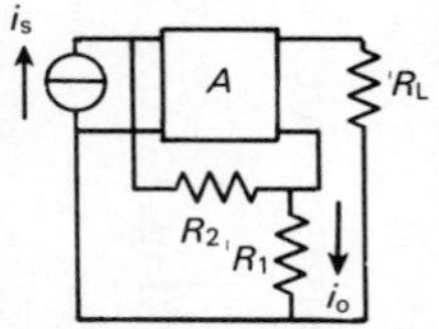 i_s, A, R_L, R_2, R_1, i_o	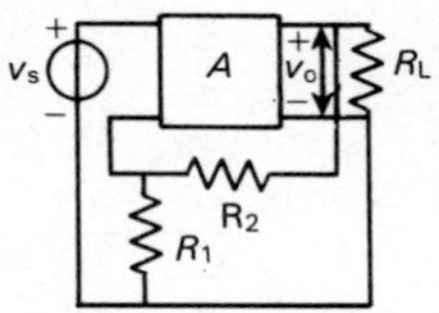 v_s, +, −, A, v_o, R_L, R_2, R_1	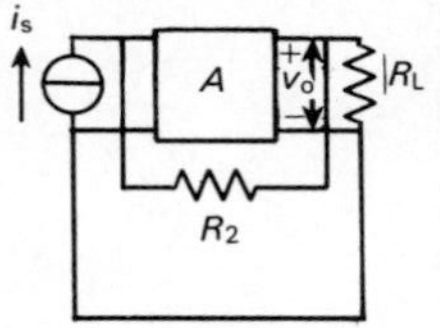 i_s, A, v_o, R_L, R_2	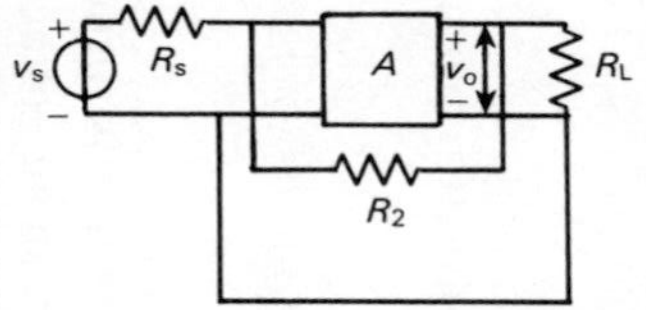 v_s, +, −, R_s, A, v_o, R_L, R_2
$A_{f,\infty}$	$\frac{i_o}{v_s} = \frac{-1}{R_1}$	$\frac{i_o}{i_s} = 1 + \frac{R_2}{R_1}$	$\frac{v_o}{v_s} = 1 + \frac{R_2}{R_1}$	$\frac{v_o}{i_s} = -R_2$	$\frac{v_o}{v_s} = -\frac{R_2}{R_s}$
$r_{i,f}$	$r_i(1 - A_r)$	$\frac{r_i}{1 - A_r}$	$r_i(1 - A_r)$	$\frac{r_i}{1 - A_r}$	$R_s + \frac{r_i}{1 - A_r}$
$r_{o,f}$	$r_o(1 - A_r)$	$r_o(1 - A_r)$	$\frac{r_o}{1 - A_r}$	$\frac{r_o}{1 - A_r}$	$\frac{r_o}{1 - A_r}$

Fig. 13.2 Five feedback methods and relevant formulas

13.5 NONIDEAL AMPLIFIERS

Example 1

Single-stage amplifier (Fig. 13.3): the transistor parameters are $h_{ie} = 2\ \text{k}\Omega$, $h_{fe} = 100$, $h_{re} = h_{oe} = 0$.

(a) The amplifier is a transimpedance amplifier (Fig. 13.2).

(b) $A_{f,\infty} = -R_f = -100\ \text{V}\,\text{mA}^{-1}$.

(c) In order to calculate A_r the signal source is turned off. Somewhere in the feedback loop we now impress a small voltage v_1 or current i_1 and we follow the gain and/or attenuation of this signal around the loop (Fig. 13.4). Then

$$\frac{i_2}{i_1} = h_{fe}, \quad \frac{i_3}{i_2} = \frac{-R_c}{R_c + R_f + h_{ie}}$$

Thus the loop gain is

$$A_r = \frac{-h_{fe}R_c}{R_c + R_f + h_{ie}}$$

$$= \frac{-1000}{112} = -8.93$$

(The minus sign of A_r indicates negative feedback.)

(d) $A_f = -100\,\dfrac{8.93}{9.93} \simeq -90$

Example 2

Two-stage amplifier (Fig. 13.5): the transistors are assumed identical with $h_{ie} = 2\ \text{k}\Omega$, $h_{fe} = 50$, $h_{re} = h_{oe} = 0$.

(a) The amplifier is a voltage amplifier (Fig. 13.2).

(b) $A_{f,\infty} = 1 + \dfrac{R_2}{R_1} = 11$

(c) A small voltage v_1 is injected into the loop (Fig. 13.6). Following v_1 around the loop, we obtain:

$$\frac{v_2}{v_1} = \frac{R_{c1}h_{ie}}{R_{c1} + h_{ie}}\, g_m = 37.5 \quad \left(g_m = \frac{h_{fe}}{h_{ie}}\right)$$

$$\frac{v_3}{v_2} = \left[-R_{c2}\left(R_2 + \frac{R_1}{1 + g_m R_1}\right)\right] \Big/ \left(R_{c2} + R_2 + \frac{R_1}{1 + g_m R_1}\right) = -8.58$$

$$\frac{v_4}{v_3} = \left(\frac{R_1}{1 + g_m R_1}\right) \Big/ \left(\frac{R_1}{1 + g_m R_1} + R_2\right) = 0.0426$$

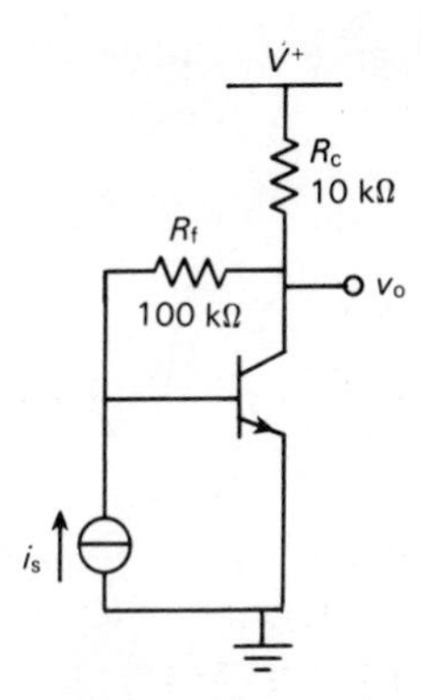

Fig. 13.3 Example of single-stage feedback amplifier

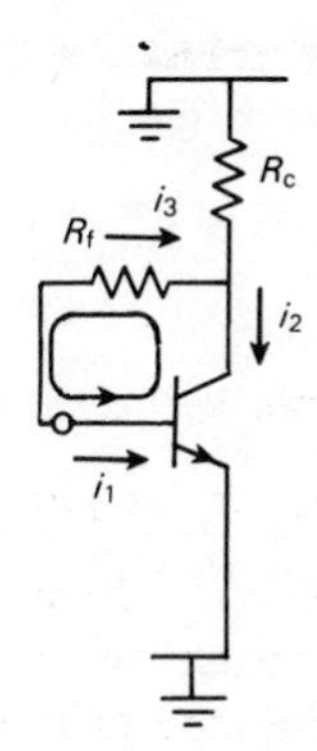

Fig. 13.4 AC equivalent circuit of Fig. 13.3

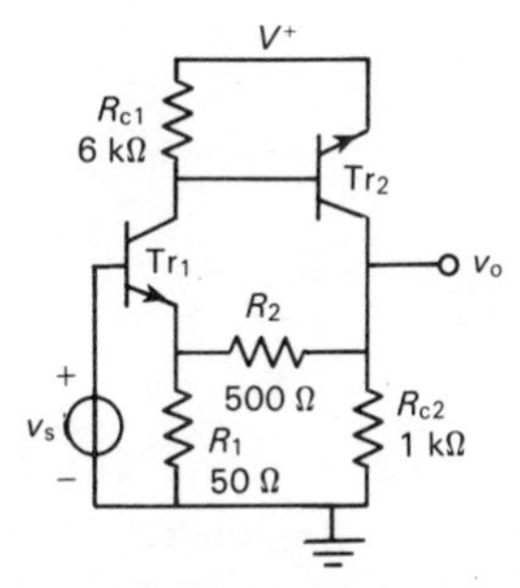

Fig. 13.5 Example of two-stage feedback amplifier

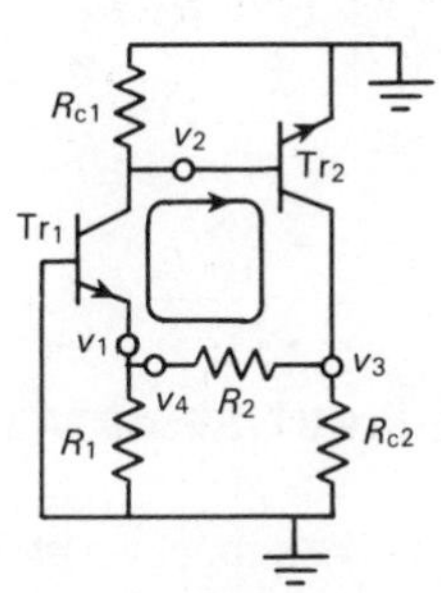

Fig. 13.6 AC equivalent circuit of Fig. 13.5

Thus $A_r = (37.5)(-8.58)(0.0426) = -13.7$

(d) $A_f = 11 \dfrac{13.7}{14.7} = 10.3$

13.6 INPUT RESISTANCE $r_{i,f}$

(a) For the current amplifier and transimpedance amplifier, employing input shunt feedback

$$r_{i,f} = \frac{r_i}{1 - A_r}$$

where r_i is the input resistance of the same amplifier having $A = 0$.

(b) For the voltage amplifier and transadmittance amplifier, employing input series feedback

$$r_{i,f} = r_i(1 - A_r)$$

Example 1

(See Example 1 of Section 13.5, Fig. 13.7.)

$$r_i = \frac{h_{ie}(R_f + R_c)}{h_{ie} + R_f + R_c} = 1.96 \text{ k}\Omega$$

$$r_{i,f} = \frac{1.96 \text{ k}\Omega}{9.93} \simeq 0.198 \text{ k}\Omega$$

Example 2

(See Example 2 of Section 13.5, Fig. 13.8.)

$$r_i = h_{fe}\left[\frac{1}{g_m} + \frac{R_1(R_2 + R_{c2})}{R_1 + R_2 + R_{c2}}\right] = 4.42 \text{ k}\Omega$$

$$r_{i,f} = 4.42 \text{ k}\Omega \times 14.7 \simeq 65 \text{ k}\Omega$$

13.7 OUTPUT RESISTANCE $r_{o,f}$

(a) For the voltage amplifier and transimpedance amplifier, employing output shunt feedback

$$r_{o,f} = \frac{r_o}{1 - A_r}$$

where r_o is the output resistance of the same amplifier having $A = 0$.

(b) For the current amplifier and transadmittance amplifier, employing output series feedback

$$r_{o,f} = r_o(1 - A_r)$$

Example 1

(See Example 1 of Section 13.5, Fig. 13.9.)

$$r_o = \frac{R_c(R_f + h_{ie})}{R_c + R_f + h_{ie}} = 9.11 \text{ k}\Omega$$

$$r_{o,f} = 9.11 \text{ k}\Omega \times 9.93 \simeq 90 \text{ k}\Omega$$

Example 2

(See Example 2 of Section 13.5, Fig. 13.10.)

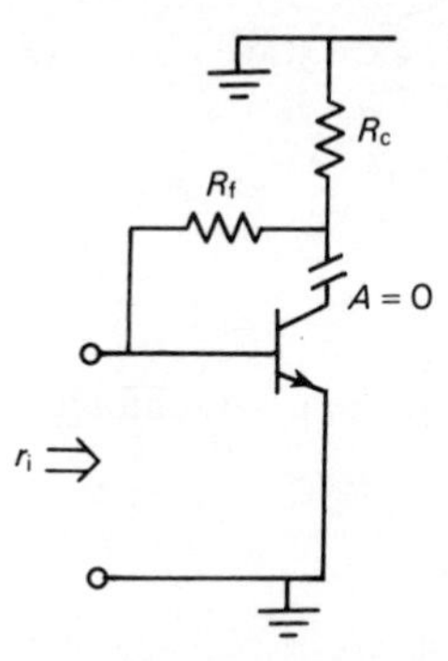

Fig. 13.7 Circuit to determine the input resistance r_i of the single-stage amplifier of Fig. 13.3 with $A = 0$

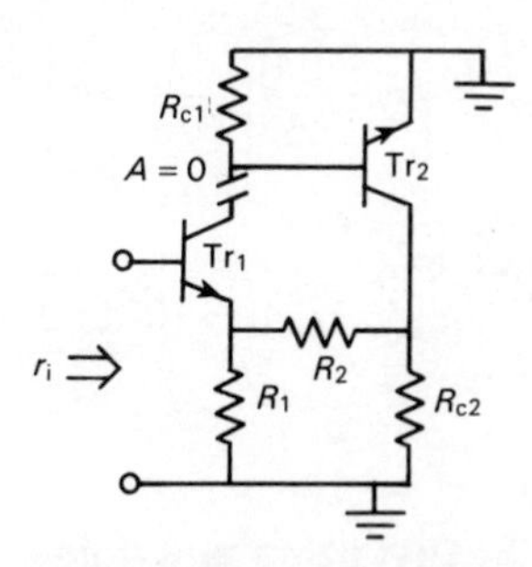

Fig. 13.8 Circuit to determine the input resistance r_i of the two-stage amplifier of Fig. 13.5 with $A = 0$

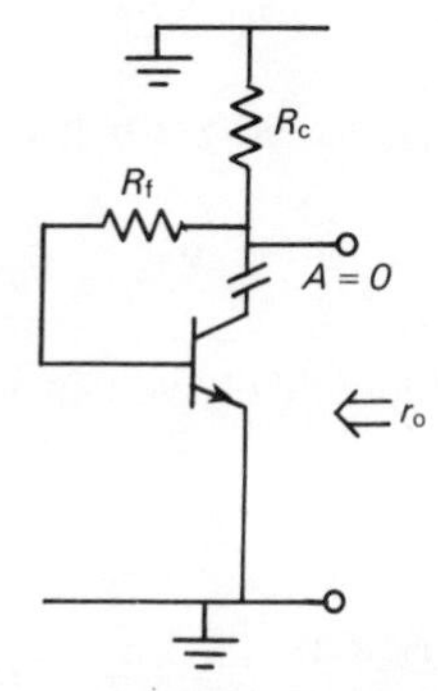

Fig. 13.9 Circuit to determine the output resistance r_o of the single-stage amplifier of Fig. 13.3 with $A = 0$

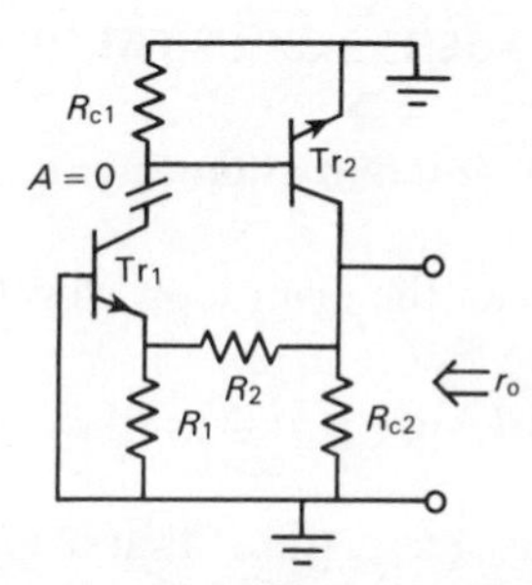

Fig. 13.10 Circuit to determine the output resistance r_o of the two-stage amplifier of Fig. 13.5 with $A = 0$

$$r_o = \left[R_{c2}\left(R_2 + \frac{R_1}{1 + g_m R_1}\right)\right] \Bigg/ \left(R_{c2} + R_2 + \frac{R_1}{1 + g_m R_1}\right) = 0.343 \text{ k}\Omega$$

$$r_{o,f} = \frac{0.343 \text{ k}\Omega}{14.7} \simeq 0.023 \text{ k}\Omega$$

13.8 ADDITIONAL EFFECTS OF NEGATIVE FEEDBACK

13.8.1 Constancy of gain

It follows from Section 13.1 that

$$\frac{dA_f}{A_f} = \frac{dA}{A} \frac{1}{1 - A_r}$$

In most cases $1 - A_r \gg 1$ so that

$$\left|\frac{dA_f}{A_f}\right| \simeq \left|\frac{dA}{A}\right| \frac{1}{|A_r|} \ll \left|\frac{dA}{A}\right|$$

Example
If in an amplifier $A_r = 10^3$ and A changes by 20 percent, A_f changes by only 0.02 percent.

13.8.2 Distortion and noise

Amplifier distortion and amplifier noise are decreased by the factor $1 - A_r$.

13.9 FEEDBACK IN FREQUENCY-DEPENDENT AMPLIFIERS

13.9.1 Stability conditions

Assuming the open loop gain A is frequency dependent, we can write

$$A_f(j\omega) = \frac{A(j\omega)}{1 - A_r(j\omega)}$$

If we further assume that $A(j\omega)$ can be expressed as

$$A(j\omega) = \frac{A}{P(\omega) + jQ(\omega)}$$

then it follows that

$$\frac{A_f(j\omega)}{A_f} = \frac{1 - A_r}{P(\omega) + jQ(\omega) - A_r}$$

where $A = A(\omega = 0)$, $A_f = A_f(\omega = 0)$ and $A_r = A_r(\omega = 0)$. This expression leads to the general conclusions:

- Peaks in the amplitude characteristic of $|A_f(j\omega)|$ are absent if

$$|A_r| \leqslant \frac{P(\omega)^2 + Q(\omega)^2 - 1}{2[1 - P(\omega)]}$$

- The amplifier will not oscillate if $|A_r| < P(\omega_o)$ where ω_o is the solution of $Q(\omega_o) = 0$.

Example 1
Amplifier with one time constant:

$$A(j\omega) = \frac{A}{1 + j\omega\tau}$$

Thus $P(\omega) \equiv 1$, $Q(\omega) = \omega\tau$:

- No peaks in the amplitude characteristics.
- No oscillations are possible.

Bandwidth

$$B_f = B(1 - A_r) = B\frac{A}{A_f}$$

(Fig. 13.11), so that

$$A_f B_f = AB$$

(gain-bandwidth product is constant).

Example 2

Amplifier with two time constants:

$$A(j\omega) = \frac{A}{(1 + j\omega\tau_1)(1 + j\omega\tau_2)}$$

Then $P(\omega) = 1 - \omega^2\tau_1\tau_2$, $Q(\omega) = \omega(\tau_1 + \tau_2)$:

- No peaks in the amplitude response occur if

$$|A_r| \leqslant \frac{1}{2}\left(\frac{\tau_1}{\tau_2} + \frac{\tau_2}{\tau_1}\right)$$

In the case where $\tau_1 \gg \tau_2$, the condition for no peaks can be written as

$$\frac{\omega_2}{\omega_1} \geqslant 2|A_r|$$

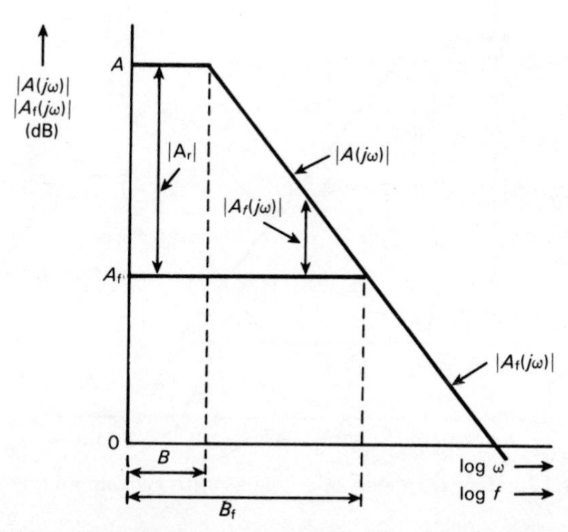

Fig. 13.11 Gain response of amplifier with one time constant

where $\omega_1 = 1/\tau_1$ and $\omega_2 = 1/\tau_2$. If $\omega_2/\omega_1 = 2\,|A_r|$, the amplitude response is maximally flat (Fig. 13.12):

- No oscillations are possible.

Example 3

Amplifier with three time constants:

$$A(j\omega) = \frac{A}{(1 + j\omega\tau_1)(1 + j\omega\tau_2)(1 + j\omega\tau_3)}$$

Then $P(\omega) = 1 - \omega^2(\tau_1\tau_2 + \tau_1\tau_3 + \tau_2\tau_3)$, $Q(\omega) = \omega(\tau_1 + \tau_2 + \tau_3) - \omega^3\tau_1\tau_2\tau_3$:

- No peaks in the amplitude response if

$$|A_r| \leqslant \frac{\tau_2^2 + \tau_2^2 + \tau_3^2}{2(\tau_1\tau_2 + \tau_1\tau_3 + \tau_2\tau_3)}$$

- Oscillations are possible at $\omega = \omega_o$ where

$$\omega_o^2 = \frac{\tau_1 + \tau_2 + \tau_3}{\tau_1\tau_2\tau_3}$$

- No oscillations will occur if

$$|A_r| < 2 + \frac{\tau_1}{\tau_2} + \frac{\tau_1}{\tau_3} + \frac{\tau_2}{\tau_1} + \frac{\tau_2}{\tau_3} + \frac{\tau_3}{\tau_1} + \frac{\tau_3}{\tau_2}$$

(Fig. 13.13).

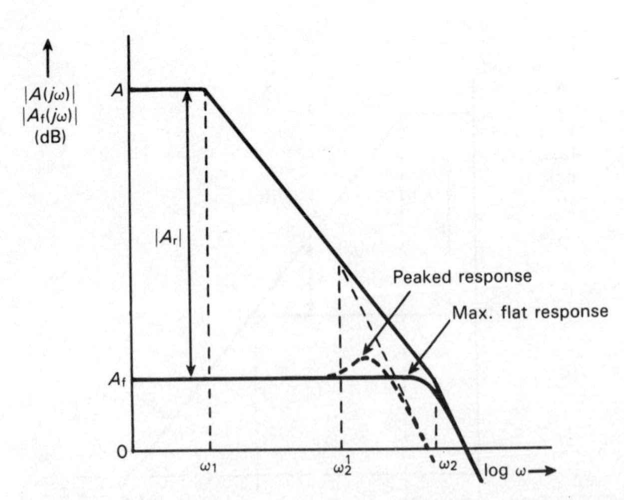

Fig. 13.12 Gain response of amplifier with two time constants

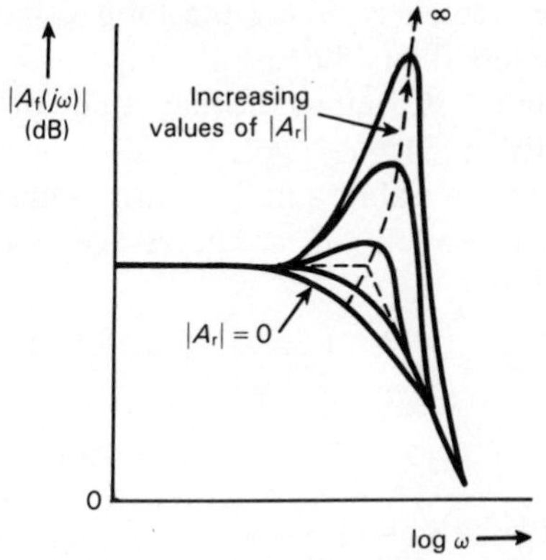

Fig. 13.13 Possible responses of amplifier with three time constants

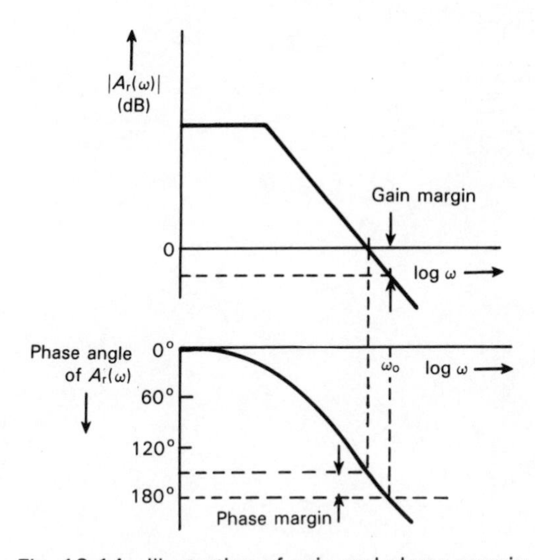

Fig. 13.14 Illustration of gain and phase margin

13.9.2 Gain and phase margin

In order that feedback amplifiers operate without the risk of oscillations, the concepts of gain and phase margins were introduced.

If (Fig. 13.14) ω_o is the frequency where oscillations are likely to occur, the design should be such that the loop gain at this frequency is less than 1. The **gain margin** is the value of $|A_r(j\omega_o)|$ in dB. A negative value is required for stable operation.

Similarly, at the frequency where the loop gain becomes unity, the phase shift should be less than 180°.

The **phase margin** is 180° minus the magnitude of the angle of $A_r(j\omega)$ at the frequency at which $|A_r(j\omega)| = 1$.

Typical figures for a stable amplifier are: gain margin $\simeq -10$ dB, phase margin $\simeq 50°$. These margins are necessary to allow for component tolerance, drift, etc.

14
DC Amplifiers

14.1 DIFFERENTIAL AMPLIFIERS

14.1.1 Discrimination and rejection

The importance of a differential amplifier is its property to provide high voltage gain to a differential signal between its inputs while suppressing signals common to the two inputs.

Such common mode signals frequently result from noise pickup or hum. The circuit of Fig. 14.1(a) is a model of an ideal differential amplifier (identical collector resistors, identical emitter resistors, identical transistors) which is biased by an ideal DC current source I. A current $\frac{1}{2}I$ then flows through each branch.

If a common mode signal v_c is applied at both inputs, the signal voltage at point X will be v_c and the currents in each branch remain unchanged. Therefore, no common mode output signal at the collectors will appear. If a differential mode signal is applied between the inputs, e.g. $+\frac{1}{2}v_d$ and $-\frac{1}{2}v_d$, no signal voltage will be present at X. Therefore, both transistors amplify the differential mode signal equally and in opposite phase.

According to the principle of superposition, if both signal types are present at the inputs, the differential mode signal will be amplified, the common mode signal will be completely suppressed.

In Figs. 14.1(b) and 14.1(c) practical (nonideal) differential amplifiers are shown with bipolar transistors and JFETs respectively. When assuming the transistors to be ideal current sources at the outputs ($h_{oe} = 0$, $y_{oe} = 0$) and further assuming that $R_e \gg R_a$, $R_e \gg R_b$ the following definitions of voltage gain can be derived:

- Differential gain A_d of a differential signal:

$$A_d = \frac{v_3 - v_4}{v_1 - v_2} \simeq \frac{R_1 + R_2}{R_3 + R_4}$$

- Differential gain A_c of a common mode signal:

$$A_c = \frac{v_3 - v_4}{\frac{1}{2}(v_1 + v_2)} \simeq \frac{R_2 R_3 - R_1 R_4}{R_e(R_3 + R_4)}$$

R_3 and R_4 include source resistances and transconductances.

- Common mode gain A_b of a common mode signal:

$$A_b = \frac{\frac{1}{2}(v_3 + v_4)}{\frac{1}{2}(v_1 + v_2)} \simeq \frac{R_2R_3 + R_1R_4}{2R_e(R_3 + R_4)}$$

From these expressions the **discrimination ratio** F is defined as

$$F = \frac{A_d}{A_b} = 2R_e \frac{R_1 + R_2}{R_2R_3 + R_1R_4}$$

The **rejection ratio** H or **common mode rejection ratio** (CMRR) is defined as

$$H = \frac{A_d}{A_c} = R_e \frac{R_1 + R_2}{R_2R_3 - R_1R_4}$$

The discrimination ratio F is a figure of merit expressing the magnitude of the common mode signals at each output.

The rejection ratio H expresses the difference of the common mode signals between both outputs. It is a measure of the asymmetry of a differential amplifier. Such asymmetry can be caused by mismatches in signal source resistances, collector capacitances, collector resistances, emitter-base junctions, etc. Due to these asymmetries, a common mode signal produces slightly different output signals. When connecting these output signals to a succeeding differential amplifier, the difference is a differential input signal to this second differential amplifier which is amplified by A_d of the second stage.

The final output stage thus contains an unwanted error voltage. Obviously, large values of F and H are obtained when R_e has a large value. If R_e is a fixed resistor, the differential amplifier is often referred to as a **long-tailed pair.** Much better performance is achieved when R_e is replaced by a current source so that R_e becomes the high output resistance of the current source.

Example 1

In the circuit of Fig. 14.2 it is assumed that $h_{fe} = 50$, $h_{re} = h_{oe} = 0$ for both transistors.

Since 0.5 mA flows through each transistor, $r_{e1} = r_{e2} = 0.05\ \text{k}\Omega$. The source resistors transferred to the emitters become 0.04 kΩ and 0.02 kΩ for the left and right branches respectively. Thus $R_3 = 0.09\ \text{k}\Omega$, $R_4 = 0.07\ \text{k}\Omega$. Then

$$H = 800 \frac{198}{100(0.09) - 98(0.07)} \simeq 74\,000 \text{ (97 dB)}$$

$$F = 1600 \frac{198}{100(0.09) + 98(0.07)} \simeq 20\,000 \text{ (86 dB)}$$

$$A_d \simeq 1238, \quad A_c \simeq 0.0167, \quad A_b \simeq 0.0620$$

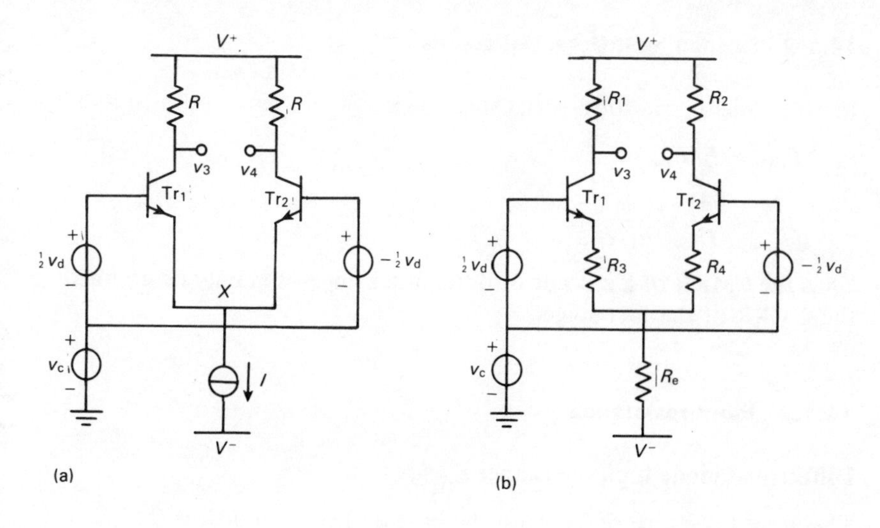

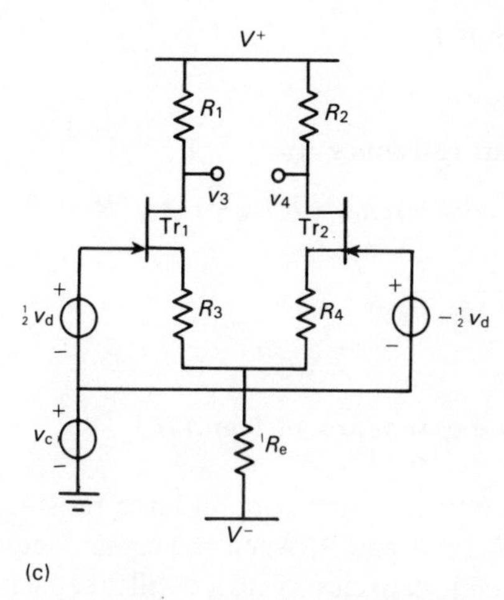

Fig. 14.1 (a) Model of ideal differential amplifier (b) Actual differential amplifier with bipolar transistors (c) Actual differential amplifier with JFETs

If, for example, the common mode voltage $v_c = 1$ V, the average of the two output signals is $v_c A_b = 62$ mV. The difference of the two signals is $v_c A_c = 16.7$ mV. Thus one signal is 70 mV, the other 54 mV.

Example 2

When considering the two voltage dividers of Fig. 14.3 as a differential stage, we find

$$H = 480$$
$$F = 1$$

14.1.2 Cascading differential stages

In a cascade of two differential stages (Fig. 14.4) it can be shown that

$$F_{\text{casc}} \simeq F_1 F_2$$

$$\frac{1}{H_{\text{casc}}} \simeq \frac{1}{H_1} + \frac{1}{F_1 H_2}$$

Thus the CMRR of a cascade of differential stages is mainly determined by the CMRR of the first stage.

14.1.3 Input resistance

Differential mode input resistance $r_{i,d}$

The input resistance of Tr_1 with the base of Tr_2 grounded, is

$$r_{i,d} \simeq h_{fe,1}(R_3 + R_4)$$

Common mode input resistance $r_{i,c}$

Assuming $h_{fe,1} = h_{fe,2} = h_{fe}$, $h_{oe,1} = h_{oe,2} = h_{oe}$, $R_1 = R_2 = R$,

$$r_{i,c} \simeq h_{fe} R_e \frac{1 + h_{oe} R}{1 + 2 h_{oe} R_e}$$

14.1.4 Frequency dependence of *F* and *H*

Since differential amplifiers often contain large resistance values, parasitic capacitances will affect F and H when the signal frequency increases. To account for the parasitic capacitances C_{cb} (Miller capacitances) and C_{R_e}, the expressions for F and H should be modified by replacing R_1 by Z_1, R_2 by Z_2 and R_e by Z_e (Fig. 14.5).

As the Miller capacitances are largely cancelled out in F and H, C_{R_e} remains the main cause for the deterioration of F and H (Fig. 14.6).

Example

If $R_e = 1\ \text{M}\Omega$, $C_{R_e} = 5\ \text{pF}$, a 3 dB decrease of F and H will occur at $f \simeq 30\ \text{kHz}$. The common mode input impedance decreases at the same rate.

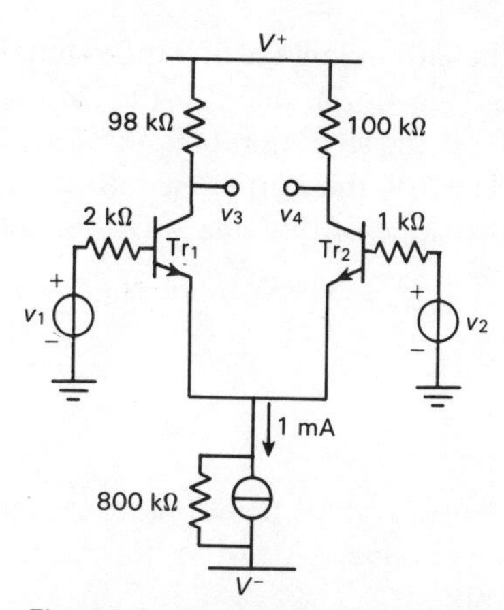

Fig. 14.2 Example of differential amplifier to calculate F and H

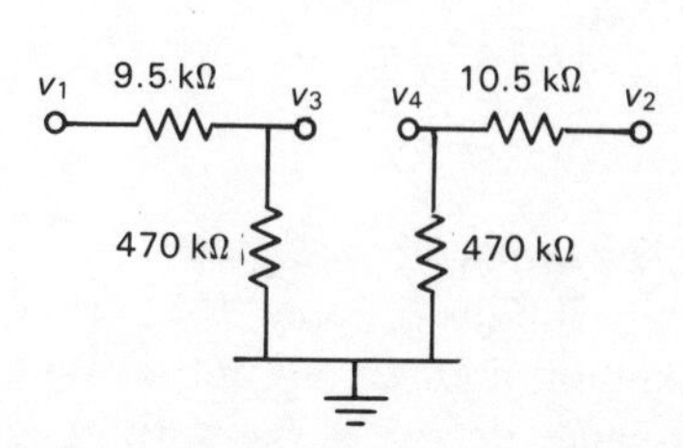

Fig. 14.3 A voltage divider as a differential stage

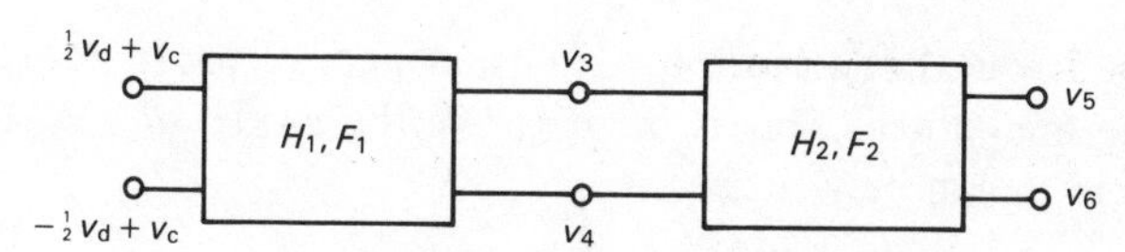

Fig. 14.4 Cascade of two differential stages

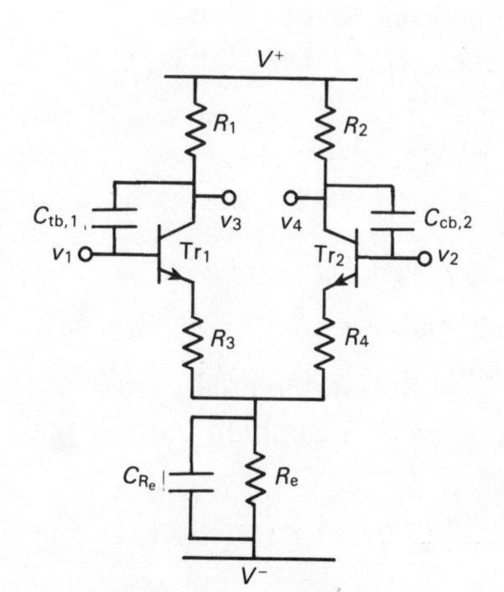

Fig. 14.5 Parasitic capacitances in a differential amplifier

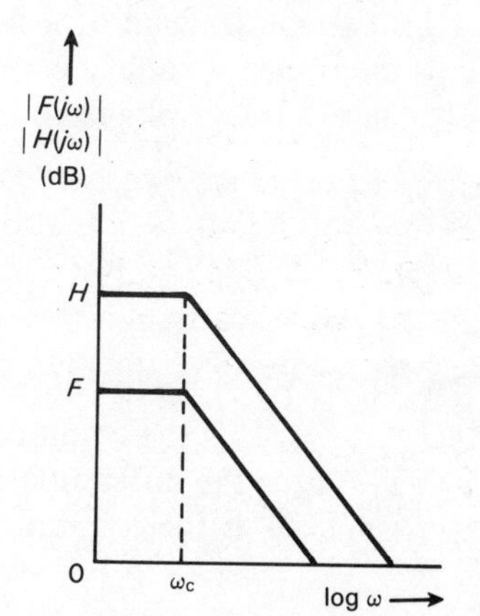

Fig. 14.6 Frequency response of F and H

14.1.5 Slewing rate limit

If the differential signal is a step voltage, the output voltage of a differential amplifier follows an exponential curve (Fig. 14.7) due to the Miller capacitances C_{cb}. The tangent of this curve is the **slewing rate limit** or **slew rate** S. If V^+ is the maximum voltage level of the output, a maximum sinusoidal output signal is defined since the slope of this sine wave cannot exceed S. The frequency of this signal is

$$f_{max} = \frac{S}{2\pi V^+}$$

Example

If $S = 3\ V\,(\mu s)^{-1}$ and $V^+ = 12$ V, $f_{max} \simeq 40$ kHz. At this frequency the maximum undistorted output signal can be obtained. A further increase in frequency decreases the output signal amplitude.

14.1.6 Single-ended output

In differential stages the two outputs are usually converted to a single-ended output in the final stage. This is indicated in the model of Fig. 14.8. The output voltage v_o can be written as

$$v_o = A_+ v_2 - A_- v_1$$

where

A_+ is the voltage gain of a signal connected to the +input
A_- is the voltage gain of a signal connected to the −input
v_1 is an arbitrary voltage connected to the −input
v_2 is an arbitrary voltage connected to the +input

Rewriting v_o as

$$v_o = \tfrac{1}{2}(A_+ + A_-)(v_2 - v_1) + (A_+ - A_-)\tfrac{1}{2}(v_2 + v_1)$$

where

$\frac{1}{2}(A_+ + A_-) = A_d$ is the differential gain of a differential signal
$A_+ - A_- = A_c$ is the common mode gain of a common mode signal
$v_2 - v_1 = v_d$ is the differential voltage
$\frac{1}{2}(v_2 + v_1) = v_c$ is the common mode voltage.

Then

$$v_o = A_d v_d + A_c v_c$$

$$= A_d\left(v_d + \frac{1}{H} v_c\right)$$

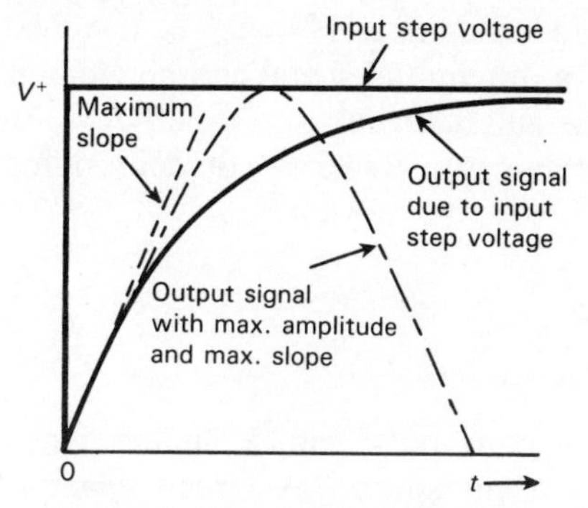

Fig. 14.7 The slewing rate limit

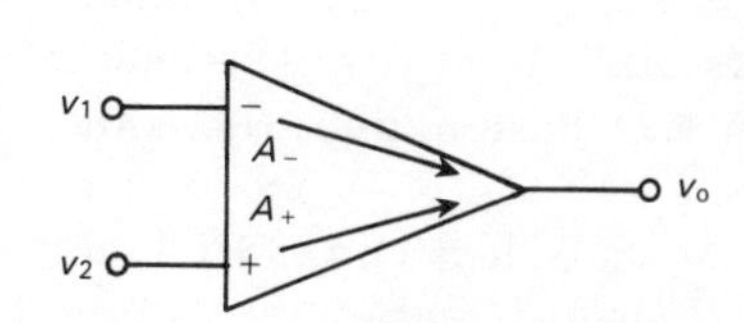

Fig. 14.8 Single-ended differential amplifier

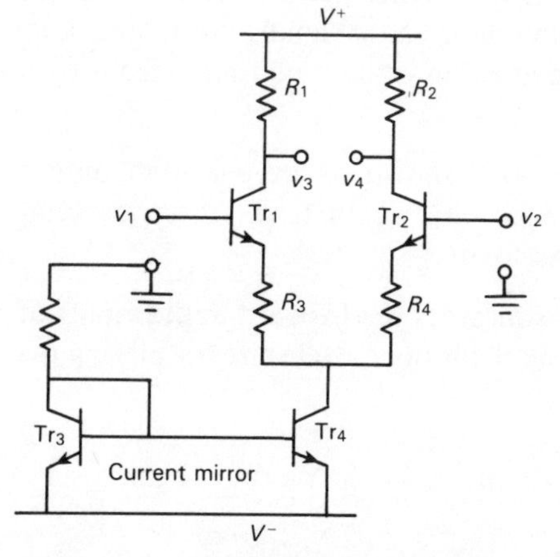

Fig. 14.9 Differential amplifier with current mirror

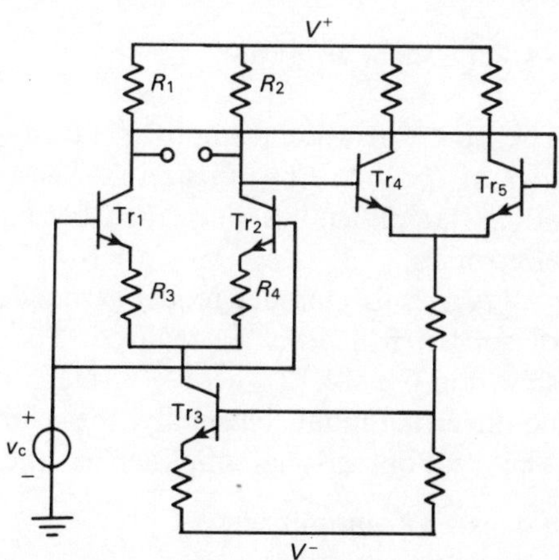

Fig. 14.10 Differential amplifier with feedback to reduce common-mode effects

Again, the common mode rejection ratio is defined as $H = A_d/A_c$. It can be seen as the ratio of the amplitudes of a differential signal and a common mode signal, both resulting in the same output signal.

It should be noted that the discrimination ratio is not defined for single-ended outputs.

14.1.7 Further improvements

- As stated in Section 14.1.1 large values of H and F can be obtained when R_e is the output resistance of a current source. A circuit example employing a current mirror is shown in Fig. 14.9.
- Feedback is sometimes employed to further reduce common-mode signals. Figure 14.10 shows how common mode signals are fed back to the current source. In this way a reduction of common mode gain by a factor of 3 can often be obtained.
- The cascode circuit (Section 11.5) was shown to have a high output resistance and virtually no Miller effect. In Fig. 14.11 the cascode circuit is used for both current source and load.

The common mode signal at the two sources is fed forward to the inputs of Tr_3 and Tr_4 at the same time providing the proper level shift for biasing the cascode FETs.

14.2 OPERATIONAL AMPLIFIERS

14.2.1 Ideal amplifier

The term 'operational amplifier' first appears in 1947 in an article (*Proceedings of the IRE*) by Ragazzini. The present day integrated circuit operational amplifier has since then become a basic building block in analog electronics.

A useful reference model is the idealized operational amplifier which, of course, will never be realized. The symbol for such an ideal model is shown in Fig. 14.12 where '−' refers to the inverting input and '+' to the noninverting input. Normally, the required supply voltages are not shown. The ideal operational amplifier has the following properties:

- It is a DC amplifier.
- The amplifier has a differential input.
- The voltage gain A is infinite.

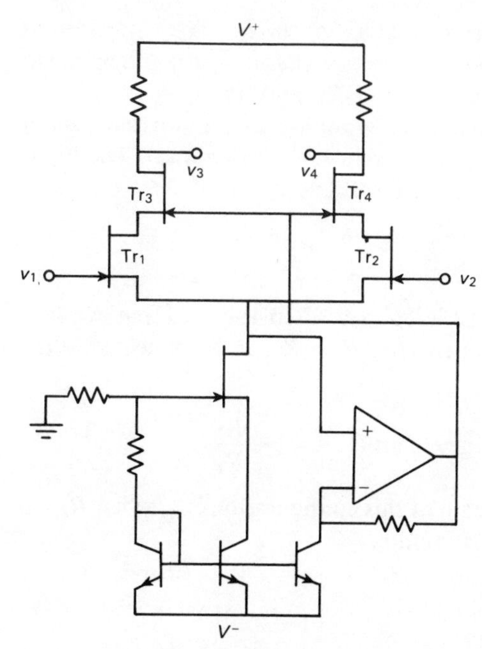

Fig. 14.11 Cascode differential amplifier to eliminate Miller effect

Fig. 14.12 Symbol of operational amplifier

- The bandwidth is infinite.
- The input resistance is infinite.
- The output resistance is zero.
- The output voltage is zero when the two input voltages are zero.

It follows from these properties that:

- No current flows in either input terminal since $r_i = \infty$.
- There is no voltage difference between v_1 and v_2 since any output signal can be produced by an infinitesimally small input signal.

It is comforting to know that many operational amplifiers in their applications may be treated as ideal models, the difference between ideal and actual often being negligible.

In the majority of applications negative feedback is employed and the presence of two input terminals allows a choice between two basic amplifier configurations.

Inverting amplifier

In Fig. 14.13 the + input is grounded and the input signal v_1 is applied through R_1. A positive voltage v_1 causes a current $i_1 = v_1/R_1$ to flow as shown. The − input terminal remains at zero volts so that a voltage drop $-i_1R_2$ develops across R_2. Thus $v_o = -i_1R_2$ and the voltage gain is $v_o/v_1 = -R_2/R_1$. The − input is called the **summing point** or **virtual ground** (it is not a real ground since no signal current flows to ground). The input resistance of the inverting amplifier is thus R_1.

Noninverting amplifier

In Fig. 14.14 the input voltage v_2 is connected to the + input while a fraction of the output voltage, namely $[R_1/(R_1 + R_2)]v_o$ is connected to the − input. Thus

$$v_2 = \frac{R_1}{R_1 + R_2} \cdot v_o \text{ and the voltage gain is } \frac{v_o}{v_2} = 1 + \frac{R_2}{R_1}$$

Evidently, the minimum voltage gain of this configuration is 1, when $R_2 = 0$ or $R_1 = \infty$. The input resistance is infinite.

14.2.2 Nonideal amplifier

The nonideal operational amplifier has a number of departures from the ideal model which can be divided into static and dynamic errors.

Static errors

The practical operational amplifier has a nonzero output voltage when the input terminals are grounded. This nonzero output voltage is called **output offset voltage** (ΔV_o). This offset voltage is due to a number of sources which are referred to the input terminals.

Input offset current. The input stage of an operational amplifier, either with bipolars or with FETs, is a differential amplifier. With bipolars, input base currents are required for the input transistors. These transistors never match perfectly and their current amplification factors will not be the same. Hence the two input currents I_- and I_+ will be different (Fig. 14.15).

External to the amplifier, these currents flow into resistances R_c and $R_1//R_2$ causing different voltage drops. This difference voltage is amplified and results in an offset voltage at the output. When making $R_c = R_1//R_2$, a first order balance is achieved, leaving an output offset voltage $\Delta V_o = |I_- - I_+|R_2$.

With FET inputs, it is the difference in leakage currents that causes an output offset voltage. Thus this voltage is smaller when the resistance values at the input terminals are smaller. Obviously, FET-input units allow much larger resistor values to be used.

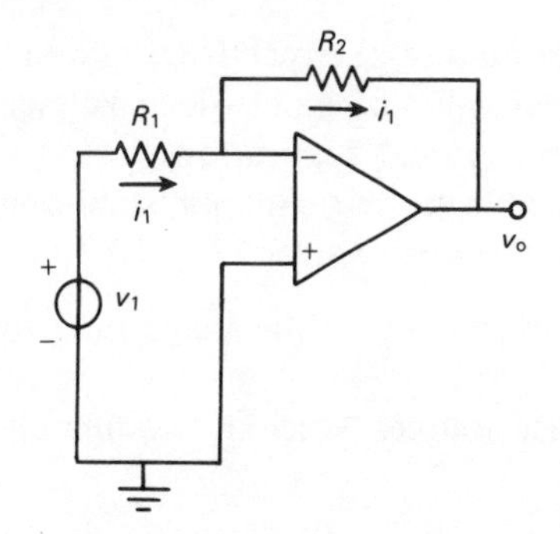

Fig. 14.13 The inverting operational amplifier

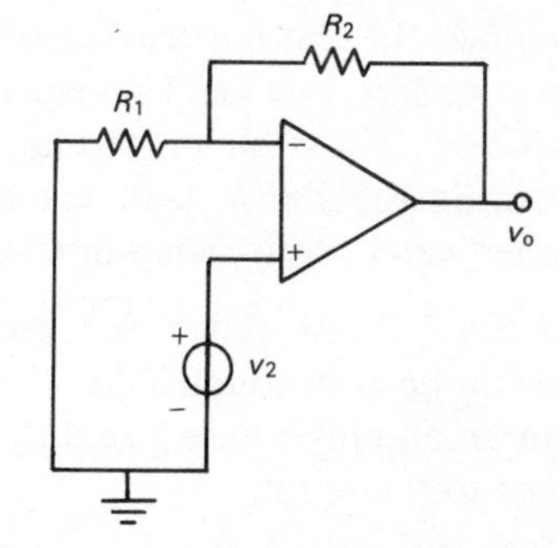

Fig. 14.14 The noninverting operational amplifier

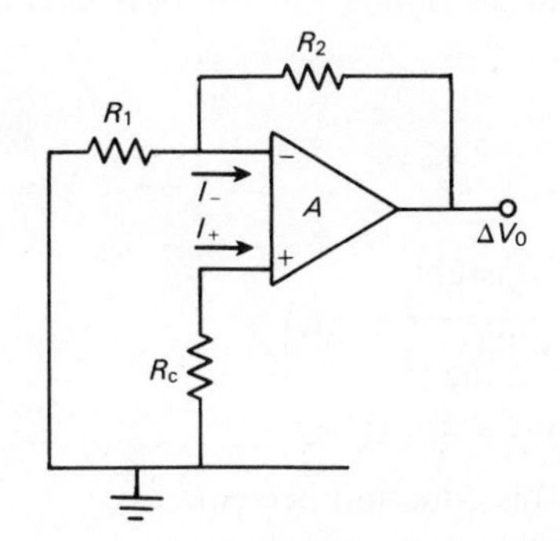

Fig. 14.15 Input currents in an operational amplifier causing an output offset voltage

The **input offset current** is defined as $I_{os} = | I_- - I_+ |$. The average value of the two input currents is defined as the **input bias current,** thus $I_b = \frac{1}{2}(I_- + I_+)$.

Input offset voltage. This offset is also due to imperfect matching of the input transistors. With bipolars, the voltage drops across the forward biased *pn* junctions will not be equal, with FETs the values of gate-source voltages will be slightly different.

The **input offset voltage** is defined as the voltage which has to be applied to one of the inputs to obtain zero output voltage.

Many operational amplifiers have leads brought out between which a nulling potentiometer can be connected to cancel the output offset voltage. Unfortunately, voltage and current offset are both temperature dependent.

Drift. With grounded inputs and disregarding voltage and current offset, the output voltage will show variations due to changes in temperature and aging of materials. These slow changes are termed **drift.**

Offset errors are very objectionable and many efforts have been made to minimize them as much as possible. After nulling, thermal drifts of input voltages and currents still remain. A typical value of input offset voltage drift is 5 μV $^{\circ}$C^{-1} which is too large in some critical applications.

Considerably less drift can be realized with the **chopper stabilized** amplifier which is discussed in Chapter 22.

Noise. Noise sets a fundamental lower limit to the magnitude of signals that can be amplified.

In an amplifier there are different noise sources which are commonly referred to the input.

- External noise sources associated with the resistors connected to the input terminals. They generate **thermal noise** due to the random motion of charges. The noise of an ohmic resistor R is expressed by the Nyquist equation

$$\overline{e_n^2} = 4kTRB \text{ (V}^2\text{)}$$

where

k is Boltzmann's constant
T is the absolute temperature (K)
R is the resistance value (Ω)
B is the system bandwidth (Hz)

At room temperature this equation becomes

$$\overline{e_n^2} \simeq 1.6 \times 10^{-20} RB \text{(V}^2\text{)}$$

- Internal current and voltage noise sources. The model of an amplifier with its noise sources is shown in Fig. 14.16. From this model the total input noise voltage is found as

$$\overline{e_t^2} = [(\overline{e_{n,-}^2} + \overline{e_{n,+}^2}) + \overline{i_{n,-}^2}(R_1//R_2)^2 + \overline{i_{n,+}^2} R_c^2 + 4kT(R_c + R_1//R_2)]B$$

Example

If $R_1 = R_c = 100\ \Omega$, $R_2 = 10$ kΩ, $\overline{i_{n,-}^2} = \overline{i_{n,+}^2} = 9$ (pA)^2 Hz^{-1}, $\overline{e_{n,-}^2} = \overline{e_{n,+}^2} = 0.01$ (μV)^2 Hz^{-1} and $B = 10$ Hz, it follows that

$$\sqrt{\overline{e_t^2}} \simeq 0.6\ \mu\text{V}$$

A further discussion of noise and its definitions is presented in Chapter 20.

Dynamic errors

Gain-bandwidth considerations. In most operational amplifiers the slope or rolloff of the open loop gain characteristic starts at a low frequency (5–100 Hz) and continues at a rate of 6 dB per octave. The maximum phase shift which accompanies such a slope is 90°. According to Chapter 13 there

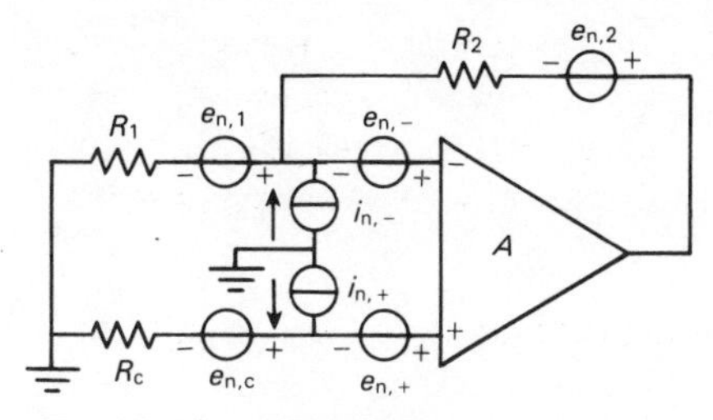

Fig. 14.16 Noise sources in an operational amplifier

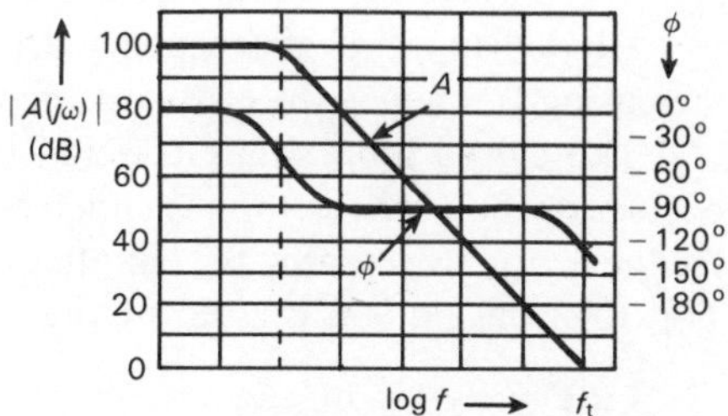

Fig. 14.17 Typical gain and phase response of an operational amplifier

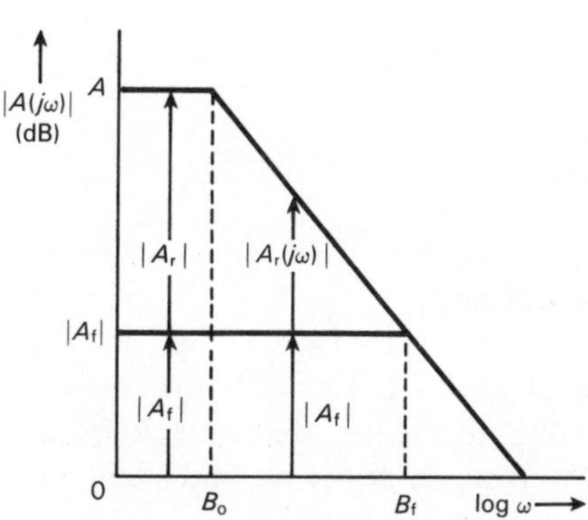

Fig. 14.18 Gain characteristics of an operational amplifier

is a 90° phase margin and no danger of oscillations. The amplifier is said to have **closed loop stability**.

With increasing frequency the gain approaches unity and additional phase shift may accumulate (Fig. 14.17) but at the **unity gain frequency** f_t, where $|A(j\omega)| = 1$, the phase shift is less than 180°.

Such a built-in stability is achieved by integrating a small capacitor at some strategic point inside the amplifier which is called **internally compensated.** Other amplifier types have leads brought out allowing compensation by external components.

When feedback is applied, the closed loop gain A_f of the **inverting** amplifier can be derived using the feedback theory of Chapter 13.

The asymptotic gain is $A_{f,\infty} = -R_2/R_1$, the loop gain

$$A_r = -A[R_1/(R_1 + R_2)]$$

so that the closed loop gain is

$$A_f = A_{f,\infty} \frac{-A_r}{1 - A_r}$$

$$= -\frac{R_2}{R_1} \frac{1}{1 + \frac{1}{A}\left(1 + \frac{R_2}{R_1}\right)}$$

For large values of closed loop gain ($R_2 \gg R_1$): $A_r \simeq -A(R_1/R_2)$ so that $|A_r||A_f| \simeq A$. This relation is indicated by arrows in Fig. 14.18.

Obviously, A_r is as frequency dependent as A which means that with increasing frequency the error between A_f and $A_{f,\infty}$ increases.

The closed loop gain characteristic thus follows a straight line until open loop and closed loop characteristics meet. Application of negative feedback thus decreases the amplifier gain but increases the bandwidth. This relationship is expressed by the **gain-bandwidth product** GBP:

$$\text{GBP} = B_o A = B_f |A_f|$$

where

B_o is the open loop bandwidth
B_f is the closed loop bandwidth

Hence the gain-bandwidth product is constant.

Since $B_f = B_o(A/|A_f|)$ and $A/|A_f| = |A_r|$, the closed loop bandwidth $B_f = B_o |A_r|$ where A_r is the loop gain at $f = 0$ Hz. Feedback thus increases the amplifier bandwidth by a factor equal to the low-frequency loop gain. For the closed loop gain of the **noninverting amplifier** we find similarly

$$A_f = \left(1 + \frac{R_2}{R_1}\right) \frac{1}{1 + \frac{1}{A}\left(1 + \frac{R_2}{R_1}\right)}$$

Impedances. According to Chapter 13, negative feedback reduces the output resistance of the operational amplifier by a factor equal to the loop gain, thus

$$r_{o,f} \simeq \frac{r_o}{|A_r|}$$

where r_o is the output resistance without feedback.

The input resistance of the inverting amplifier when looking at the virtual ground, is

$$r_i \simeq r_{i,d} // 2r_{i,c} // \frac{R_2}{|A_r|}$$

where $r_{i,d}$ is the differential mode input resistance of the amplifier and $r_{i,c}$

the common mode resistance. In practical circuits the term $R_2/|A_r|$ is much smaller than the other two so that

$$r_i \simeq \frac{R_2}{|A_r|}$$

(Fig. 14.19(a)).

In the noninverting amplifier the input resistance increases by feedback (Fig. 14.19(b)). Resistance $r_{i,d}$ increases by the factor $|A_r|$. Parallel to this resistance is the resistance $2r_{i,c}$ where $r_{i,c}$ is the common mode input resistance of the amplifier. Thus

$$r_i \simeq r_{i,d}|A_r|//2r_{i,c}$$

Both resistance values are normally very large (hundreds of MΩs) and other effects will determine the actual value of the input resistance. The values of output and input resistances depend strongly on the frequency since all expressions contain $|A_r|$ as a factor.

CMRR. The inverting amplifier shows no common mode effect since the + input is at ground.

In the noninverting amplifier the − input follows the + input. When

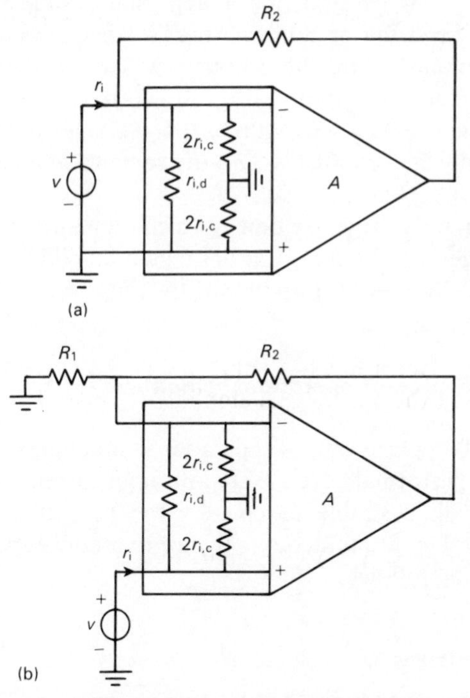

Fig. 14.19 (a) Input impedance of the noninverting amplifier (b) Input impedance of the inverting amplifier

connecting a signal source v_s to the + input, the common mode voltage equals v_s. If, for example, the amplifier has a CMRR = 10^4, amplification of noninverting signals is accurate to within 0.01 percent.

The CMRR is not a constant parameter but its value starts to roll off at about 100 Hz with increasing frequency.

Slew rate limiting. The slew rate was introduced in Section 14.1.5; its effect becomes apparent when applying a large step voltage at the input. The reasons for this effect are internal to the amplifier, often it is the frequency compensation capacitor. With high rates of change of signals insufficient current is available to charge or discharge this capacitor.

Externally, slew rate limiting is noticeable when an expected sine wave output signal becomes triangular in shape.

The frequency at which a maximum sinusoidal output signal can be obtained is called the **power bandwidth** f_p and is related to the slew rate S by

$$f_p = \frac{S}{2\pi V_p}$$

where V_p is the maximum possible excursion of the output signal.

Settling time. When applying a step input voltage to a unity-gain amplifier (either inverting or noninverting) so that the output signal just reaches its maximum value, the output will show a damped oscillation before the final value is reached (Fig. 14.20(a)). The **settling time** t_s is defined as the time it takes for the output voltage to reach a specified percentage (usually 0.1 or 0.01 percent) of the final value, after application of a step input.

A related quantity is the **rise time** t_r which is measured when applying a small step voltage (e.g. 20 mV) to a unity-gain amplifier. The rise time is defined as the 10–90 percent response of the amplifier (Fig. 14.20(b)).

14.2.3 Classification of operational amplifiers

The diversity of applications of operational amplifiers entails different requirements on performance (e.g. very low input current, wide bandwidth, etc.). Since not all desirable parameters can be met simultaneously, a trade-off is inevitable. Manufacturers therefore produce special types which can be classified as follows.

General purpose types

These have moderate drift and bandwidth specifications, for example bias current drift: 0.2–2 nA °C^{-1}, input offset voltage drift: 2–20 μV °C^{-1}. Bandwidths are usually less than 2 MHz.

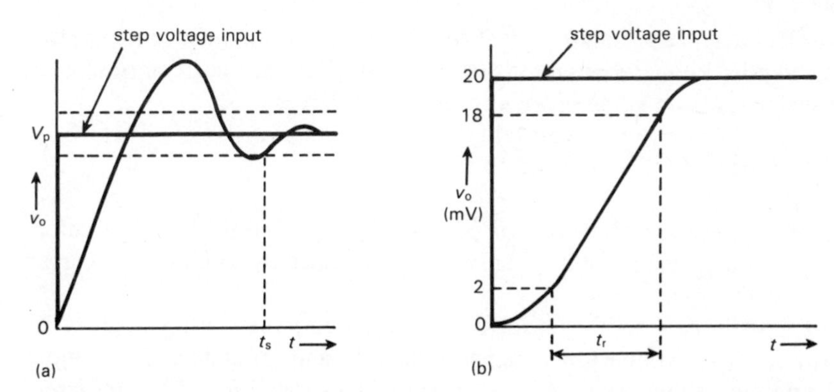

Fig. 14.20 (a) The definition of settling time (b) The definition of rise time

High input impedance types

These types have mostly an FET input stage. The use of resistor values exceeding 100 kΩ poses no problem. Bias currents may range from 1–10 pA.

Low drift types

Input offset voltage drift values are of the order of 0.2–5 $\mu V\,^\circ C^{-1}$. In chopper stabilized types this value is about 0.1 $\mu V\,^\circ C^{-1}$. Bias current drift may be as low as 1 pA $^\circ C^{-1}$.

Wide bandwidth types

Usually the GBP is at least 10 MHz, slew rates range from 25–1000 V μs^{-1}.

High output voltage and current types

Output voltages of 20–100 V can be obtained while load currents may exceed 100 mA.

Instrumentation types

Instrumentation amplifiers are precision differential amplifiers containing two or three cascaded operational amplifiers. Input resistances of 1000 MΩ are no exception. Drifts are very low ($\simeq 1\ \mu V\,^\circ C^{-1}$), CMRR is high ($>100$ dB) and open loop gains range from 10^5–10^6.

14.2.4 Operational transconductance amplifier

A later development in operational amplifiers is the operational transconductance amplifier (OTA). This is an amplifier with normal differential input terminals but the output stage is a current source. Its transfer function is therefore a transconductance:

$$\frac{i_o}{v_i} = g_m$$

where v_i is the difference voltage between the input terminals. Furthermore, g_m is not a constant but its value can be controlled by an input bias current I_{bias} as sketched in Fig. 14.21.

The OTA is an almost ideal differential amplifier with a single-ended output. When connecting a signal voltage v_1 and grounding the + input, the output voltage across R_L may be varied by varying I_{bias}. Thus the circuit has a current controlled voltage gain. Section 14.2.5 contains an application of the OTA.

14.2.5 Applications

Many handbooks containing hundreds of operational amplifier applications have appeared, not counting the avalanche of manufacturers application notes. Only a few commonly encountered circuits are therefore described. In these circuits the amplifier is treated as an ideal model unless this is not justified.

Voltage follower (Fig. 14.22)

A very high input impedance, a very low output impedance and unity voltage gain are the properties of this circuit. Hence it is an excellent buffer between a high resistance source and a low resistance load.

Constant voltage source (Fig. 14.23)

The output voltage is

$$V_o = -V_z \frac{R_2}{R_1}$$

Inverting adder (Fig. 14.24)

The output voltage is

$$v_o = -\left(v_1 \frac{R_o}{R_1} + v_2 \frac{R_o}{R_2} + v_3 \frac{R_o}{R_3}\right)$$

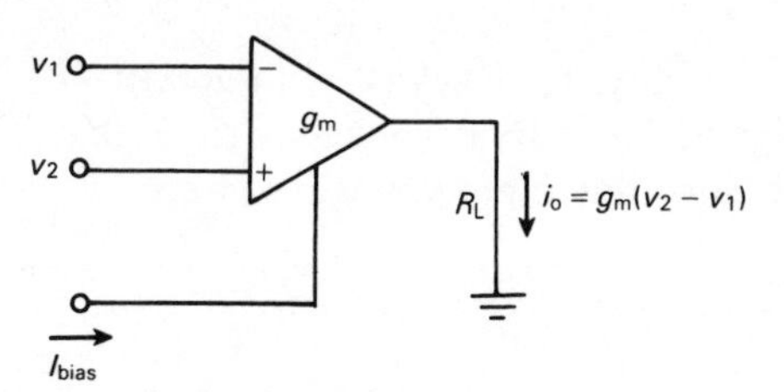

Fig. 14.21 Basic circuit of an OTA

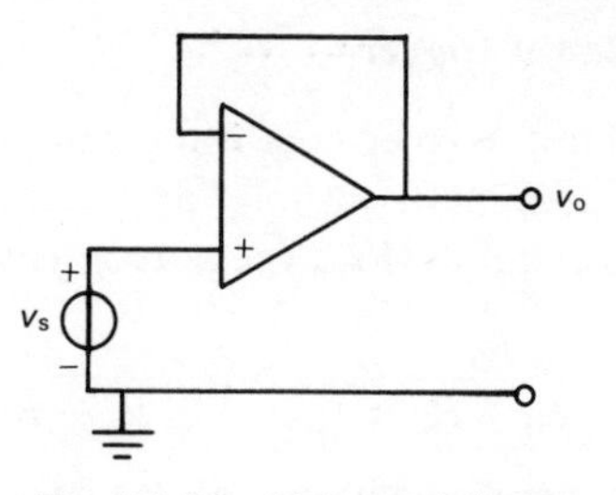

Fig. 14.22 Voltage follower

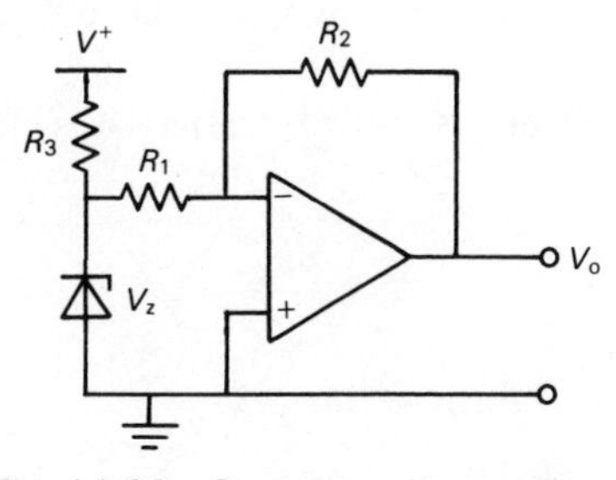

Fig. 14.23 Constant voltage source

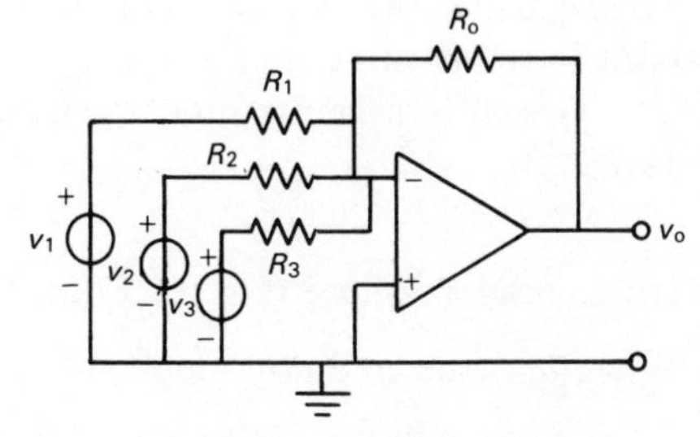

Fig. 14.24 Inverting adder

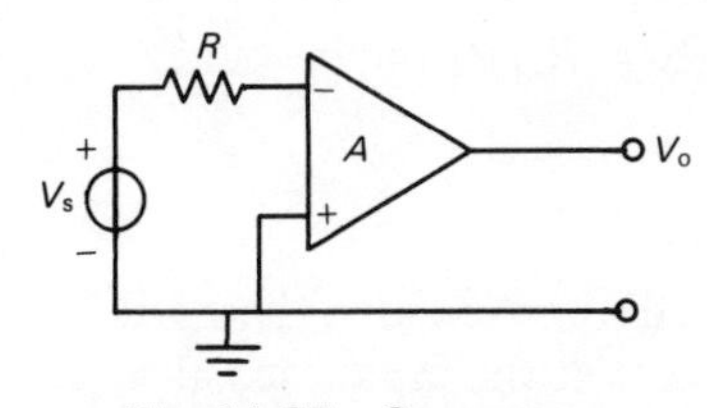

Fig. 14.25 Comparator

The signal sources $v_1, v_2, \ldots$ are isolated from each other by the virtual ground.

Comparator (Fig. 14.25)

The amplifier operates without feedback. Thus

$$V_o = V^- \quad \text{when } V_s > 0$$
$$V_o = V^+ \quad \text{when } V_s < 0$$

In an actual amplifier A is not infinite and there is a small transition region around zero volts where V_s is undetermined. The width of this region is

$$\frac{1}{A}(V^+ - V^-)$$

Schmitt trigger

Schmitt trigger operation is based on positive feedback (Fig. 14.26(a)). The input signal is usually an analog signal. The output voltage is either V_{max} (positive) or V_{min} (negative), the noninverting input voltage is either

$$\frac{R_1}{R_1 + R_2} V_{max} \quad \text{or} \quad \frac{R_1}{R_1 + R_2} V_{min}$$

An increasing input voltage passing the noninverting input level switches the output to the negative state (Fig. 14.26(b)). The decreasing input voltage when passing the negative noninverting input level, switches the output back to the positive state.

The difference in input switching levels is called **hysteresis** (Fig. 14.26(c)).

Differential amplifier (Fig. 14.27(a))

When v_c is the common mode voltage, the output voltage is

$$v_o = -v_1 \frac{R_2}{R_1} + v_2 \frac{R_4}{R_3} \frac{1 + (R_2/R_1)}{1 + (R_4/R_3)} + v_c \frac{(R_4/R_3) - (R_2/R_1)}{1 + (R_4/R_3)}$$

Apparently, when $R_4/R_3 = R_2/R_1$ the output is

$$v_o = \frac{R_2}{R_1}(v_2 - v_1)$$

with no common mode effects. Actually, there will be common mode errors since equal voltage sources at the inputs see unequal loads.

Differential amplifier (Fig. 14.27(b))

In this circuit the common mode effect of the previous circuit is eliminated. The differential output is

$$v_{o1} - v_{o2} = \left(1 + \frac{R_2}{R_1}\right)(v_1 - v_2)$$

Integrator (Fig. 14.28)

When long-term integration is performed the amplifier usually shows its nonideal character due to the finite values of A and r_i. The transfer function in Laplace notation is

$$\frac{v_o}{v_s} = \frac{-1}{s\tau + \frac{1}{A}\left(1 + \frac{R}{r_i} + s\tau\right)}$$

where $\tau = RC$. If A were infinite, $v_o/v_s = -1/s\tau$ so that

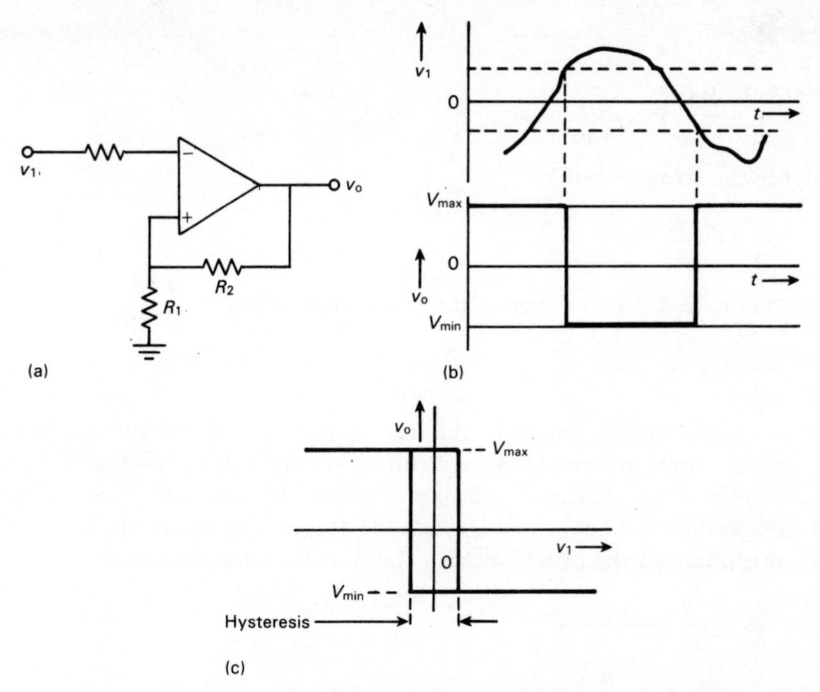

Fig. 14.26 (a) Schmitt trigger (b) Switching characteristic of Schmitt trigger (c) Hysteresis of a Schmitt trigger

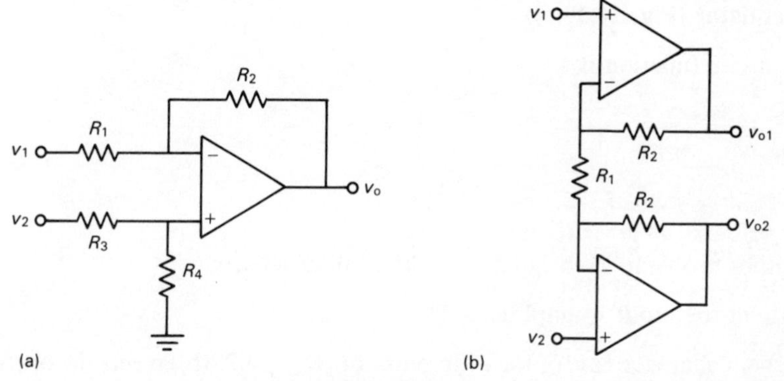

Fig. 14.27 (a) Simple differential amplifier (b) Improved differential amplifier

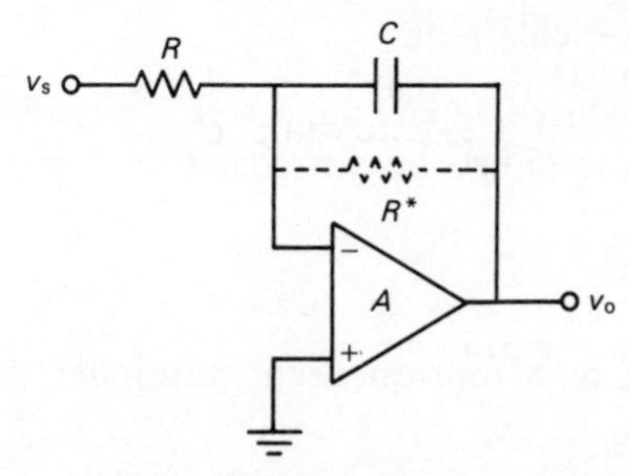

Fig. 14.28 Integrator

$$v_o = \frac{-1}{\tau} \int v_s \, dt$$

Hence the error term is

$$\frac{1}{A}\left(1 + \frac{R}{r_i} + s\tau\right)$$

Normally, $R \ll r_i$ and the error term is approximately

$$\frac{1}{A}(1 + s\tau)$$

A static error is caused by the input offset current and voltage. With grounded input, the resulting input offset voltage will be integrated. This problem is often reduced by placing a resistor R^* across C. This resistor decreases the amplifier gain at DC and low frequencies and reduces the rate of integration of the offset voltage. The transfer function is now

$$\frac{v_o}{v_s} = \frac{-1}{\tau} \frac{1}{s + \frac{1}{R^*C}}$$

The integration is thus less ideal than it should be.

Differentiator (Fig. 14.29(a))

The transfer function is

$$\frac{v_o}{v_s} = \frac{-s\tau}{1 + \frac{s\tau}{A}}$$

where $s\tau/A$ is the error term.

Other problems associated with differentiators are:

- Noise at the input is amplified.
- The gain characteristic of the differentiator (Fig. 14.29(b)) meets the open loop characteristic of the amplifier with increasing frequency. At this point the slope of the differentiator changes suddenly by 12 dB octave^{-1}. This may result in oscillations.

Both problems are reduced when connecting a resistor R^* in series with C (Fig. 14.29(c)) which modifies the differentiator characteristic as shown.

Multiplier

The use of an OTA as a four-quadrant multiplier is shown in the basic

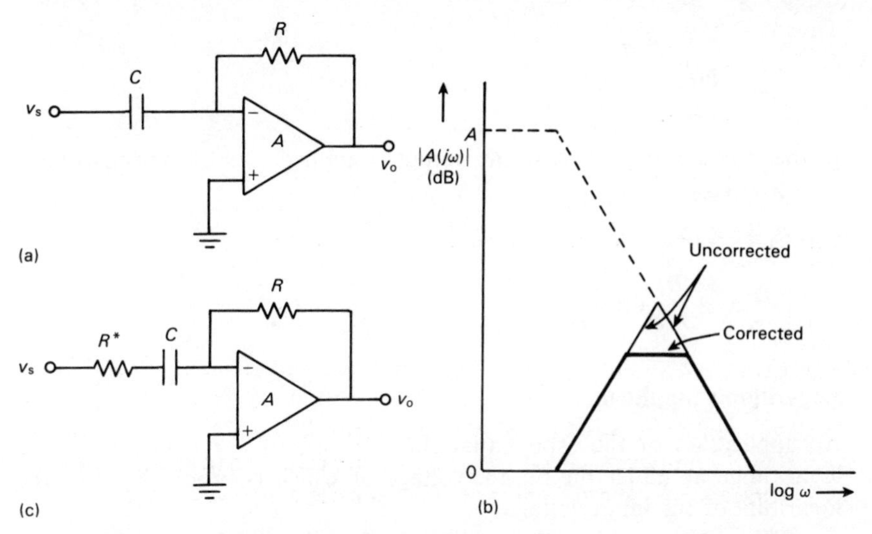

Fig. 14.29 (a) Differentiator (b) Response of differentiator (c) Corrected differentiator

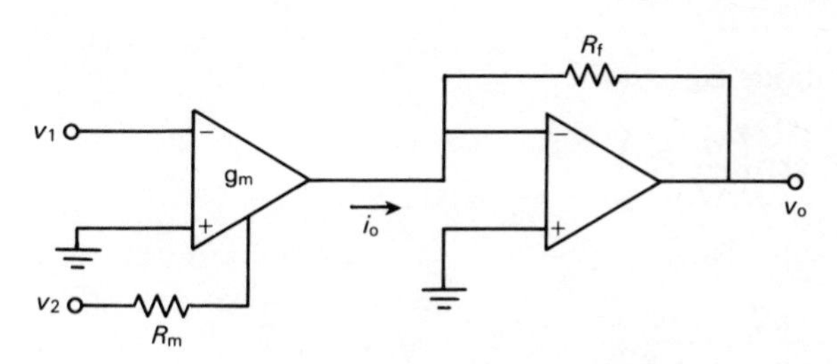

Fig. 14.30 Four-quadrant multiplier

circuit of Fig. 14.30. The bias current input is used in conjunction with the inverting input.

The output current is

$$i_o = g_m v_1$$

while

$$g_m = a \frac{v_2}{R_m}$$

where a is a constant.

Thus

$$i_o = a\,\frac{v_1 v_2}{R_m}$$

In the circuit shown the current output is applied to a current-to-voltage converter so that

$$v_o = -\,i_o R_f$$

$$= -\,a\,\frac{R_f}{R_m}\,v_1 v_2$$

Logarithmic amplifier

An application of the exponential characteristic of a *pn* junction is the logarithmic amplifier the output voltage of which is proportional to the logarithm of the input voltage.

Figure 14.31(a) shows the principle of operation. For positive values of v_s

$$\frac{v_s}{R} = i_{c1} \quad \text{and} \quad v_{be,1} = -\,v_{o1}$$

Using the diode equation

$$i_c \simeq I_s \exp\left(\frac{q}{kT}\,v_{be}\right)$$

we find

$$v_{o1} = \frac{-\,kT}{q}\ln\frac{v_s}{I_s R}$$

Transistor Tr_2 serves to compensate for leakage current. When assuming equal leakage currents in both transistors

$$v_{be,2} = \frac{kT}{q}\ln\frac{V^+}{I_s R^*}$$

while $v_{o2} = v_{o1} + v_{be,2}$ from which it follows that

$$v_{o2} = \frac{-\,kT}{q}\ln\frac{v_s R^*}{V^+ R}$$

Thus v_{o2} is proportional to the logarithm of v_s. A logarithmic amplifier characteristic, illustrating the large range of input signal voltage, is shown in Fig. 14.31(b).

Active filters

A small number of commonly used active filters is discussed in Chapter 15.

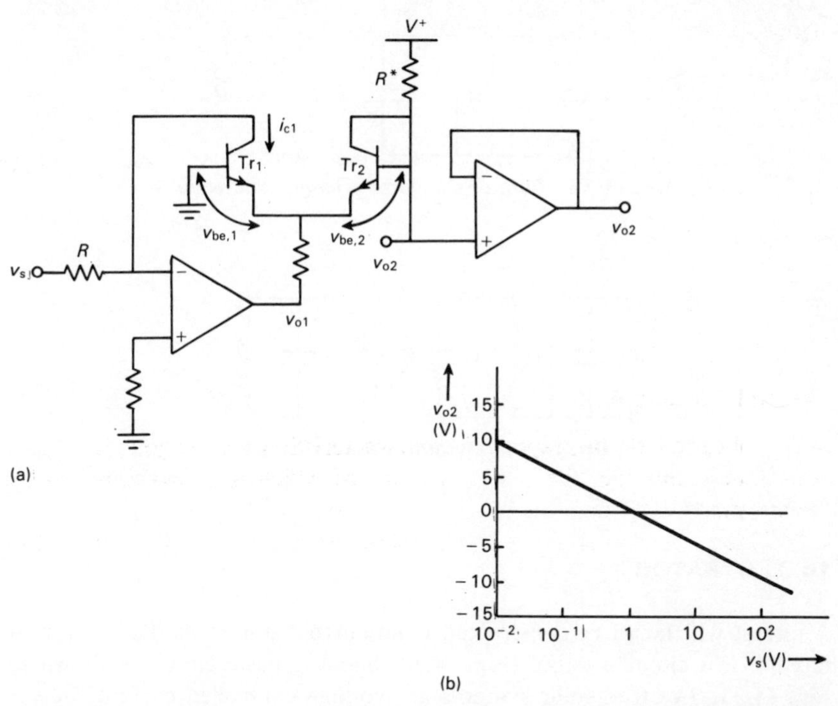

Fig. 14.31 (a) Logarithmic amplifier (b) Typical response of a logarithmic amplifier

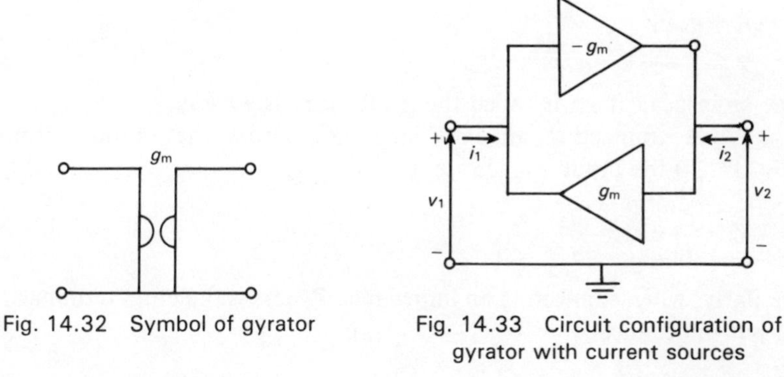

Fig. 14.32 Symbol of gyrator

Fig. 14.33 Circuit configuration of gyrator with current sources

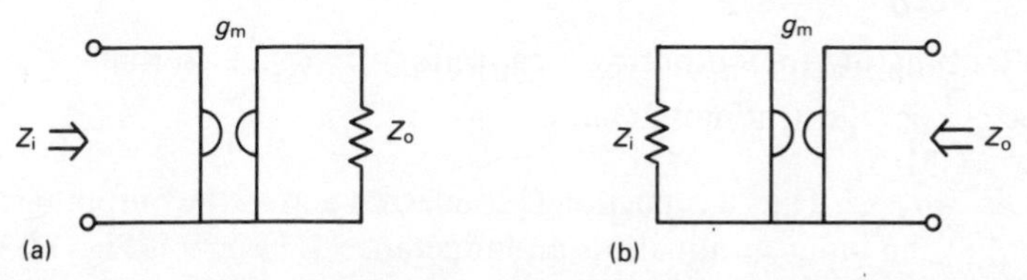

Fig. 14.34 (a) Impedance Z_o across output terminals of gyrator (b) Impedance Z_i across input terminals of gyrator

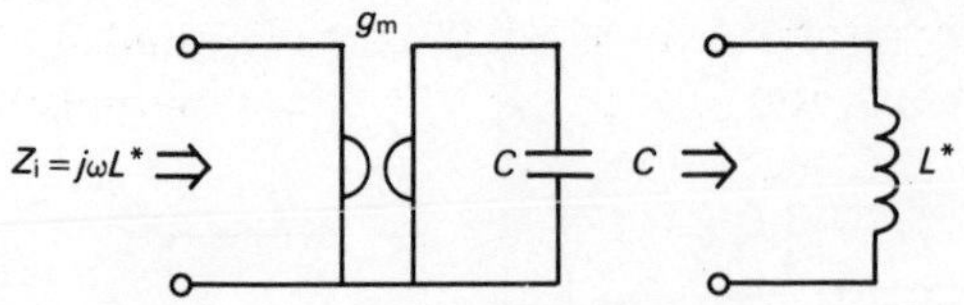

Fig. 14.35 Gyrator simulates a large inductance

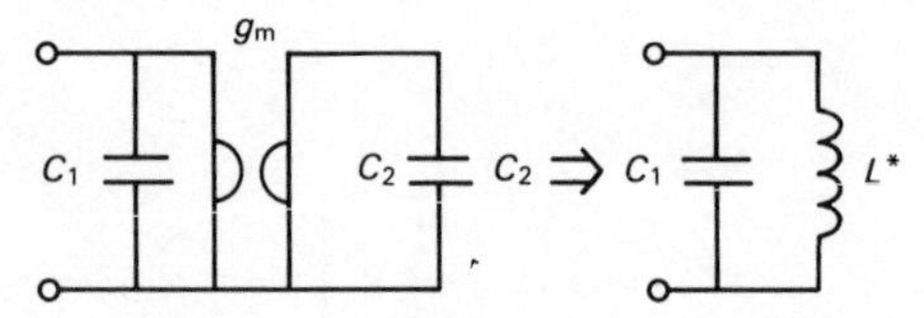

Fig. 14.36 Gyrator resonant circuit

14.3 GYRATOR

A circuit which can be constructed from current sources is the **gyrator**. It has its own circuit symbol (Fig. 14.32) and the basic circuit is shown in Fig. 14.33. The triangular symbols are voltage controlled current sources such as OTAs, one inverting, the other noninverting.

The circuit equations are

$$i_1 = g_m v_2$$
$$i_2 = -g_m v_1$$

The reciprocal of g_m is called the **gyration resistance** r_g.

When connecting an impedance Z_o across the output terminals (Fig. 14.34) the input impedance is

$$Z_i = \frac{v_1}{i_1} = r_g{}^2 \frac{-i_2}{v_2} = \frac{r_g{}^2}{Z_o}$$

Similarly, when connecting an impedance Z_i across the input terminals, the output impedance is

$$Z_o = \frac{r_g{}^2}{Z_i}$$

When terminating the output by a capacitor, $Z_o = \dfrac{1}{j\omega C}$ so that

$$Z_i = j\omega(r_g{}^2 C) = j\omega L^*$$

where $L^* = r_g{}^2 C$. Thus a capacitor C connected across the output terminals is 'seen' at the input terminals as an inductance $L^* = r_g{}^2 C$ (Fig. 14.35). In other words, a capacitor of small value can simulate a large inductance.

Integrated gyrators are capable of simulating inductance values up to 1 MH. When in addition to the capacitive output termination, the input is also terminated by a capacitor, the circuit behaves as a resonant circuit (Fig. 14.36). The resonance frequency is expressed by

$$\omega_o = \frac{1}{\sqrt{L^* C_1}} = \frac{1}{r_g \sqrt{C_1 C_2}}$$

Example

Let $r_g = 20$ kΩ, $C_1 = C_2 = 10$ μF, $\omega_o = 5$ rad^{-1} s so that the resonance frequency, $f_o \simeq 0.8$ Hz. Based on this property, very low frequency circuits can be implemented such as oscillators and low-pass filters.

It can be shown that a cascade of two gyrators simulates a DC transformer.

15
Active Filters

15.1 GENERAL

The active element in active filters is the operational amplifier, the passive elements are resistors and capacitors. Active filters find widespread use because of their small size and weight, particularly at low frequencies.

Filters can be designed from 10^{-3} Hz up to 1 MHz. At low frequencies voltage gains of 40 dB are possible. Q values may extend up to a few hundred. However, the characteristics of high-Q filters are sensitive to changes in element values.

The very large variety of filters and types prohibits a comprehensive treatment of the subject. The filters to be described are:

- Second order low-pass filter (Fig. 15.1(a)). Special types: Butterworth, Bessel, Chebyshev.
- Third order low-pass filter (Fig. 15.1(b)). Special types: Butterworth, Bessel, Chebyshev.
- Fourth order low-pass filter (Fig. 15.1(c)). Special types: Butterworth, Bessel.
- Second order high-pass filter (Fig. 15.2(a)). Special types: Butterworth, Bessel, Chebyshev.
- Third order high-pass filter (Fig. 15.2(b)). Special types: Butterworth, Bessel, Chebyshev.
- Fourth order high-pass filter (Fig. 15.2(c)). Special types: Butterworth, Bessel.

These filter designs are based on 'equal component values': resistor values are equal as well as capacitor values with the exception of the third order Chebyshev filters.

- Second order bandpass filter } special types: twin-T configurations
- Second order band-reject filter }
- All-pass filter ($0 \leqslant \phi < 180^\circ$)

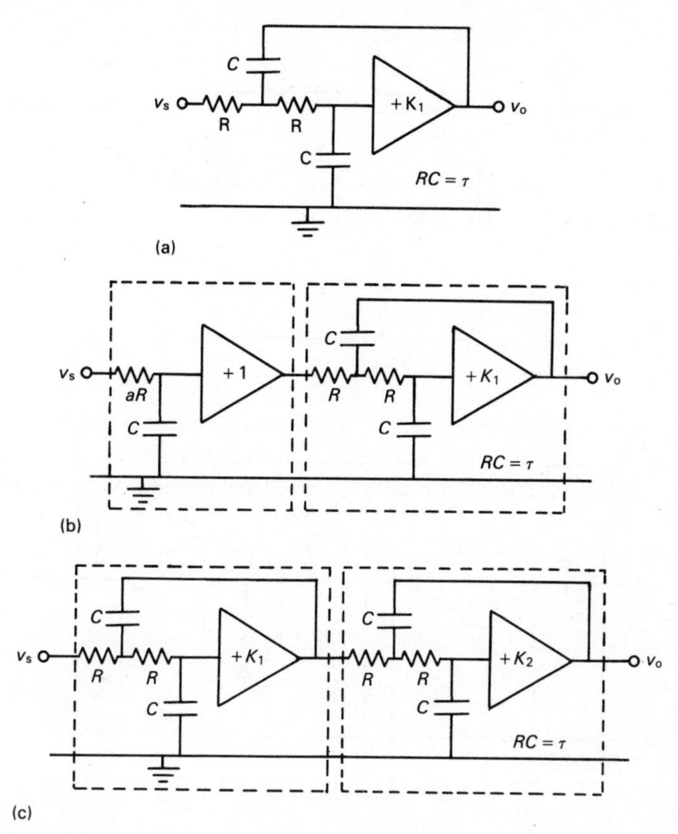

Fig. 15.1 (a) Second order low-pass filter (b) Third order low-pass filter (c) Fourth order low-pass filter

Important parameters of filter characteristics are:

- Cutoff frequency ω_o is the frequency determined by the points of intersection of the low-frequency and high-frequency asymptotes of the logarithmic amplitude characteristic (Fig. 15.3).
- -3 dB frequency ω_c is the frequency at which the amplitude of the output signal is $\frac{1}{2}\sqrt{2}$ times the amplitude of the signal in the passband (Fig. 15.3).
- Center frequency ω_m is the frequency at which the amplitude response of a bandpass (band-reject) filter is maximum (minimum).
- Group delay Δt is the time required for a signal to pass through the filter

For ease of description the following substitutions are made: $x = \omega\tau$, $x_o = \omega_o\tau$, $x_c = \omega_c\tau$, $x_m = \omega_m\tau$, where $\tau = RC$. The transfer function of a filter is $H(j\omega)$.

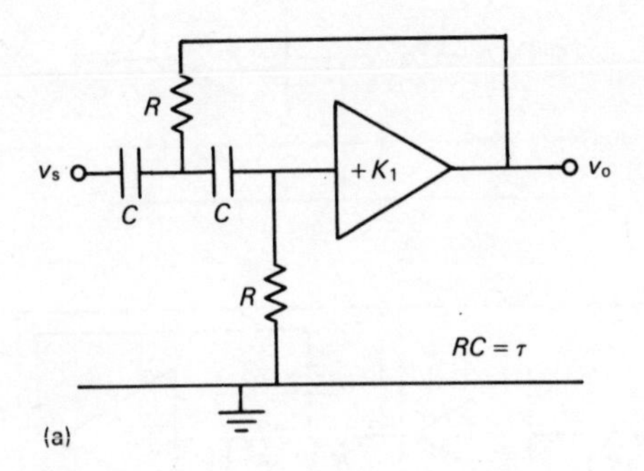

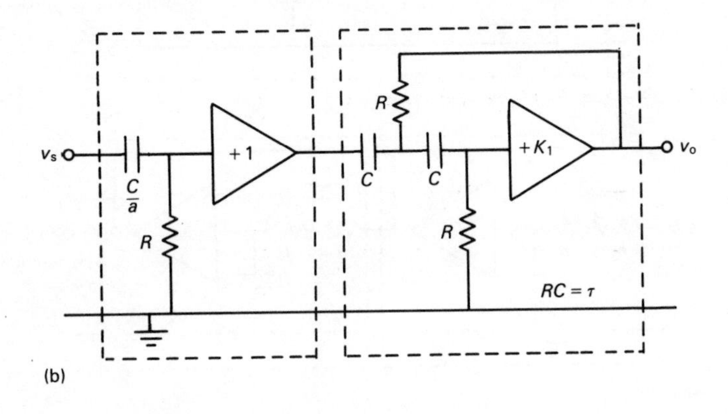

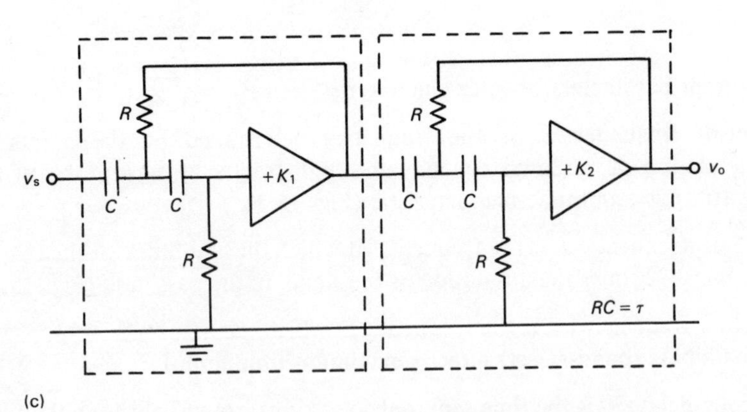

Fig. 15.2 (a) Second order high-pass filter (b) Third order high-pass filter (c) Fourth order high-pass filter

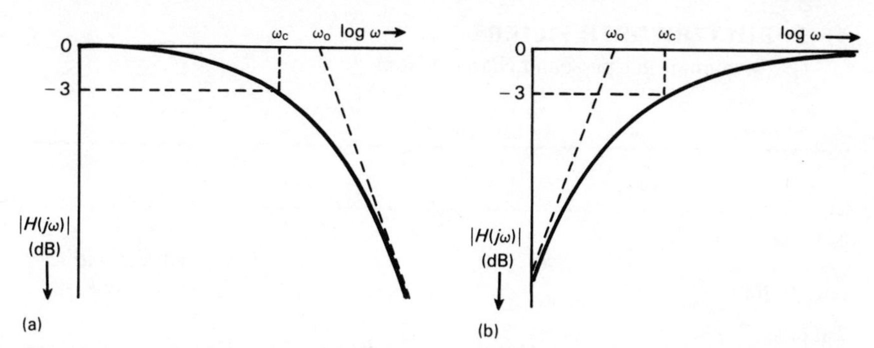

Fig. 15.3 (a) Cutoff and 3-dB frequencies of a low-pass filter (b) Cutoff and 3-dB frequencies of a high-pass filter

15.2 LOW-PASS AND HIGH-PASS FILTERS

Table 12

	$H(j\omega)$
2nd order low-pass filter	$\dfrac{K_1}{1 - x^2 + jx(3 - K_1)}$
2nd order high-pass filter	$\dfrac{K_1 x^2}{x^2 - 1 - jx(3 - K_1)}$
3rd order low-pass filter	$\dfrac{K_1}{1 - x^2[1 + a(3 - K_1)] + jx[a + 3 - K_1 - ax^2]}$
3rd order high-pass filter	$\dfrac{K_1 x^3}{x^3 - x[1 + a(3 - K_1)] + j[a - (a + 3 - K_1)x^2]}$
4th order low-pass filter	$\dfrac{K_1 K_2}{1 + x^4 - x^2[K_1 K_2 + 11 - 3(K_1 + K_2)] + jx(1 - x^2)(6 - K_1 - K_2)}$
4th order high-pass filter	$\dfrac{K_1 K_2 x^4}{1 + x^4 - x^2[K_1 K_2 + 11 - 3(K_1 + K_2)] + jx(1 - x^2)(6 - K_1 - K_2)}$

15.3 BUTTERWORTH FILTERS

(Maximum flat passband characteristic)

Table 13

	a	K_1	K_2	x_o	x_c	
2nd order low-pass filter (Fig. 15.1(a))	—	1.586	—	1	1	amplitude characteristic: Fig. 15.4, curve A
2nd order high-pass filter (Fig. 15.2(a))	—	1.586	—	1	1	
3rd order low-pass filter (Fig. 15.1(b))	1	2	—	1	1	
3rd order high-pass filter (Fig. 15.2(b))	1	2	—	1	1	
4th order low-pass filter (Fig. 15.1(c))	—	2.235	1.152	1	1	
4th order high-pass filter (Fig. 15.2(c))	—	2.235	1.152	1	1	

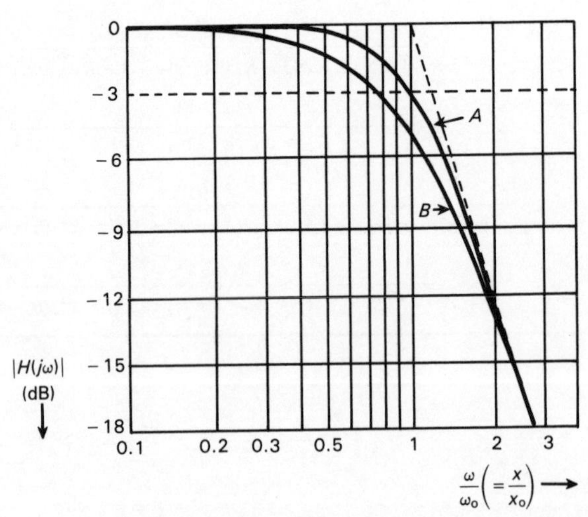

Fig. 15.4 Response of second-order Butterworth (A) and Bessel filter (B)

15.4 BESSEL FILTERS

(Constant group delay = linear phase characteristic in the passband)

Table 14

	a	K_1	K_2	x_o	x_c	$\frac{\Delta t}{\tau}$ at $x=0$	$\frac{\Delta t}{\tau}$ at $x=x_c$	*Ratio*	
2nd order low-pass filter (Fig. 15.1(a))	—	1.268	—	1	0.786	1.73	1.40	0.809	amplitude characteristic Fig. 15.4 curve B
2nd order high-pass filter (Fig. 15.2(a))	—	1.268	—	1	1.272				
3rd order low-pass filter (Fig. 15.1(b))	1.25	1.89	—	0.93	0.837	2.36	2.31	0.977	
3rd order high-pass filter (Fig. 15.2(b))	1.25	1.89	—	1.08	1.19				
4th order low-pass filter (Fig. 15.1(c))	—	1.663	1.042	1	0.624	3.30	3.20	0.970	
4th order high-pass filter (Fig. 15.2(c))	—	1.663	1.042	1	1.603				

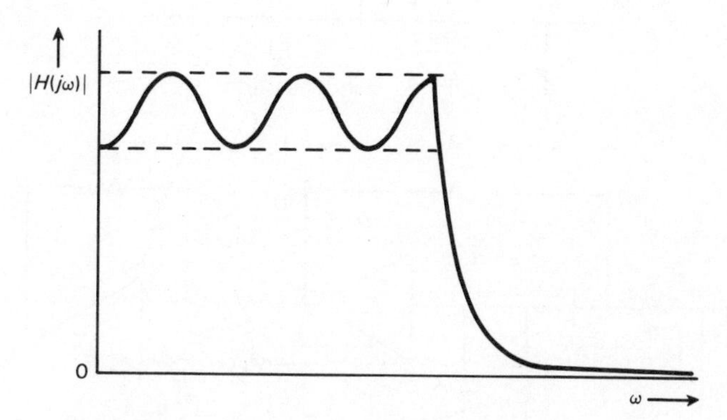

Fig. 15.5 Response of third-order Chebyshev filter

15.5 CHEBYSHEV FILTERS

(Sharp transition characteristic, equal ripple in the passband (Fig. 15.5).)

Table 15

	a	K_1	K_2	x_o	x_c	*Ripple* (dB)
2nd order	—	1.96	—	1	1.25	1
low-pass filter	—	2.12	—	1	1.33	2
(Fig. 15.1(a))	—	2.23	—	1	1.39	3
2nd order	—	1.96	—	1	0.80	1
high-pass filter	—	2.12	—	1	0.75	2
(Fig. 15.2(a))	—	2.23	—	1	0.72	3
3rd order	1.27	2.13	—	0.92	1.28	1
low-pass filter	2.19	2.39	—	0.77	1.16	2
(Fig. 15.1(b))	3.47	2.95	—	0.66	1.05	3
3rd order	1.27	2.13	—	1.08	0.78	1
high-pass filter	2.19	2.39	—	1.30	0.87	2
(Fig. 15.2(b))	3.47	2.95	—	1.51	0.95	3

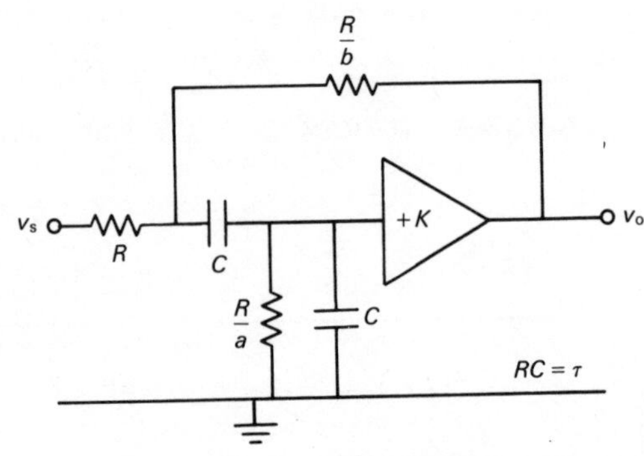

Fig. 15.6 Bandpass filter

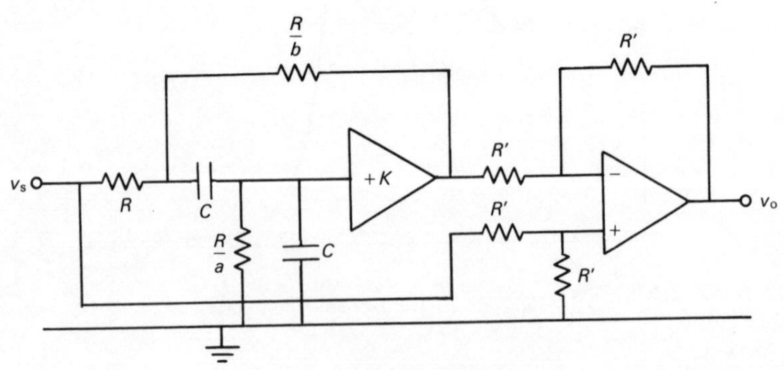

Fig. 15.7 Band-reject filter

15.6 BAND FILTERS

Table 16

	$H(j\omega)$	x_{m}	$\lvert H(j\omega_{\mathrm{m}})\rvert$	Q	B
Bandpass filter (Fig. 15.6)	$\dfrac{Kjx}{a(b+1) - x^2 + jx[a+2+b(2-K)]}$	$\sqrt{a(b+1)}$	$\dfrac{K}{a+2+b(2-K)}$	$\dfrac{\sqrt{a(b+1)}}{a+2+b(2-K)}$ $\left[K < \dfrac{1}{a}(a+2b+2)\right]$	$\dfrac{a+2+b(2-K)}{2\pi\tau}$
Band-reject filter (Fig. 15.7)	$\dfrac{a(b+1) - jx^2 + jx[a+(b+1)(2-K)]}{a(b+1) - x^2 + jx[a+2+b(2-K)]}$	$\sqrt{a(b+1)}$	$\dfrac{a+(b+1)(2-K)}{a+2+b(2-K)}$	$\dfrac{\sqrt{a(b+1)}}{a+2+b(2-K)}$ $\left[K < \dfrac{1}{a}(a+2b+2)\right]$	$\dfrac{a+2+b(2-K)}{2\pi\tau}$
Twin-T bandpass filter (Fig. 15.8)	$\dfrac{4jx(1-K)}{1-x^2+4jx(1-K)}$	1	1	$\dfrac{1}{4(1-K)}$ $(K<1)$	$\dfrac{2(1-K)}{\pi r}$ $(K<1)$
Twin-T bandpass filter (Fig. 15.9)	$\dfrac{-jx}{1-x^2+\dfrac{1}{\alpha}(1+jx)}$	$1+\dfrac{1}{\alpha}$	α	$\sqrt{\alpha(\alpha+1)}$	$\dfrac{1}{2\alpha\pi\tau}$
Twin-T band-reject filter (Fig. 15.10)	$\dfrac{1-x^2}{1-x^2+4jx(1-K)}$	1	0	$\dfrac{1}{4(1-K)}$ $(K<1)$	$\dfrac{2(1-K)}{\pi\tau}$ $(K<1)$

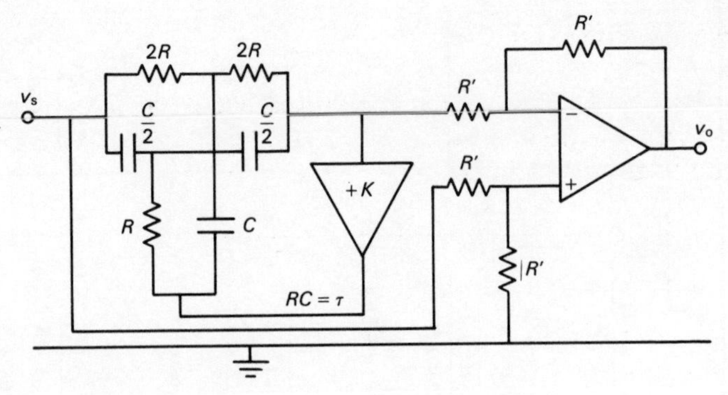

Fig. 15.8 Twin-T bandpass filter

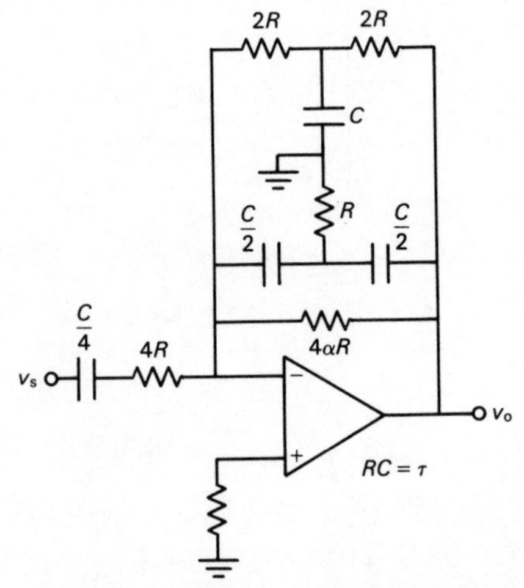

Fig. 15.9 Alternate twin-T bandpass filter

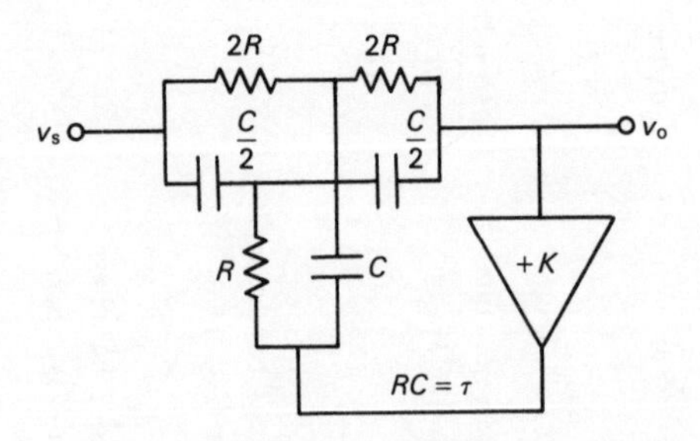

Fig. 15.10 Twin-T band-reject filter

15.7 ALL-PASS FILTER

All-pass filters have a flat amplitude response but a frequency dependent phase shift. An example is shown in Fig. 15.11. Phase shift and corresponding time delay are shown in the table below.

$H(j\omega)$	$\tan\phi$	$\frac{\Delta t}{\tau}$
$\frac{1-jx}{1+jx}$	$\frac{-2x}{1-x^2}$	$\frac{2x^2}{1+x^4}$

Example 1

Design of a fourth order Bessel filter with $f_c = 10$ kHz.

The circuit is that of Fig. 15.1(c) and Table 14 shows $K_1 = 1.663$, $K_2 = 1.042$, $x_c = 0.624$ so that $\omega_c\tau = 2\pi 10^4$ and $\tau \simeq 10\ \mu$s.

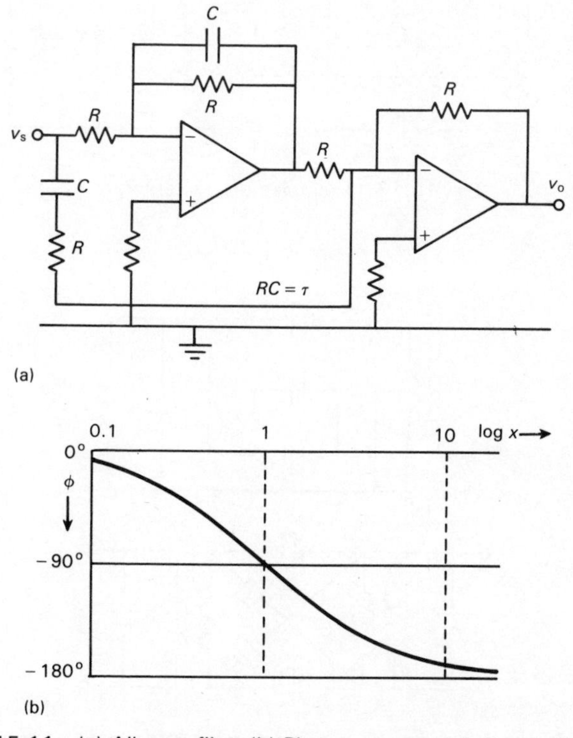

Fig. 15.11 (a) All-pass filter (b) Phase response of all-pass filter

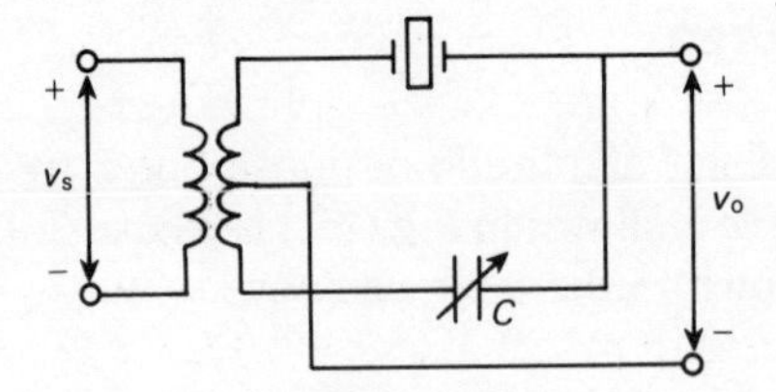

Fig. 15.12 Crystal filter with capacitive tuning

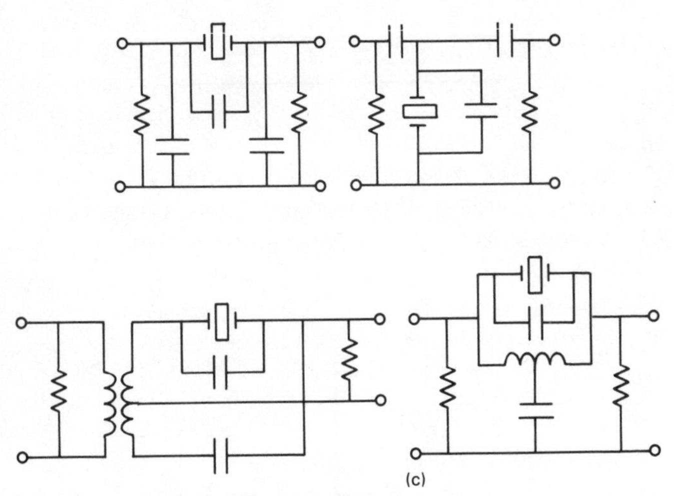

Fig. 15.13 Examples of crystal filters

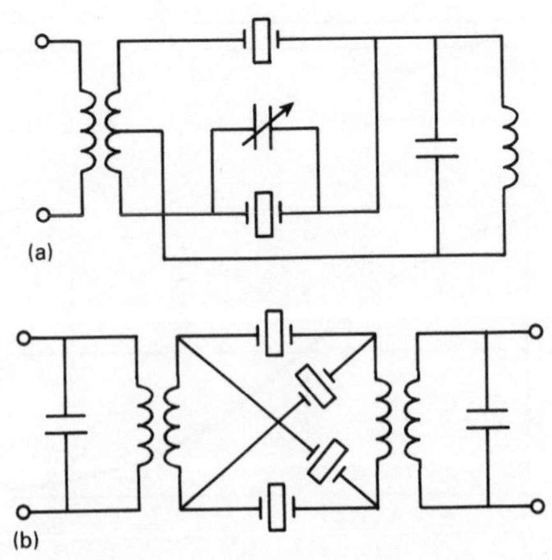

Fig. 15.14 (a) Half-lattice crystal filter (b) Full-lattice crystal filter

Example 2

Design of a bandpass filter with $Q = 10$, $f_m = 5$ kHz, $|H(j\omega_m)| = 20$. Table 16 (Fig. 15.6) shows that $|H(j\omega_m)| = KQ/x_m$. Choosing $\tau = 10^{-4}$ s, we find $x_m = \pi = 3.14$ so that $K = 2\pi = 6.28$. Also, $a + 2 + b(2 - K) = x_m/Q$ or $a + 2 - 4.28b = 0.314$ and $a(b+1) = x_m{}^2 = 9.87$. From these two equations it follows that $a \simeq 7.22$ and $b \simeq 1.37$.

15.8 CRYSTAL FILTERS

The reactance characteristic of a crystal (see Chapter 7) can be changed by placing an inductance in series or in parallel with the crystal.

In the series connection the parallel resonant frequency remains unchanged while f_s becomes lower. In the parallel connection f_s remains unaltered while f_p goes higher. This frequency spreading property allows the use of a crystal as a bandpass filter, for example in intermediate frequency amplifiers and single-sideband filters.

In addition, capacitive tuning can be used to construct bandpass filters. In Fig. 15.12 capacitor C can be adjusted so that only a narrow band around the crystal frequency is passed to the output.

Figure 15.13 shows a few more configurations of crystal bandpass filters. Wider bandwidths can be obtained when using more crystals as shown in Fig. 15.14. Figure 15.14(a) is a half-lattice filter with two slightly different crystals. A full-lattice filter is shown in Fig. 15.14(b) where the crystals in the horizontal legs are identical. The crystals in the diagonals are also identical but different from the other two.

16
Power Amplifiers

16.1 GENERAL

Normally, an amplifying system consists of a number of amplifying stages in cascade. These stages amplify the small input signals to a large enough level. The final stage is the power amplifier which is designed to deliver power to a generally low impedance load (loudspeaker, servo motor, etc.). With these large signals and high power levels emphasis should be placed on matters like efficiency, distortion (Chapter 20) and dissipation.

Power amplifiers can be classified in several ways, e.g. voltage, current or power amplifiers and according to their frequency range: DC amplifiers, audio amplifiers (20 Hz–20 kHz), video amplifiers (up to a few MHz), radio frequency amplifiers (up to hundreds of MHz), ultrahigh-frequency amplifiers (up to some GHz).

We will restrict ourselves to low-frequency power amplifiers and classify them according to their mode of operation:

- Class A amplifiers: this amplifier is biased so that bias current flows at all times.
- Class B amplifiers: biasing is such that the amplifier is just cut off so that the operating point is at an extreme end of its load line.
- Class AB amplifiers: biasing is just above cutoff.

16.2 CLASS A AMPLIFIERS

Fig. 16.1(a) shows an output stage with resistive load R_L. To obtain maximum output signal, biasing is at the midpoint of the load line. The average current is constant, whether or not an AC signal is present at the output (Fig. 16.1(b)). Thus the AC power in R_L is

$$P_L = \tfrac{1}{4} V^+ I_b$$

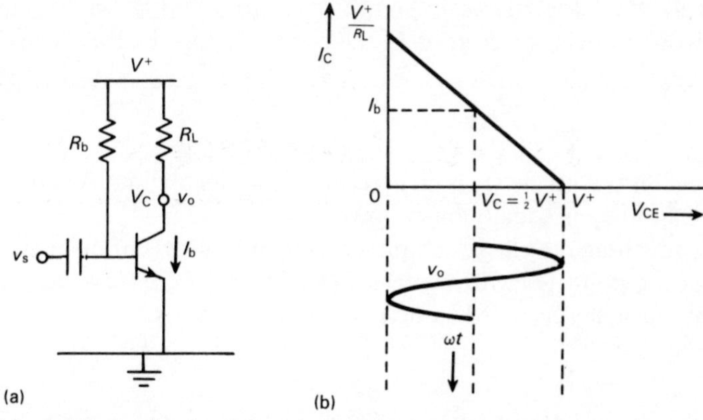

Fig. 16.1 (a) Basic circuit of class A amplifier (b) Load line of class A amplifier with resistive load

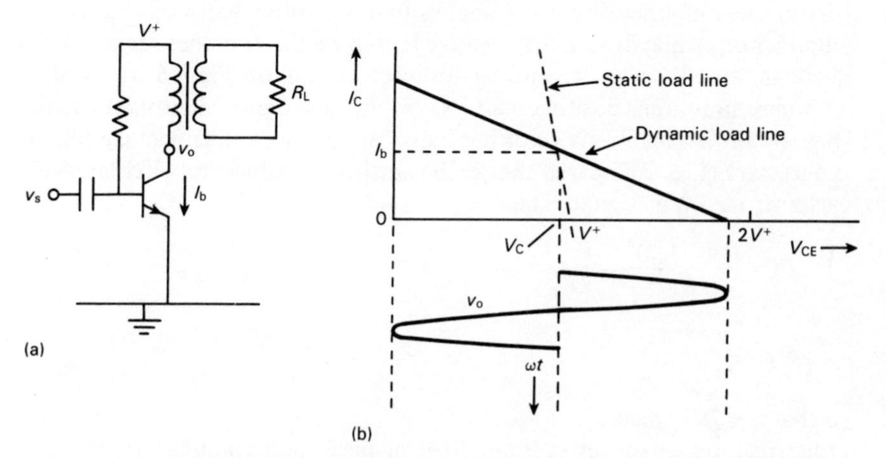

Fig. 16.2 (a) Class A amplifier with inductive load (b) Load line of class A amplifier with inductive load

The power supply delivers a DC power

$$P_S = V^+ I_b$$

The **efficiency** is defined as

$$\eta = \frac{P_L}{P_S} = \frac{\text{AC power in load at maximum swing}}{\text{DC power delivered by power supply}} \times 100\%$$

Clearly, in this circuit $\eta = 25$ percent.

Improved power transfer is obtained when using an inductive load. In the circuit of Fig. 16.2(a) a transformer is used to couple the output signal to the load. Here the operating point is close to V^+ (Fig. 16.2(b)) and the output swing is twice as large as in the circuit of Fig. 16.1(a). Thus

$$P_L = V^+ \tfrac{1}{2}\sqrt{2}\, I_b \tfrac{1}{2}\sqrt{2} = \tfrac{1}{2} V^+ I_b$$

$$P_S = V^+ I_b$$

Hence the limit of efficiency is 50 percent. If the signal input is variable, the average efficiency is usually much lower.

The advantage of class A amplifiers is their low level of distortion since the operating point is in the center of the linear range. However, the low values of efficiency have led to designs of other classes.

16.3 CLASS B AMPLIFIERS

Biasing the amplifier at cut-off implies that a positive half-cycle turns the amplifier on, while during the negative half-cycle the amplifier remains off. Such an amplifier is the emitter-follower circuit of Fig. 16.3(a) which amplifies only the positive half waves of an input sinusoidal signal (Fig. 16.3(b)). To amplify the other halves as well, a complementary circuit is necessary (Fig. 16.3(c)) so that each transistor conducts on alternate half cycles of the input signal. Thus

$$P_L = \frac{(\tfrac{1}{2} V^+ \tfrac{1}{2}\sqrt{2})^2}{R_L} = \frac{V^{+2}}{8R_L}$$

$$P_S = \frac{V^{+2}}{2\pi R_L}$$

so that $\eta = 78$ percent.
Other features of this class B amplifier or **push–pull amplifier** are:

- Quiescent power drain is negligible.
- Two active devices share the power dissipation.
- There is considerable **crossover distortion** at small output signals.

The crossover distortion is illustrated in Fig. 16.3(d) and is due to the nonlinear characteristic of a forward biased *pn* junction at low current levels. Analysis shows that the even harmonic distortion is largely canceled out. No crossover distortion would occur if the bases were driven by sinusoidal current sources. However, since β is a function of output current level, another type of distortion would appear instead.

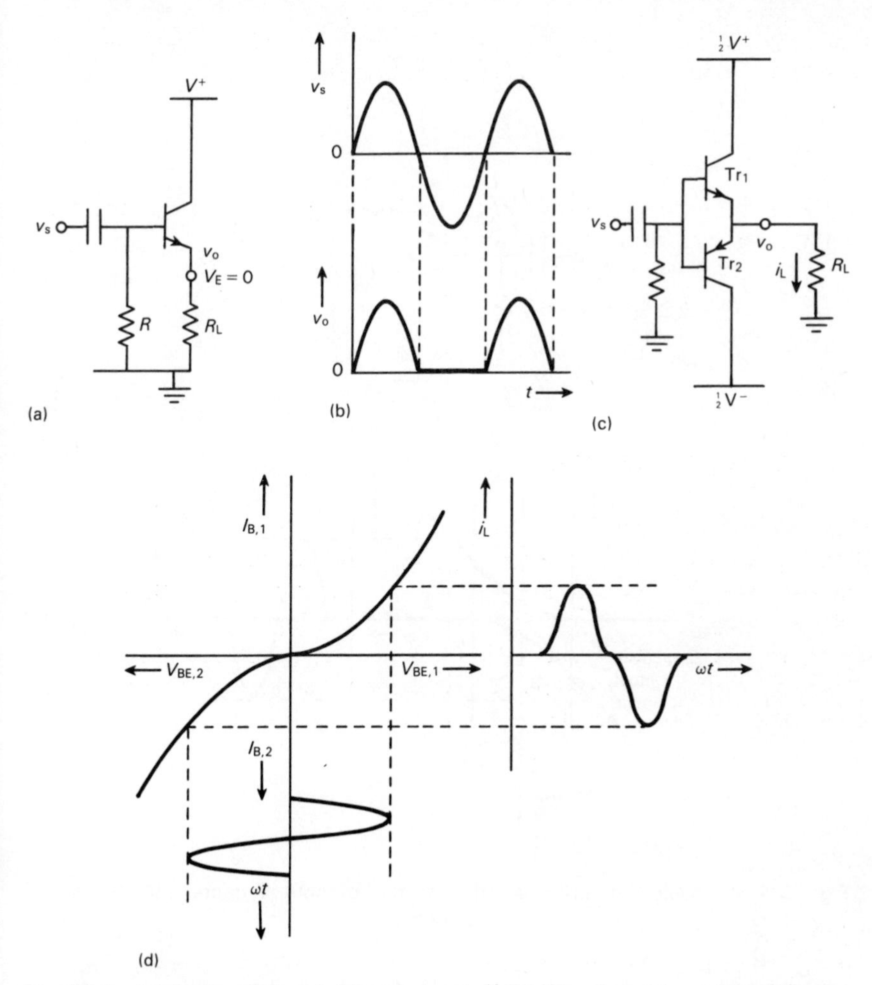

Fig. 16.3 (a) Emitter follower biased at cutoff (b) Waveforms in emitter follower circuit (c) Basic push–pull amplifier circuit (d) Input characteristic of push–pull circuit

16.4 CLASS AB AMPLIFIERS

A great deal of the crossover distortion of a class B push–pull amplifier can be eliminated by biasing both transistors just above the *pn* threshold voltage (Fig. 16.4(a)). The resulting composite transfer characteristic is shown in Fig. 16.4(b). This circuit is called the **complementary push–pull stage.** As it is difficult to obtain matched complementary transistors, some even harmonic distortion will be left uncanceled.

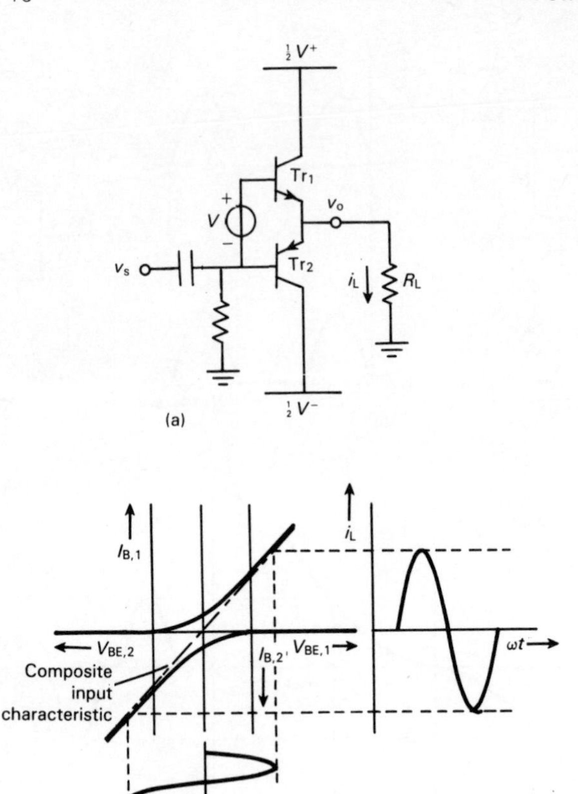

Fig. 16.4 (a) Basic circuit of class AB amplifier (b) Input characteristic of class AB amplifier

A solution to this problem is the **quasi-complementary stage** or **single-ended push-pull** circuit (SEPP, Fig. 16.5(a)).

The two transistors are of the same type and will usually match more closely. The circuit arrangement for supplying the small idling current is mostly either a diode–potentiometer arrangement (Fig. 16.5(b)) or a transistor with an adjustable value of V_{CE} (Fig. 16.5(c)).

The efficiency of the class AB amplifier is less than 78 percent due to the required bias current. It can be shown that maximum dissipation in the transistors occurs when the output swing is 63 percent of its maximum value.

Often small resistors ($< 1\ \Omega$) are placed in the output circuit as shown in Fig. 16.5(c). This eases bias adjustment and provides a small local feedback. Single-supply operation is possible when amplifier output and load are connected by a coupling capacitor C_L.

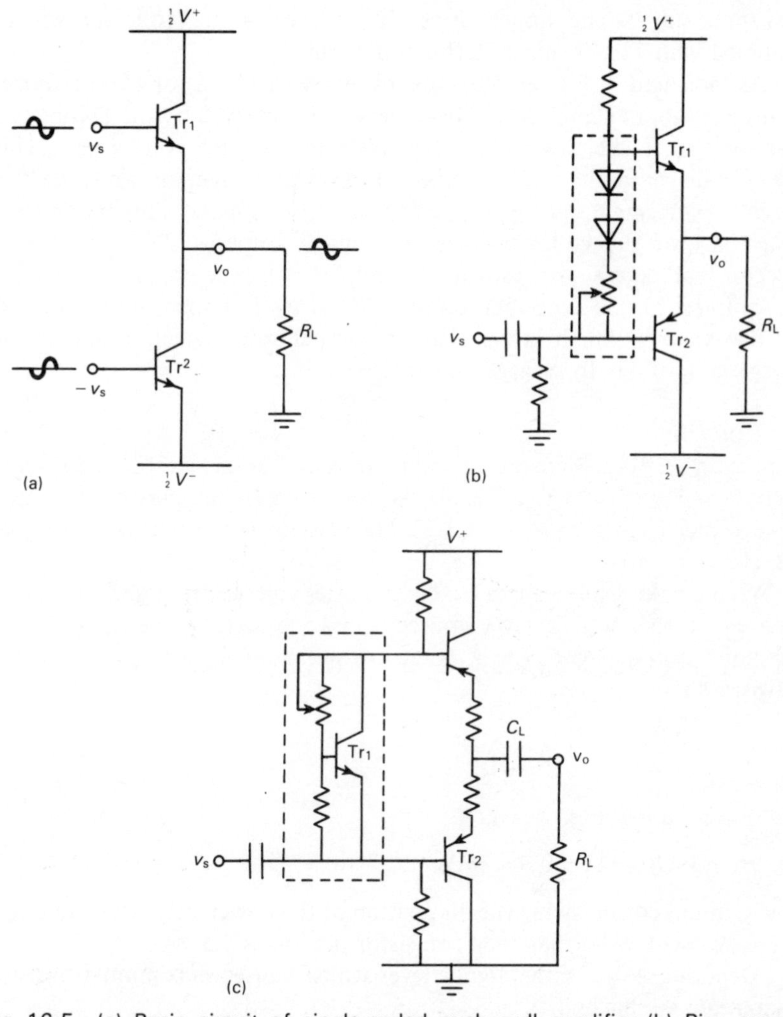

Fig. 16.5 (a) Basic circuit of single-ended push–pull amplifier (b) Bias current arrangement of SEPP amplifier (c) Alternate bias current arrangement of SEPP amplifier

16.5 DRIVER CIRCUITS

In case the output transistors are bipolars, the base currents for these transistors, in view of the large output currents, are not negligible and can reach values of tens of mA.

These base currents are supplied by one or more medium-power transistors, called **drivers**. Figure 16.6 shows the drivers for a single-ended push–pull stage. The upper driver Tr_3 is an emitter follower which, combined with Tr_1, forms a Darlington circuit.

As indicated in Fig. 16.5(a), the lower power transistor (Tr_2) requires the inverted input signal $-v_s$. This inversion is achieved by the Darlington variation circuit where two inversions take place with unit voltage gain. This way both driver inputs can be connected to the same signal v_s while the bias current arrangement can be applied to the driver inputs. This bias circuit forms part of the collector load of voltage amplifier Tr_5. The actual collector load consists of resistors R_1 and R_2 where the upper end of R_2 is AC coupled to the amplifier output. Thus R_2 is bootstrapped and its effective value is much larger than the actual value which enhances the voltage gain of Tr_5 by a large amount.

Example

Assume an SEPP power stage is required to deliver 25 W into an 8 Ω load. Then $P_L = V_{rms}^2/R_L$ where V_{rms} is the rms value of the maximum output signal. Thus $V_{rms} = \sqrt{P_L R_L} = 14.1$ V. The peak-to-peak value of this signal is $2\sqrt{21} \times 4.1 = 40$ V.

When making allowances for base-emitter voltage drops and saturation voltages of the drivers the required supply voltage is set at 45 V. At maximum output signal the average current flowing through the power transistors is

$$I_{av} = \frac{\sqrt{2}}{\pi}\frac{V_{rms}}{R_L} \simeq 0.8 \text{ A}$$

The power supply thus delivers

$$P_S = V^+ I_{av} = 45 \times 0.8 = 36 \text{ W}$$

At maximum output swing the dissipation of the power transistors amounts to $36 - 25 = 11$ W so that each transistor dissipates 5.5 W.

Denoting V_{rms}^* as the signal level which causes maximum transistor dissipation, we have

$$P_{trans} = P_S - P_L = \frac{\sqrt{2}}{\pi}\frac{V_{rms}^*}{R_L} V^+ - \frac{V_{rms}^{*\,2}}{R_L}$$

the maximum value of which is

$$P_{trans,\,max} = \frac{V^{+2}}{2\pi^2 R_L} = 12.8 \text{ W}$$

which occurs at

$$V_{rms}^* = \frac{\sqrt{2}}{2\pi} V^+ = 10.1 \text{ V}$$

Maximum dissipation in each transistor is thus 6.4 W.

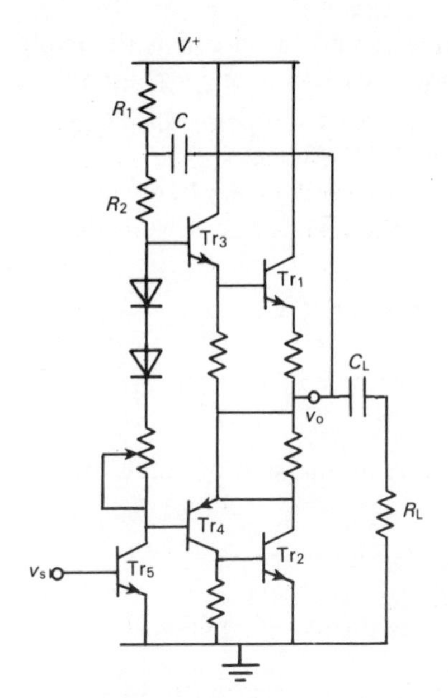

Fig. 16.6 Example of output stage and drivers of SEPP amplifier

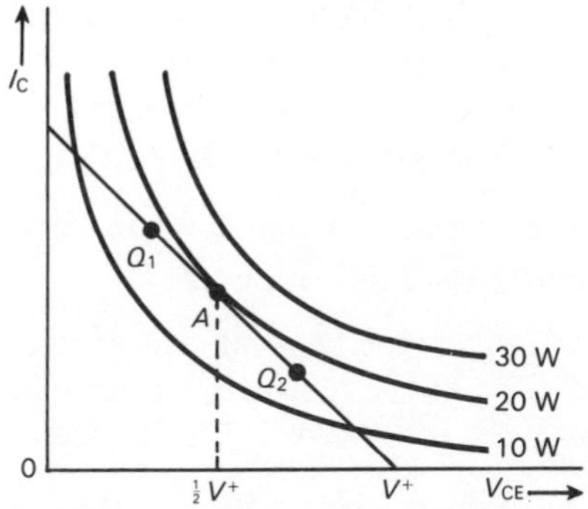

Fig. 16.7 Constant power hyperbolas and load line of power amplifier

The efficiency value is

$$\eta = \frac{25}{36} \times 100\% \simeq 69\%$$

16.6 THERMAL STABILITY

With bipolar transistors in the output stage, the temperature of a base-collector junction may rise either by ambient temperature rise or by the transistors own power dissipation. Junction heating increases the collector current. If this increasing current lowers the collector–emitter voltage sufficiently, power dissipation decreases and the circuit is thermally stable. If V_{CE} does not decrease or increases, power dissipation increases and **thermal runaway** will occur.

Figure 16.7 shows the load line of a power transistor in an $I_C - V_{CE}$ diagram in addition to three constant power hyperbolas of which one is tangent to the load line at point A.

If Q_1 is the operating point and I_C increases, Q_1 moves upward to the lower value hyperbola. The circuit is then thermally stable. If, however, Q_2 is the operating point, an increase in I_C makes Q_2 move upward to the higher value hyperbola. Dissipation increases and the circuit is thermally unstable. Hence the operation point should be less than $\frac{1}{2}V^+$ for stable operation.

'Ohms law' expressed in thermal quantities is

$$\Delta T = T_j - T_a = \theta P$$

where

T_j is the collector–base junction temperature (°C),
T_a is the ambient temperature (°C),
θ is the thermal resistance between junction and ambient (°C W^{-1}),
P is the power dissipation at the junction (W).

The thermal resistance θ can be considered as the sum of three thermal resistances, shown in Fig. 16.8:

$\theta_{j,c}$ is the thermal resistance between junction and case,
$\theta_{c,s}$ is the thermal resistance between case and heat sink,
$\theta_{s,a}$ is the thermal resistance between heat sink and ambient.

Of these, $\theta_{j,c}$ is specified by the manufacturer.

A low value of $\theta_{c,s}$ requires good thermal contact between case and heat sink while a low value of $\theta_{s,a}$ depends upon shape, size and position of the heat sink.

When appreciable power is dissipated, heat sinks of the shape of Fig. 16.9(a) are often employed. Curves of ΔT against P are shown in Fig. 16.9(b) for horizontal and vertical position of the heat sink.

Example

Maximum allowable power dissipation is expressed by the thermal law

$$P_{max} = \frac{T_{j,\,max} - T_{a,\,max}}{\theta}$$

where $T_{j,\,max}$ is specified by the manufacturer.

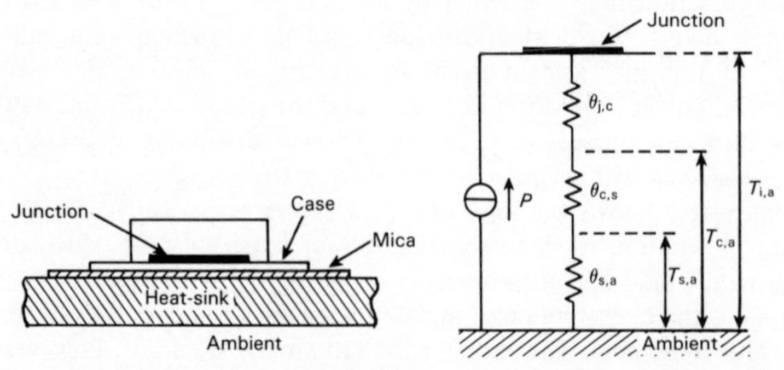

Fig. 16.8 Power transistor mounted on heat sink and equivalent circuit

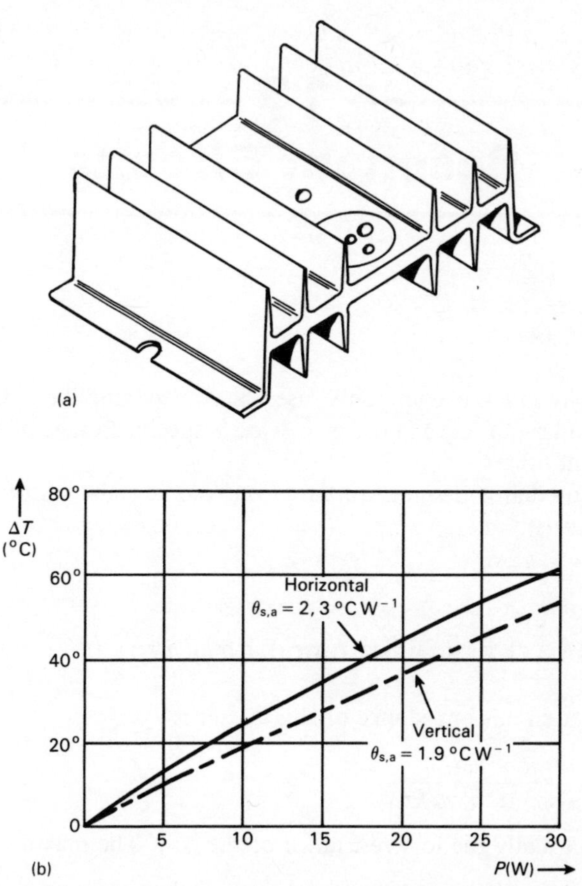

Fig. 16.9 (a) Example of heat sink construction (b) Temperature characteristics of heat sink of (a)

If, for example, $T_{j,\,max} = 200\,^\circ\mathrm{C}$, $T_{a,\,max} = 40\,^\circ\mathrm{C}$ and $P_{max} = 80$ W, a total thermal resistance of less than $2\,^\circ\mathrm{C}\;W^{-1}$ is required.

The use of power FETs instead of bipolars has gained increased popularity for two main reasons:

- There is no risk of thermal runaway. A temperature rise decreases the mobility of the majority carriers. Hence, drain current I_D decreases (unlike collector current behavior). Therefore, FETs are thermally stable devices.
- Since gate currents are very small ($\leqslant 100$ nA), driving power FETs is much easier than driving bipolars.

17 Resonant Circuits

17.1 GENERAL

Resonant circuits are commonly used in tuned amplifiers (e.g. radio, television and radar, etc.) in order to select a specified range of frequencies and reject all others.

Many oscillator designs are based on resonant circuits of which two basic types exist.

17.2 PARALLEL RESONANT CIRCUIT (Fig. 17.1)

The complex circuit impedance of this circuit is

$$Z_p = \frac{R + j\omega L}{1 - \omega^2 LC + j\omega RC}$$

where R is usually the loss resistance of the coil. The maximum value of $|Z_p|$ is

$$Z_{p,\,max} = \frac{L}{RC}$$

This maximum value occurs at the frequency

$$\omega_o = 2\pi f_o = \frac{1}{\sqrt{LC}}\sqrt{1 - \frac{R}{Z_{p,\,max}}} = \sqrt{\frac{1}{LC} - \frac{R^2}{L^2}}$$

The frequency f_o is the **resonance frequency**.

An important characteristic of a resonant circuit is its **quality factor** Q which is directly related to the loss resistance R and is defined as

$$Q = \sqrt{\frac{Z_{p,\,max}}{R}} = \frac{1}{R}\sqrt{\frac{L}{C}} \simeq \frac{\omega_o L}{R} \simeq \frac{1}{\omega_o RC}$$

Consequently ω_o can be expressed as

$$\omega_o = \sqrt{\frac{1}{LC}\left(1 - \frac{1}{Q^2}\right)}$$

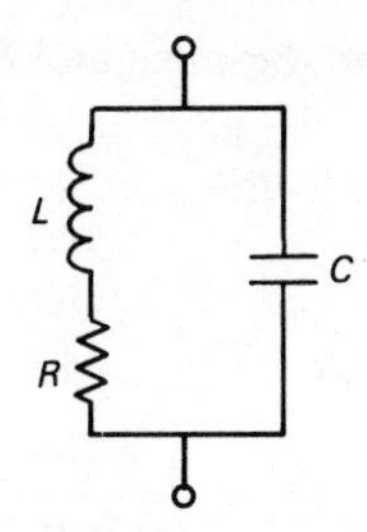

Fig. 17.1 Parallel resonant circuit

From the foregoing relations the bandwidth of $|Z_p|$ is found as

$$B = \frac{f_o}{Q} \quad \text{(at } -3 \text{ dB points)}$$

$$B^* \simeq 10\,\frac{f_o}{Q} \quad \text{(at } -20 \text{ dB points)}$$

A convenient parameter in calculations is the **relative detuning** γ, defined as

$$\gamma = \frac{\omega}{\omega_o} - \frac{\omega_o}{\omega} = \frac{f}{f_o} - \frac{f_o}{f}$$

Then

$$Z_p = \frac{Z_{p,\,\max}\left(1 - j\dfrac{1}{Q}\dfrac{\omega_o}{\omega}\right)}{1 + j\gamma Q}$$

It is common practice to express the behavior of the circuit around the resonance frequency, where $\omega \simeq \omega_o$ and (normally) $Q \gg 1$, in terms of the parameters γ and Q. Then

$$\frac{|Z_p|}{Z_{p,\,\max}} \simeq \frac{1}{\sqrt{1 + (\gamma Q)^2}}$$

Graphically, this is the **resonance curve, selectivity curve** or **relative response curve** (Fig. 17.2).
The phase angle between voltage and current is

$$\tan \phi = -\gamma Q$$

Example 1
A parallel resonant circuit with $Q = 100$ is resonant at 500 kHz. Determine

the bandwidth and the relative response at 505 kHz.

$$B = \frac{f_o}{Q} = \frac{500\text{ kHz}}{100} = 5\text{ kHz}$$

$$\gamma = \frac{505}{500} - \frac{500}{505} = 0.0199$$

$$\frac{|Z_p|}{Z_{p,\max}} = \frac{1}{\sqrt{1+(\gamma Q)^2}} = \frac{1}{\sqrt{1+1.99^2}} \simeq 0.45$$

The series resistance R in the circuit of Fig. 17.1 may be replaced by the equivalent parallel resistance R_p of Fig. 17.3 where R_p has the value

$$R_p = \frac{L}{RC} \text{ and } Q = R_p\sqrt{\frac{C}{L}}$$

In practical situations the resonant circuit is connected to a signal source via a resistor R_s (Fig. 17.4(a)). As this circuit is equivalent to the circuit of Fig. 17.4(b), the resistance R_s appears in parallel with R_p so that the resultant Q is

$$Q^* = \frac{R_pR_s}{R_p+R_s}\sqrt{\frac{C}{L}} = \frac{1}{\dfrac{\omega_oL}{R_s} + \dfrac{R}{\omega_oL}} \quad \text{or} \quad \frac{1}{Q^*} = \frac{1}{Q} + Q\frac{R}{R_s}$$

Example 2

The resonant circuit of Example 1 has a loss resistance $R = 5\ \Omega$ and is connected to a voltage source via a series resistance $R_s = 100\text{ k}\Omega$. Then

$$\frac{1}{Q^*} = \frac{1}{100} + 100\frac{5}{10^5} = \frac{3}{200}$$

so that $Q^* \simeq 67$.

17.3 SERIES RESONANT CIRCUIT (Fig. 17.5)

The circuit impedance here is

$$Z_s = R + j\left(\omega L - \frac{1}{\omega C}\right)$$

The minimum value of $|Z_s|$ is

$$Z_{s,\min} = R$$

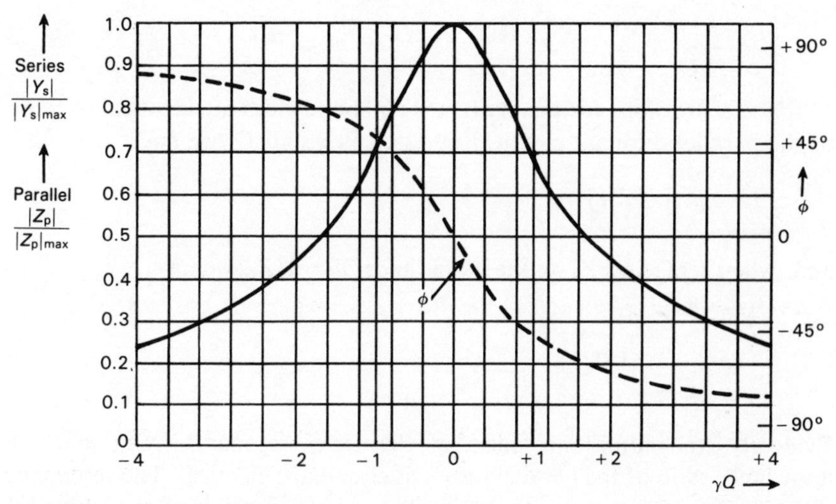

Fig. 17.2 Universal amplitude and phase response curves of resonant circuits

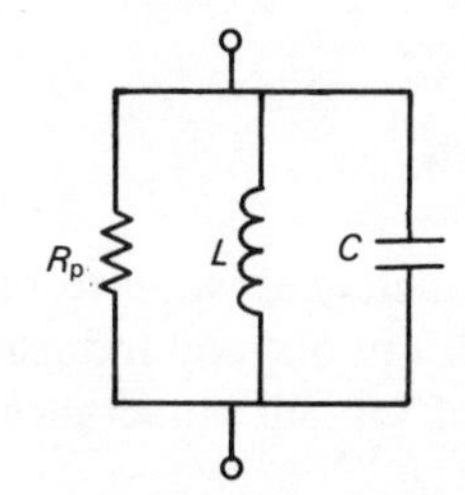

Fig. 17.3 Transformation of a series resistance to a parallel resistance

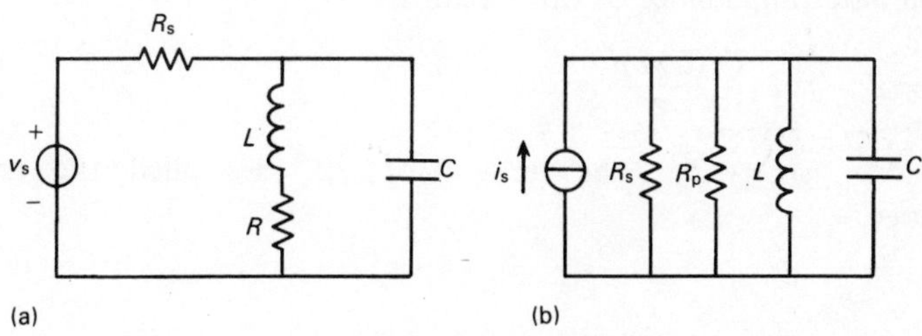

Fig. 17.4 (a) Parallel resonant circuit connected to source (b) Equivalent circuit of (a)

which occurs at the resonance frequency

$$\omega_o = \frac{1}{\sqrt{LC}}$$

The expressions for quality factor and bandwidth are identical to those of the parallel resonant circuit. Expressed in γ and Q, we find

$$\frac{|Z_s|}{Z_{s,\,min}} = \sqrt{1 + (\gamma Q)^2}$$

and, when replacing Z_s by the admittance $Y_s = 1/Z_s$, we have

$$\frac{|Y_s|}{Y_{s,\,max}} = \frac{1}{\sqrt{1 + (\gamma Q)^2}}$$

$$\tan \phi = \gamma Q$$

Evidently, the admittance ratio of the series resonant circuit and the impedance ratio of the parallel resonant circuit are identical. The resonance curve of Fig. 17.2 therefore is applicable to both circuits and is a **universal resonance curve.**

17.4 COUPLED CIRCUITS

The concept of mutual inductance was introduced in Chapter 3. In the circuit of Fig. 17.6 a current i in one coil induces a voltage $M(\mathrm{d}i/\mathrm{d}t)$ in the other coil. The coefficient of mutual inductance M is defined as

$$M = k\sqrt{L_1 L_2}$$

where k is the **coupling factor**. Its value depends on the mutual position of L_1 and L_2 ($0 \leqslant k \leqslant 1$).

The input impedance of this circuit is

$$\frac{v_1}{i_1} = Z = j\omega L_1 + \frac{(\omega M)^2}{Z^*}$$

where $Z^* = j\omega L_2 + Z_2$. The term $(\omega M)^2/Z^*$ is called the **reflected impedance.**

17.5 MISCELLANEOUS METHODS OF COUPLING

In addition to mutual inductance coupling, a large number of other coupling methods exist. The choice of a particular circuit depends upon

Fig. 17.5 Series resonant circuit

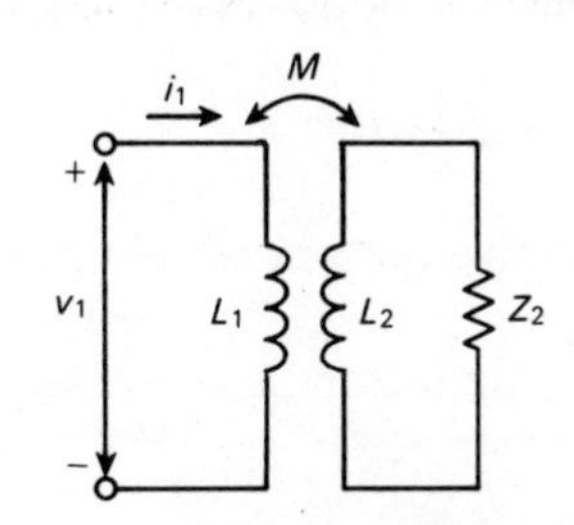

Fig. 17.6 A circuit coupled by mutual inductance

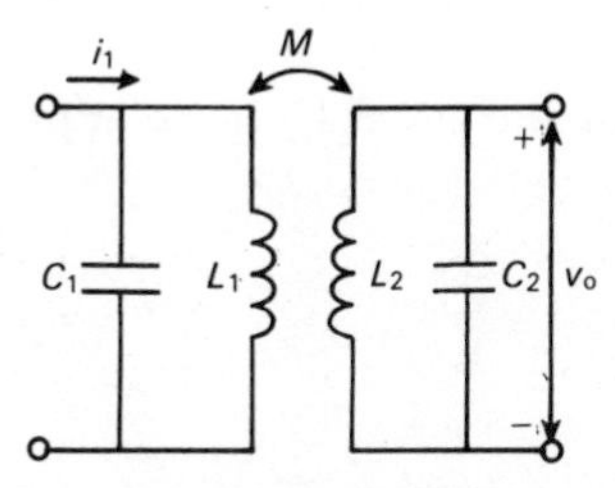

Fig. 17.7 Two resonant circuits coupled by mutual inductance

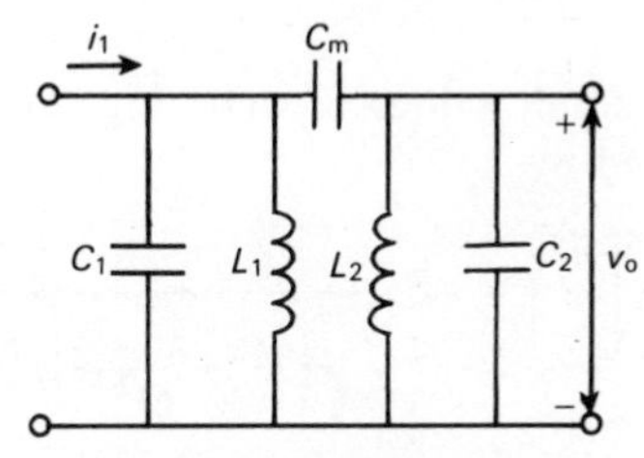

Fig. 17.8 Top capacitive coupling of two resonant circuits

requirements of selectivity, gain, bandwidth constancy with varying resonance frequency, etc.

- Mutual inductance coupling of resonant circuits (Fig. 17.7)

  coupling factor: $k = \dfrac{M}{\sqrt{L_1 L_2}}$

- Top capacitive coupling (Fig. 17.8)

  coupling factor: $k = \dfrac{C_m}{\sqrt{(C_1 + C_m)(C_2 + C_m)}}$

  This coupling is often used with antennas, RF coils and IF transformers.

- Bottom capacitive coupling (Fig. 17.9)

  coupling factor: $k = \dfrac{\sqrt{(C_1 + C_m)(C_2 + C_m)}}{C_m}$

If L_1 and L_2 are mutually coupled, a reasonably constant bandwidth over a tuning range of 2–3 is obtained.

- Top inductive coupling (Fig. 17.10)

  coupling factor: $k = \sqrt{\dfrac{L_1 L_2}{(L_1 + L_m)(L_2 + L_m)}}$

- Capacitive bottom coupling (Fig. 17.11)

$$\text{coupling factor: } k = \sqrt{\frac{C_1 C_2}{(C_1 + C_m)(C_2 + C_m)}}$$

- Inductive bottom coupling (Fig. 17.12)

$$\text{coupling factor: } k = \frac{L_m}{\sqrt{(L_1 + L_m)(L_2 + L_m)}}$$

17.6 RESPONSE OF COUPLED CIRCUITS

For small values of detuning the response of the circuits can be approximated by

$$\left|\frac{v_o}{i_1}\right| = |Z| \frac{k}{\sqrt{\left(k^2 + \frac{1}{Q^2} - \gamma^2\right)^2 Q^2 + 4\gamma^2}}$$

where

$|Z| = \sqrt{|Z_1 Z_2|}$ is the geometric mean of the impedances of the resonant circuits

$Q = \sqrt{Q_1 Q_2}$ is the geometric mean of the quality factors of the resonant circuits

The value of k determines the shape of the response curve:

- Overcritical coupling: $kQ > 1$ (Fig. 17.13)

$$\left|\frac{v_o}{i_1}\right|_{max} = \tfrac{1}{2}|Z| \quad \text{when } \gamma = \gamma_o = \pm\sqrt{k^2 - \frac{1}{Q^2}}$$

bandwidth between peaks:

$$B^* = f_o \sqrt{k^2 - \frac{1}{Q^2}} = f_o \gamma_o$$

bandwidth (−3 dB points):

$$B = f_o \gamma_o \sqrt{2} = B^* \sqrt{2}$$

peak-to-valley ratio:

$$\delta = \frac{1 + k^2 Q^2}{2kQ}$$

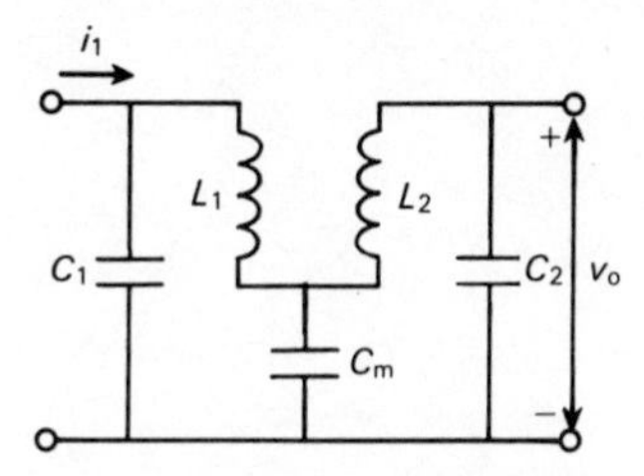

Fig. 17.9 Bottom capacitive coupling of two resonant circuits

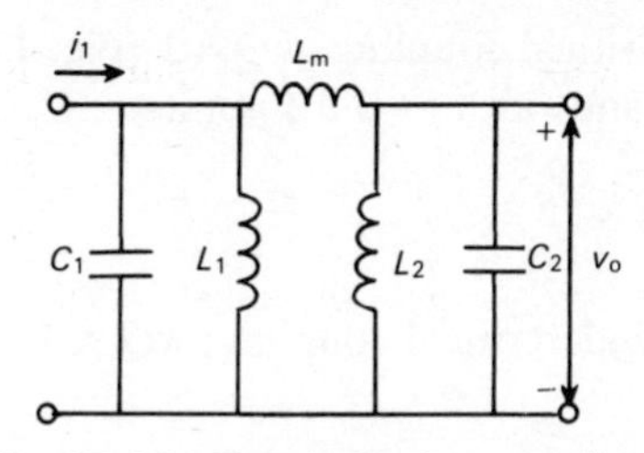

Fig. 17.10 Top inductive coupling of two resonant circuits

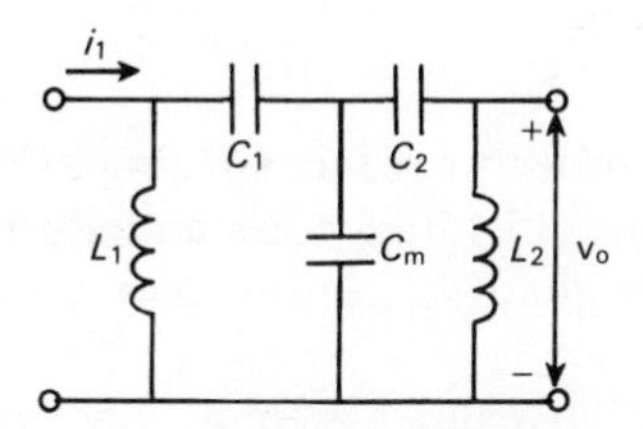

Fig. 17.11 Capacitive bottom coupling of two resonant circuits

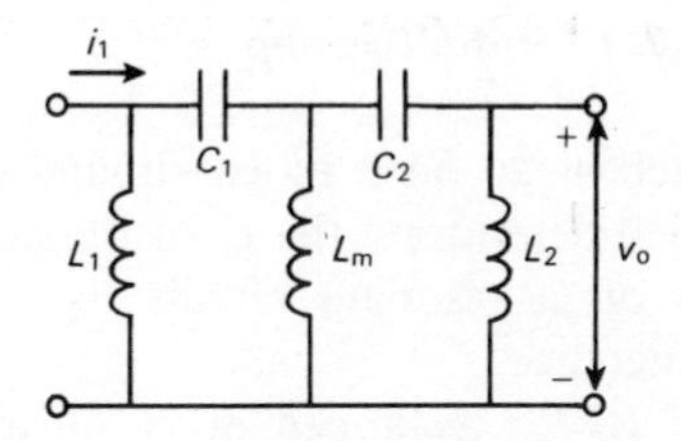

Fig. 17.12 Inductive bottom coupling of two resonant circuits

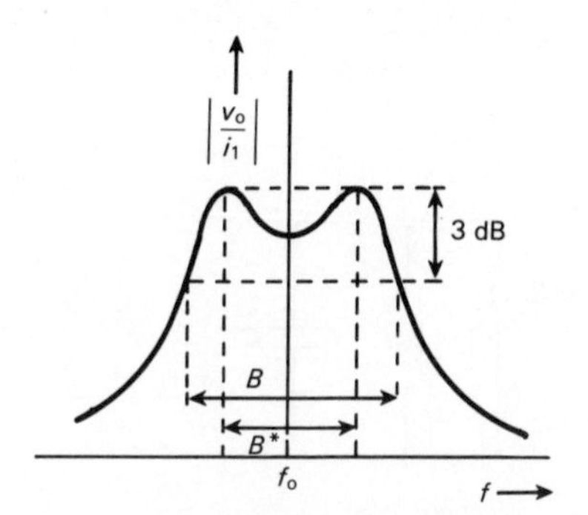

Fig. 17.13 Overcritical coupling of two resonant circuits

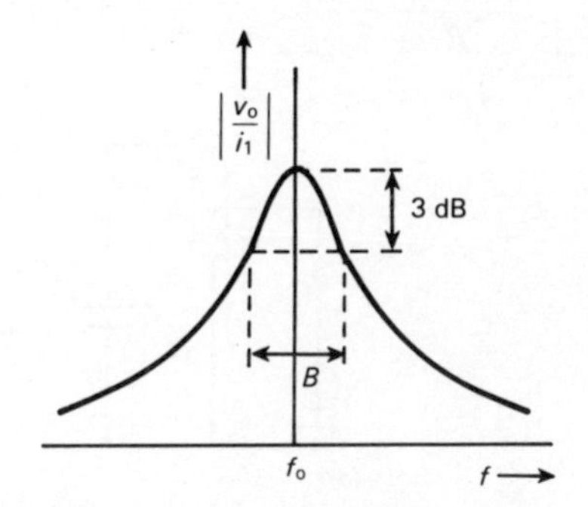

Fig. 17.14 Critical coupling of two resonant circuits

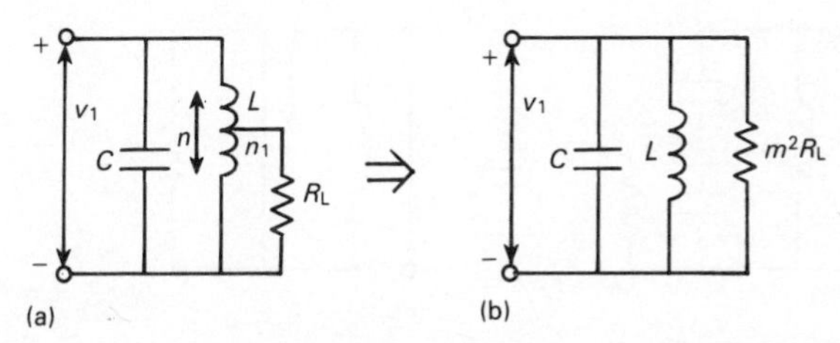

Fig. 17.15 Load reduction by inductive tap and equivalent circuit

- Critical coupling: $kQ = 1$ (Fig. 17.14)
  bandwidth (−3 dB points):

$$B = \frac{f_o}{Q}\sqrt{2}$$

- Undercritical coupling: $kQ < 1$.

17.7 REDUCTION OF LOADING EFFECTS

17.7.1 Inductive tap

Often a tap on a tuned circuit is used in order to reduce loading effects which deteriorate the Q of the circuit. Figure 17.15(a) shows an inductive tap on a resonant circuit to connect the load R_L across part of the inductance.

If R_L is tapped at n_1 turns, the effective load seen by the input terminals is $m^2 R_L$ where m is the ratio n/n_1. The equivalent circuit is shown in Fig. 17.15(b). The effective impedance of the circuit is

$$\frac{m^2 R_L Z_{p,\,max}}{m^2 R_L + Z_{p,\,max}}$$

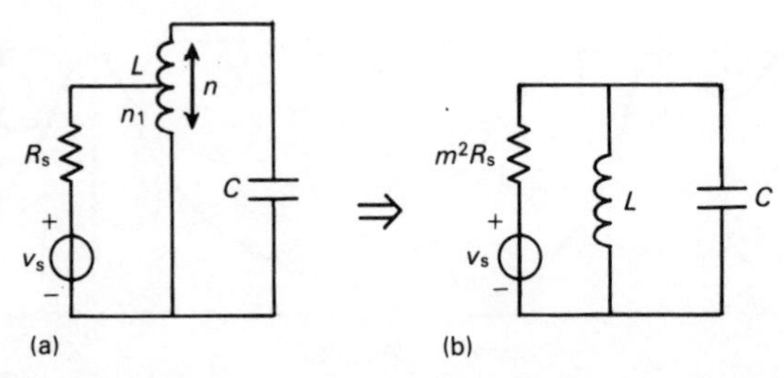

Fig. 17.16 Source reduction by inductive tap and equivalent circuit

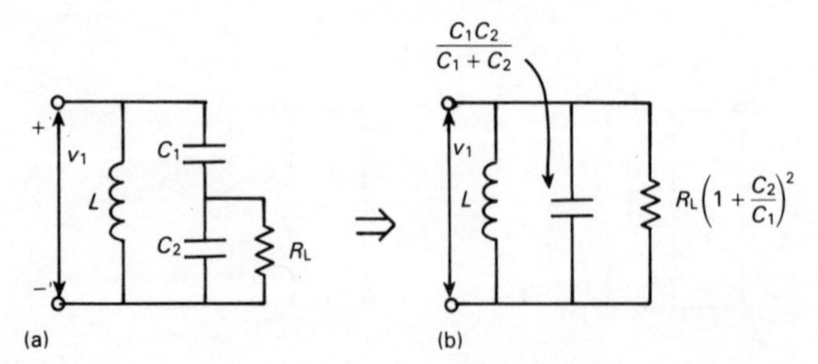

Fig. 17.17 Load reduction by capacitive tap and equivalent circuit

where $Z_{p,\,max}$ is the unloaded impedance of the circuit at resonance. The Q of the circuit is then reduced by the factor

$$\frac{m^2 R_L}{m^2 R_L + Z_{p,\,max}}$$

Hence the loaded Q is

$$Q^* = Q \frac{1}{1 + \dfrac{Z_{p,\,max}}{m^2 R_L}}$$

The source can be tapped to the circuit as shown in Fig. 17.16(a). It can be shown that the effective source resistance is stepped up by the factor m^2. The equivalent circuit is shown in Fig. 17.16(b).

17.7.2 Capacitive tap

Arrangements similar to those in the previous paragraph can be made by tapping the capacitance of the tuned circuit (Fig. 17.17(a)). If the tap is to be effective, $\omega C_2 R_L \gg 1$; if so the equivalent circuit of Fig. 17.17(b) is valid. The tuning capacitance remains the same while the effective load is

$$R_L' = R_L \left(1 + \frac{C_2}{C_1}\right)^2$$

Obviously, tapping reduces the output voltage across R_L.

18
Sine Wave Oscillators

18.1 GENERAL

If in a linear feedback amplifier the output signal is identical to the input signal, the amplifier will oscillate when input and output are connected. The identity condition of the input and output signals implies that the magnitude of the loop gain $|A_r| = 1$ while the phase shift is 0° or an integral multiple of 2π.

The necessary condition for oscillations to occur can therefore be expressed by the complex loop gain

$$A_r(j\omega) = 1$$

This is the **Barkhausen criterion** which, as outlined above, can be written as two conditions which must be satisfied simultaneously:

$$\mathrm{Re}[A_r(j\omega)] = 1$$
$$\mathrm{Im}[A_r(j\omega)] = 0$$

The condition $\mathrm{Re}[A_r(j\omega)] = 1$ is the starting condition which says that sufficient output signal must be fed back to the input. In practical situations this precise equality can never be satisfied due to small changes in component values in the oscillator. If, because of such drift, $\mathrm{Re}[A_r(j\omega)]$ becomes slightly less than unity, oscillations will stop. Therefore, to be sure that oscillations will always occur, it is necessary to arrange that $\mathrm{Re}[A_r(j\omega)] > 1$. Then the amplitude of the oscillations will grow until they are limited by nonlinearities in the circuit.

Consider an oscillator which has not yet started its oscillations. As in all electronic systems, noise is present everywhere. One of the noise components will have the same frequency as the oscillation frequency. This component is amplified, passed by the frequency selective network, again amplified, etc., so that oscillations have started.

This means that an oscillator can be considered as a noise amplifier. The selectivity bandwidth of the oscillator is usually narrow but not zero, so that some of the adjacent noise frequency components will also be passed by the network giving rise to small fluctuations in phase and amplitude of the output signal.

These fluctuations will be less with higher values of the Q of the network. The phase shift condition $\mathrm{Im}[A_r(j\omega)] = 0$ determines the oscillation frequency.

18.2 FREQUENCY STABILITY

Frequency stability of an oscillator can be loosely defined as the ability to maintain a fixed frequency over a long period of time.

Changes in **temperature** will affect the oscillation frequency to some degree. The temperature stability is defined as

$$\frac{1}{\omega_o}\left(\frac{d\omega}{dT}\right)_{\omega=\omega_o} \quad \text{ppm } ^\circ\text{C}^{-1}$$

where $\omega_o/2\pi$ is the oscillation frequency.

Obviously, the frequency determining components in the feedback network should be very stable. High stability capacitor types are silver mica, polystyrene and teflon. Self inductance and resistivity of an inductor both increase with temperature. To offset this increase, negative temperature coefficient capacitors are sometimes used but tracking is difficult. The most stable components are crystals. Resonance frequencies are 10^3–10^4 times more stable than those of LC circuits.

Changes in **amplifier parameters** will cause some frequency instability. Ideally, only the feedback network should determine the oscillator parameters. Actually, input impedance Z_i and output impedance Z_o of the amplifier load the feedback network and will have some effect on oscillator performance. The bulk of the required phase shift is supplied by the feedback network but parasitic capacitances of the amplifier will cause a parasitic phase shift. The total phase shift can thus be written as $\phi = 0 = \phi_n + \phi_p$ where ϕ_n is the network phase shift and ϕ_p the parasitic phase shift. If ϕ_p changes by $d\phi_p$, ϕ_n must change by $d\phi_n = -d\phi_p$. A change $d\phi_n$ causes a relative frequency change $d\omega/\omega_o$ the magnitude of which is determined by the slope of the phase characteristic of the feedback network.

Therefore, frequency stability with regard to parameter changes is defined as

$$\omega_o\left(\frac{d\phi}{d\omega}\right)_{\omega=\omega_o}$$

Two methods of minimizing the influence of circuit parameters are as follows:

- Select the values of the frequency determining components so that error terms are minimized. For example, in the Colpitts oscillator

(Section 18.4.1) choose

$$\frac{1}{C_1C_2r_ir_o} \ll \frac{1}{L}\left(\frac{1}{C_1}+\frac{1}{C_2}\right) \quad \text{or} \quad C_1 + C_2 \gg \frac{L}{r_ir_o}$$

- Swamp out part of the active parameter by putting appropriate values of resistances in parallel with Z_i and Z_o.

18.3 AMPLITUDE STABILITY

In a linear oscillator the amplitude of the signal is undetermined. Since the magnitude of the loop gain is slightly larger than unity, the amplitude of the signal will grow until the limits of the linear range are reached.

LC oscillators usually do not require amplitude stabilization due to the selectivity of the LC circuit.

In RC oscillators the amplitude can be stabilized by utilizing various nonlinear devices:

- NTC and PTC resistors (incandescent lamps),
- the FET as a variable channel resistance, controlled by V_{GS},
- the transconductance g_m of a bipolar transistor, which is controlled by bias current,
- Zener diodes.

18.4 LC OSCILLATORS

Most LC oscillators are Π-type feedback oscillators containing three impedances in the feedback network (Fig. 18.1). When neglecting input and output capacitances, $Z_i = r_i$ and $Z_o = r_o$. If

$$\frac{1}{Z_1} = \frac{1}{Z_1^*} + \frac{1}{r_i}, \qquad \frac{1}{Z_2} = \frac{1}{Z_2^*} + \frac{1}{r_o}$$

the loop gain is

$$A_r(j\omega) = \frac{-g_mZ_1Z_2}{Z_1 + Z_2 + Z_3}$$

When writing $Z_1 = R_1 + jX_1$, $Z_2 = R_2 + jX_2$ and $Z_3 = R_3 + jX_3$, the two oscillation conditions are

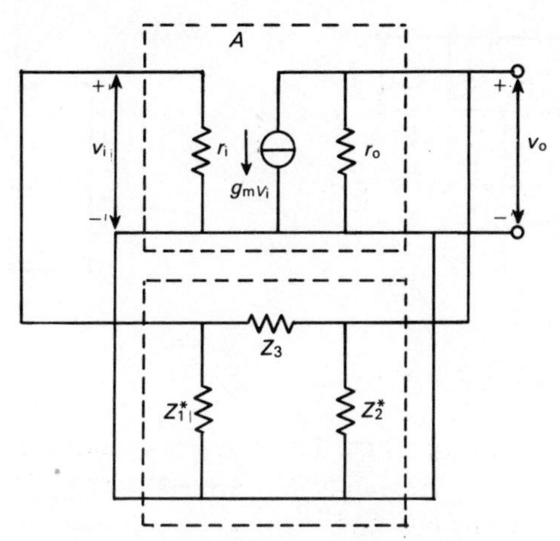

Fig. 18.1 General model of oscillator circuit

$$g_m(X_1X_2 - R_1R_2) \geqslant R_1 + R_2 + R_3 \quad \text{(starting condition)}$$

$$X_1(1 + g_mR_2) + X_2(1 + g_mR_1) + X_3 = 0 \quad \text{(oscillation frequency)}$$

These equations lead to a number of classical oscillator configurations.

18.4.1 Colpitts oscillator (Fig. 18.2)

Here

$$\frac{1}{Z_1} = j\omega C_1 + \frac{1}{r_i}, \quad \frac{1}{Z_2} = j\omega C_2 + \frac{1}{r_o}, \quad Z_3 = j\omega L + R_3$$

where R_3 is the loss resistance of the coil.

The starting condition requires a transconductance:

$$g_m \geqslant \frac{C_2}{r_iC_1} + \frac{C_1}{r_oC_2} + \frac{R_3}{L}(C_1 + C_2)$$

The oscillation frequency is determined by

$$\omega_o{}^2 = \frac{1}{L}\left(\frac{1}{C_1} + \frac{1}{C_2}\right) + \frac{1}{C_1C_2r_ir_o} + \frac{R_3}{L}\left(\frac{1}{C_1r_i} + \frac{1}{C_2r_o}\right)$$

18.4.2 Clapp oscillator (Fig. 18.3)

A capacitor C_3 is connected in series with L so that $C_3 \ll C_1$ and $C_3 \ll C_2$. The starting condition is the same as in the Colpitts oscillator. The

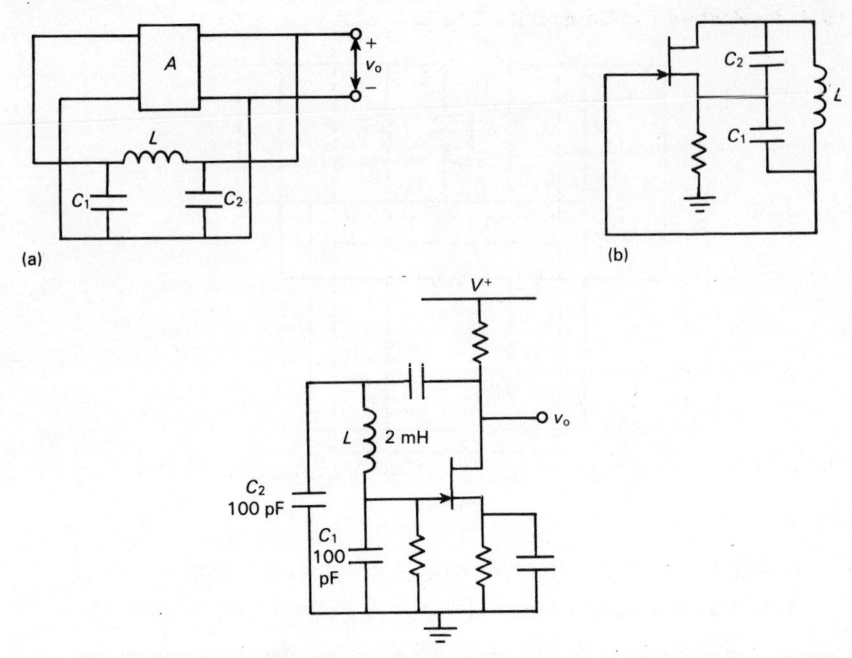

Fig. 18.2 (a) Colpitts oscillator configuration (b) Basic circuit example of Colpitts oscillator (c) Circuit example of 500 kHz Colpitts oscillator

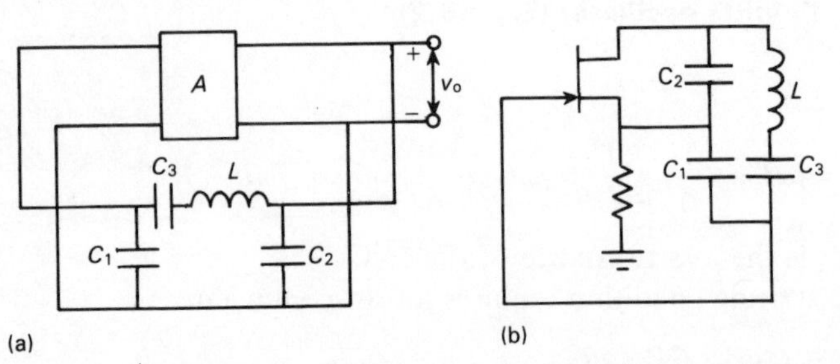

Fig. 18.3 (a) Clapp oscillator configuration (b) Basic Clapp oscillator circuit

oscillation frequency is

$$\omega_o{}^2 = \frac{1}{L}\left(\frac{1}{C_1} + \frac{1}{C_2} + \frac{1}{C_3}\right) \simeq \frac{1}{LC_3}$$

The oscillation frequency is thus largely determined by C_3. The combination of L and C_3 can be considered as a self inductance

$$L^* \simeq LC_3\left(\frac{1}{C_1} + \frac{1}{C_2}\right)$$

This has the effect of making ω_o insensitive to circuit parameter changes.

18.4.3 Hartley oscillator (Fig. 18.4)

$$\frac{1}{Z_1} = \frac{1}{j\omega L_1} + \frac{1}{r_i}, \quad \frac{1}{Z_2} = \frac{1}{j\omega L_2} + \frac{1}{r_o}, \quad Z_3 = \frac{1}{j\omega C}$$

The starting condition is

$$g_m \geqslant \frac{1}{r_i}\frac{L_1 + M}{L_2 + M}$$

and the oscillation frequency is

$$\omega_o{}^2 = \frac{1}{(L_1 + L_2 + 2M)C + \dfrac{L_1 L_2 - M^2}{r_i r_o}}$$

where M is the coefficient of mutual inductance between L_1 and L_2.

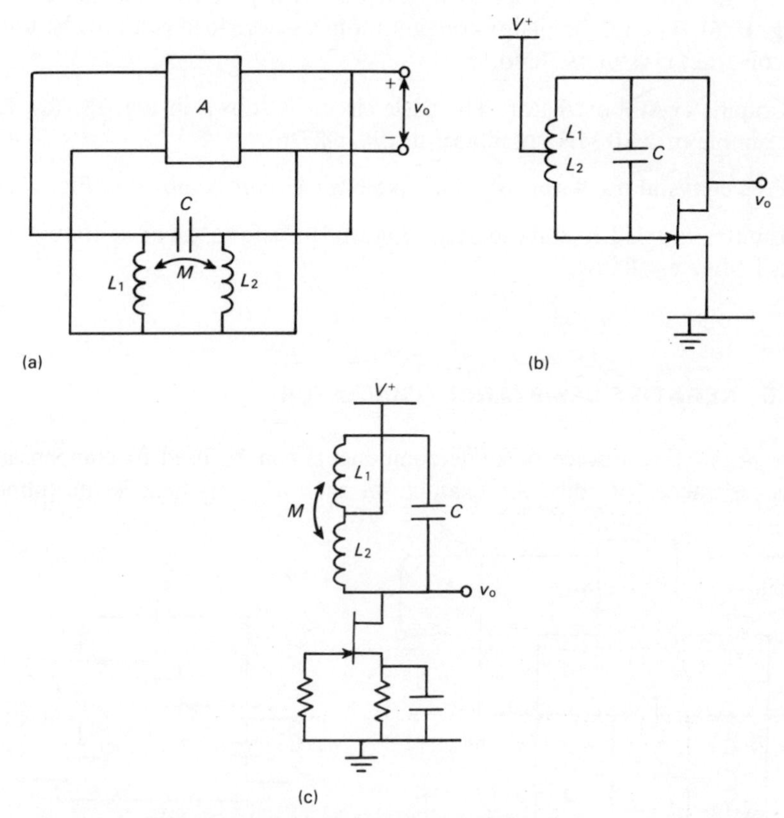

Fig. 18.4 (a) Hartley oscillator configuration (b) Example of Hartley oscillator (c) Example of Hartley oscillator

18.4.4 Tuned plate-tuned grid (TPTG) oscillator (Fig. 18.5)

The TPTG oscillator is a variation of the Hartley oscillator where the inductances are replaced by resonant circuits.

When both tuned circuits are identical and loss resistances ignored, the oscillation frequency is

$$\omega_o^2 \simeq \frac{1}{LC\left(1 + 2\dfrac{C_2}{C}\right)}$$

Thus both resonance circuits must be tuned to the inductive side of resonance.

18.4.5 Crystal oscillator

A crystal can operate either as a series or as a parallel resonant circuit (Fig. 18.6). The LC oscillator configurations as described can thus be used to construct crystal oscillators:

- Colpitts crystal oscillator. The basic circuit is shown in Fig. 18.7(a); an example of a 10 MHz oscillator in Fig. 18.7(b).
- Pierce crystal oscillator, of which the basic circuit is shown in Fig. 18.8.
- Emitter coupled crystal oscillator. Figure 18.9 shows the basic circuit and a 1 MHz oscillator.

18.5 NEGATIVE RESISTANCE OSCILLATOR

The negative resistance of some components can be used to compensate loss resistances of coils. An example of such a component is the tunnel

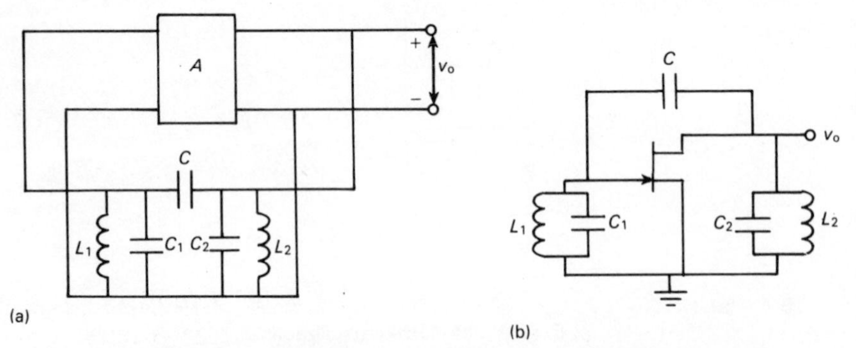

Fig. 18.5 (a) TPTG oscillator configuration (b) TPTG oscillator circuit

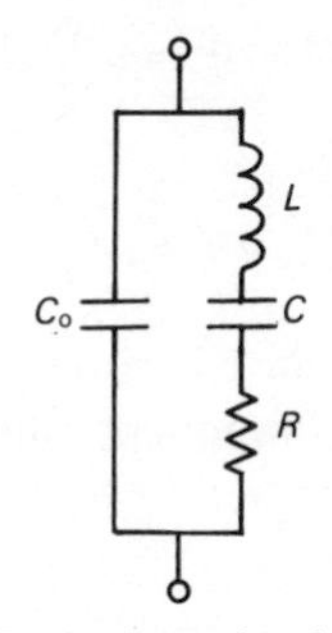

Fig. 18.6 Equivalent circuit of a crystal

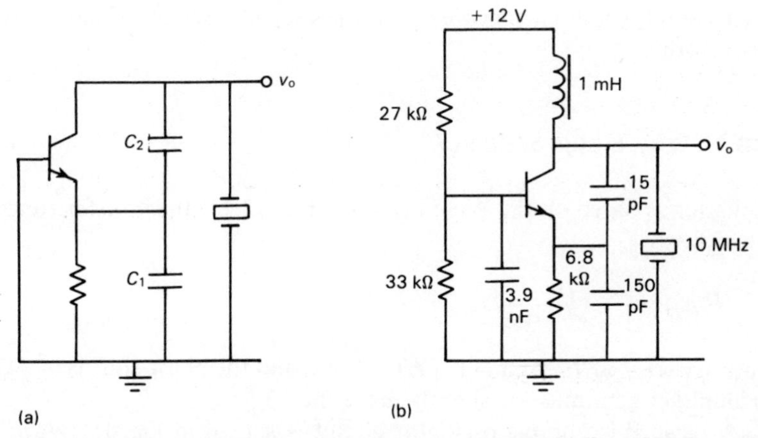

Fig. 18.7 (a) Basic Colpitts crystal oscillator (b) Example of 10 MHz crystal oscillator

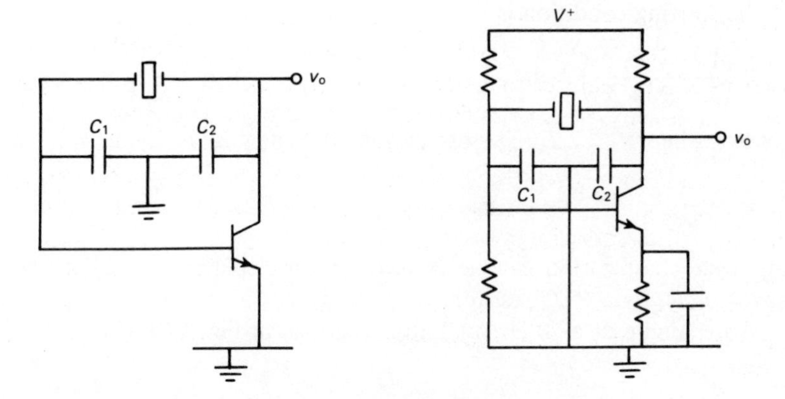

Fig. 18.8 Basic Pierce crystal oscillator

diode. The basic circuit is shown in Fig. 18.10(a) of which the oscillation frequency is

$$\omega_o^2 = \frac{1}{C + C_d}\left(\frac{1}{L} - \frac{1}{C_d r_d^2}\right)$$

An example of a 1 MHz oscillator is shown in Fig. 18.10(b). The tunnel diode, being a high-frequency component, allows oscillation frequencies up to several GHz.

18.6 RC OSCILLATORS

When stability requirements are not too severe, RC oscillators may perform satisfactorily.

18.6.1 Wien bridge oscillator

Oscillation is based on the Wien network (Fig. 18.11) the transfer function of which is

$$H(j\omega) = \frac{jx}{1 - x^2 + 3jx}$$

where $x = \omega\tau = \omega RC$. At $x = 1$, $|H(j\omega)| = \frac{1}{3}$ and the phase shift is 0°. Thus the amplifier gain must be slightly more than 3.

A basic Wien-bridge oscillator circuit is shown in Fig. 18.12(a). The loop gain at resonance is

$$A_r(j\omega_o) = A\left(\frac{1}{3} - \frac{R_1}{R_1 + R_2}\right)$$

and the starting condition is

$$A\left(\frac{1}{3} - \frac{R_1}{R_1 + R_2}\right) > 1$$

If $A \gg 1$, then $R_2 = 2R_1$. The oscillation frequency is

$$\omega_o = \frac{1}{RC}$$

Amplitude stabilization can be achieved when replacing R_1 by an NTC resistor or R_2 by a PTC resistor.

An example of a 70 Hz oscillator is shown in Fig. 18.12(b).

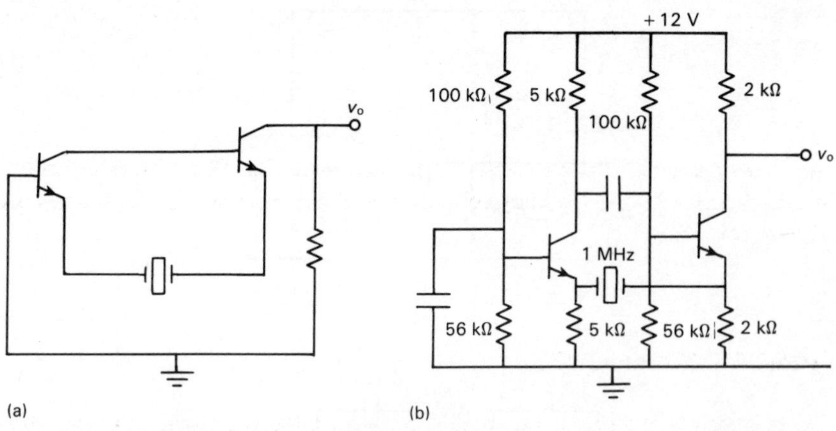

Fig. 18.9 (a) Emitter coupled crystal oscillator (b) Example of 1 MHz emitter coupled oscillator

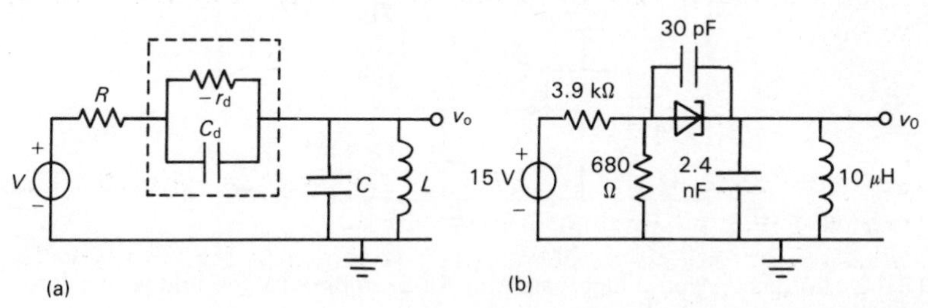

Fig. 18.10 (a) Basic tunnel diode oscillator circuit (b) Example of tunnel diode oscillator circuit

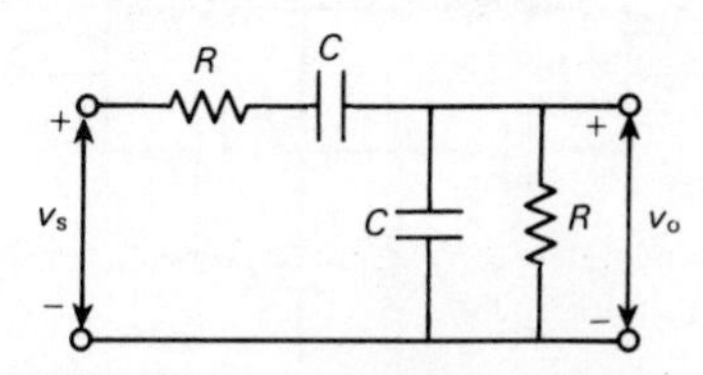

Fig. 18.11 The Wien network

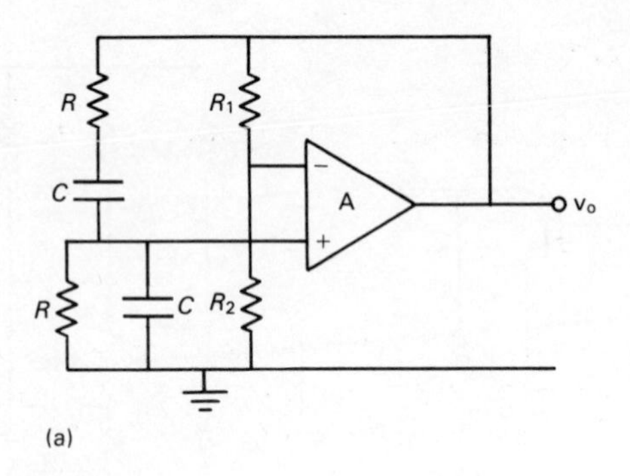

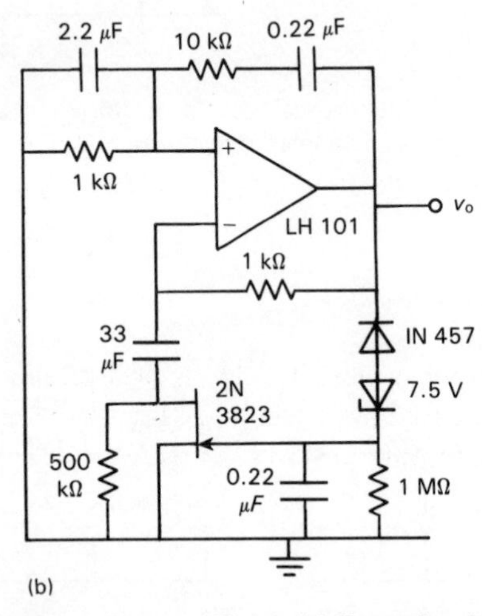

Fig. 18.12 (a) Basic Wien bridge oscillator (b) Example of Wien bridge oscillator

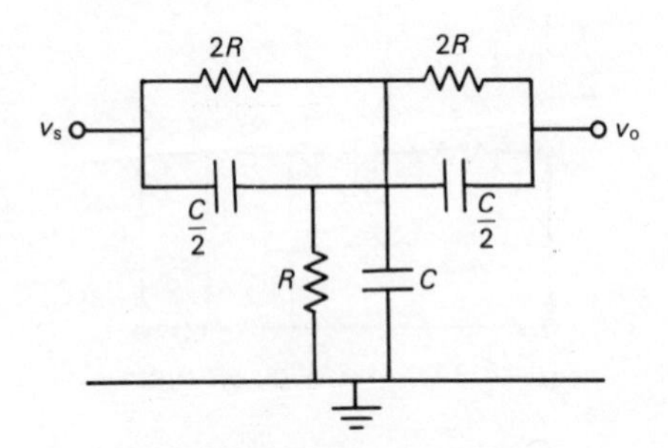

Fig. 18.13 Twin-T network

18.6.2 Twin-T oscillator

This oscillator is based on the twin-T network (Fig. 18.13) with transfer function

$$H(j\omega) = \frac{1 - x^2}{1 - x^2 + 4jx}$$

At $x = 1$, $|H(j\omega)| = 0$ and $\phi = 0^\circ$. When placing the network in the feedback loop of an amplifier, the circuit will oscillate when $x = 1$.

The basic oscillator circuit is that of Fig. 18.14 having an oscillation frequency

$$\omega_o = \frac{1}{RC}$$

The amplitude of the output signal is controlled by the choice of appropriate values of R_1 and R_2.

18.6.3 Phase shift oscillators

In these oscillators a phase shift of 180° is accumulated by RC networks. The remaining 180° is provided by an inverting amplifier. The RC networks can be low-pass or high-pass networks.

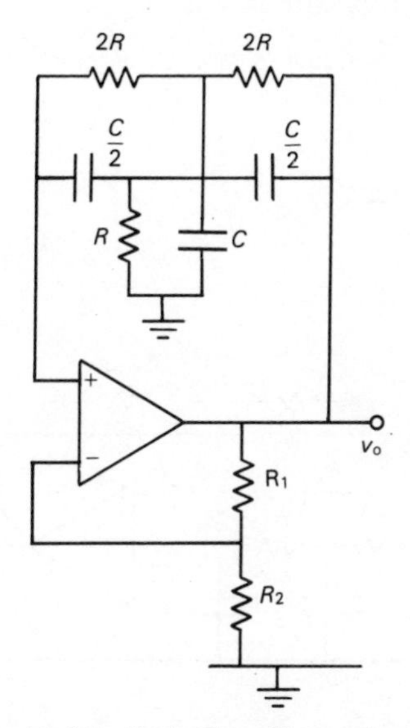

Fig. 18.14 Twin-T oscillator circuit

Oscillator with low-pass sections (Fig. 18.15)

The network attenution is $\frac{1}{29}$ thus the amplifier gain must be at least 29. The oscillation frequency is

$$\omega_o = \frac{\sqrt{6}}{RC}$$

Oscillator with high-pass sections (Fig 18.16)

The network attenuation is also $\frac{1}{29}$, the oscillation frequency is

$$\omega_o = \frac{1}{RC\sqrt{6}}$$

18.6.4 Quadrature oscillator (sine–cosine oscillator)

The quadrature oscillator produces simultaneously a sine and a cosine waveform. The oscillator is usually used for fixed-frequency operation. A functional schematic is shown in Fig. 18.17(a). Its transfer function is described by the differential equation

$$\frac{d^2 v_o}{dt^2} + \frac{1}{\tau_1 \tau_2} v_o = 0$$

where τ_1 and τ_2 are the time constants of the low-pass sections. The steady-state solution of this equation is

$$v_o = V_o \sin \frac{t}{\sqrt{\tau_1 \tau_2}}$$

and the oscillation frequency is

$$\omega_o = \frac{1}{\sqrt{\tau_1 \tau_2}}$$

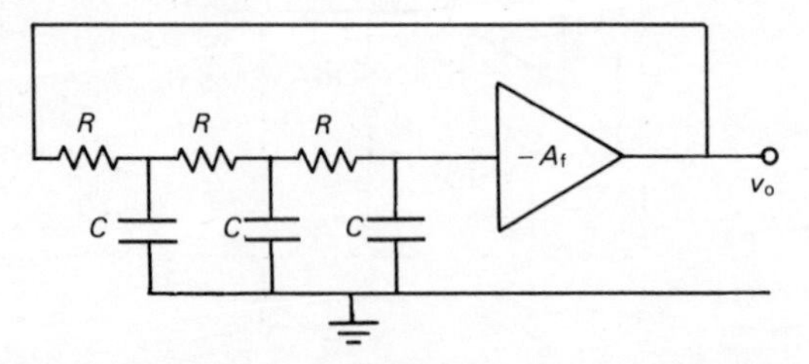

Fig. 18.15 Phase shift oscillator with low-pass sections

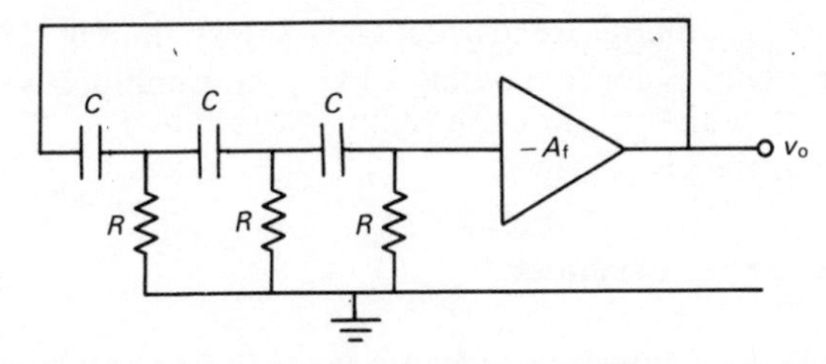

Fig. 18.16 Phase shift oscillator with high-pass sections

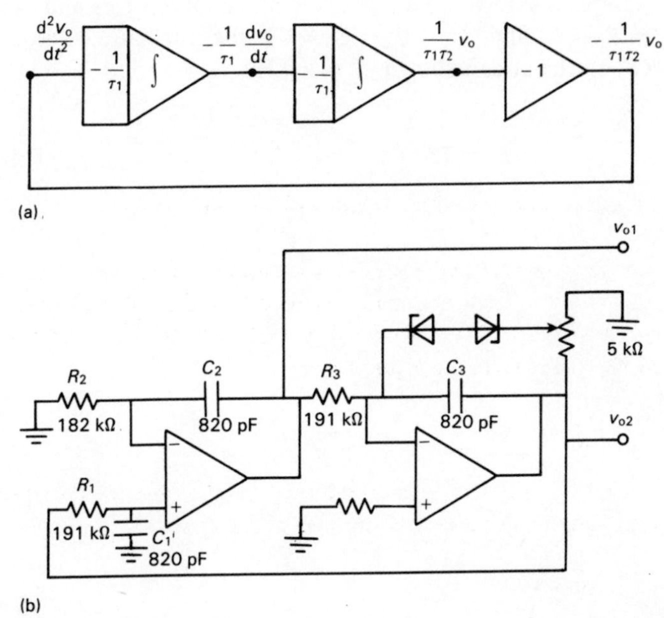

Fig. 18.17 (a) Basic quadrature oscillator (b) Example of quadrature oscillator

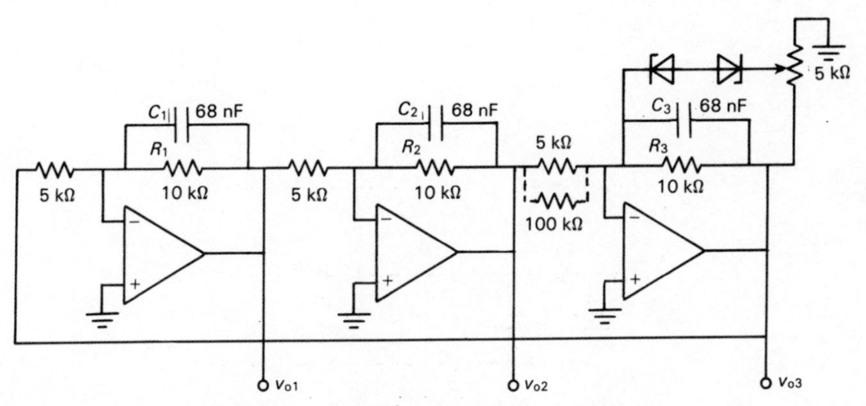

Fig. 18.18 Example of three-phase oscillator

An example of a quadrature oscillator is shown in Fig. 18.17(b). The oscillation frequency is approximately 1 kHz, the amplitude is stabilized by Zener diodes.

18.6.5 Three-phase oscillator

An extension of the quadrature oscillator is the three-phase oscillator which produces three output sine waves with a phase difference of 120°.

An example is shown in Fig. 18.18. With equal resistors and capacitors ($R_1 = R_2 = R_3 = R$ and $C_1 = C_2 = C_3 = C$) the oscillation frequency is $\omega_o = \sqrt{3}/RC$. In the circuit shown $f_o \simeq 400$ Hz.

19 Rectification; Voltage and Current Sources

19.1 GENERAL

Almost all electronic circuits and systems operate on one or more DC voltages. The AC line voltage is usually converted to a suitable value of DC voltage. Such a conversion system is a power supply which normally consists of:

- A line transformer.
- A rectifying circuit.
- A smoothing circuit.
- A voltage stabilizing circuit.

Since AC voltages, currents and powers are involved it seems useful to first recapitulate a number of AC terms.

19.2 AC TERMS AND DEFINITIONS

19.2.1 Voltages and currents

The AC voltage is a periodic and sinusoidal waveform $v(t)$ which is described by a number of quantities:

- The frequency f.
- The instantaneous value; this is the value of $v(t)$ at a certain instant t.
- The amplitude or peak value; this is the maximum value in one period; notation V_m.
- The peak-to-peak value; this is the difference between the largest positive and the largest negative value in one period; notation V_{pp}.

- The average value, defined as

$$V_{av} = \frac{1}{T} \int_0^T v(t)\, dt$$

If the voltage waveform has equal positive and negative areas, $V_{av} = 0$. In that case the average value is taken over one half period, thus

$$V_{av} = \frac{2}{T} \int_0^{T/2} v(t)\, dt$$

- The rms value. This is the value of a DC voltage which produces in a resistor R the same heat as the AC voltage, when expressed in its rms value, given by:

$$V_{rms} = \sqrt{\frac{1}{T} \int_0^T v^2(t)\, dt}$$

- The form factor:

$$f_v = \frac{V_{rms}}{V_{av}}$$

- The crest factor:

$$f_c = \frac{V_m}{V_{rms}}$$

If the waveform is sinusoidal:

- $V_{av} = \frac{2}{\pi} V_m$
- $V_{rms} = \frac{1}{2}\sqrt{2}\, V_m$
- $f_v = \frac{\pi}{2\sqrt{2}} = 1.11$
- $f_c = \sqrt{2} = 1.41$

The foregoing definitions and formulas remain the same when 'voltage' is replaced by 'current'.

19.2.2 Power

If a sinusoidal voltage is expressed as

$$v(t) = V_m \sin \omega t$$

and the current as

$$i(t) = I_m \sin(\omega t + \phi)$$

the instantaneous power is

$$p(t) = v(t)i(t)$$

and the average power is

$$P = \frac{1}{T}\int_0^T p(t)\,\mathrm{d}t = V_m I_m \cos\phi$$

Power can be represented in a phasor diagram (Fig. 19.1) to which the following definitions apply:

$V_m I_m \cos\phi$ is the real power (W)
$V_m I_m \sin\phi$ is the reactive power (VAr = volt-amperes reactive); the average value of this component is zero; it is therefore not capable of useful work
$V_m I_m$ is the apparent power (VA = volt-amperes)
$\cos\phi$ is the power factor

19.3 THE LINE TRANSFORMER

Transformer operation is based on the mutual inductance between two magnetically coupled coils. It converts one AC voltage level to another. The supply voltage is connected to the primary winding, the load to the secondary winding. The mutual inductance is achieved by an iron core which is common to both windings.

The ideal transformer (ignoring all losses) is shown in Fig. 19.2. The basic equation which applies to both coils is

$$v(t) = n\frac{\mathrm{d}\Phi}{\mathrm{d}t}$$

(Chapter 3). Thus if n_p and n_s are the numbers of turns for primary and secondary windings

$$\frac{V_p}{V_s} = \frac{n_p}{n_s}$$

Since the power loss is assumed zero

$$V_p I_p = V_s I_s$$

In Chapter 3 it was shown that the value of the self-inductance is proportional to the square of the number of turns. If follows then that

$$\frac{V_p}{V_s} = \frac{I_s}{I_p} = \frac{n_p}{n_s} = \sqrt{\frac{L_p}{L_s}}$$

The input impedance of the transformer is

$$Z_p = \frac{V_p}{I_p} = \left(\frac{n_p}{n_s}\right)^2 \frac{V_s}{I_s} = \left(\frac{n_p}{n_s}\right)^2 R_L$$

The nonideal transformer has two kinds of losses: **iron losses** and **copper losses.** Iron loss or core loss is the power dissipated in the iron core. This loss is virtually independent of the load current. Therefore it is also called **no-load loss.**

Core losses consist of two types: **hysteresis loss** and **eddy current loss.** Hysteresis loss is due to the reversal of the magnetic state of the iron in alternate half cycles (Chapter 3). Eddy current loss follows from Faraday's law which states that an electromagnetic force is induced in a conductor when the associated flux changes. Due to this electromagnetic force, a current results which opposes the change of magnetic flux.

The magnitude of the eddy currents is decreased by using core laminations which are isolated from one another by oxide, varnish or both. These laminations increase the effective magnetic resistance of the iron. The magnitude of eddy current loss depends on a number of parameters and can be expressed as

$$P_e = af^2 B_m{}^2 d^2$$

where

a is a constant
f is the frequency
B_m is the maximum flux density
d is the thickness of the laminations

Copper loss results from the ohmic resistance of the windings and can be written as

$$P_{Cu} = I_p{}^2 R_p + I_s{}^2 R_s$$

The equivalent circuit of a transformer at low frequencies is shown in Fig. 19.3. Because of nonideal coupling of primary and secondary ($k < 1$) not all the primary flux will link the secondary and vice versa. These effects are accounted for by the **leakage inductances** L_1 and L_2.

Transformer losses are expressed by the efficiency

$$\eta = \frac{\text{output power}}{\text{input power}} \times 100\%$$

Figure 19.4 shows a common construction for small power transformers assembled from E- and I-type iron laminations. Such transformers thus convert an AC voltage to another AC voltage (higher or lower) of suitable value, at the same time providing isolation between secondary circuit and line.

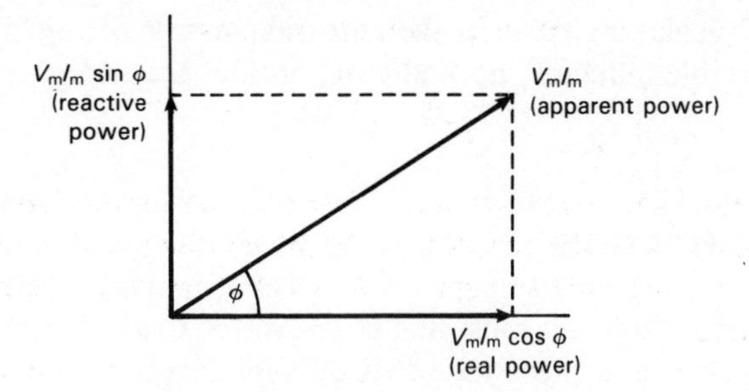

Fig. 19.1 Power phasor diagram

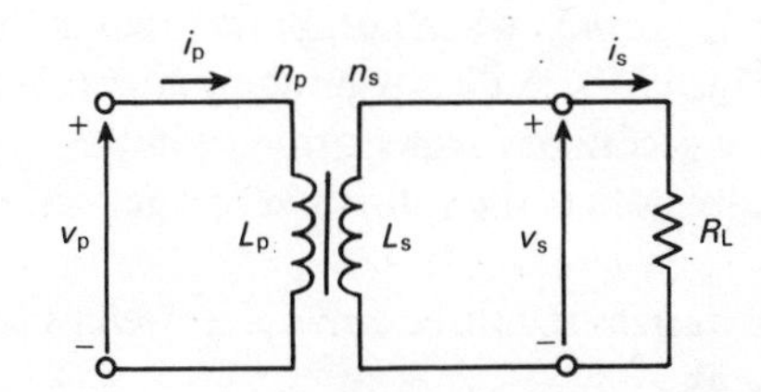

Fig. 19.2 Equivalent circuit of ideal transformer

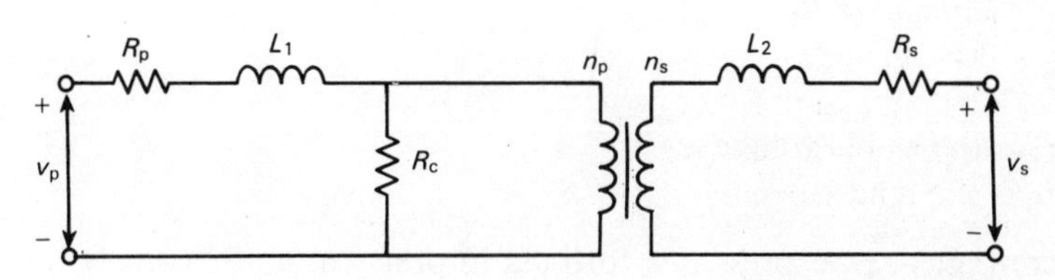

Fig. 19.3 Low frequency equivalent circuit of actual transformer

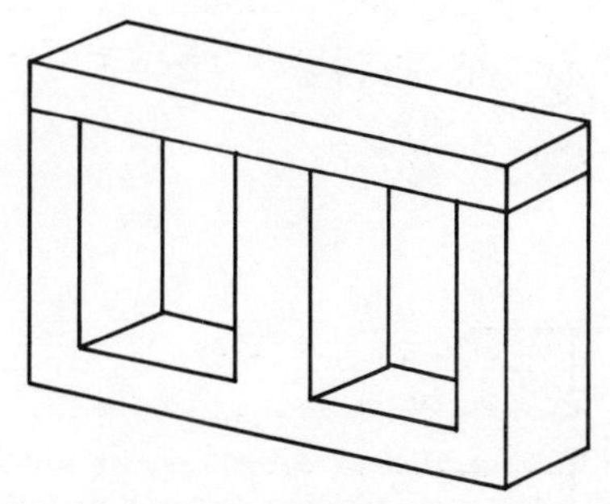
Fig. 19.4 Core construction of small transformer

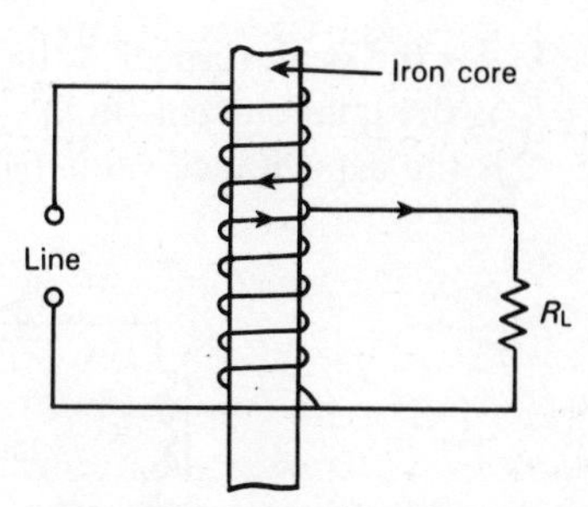

Fig. 19.5 Autotransformer

No such isolation exists in the **autotransformer** of Fig. 19.5 which is merely a variable inductor, normally of toroidal shape.

19.4 RECTIFICATION

The secondary AC voltage must first be converted to a unipolar voltage by a rectifier circuit and is then smoothed by some form of filter circuit. The simplest rectifier is the half-wave rectifier (Fig. 19.6) consisting of a single diode followed by a smoothing capacitor. This circuit is rarely used because of its inefficiency, since only one-half of each cycle is utilized.

Improved efficiency results when combining two half-wave circuits into a full-wave circuit (Fig. 19.7). A disadvantage sometimes is the requirement of a center tap on the secondary transformer winding.

Almost universally used is the full-wave bridge rectifier (Graetz circuit) of Fig. 19.8.

Figure 19.9 summarizes the three circuits as well as the relevant voltage and current waveforms.

From a rather extensive analysis of rectified waveforms a ripple voltage formula can be approximated:

$$V_{\mathrm{pp}} \simeq \frac{I_{\mathrm{L}}}{f_{\mathrm{r}}C} = \frac{V_{\mathrm{o}}}{f_{\mathrm{r}}R_{\mathrm{L}}}$$

where

f_{r} is the ripple frequency
I_{L} is the load current

It is convenient to express this formula in practical units, namely

$$V_{\mathrm{pp}} \simeq 1000\,\frac{I_{\mathrm{L}}}{f_{\mathrm{r}}C}$$

where

V_{pp} is the peak-to-peak value of ripple voltage (V)
I_{L} is the load current (mA)
C is the capacitance value (μF)

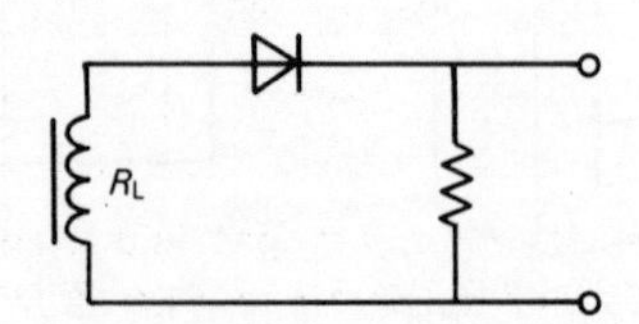

Fig. 19.6 Half-wave rectification

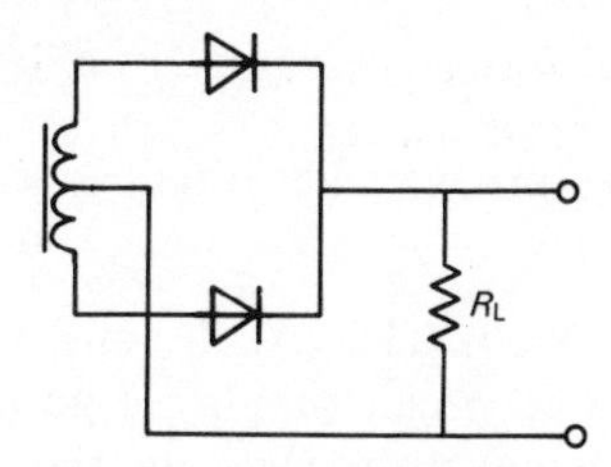

Fig. 19.7 Full-wave rectification with center tap

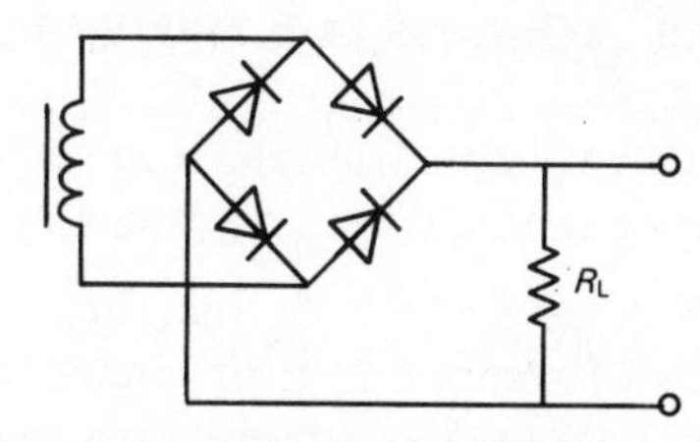

Fig. 19.8 Full-wave rectification without center tap

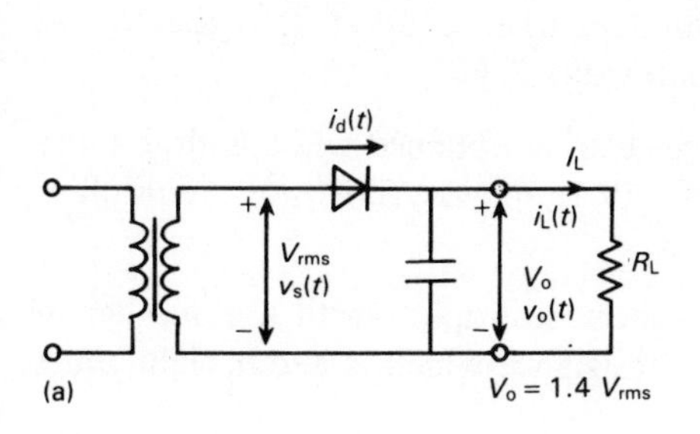

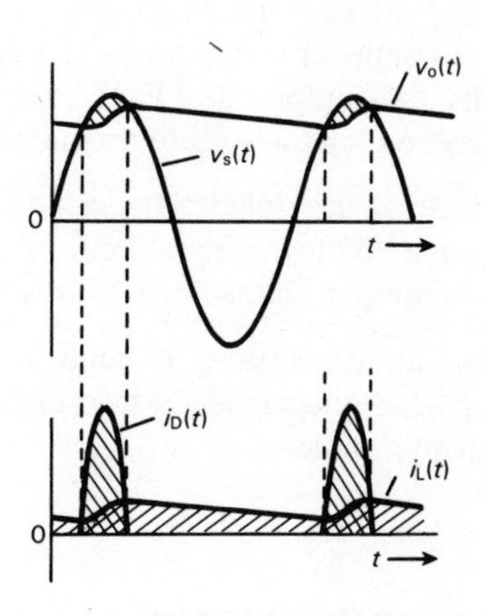

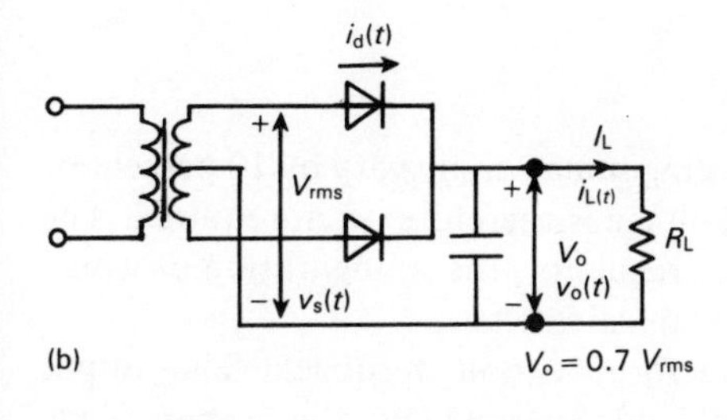

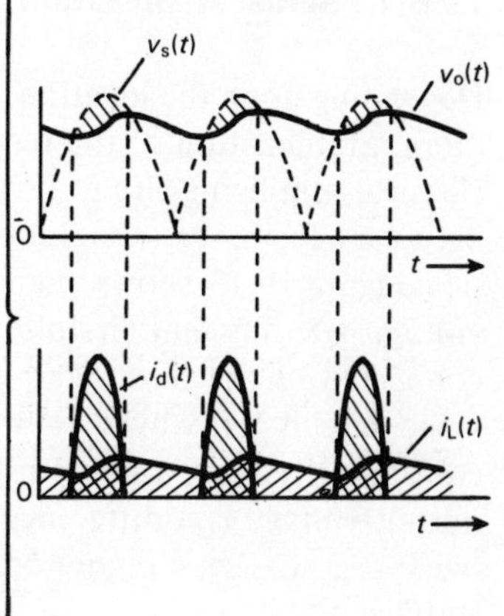

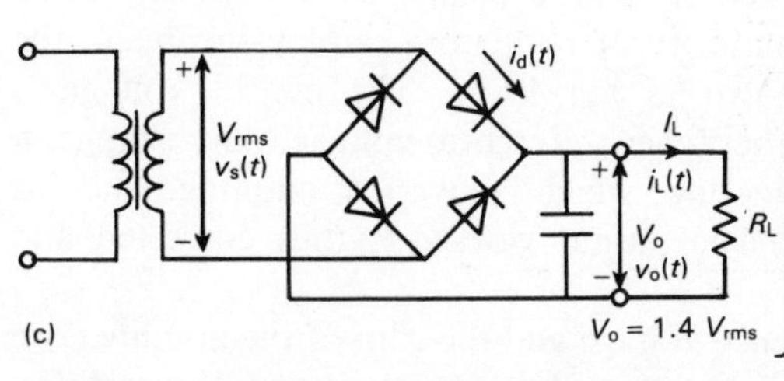

Fig. 19.9 (a) Half-wave rectification with load and relevant waveforms (b) Full-wave rectification with load and relevant waveforms (c) Full-wave rectification with load and relevant waveforms

19.5 VOLTAGE MULTIPLIERS

Higher voltages from the secondary of a transformer can be obtained when adding more diodes and capacitors.

- Voltage doubler (Greinacher circuit, Fig. 19.10): This circuit consists of two half-wave rectifier circuits of which one circuit supplies a positive voltage, the other circuit a negative voltage. At the output the two voltages are added.
- Voltage doubler (Villard circuit, Fig. 19.11): This circuit is a cascade of two capacitor-diode sections. Capacitor C_1 is charged, leaving diode D_1 just short of conducting. The first section thus raises the input signal by its own amplitude (V_m). This voltage, when connected to the second section, charges C_2 to its maximum value ($2V_m$).
- Voltage multiplier: Higher voltages can be obtained when adding more stages to the previous circuit. Figure 19.12 shows a three-stage multiplier, the output voltage of which is $6V_m$.

The output resistance of multipliers increases rapidly with the number of stages. A quadrupler has an output resistance which is about eight times that of a doubler.

19.6 STABILIZATION

19.6.1 Series stabilization

Depending upon the location, line voltages may easily vary by 10 percent or more. Stabilization of the rectified voltage is therefore often required. The classical stabilizing circuit is the series regulator, the configuration of which may vary from very simple to very sophisticated.

Figure 19.13 shows the simplest form without feedback. The output voltage of the emitter-follower is held constant by the Zener diode connected to the base. The majority of regulators employ feedback, the basic circuit of which is shown in Fig. 19.14. The output voltage is continuously compared with the Zener (reference) voltage. Any change in output voltage is a difference voltage which is inverted, amplified and fed back to the series regulator. The output voltage is thus corrected and stabilized.

The stability of the reference voltage and the gain of the amplifier are important parameters for the quality of the output voltage. Generally, a

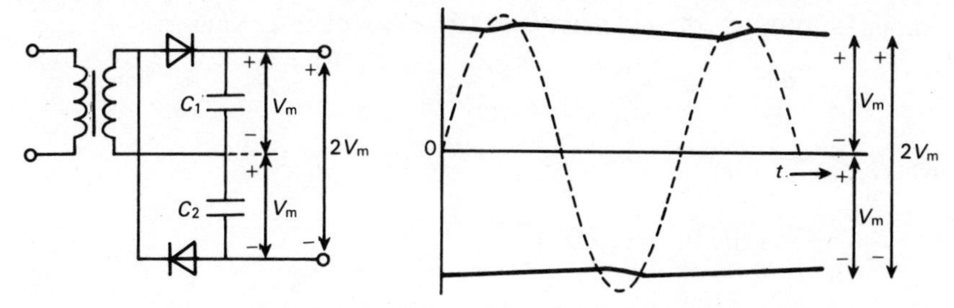

Fig. 19.10 Greinacher voltage doubler and waveforms

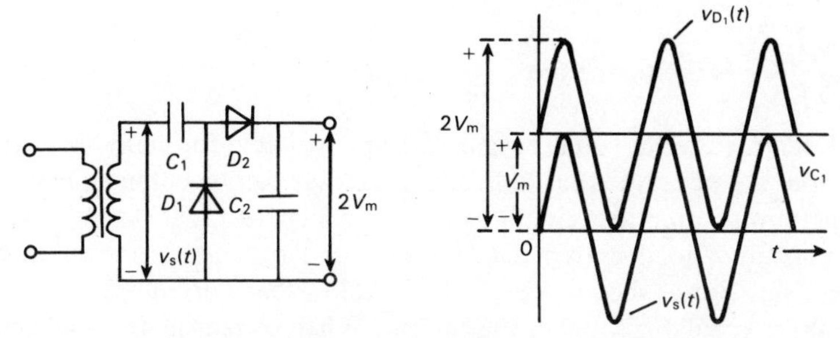

Fig. 19.11 Villard voltage doubler and waveforms

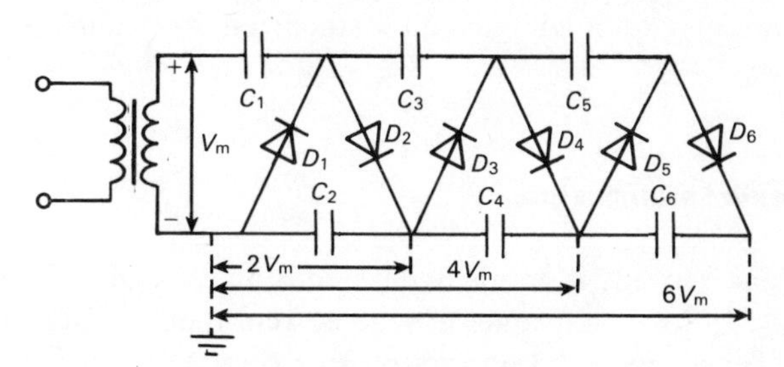

Fig. 19.12 Voltage multiplier

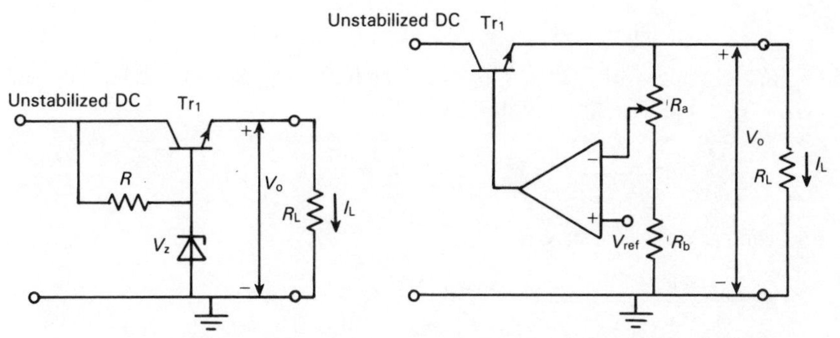

Fig. 19.13 Simple series regulator

Fig. 19.14 Series regulator with adjustable output voltage

change in output voltage may result from any of three causes:

$$\Delta V_o = \frac{\partial V_o}{\partial V_i} \Delta V_i + \frac{\partial V_o}{\partial I_L} \Delta I_L + \frac{\partial V_o}{\partial T} \Delta T$$

where

$$\frac{\partial V_o}{\partial V_i} = s \text{ is the line regulation}$$

$$\frac{-\partial V_o}{\partial I_L} = r_o \text{ is the output resistance}$$

$$\frac{\partial V_o}{\partial T} = \tau \text{ is the temperature coefficient}$$

Integrated series voltage regulators are available for many fixed values of output voltage. An adjustable output voltage can be obtained by adding an operational amplifier circuit as shown in Fig. 19.15.

Maximum load current can be increased by connecting an external power transistor as shown in Fig. 19.16. At low load currents, Tr_1 is off and the voltage regulator supplies the current. When increasing the load current up to a certain threshold value, determined by resistor R, Tr_1 begins to conduct. From here on it will pass all load current exceeding the threshold current.

19.6.2 Parallel stabilization

When the load current is more or less constant, a parallel regulator is sometimes used. For small values of load current the regulator may consist of merely a Zener diode and a series resistor (Fig. 19.17).

A regulator with feedback is shown in Fig. 19.18 the output voltage of which can be expressed as

$$V_o \simeq V_i - (I_C + I_L)R$$

If V_o changes by ΔV_o, V_{BE} will change by $\Delta V_{BE} = \Delta V_o$ so that $\Delta I_C = g_m \Delta V_o$. The line regulation is thus

$$s = \frac{\partial V_o}{\partial V_i} = \frac{1}{1 + g_m R}$$

and the output resistance is

$$r_o = \frac{-\partial V_o}{\partial I_L} = \frac{R}{1 + g_m R}$$

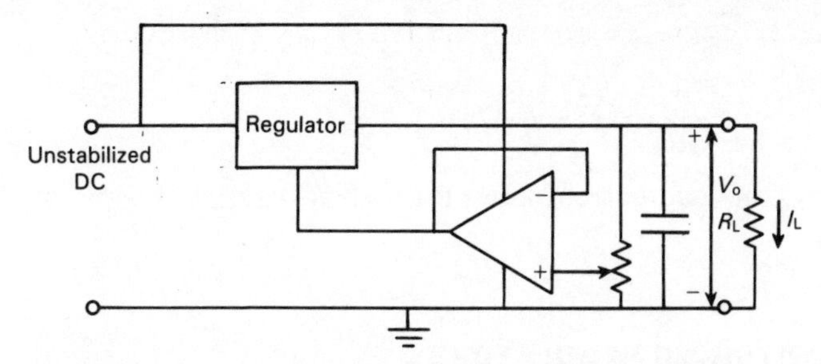

Fig. 19.15 Series stabilization with integrated regulator

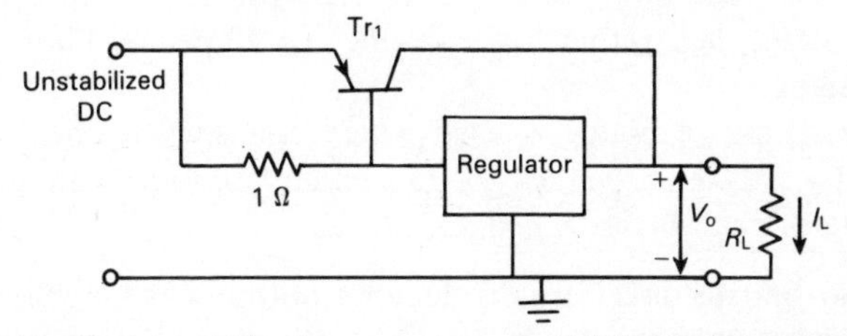

Fig. 19.16 Increasing the output current of the series regulator

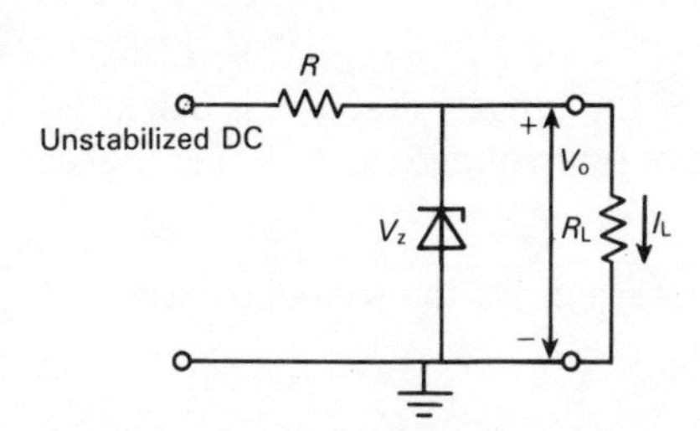

Fig. 19.17 Parallel regulator with Zener diode

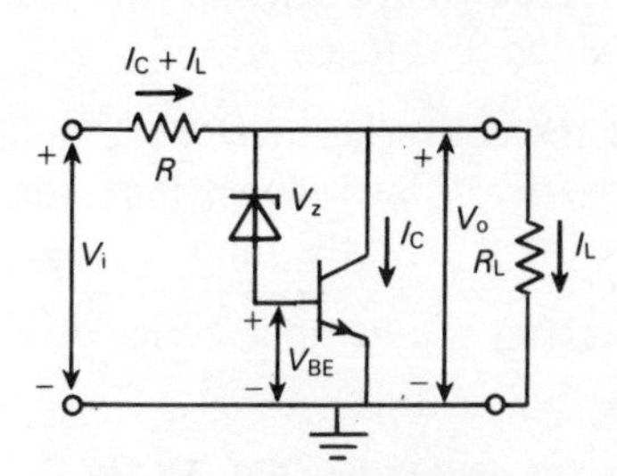

Fig. 19.18 Parallel regulator with bipolar transistor

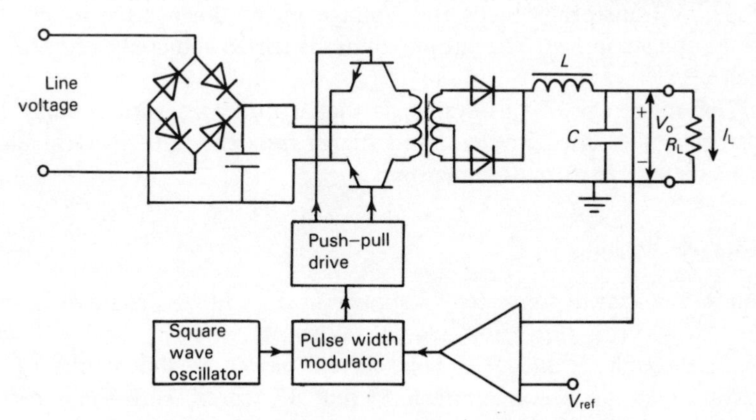

Fig. 19.19 Switching regulator (primary switching)

so that

$$\Delta V_o = \frac{1}{1 + g_m R} (\Delta V_i - R \Delta I_L)$$

Obviously, the load current flows through R, causing dissipation.

19.7 SWITCHING REGULATORS

Although series regulators with excellent performance characteristics are used everywhere, their rather low efficiency ($\simeq 50$ percent) has led to more efficient designs.

The **switching regulator** is used where efficiency is one of the prime requirements. Efficiency values of 75 percent or more can be achieved. These regulators can be classified as follows:

- **Step-up** or **boost** converters produce an output voltage which is higher than the input voltage.
- **Step-down** or **buck** converters have output voltages which are lower than the input voltage.
- If the converter inverts the polarity of the input voltage, it is called an **inverter** or **buck-boost** converter. The output voltage can be either higher or lower than the input voltage.

Switching can be done either on the primary or on the secondary side.

Primary switching (Fig. 19.19)

The line voltage is directly rectified and converted to a high frequency AC voltage. A transformer steps this voltage up or down after which it is rectified and smoothed. The output voltage is sensed and used to control the switch.

The advantage of this system is that it dispenses with a bulky line transformer since the conversion to a higher frequency (20–50 kHz) allows much smaller transformer dimensions.

Secondary switching

A basic step-down converter is shown in Fig. 19.20. The switch is a transistor which is turned on and off. When it turns on, a current starts flowing through L and R_L. This current increases slowly due to the inductance L. C_2 is now charged so that V_o increases. When a certain threshold voltage is reached, the transistor turns off and the current

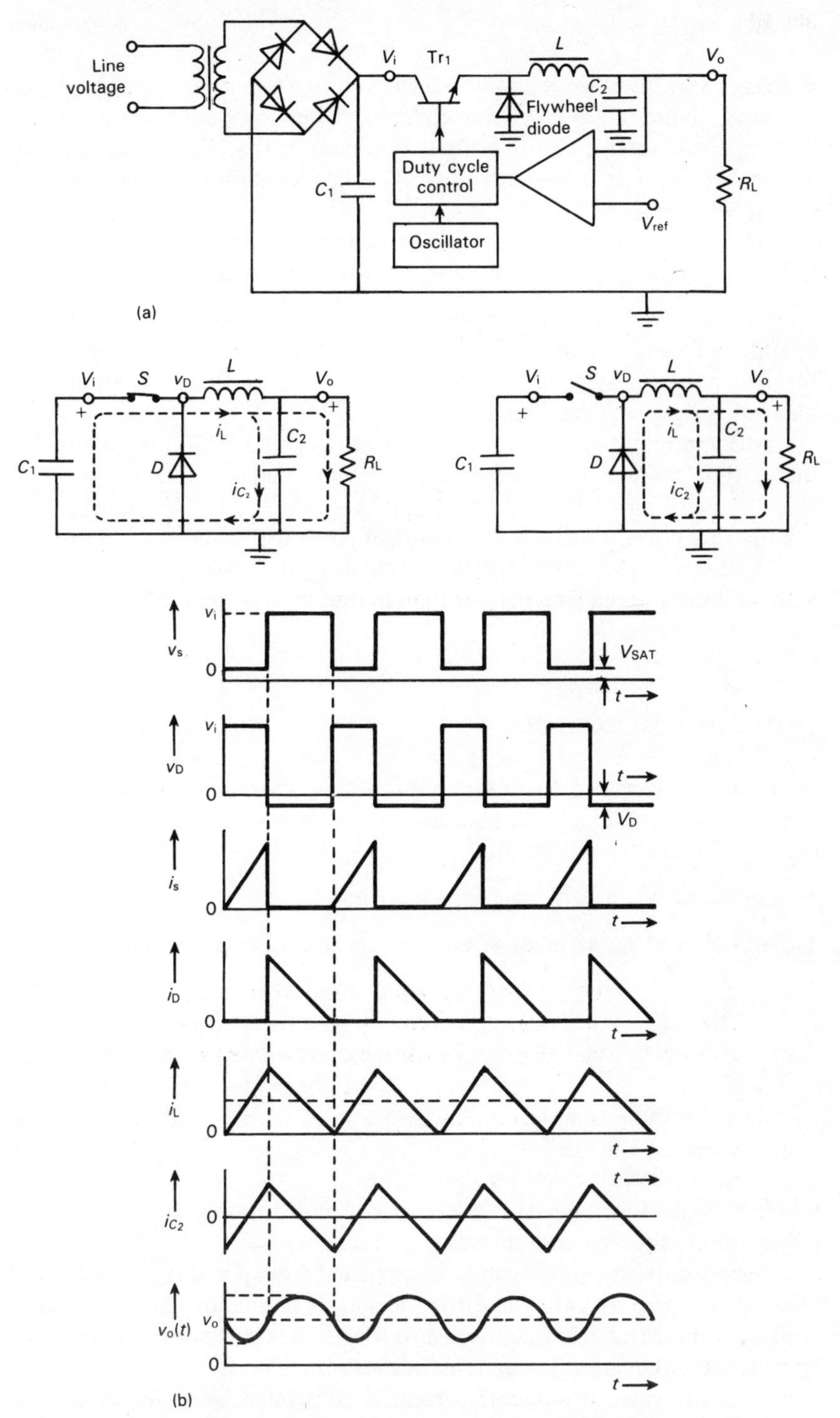

Fig. 19.20 (a) Switching regulator (secondary switching) (b) Currents and waveforms in secondary switching regulator

decreases slowly. Since the only return path of this current is through the 'flywheel'-diode, C_2 is still being charged. When the current becomes less than the load current, C_2 will begin to discharge so that V_o decreases. When V_o reaches a certain reference voltage V_{ref} the transistor turns on again and the cycle repeats.

The basic circuits of a step-up converter and a voltage inverter (buck-boost) are shown in Figs. 19.21 and 19.22 respectively.

The main advantages of switching regulators result from the higher operating frequency, allowing a smaller smoothing capacitor and transformer, and the switching mode which decreases the dissipation thus alleviating heat sink requirements.

Modern integrated switching regulators use MOSFETs. Regulation of the output voltage is achieved by different methods like pulse width modulation or pulse frequency modulation (Chapter 23). The inductors are usually ring cores, potcores or cylindrical cores with values of 0.5–1 mH.

A disadvantage of the switching regulator is its rather slow response to load variations and a larger ripple than in the conventional series regulator.

19.8 CURRENT SOURCES

Although not intended for power applications, current sources find many uses in small signal circuits, especially in integrated circuits. Examples of current source applications are:

- Generation of linearly increasing voltages.
- Simulation of high values of emitter resistance in differential amplifiers (Chapter 14).

The basic model of a current source is shown in Fig. 19.23(a). From the equivalent circuit (Fig. 19.23(b)) the output resistance is found as

$$r_o \simeq r_c \frac{R_e + r_e + \dfrac{R_s}{\beta}}{R_e + r_e + R_s}$$

where r_c is the dynamic resistance from collector to base, due to the Early effect, and r_e the dynamic emitter resistance.

Simple current sources can be constructed from a voltage follower and a transistor (Fig. 19.24). The effect of V_{BE} is eliminated since the input control voltage is directly connected to the emitter and not to the base. In these circuits the current source output resistance is r_c.

In many cases the output current is controlled by an input current rather than an input voltage. The input resistance should then be as low as

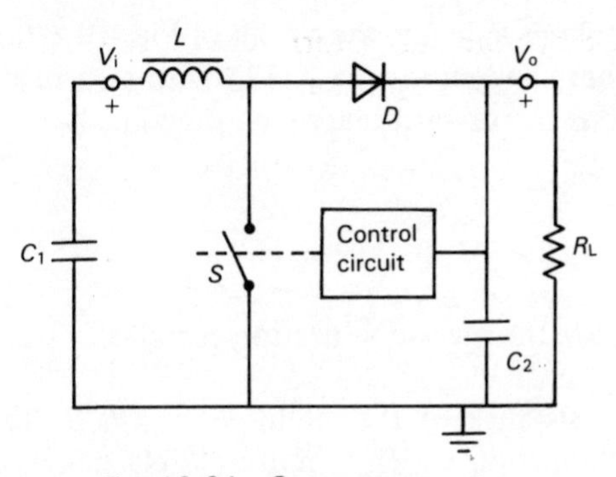

Fig. 19.21 Step-up converter

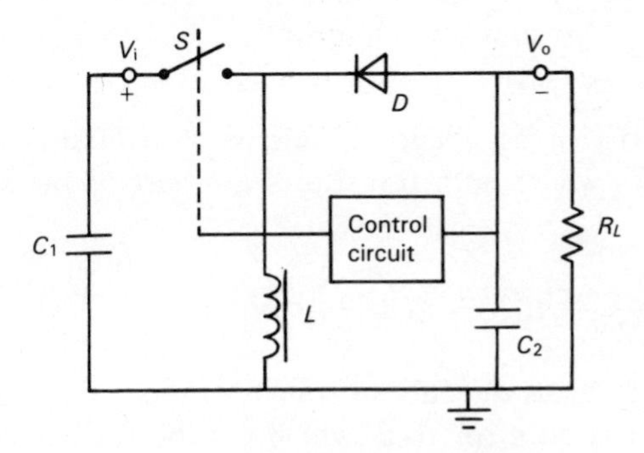

Fig. 19.22 Voltage inverter

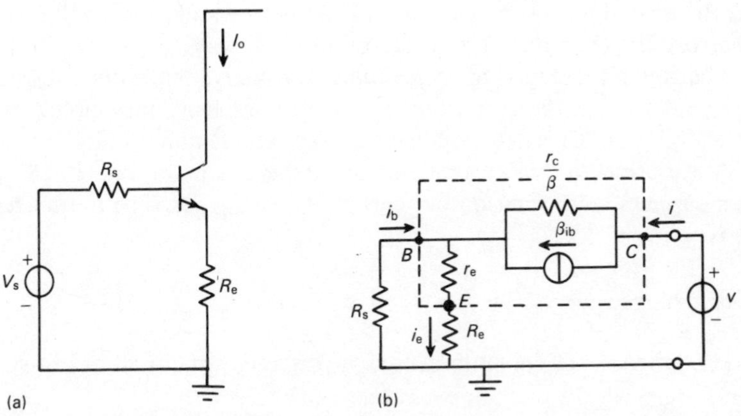

Fig. 19.23 (a) Basic current source with bipolar transistor (b) Equivalent circuit of current source to determine output resistance

possible. When making $R_e = 0$ in the circuit of Fig. 19.23(a) and connecting a diode between base and emitter (Fig. 19.25), the current transfer ratio is (using the equations of forward biased *pn* junctions)

$$\frac{I_o}{I_s} \simeq \frac{I_{s,t}}{I_{s,d} + \dfrac{I_{s,t}}{\beta}}$$

where $I_{s,t}$ and $I_{s,d}$ are the reverse saturation currents of transistor and diode respectively.

The output resistance of the circuit is $r_o \simeq r_c/\beta$. The diode can be replaced by a *pn* junction which is more closely matched to that of the transistor when using a similar transistor and connecting it as a diode (Fig. 19.26). The current transfer ratio is now

$$\frac{I_o}{I_s} \simeq \frac{I_{s,t}}{I_{s,d}\left(1 + \dfrac{1}{\beta_2}\right) + \dfrac{I_{s,t}}{\beta_2}}$$

This circuit is known as a **current mirror**. It is frequently applied in integrated circuits where both transistors are part of the same IC. Then $I_{s,t} \simeq I_{s,d}$ so that

$$\frac{I_o}{I_s} \simeq 1 - \frac{2}{\beta + 2}$$

Thus with high β values the current transfer ratio is approximately unity. The second term is an error term which can be further decreased when adding another transistor (Fig. 19.27). In this circuit

$$\frac{I_o}{I_s} \simeq 1 - \frac{2}{\beta^2 + 2\beta + 2}$$

Thus the second term is much smaller than before and even small β values give error terms less than 1 percent.

The input resistance of this circuit is $r_i \simeq 2r_e$, the output resistance $r_o \simeq \frac{1}{2} r_c$. Note that the current mirror is the complementary circuit of the emitter follower, where the voltage transfer ratio is about unity.

A variation of the current mirror is the circuit of Fig. 19.28. This circuit supplies a constant output current $I_o \simeq V_{BE,1}/R$. The output resistance is

$$r_o \simeq r_c \frac{R}{R + r_{e,1}}$$

Current mirrors are sometimes used in discrete circuits allowing simple

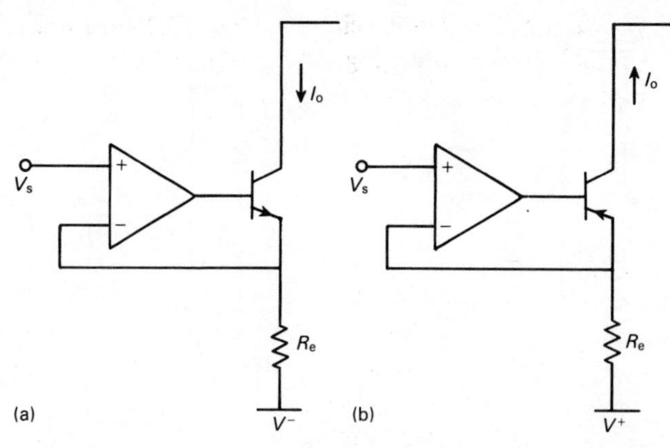

Fig. 19.24 (a) Current sink with operational amplifier (b) Current source with operational amplifier

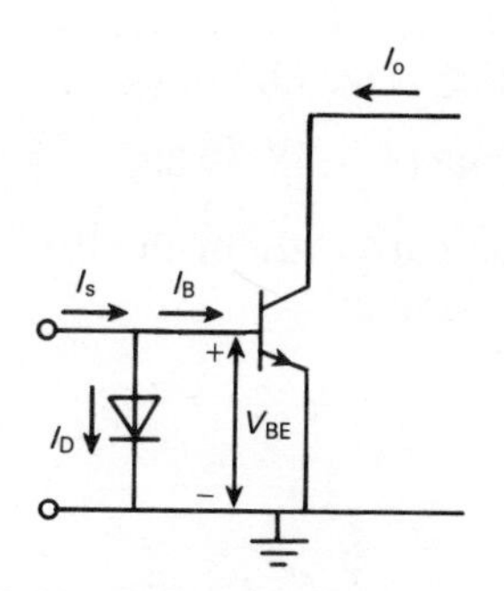

Fig. 19.25 Circuit to calculate current transfer ratio

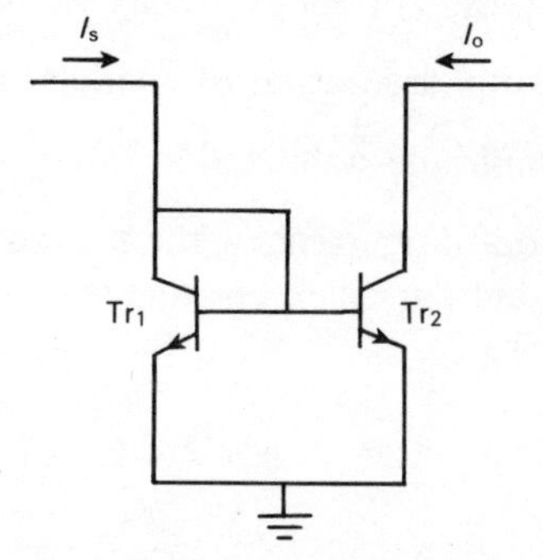

Fig. 19.26 Simple current mirror

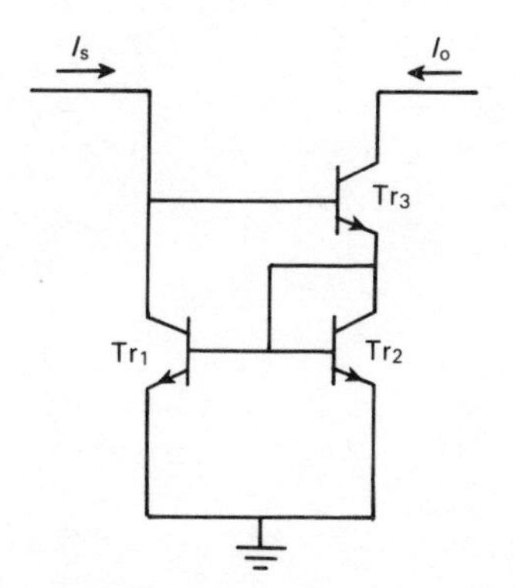

Fig. 19.27 Improved current mirror

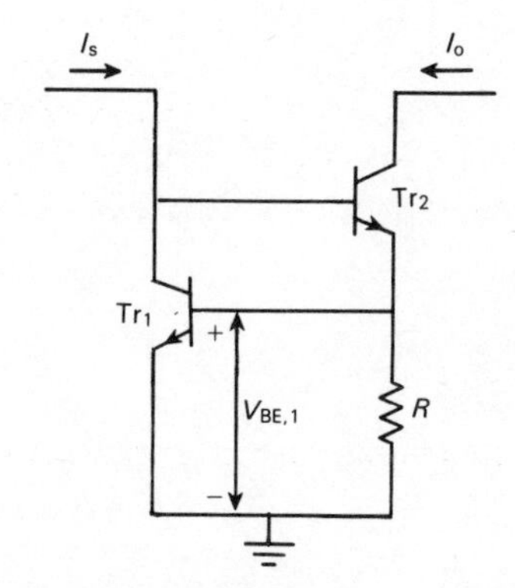

Fig. 19.28 Constant current source

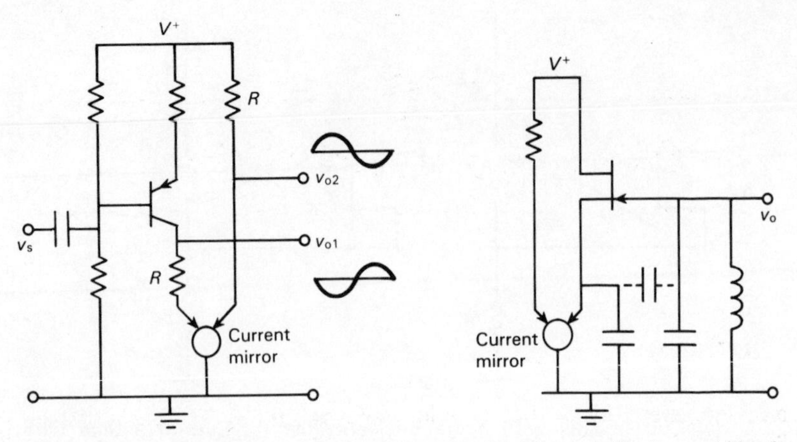

Fig. 19.29 Two-phase amplifier using current mirror

Fig. 19.30 Amplitude stabilization in an oscillator

solutions. The following are examples:

- Two-phase output of a single-stage amplifier (Fig. 19.29).
- Amplitude stabilization in a sine wave oscillator (Fig. 19.30).

The output current of the mirror is the drain current I_D of the FET. This constant current keeps the gain of the FET constant.

20
Distortion and Noise

20.1 GENERAL

If amplifiers were ideal they would have exactly linear chracteristics. Application of a sine wave input signal would result in a sinusoidal and undistorted output signal. Unfortunately, no such amplifiers exist and the output signal will not be an exact replica of the sine wave input signal. The difference is termed **distortion** which can be classified in different types.

Linear distortion

This is subdivided into **amplitude distortion** and **phase distortion** or **delay distortion.**

Amplitude distortion. When signal components of different frequencies are amplified differently (Fig. 20.1(a)) the output signal will be distorted. This distortion is caused by the fact that the amplitude characteristic of an amplifier is not a straight line but rolls off in certain frequency ranges. In these ranges amplitude distortion occurs and it is due to reactive components in the amplifier. Expressed mathematically: if the input signal is

$$v_s(t) = V_1 \sin \omega_1 t + V_2 \sin \omega_2 t$$

and the output signal

$$v_o(t) = A(\omega_1)\sin(\omega_1 t + \phi_1) + A(\omega_2)\sin(\omega_2 t + \phi_2)$$

the amplifier has amplitude distortion when $A(\omega_1) \neq A(\omega_2)$.

Phase Distortion or *Delay Distortion* This occurs when different frequency components have unequal phase shifts. If the output signal is

$$v_o(t) = AV_s(\cos \omega t - \phi) = AV_s \cos \omega\left(t - \frac{\phi}{\omega}\right)$$

the signal delay is $\Delta t = \phi/\omega$. Absence of delay distortion is thus expressed as $d\phi/d\omega = 0$. Delay distortion arises because the phase angle of $A(j\omega)$ depends upon the frequency.

Nonlinear distortion, frequency distortion or harmonic distortion

This most objectionable distortion is due to the nonlinear characteristics of the amplifier which produce new frequencies (harmonics of the fundamental) in the output signal that are not present in the input signal.

In linear amplifiers (for example high-quality audio amplifiers) reduction of this distortion requires careful and detailed circuit design.

The amount of distortion in a signal is expressed by the **distortion factor** D which is defined as

$$D = \sqrt{\frac{V_2{}^2 + V_3{}^2 + V_4{}^2 + \cdots}{V_1{}^2}}$$

where $V_2, V_3, V_4, \ldots$ are the amplitudes of second, third, fourth, ... harmonics and V_1 is the amplitude of the fundamental.

Distortion is often measured and expressed as a percentage:

$$d = D \times 100\%$$

If the input signal is $v_s = V_s \cos \omega t$, the output signal can be expanded in the series

$$v_o = a_1 v_s + a_2 v_s{}^2 + a_3 v_s{}^3 + \ldots$$

Application of trigonometric identities yields:

$$v_o = V_o + V_1 \cos \omega t + V_2 \cos 2\omega t + V_3 \cos 3\omega t + \cdots$$

where

$$V_o = \tfrac{1}{2} a_2 V_s{}^2 + \tfrac{3}{8} a_4 V_s{}^4 + \cdots$$
$$V_1 = a_1 V_s + \tfrac{3}{4} a_3 V_s{}^3 + \tfrac{5}{8} a_5 V_s{}^5 + \cdots$$
$$V_2 = \tfrac{1}{2} a_2 V_s{}^2 + \tfrac{1}{2} a_4 V_s{}^4 + \cdots$$
$$V_3 = \tfrac{1}{4} a_3 V_s{}^2 + \tfrac{5}{16} a_5 V_s{}^5 + \cdots$$

The separate distortion factors are thus

$$D_2 = \frac{|V_2|}{|V_1|} \quad \text{(second harmonic distortion)}$$

Figure 20.1(b) shows this distortion for a 20 percent second harmonic.

$$D_3 = \frac{|V_3|}{|V_1|} \quad \text{(third harmonic distortion)}$$

Figure 20.1(c) shows a signal with 20 percent third harmonic in phase with the fundamental; Fig. 20.1(d) shows a signal with 20 percent third harmonic in opposite phase with the fundamental.

Intermodulation distortion

This is a harmonic distortion which results when the input signal consists of different frequency components.

When the input signal consists of only two frequency components, namely $v_1 = V_1 \cos \omega_1 t$ and $v_2 = V_2 \cos \omega_2 t$, the output signal will contain all frequencies of the form $p\omega_1 + q\omega_2$ where p and q are integers.

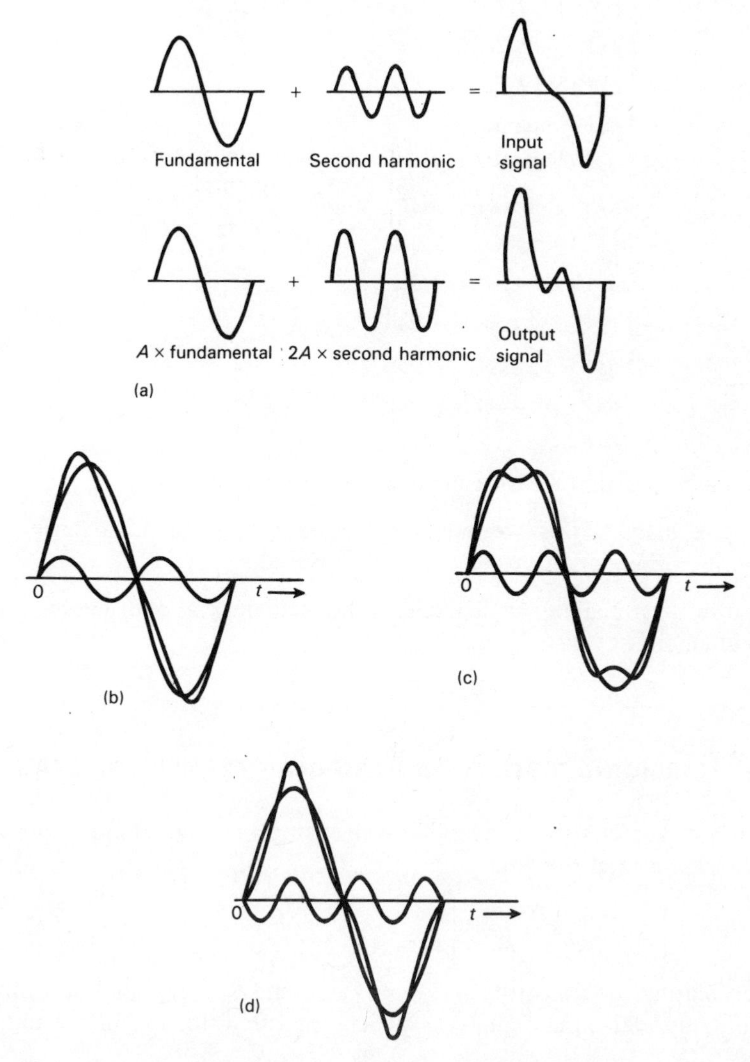

Fig. 20.1 (a) Example of amplitude distortion (b) Example of 20 percent second harmonic distortion (c) Example of 20 percent third harmonic in phase distortion (d) Example of 20 percent third harmonic opposite phase distortion

The output signal is thus

$$v_o = V_o + a_1(v_1 + v_2) + a_2(v_1 + v_2)^2 + a_3(v_1 + v_2)^3 + \cdots$$

A few mathematical manipulations yield the following frequency components in the output signal:

$$\begin{aligned}
v_o = V_o &+ \tfrac{1}{2} a_2 V_1^2 + \tfrac{1}{2} a_2 V_2^2 \\
&+ (a_1 V_1 + \tfrac{3}{4} a_3 V_1^3 + \tfrac{3}{2} a_3 V_1 V_2^2)\cos \omega_1 t \\
&+ (a_1 V_2 + \tfrac{3}{4} a_3 V_2^3 + \tfrac{3}{2} a_3 V_1^2 V_2)\cos \omega_2 t \\
&+ \tfrac{1}{2} a_2 V_1^2 \cos 2\omega_1 t \\
&+ \tfrac{1}{2} a_2 V_2^2 \cos 2\omega_2 t \\
&+ \tfrac{1}{4} a_3 V_1^3 \cos 3\omega_1 t \\
&+ \tfrac{1}{4} a_3 V_2^3 \cos 3\omega_2 t \\
&+ a_2 V_1 V_2 \cos(\omega_1 + \omega_2)t \\
&+ a_2 V_1 V_2 \cos(\omega_1 - \omega_2)t \\
&+ \tfrac{3}{4} a_3 V_1 V_2^2 \cos(\omega_1 + 2\omega_2)t \\
&+ \tfrac{3}{4} a_3 V_1 V_2^2 \cos(\omega_1 - 2\omega_2)t \\
&+ \tfrac{3}{4} a_3 V_1^2 V_2 \cos(2\omega_1 + \omega_2)t \\
&+ \tfrac{3}{4} a_3 V_1^2 V_2 \cos(2\omega_1 - \omega_2)t \\
&\ \ \vdots
\end{aligned}$$

Harmonic distortion measurements can be carried out using:

- A wave analyzer: this is a tuned detector permitting individual harmonics and intermodulation products to be measured.
- A distortion analyzer which rejects the fundamental and measures the total amount of distortion.

20.2 HARMONIC DISTORTION IN BIPOLAR TRANSISTOR STAGES

The input characteristic of a bipolar transistor stage (Fig. 20.2) is expressed by the exponential function

$$I_E \simeq I_s\left[\exp\left(\frac{q}{kT} V_{BE}\right) - 1\right]$$

When biasing the transistor at $V_{BE} = V_{BE,1}$ and $I_E = I_{E,1}$ and applying a small sinusoidal input signal $v_s = V_s \cos \omega t$ the distortion in the output

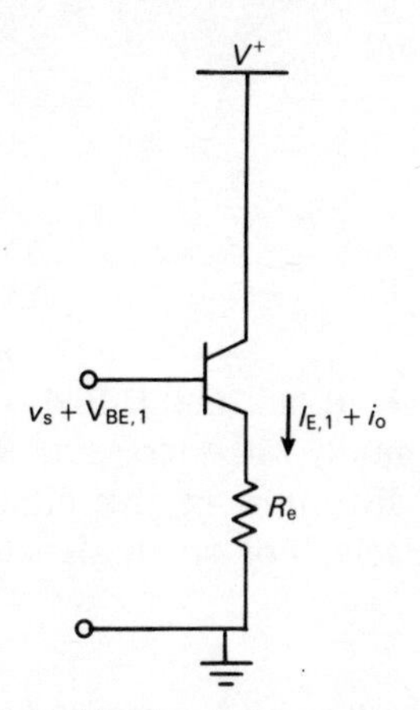

Fig. 20.2 Bipolar transistor stage to calculate harmonic distortion

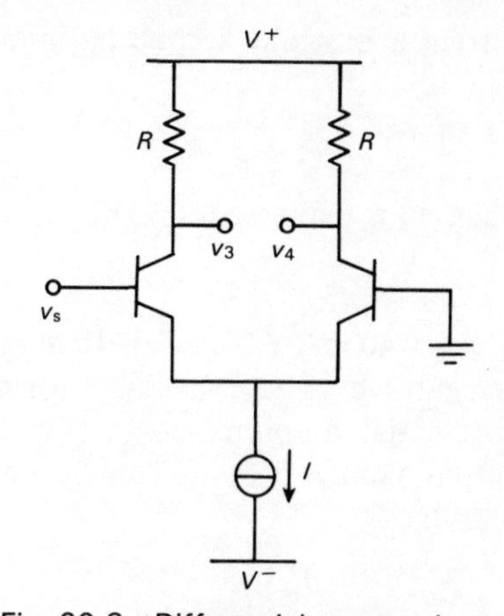

Fig. 20.3 Differential stage reduces harmonic distortion

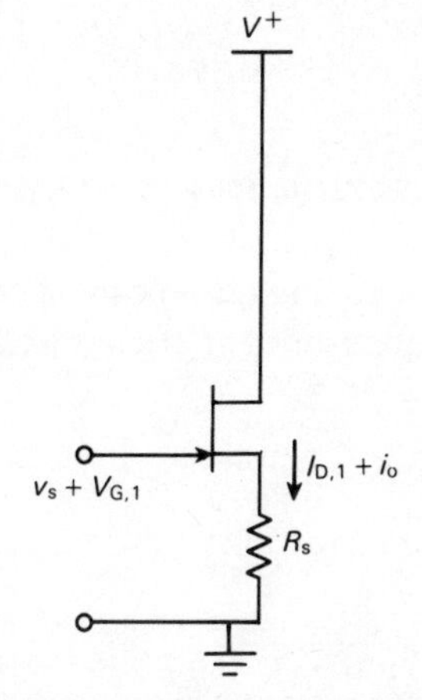

Fig. 20.4 JFET stage to calculate harmonic distortion

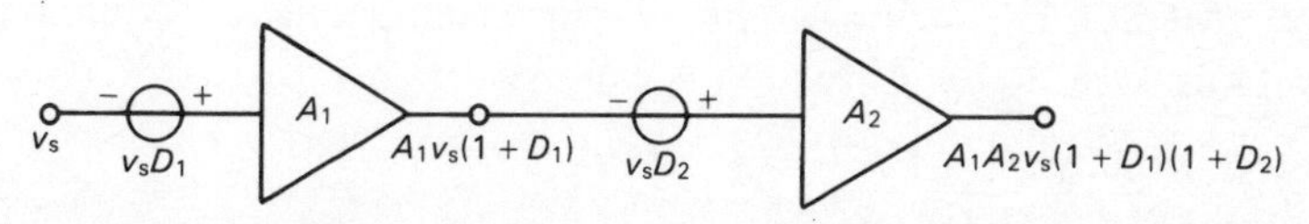

Fig. 20.5 Distortion in two cascaded stages

current is basically second harmonic distortion:

$$d \simeq d_2 \simeq \frac{V_s}{(1 + g_m R_e)^2} \ (\%)$$

when V_s is expressed in mV.

Example

If $g_m = 40 \text{ mA V}^{-1}$, $V_s = 10 \text{ mV}$, $R_e = 100\ \Omega$ it is found that $d \simeq 0.4$ percent. Obviously, the distortion decreases rapidly with increasing R_e. In a differential amplifier stage (Fig. 20.3) the distortion in the differential output voltage $v_3 - v_4$ can be derived similarly. For small signals v_s is found:

$$d \simeq d_3 \simeq \frac{V_s^2}{75} \ (\%)$$

when V_s is expressed in mV.

Apparently, distortion is reduced when applying **negative feedback** as in the circuit of Fig. 20.2 or by employing balanced stages (Fig. 20.3) where **compensation** reduces the distortion.

20.3 HARMONIC DISTORTION IN FET STAGES

The distortion in a JFET stage (Fig. 20.4) can be derived in a similar way as in the bipolar stage. The quadratic input characteristic

$$I_D = I_{DSS}\left(1 - \frac{V_{GS}}{V_P}\right)^2$$

results in second harmonic distortion:

$$d = d_2 \simeq V_s \frac{0.05 I_{DSS}}{g_m V_P^2} \frac{1}{(1 + g_m R_s)^2} \quad (\%)$$

when V_s is expressed in mV, I_{DSS} in mA, V_P in V.

Example

When $V_s = 10$ mV, $I_{DSS} = 16$ mA, $V_P = -4$ V, $g_m = 4 \text{ mA V}^{-1}$ and $R_s = 100\ \Omega$, $d_2 \simeq 0.06$ percent.

In a MOSFET, writing the input characteristic equation as $I_D = \gamma(V_{GS} - V_T)^2$, the distortion is expressed as

$$d \simeq V_s \frac{0.05\gamma}{g_m} \frac{1}{(1 + g_m R_s)^2} \ (\%)$$

20.4 DISTORTION IN A CASCADE

When considering two cascaded amplifier stages (Fig. 20.5) having distortion factors D_1 and D_2 respectively, the distortion of the output signal can be found by assuming the individual distortions to be caused by voltage sources $v_s D_1$ and $v_s D_2$ at the amplifier inputs. The amplifiers themselves are then distortion free.

The output signal distortion follows from a simple calculation:

$$D_o = \sqrt{D_1^2 + D_2^2 + D_1^2 D_2^2}$$

When distortions are not excessive, $D_1 \ll 1$, $D_2 \ll 1$ so that

$$D_o \simeq \sqrt{D_1^2 + D_2^2}$$

This is true for the percentage distortions as well. Therefore, the general equation for the distortion in a cascade of n stages is

$$d_{\text{casc}} \simeq \sqrt{d_1^2 + d_2^2 + \cdots + d_n^2}$$

20.5 NOISE AND ITS EFFECT ON LINEAR SYSTEMS

Noise is a random or stochastic signal which, unlike a deterministic signal, cannot be completely specified by a number of parameters. The knowledge of noise properties and the effect of noise on amplifiers is important since noise sets an inherent limit to the smallest signals that can be amplified and recognized. Such very small signals are encountered in areas like astronomy, radar and space technology.

The basic source of noise as it is observed on an oscilloscope or in a loudspeaker is the discrete nature of electric charges. Every component in an electric circuit produces noise, except at absolute zero temperature. As this is a somewhat inconvenient temperature to operate circuits in, designs often have to be arranged so as to minimize the effects of noise.

Noise, as a random signal, is described by statistical quantities. The average value of a noise signal is a useless parameter since it is always zero. Therefore, the magnitude of a noise signal is expressed by its mean square value. Another parameter is the frequency, since noise signals usually cover certain frequency bands.

Being an electrical signal, noise is divided into current noise and voltage noise. The notation is $\sqrt{\overline{i_n^2}}$ or i_n and $\sqrt{\overline{v_n^2}}$ or v_n respectively. The square of these rms values is the noise power dissipated in a 1 Ω resistor. The two parameters are often combined to represent the noise power per unit bandwidth, thus $A^2\,Hz^{-1}$ and $V^2\,Hz^{-1}$. This combined parameter is called the **spectral power density** or the **spectrum** for which the notations $S_i(f)$

and $S_v(f)$ are used. The total noise power in a certain frequency interval $f_1 - f_2$ is then the integral of the power density taken over this interval, thus

$$v_n = \sqrt{\int_{f_1}^{f_2} S_v(f)\,df} \quad \text{or} \quad i_n = \sqrt{\int_{f_1}^{f_2} S_i(f)\,df}$$

White noise has equal power density at all frequencies. It is a fictitious noise since infinite bandwidth implies infinite power. Nevertheless, noise is called white when it has a flat power spectrum over the frequency band of interest.

A noise voltage at the input of a linear system with transfer function $H(j\omega)$ causes an output noise voltage which equals the input noise voltage multiplied by $|H(j\omega)|$. Thus if $S_{in}(f)$ and $S_{out}(f)$ are the power densities of input and output noise,

$$\int_0^\infty S_{out}(f)\,df = \int_0^\infty S_{in}(f)\,|H(j\omega)|^2\,df$$

When assuming that $S_{in}(f)$ is constant in a certain frequency range ('band limited white noise') the output noise is

$$\int_0^\infty S_{out}(f)\,df = \overline{v_{no}^2} = S_{in} \int_0^\infty |H(j\omega)|^2\,df$$

When $H(0)$ is the transfer function at $f = 0$, the **noise bandwidth** of the system is defined as

$$B_n = \frac{1}{|H(0)|^2} \int_0^\infty |H(j\omega)|^2\,df$$

20.6 NOISE IN CONDUCTORS

20.6.1 Thermal noise

In a conductor, electrons have varying amounts of energy. When colliding with the vibrating atoms of the crystal, random motions and velocities result. Moving electrons constitute currents which produce voltages across the ends of the conductor. These noise voltages are called **thermal noise** or **Johnson noise**. The magnitude of this noise can be derived from the laws of thermodynamics and is expressed by the Nyquist equation

$$S_v = \overline{v_n^2} = 4kTRB \ (\mathrm{V}^2)$$

where

$k = 1.38 \times 10^{-23}$ J K^{-1} is Boltzmann's constant
T is the absolute temperature (K)
R is the resistance of the conductor (Ω)
B is the system bandwidth (Hz)

The equation shows that noise power is independent of a center frequency which means that thermal noise is white noise. For example, the noise voltage of a 1 MΩ resistor at room temperature, measured in a bandwidth of 10 kHz, is 13 μV.

If this resistor were the input resistor of a noiseless amplifier having a 1 MHz bandwidth, the input noise would be 1.3 mV. This could be excessive so the input resistor must be smaller. If the input noise is not to exceed, say, 10 μV the input resistor should be about 8 kΩ.

20.6.2 Excess noise

A DC current, flowing through a resistor or conductor, produces low frequency noise, the magnitude of which depends on the material of the conductor (Chapter 7). This noise is described by the equation

$$S_v = \overline{v_n^2} = \frac{KI^2R^2}{f} B \text{ (V}^2\text{)}$$

where K is a constant depending on the conductor material.

This noise is also termed $1/f$-noise or **flicker noise**, and it increases at 3 dB octave^{-1} with decreasing frequencies. Therefore in amplifiers low noise resistors are placed at critical locations.

20.6.3 Shot noise

Shot noise is caused by the fact that in a current-carrying conductor the number of charges per second which pass a cross section of the conductor fluctuates statistically.

Like thermal noise, shot noise is white noise. It is expressed by

$$S_i = \overline{i_n^2} = 2qB(I + 2I_s) \text{ (A}^2\text{)}$$

where

$q = 1.6 \times 10^{-19}$ C is the electron charge
I is the DC current (A)
I_s is the reverse saturation current (A)

Example

A forward biased diode can be represented by its dynamic resistance r_d in parallel with a noise current source i_n (Fig. 20.6). Because $r_d \simeq kT/qI_D$ the noise spectrum is

$$S_i = \frac{2kT}{r_d} B$$

If $r_d = 25\ \Omega$, the noise current in a bandwidth of 1 kHz is about 0.58 nA.

20.7 NOISE IN AMPLIFIERS

To specify and measure amplifier noise, two concepts can be used: noise factor and noise voltage, and current model.

20.7.1 Noise factor

The **noise factor** F compares noise of an actual amplifier with that of an ideal (noiseless) amplifier, both connected to the same signal source (Fig. 20.7). The output noise of an ideal amplifier is only due to the thermal noise v_{ns} or $\sqrt{\overline{v_{ns}^2}}$ of the source resistance R_s.

The definition of noise factor is

$$F = \frac{\text{noise power output of actual device}}{\text{power output due to source noise}}$$

If the amplifier has a voltage gain A, the noise voltage at the output is $\sqrt{\overline{v_{no}^2}} = A\sqrt{\overline{v_{ns}^2}}$. Across a load resistance R_L the noise factor can be written as

$$F = \frac{\overline{v_{no}^2}}{A^2\overline{v_{ns}^2}}$$

Since $\overline{v_{ns}^2} = 4kTBR_s$, the noise factor can be expressed as

$$F = \frac{\overline{v_{no}^2}}{A^2 \cdot 4kTBR_s}$$

For practical purposes the **noise figure** is often used which is defined as

$$NF = 10 \log F$$

An equivalent definition of F involves the **signal-to-noise ratio**. At the input this ratio is defined as

$$(S/N)_i = \frac{\text{signal power}}{\text{noise power}} = \frac{P_{si}}{P_{ni}} = \frac{\overline{v_s^2}}{\overline{v_{ns}^2}}$$

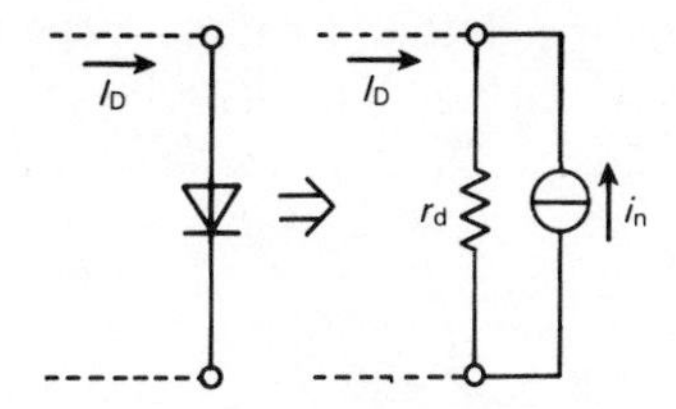

Fig. 20.6 Noise model of forward biased diode

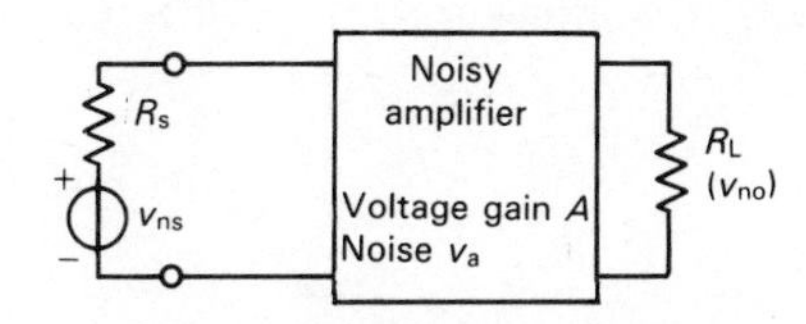

Fig. 20.7 Amplifier with load and source showing noise sources

where v_s is the input signal. Similarly, at the output

$$(S/N)_o = \frac{P_{so}}{P_{no}} = \frac{A^2 v_s^2}{\overline{v_{no}^2}}$$

The ratio of the two signal-to-noise ratios is

$$\frac{(S/N)_i}{(S/N)_o} = \frac{\overline{v_{no}^2}}{A^2 \overline{v_{ns}^2}}$$

Thus

$$F = \frac{(S/N)_i}{(S/N)_o}$$

If $\overline{v_a^2}$ is the amplifier noise, $\overline{v_{no}^2} = \overline{v_a^2} + A^2 \overline{v_{ns}^2}$ so that

$$\overline{v_a^2} = (F-1) A^2 \overline{v_{ns}^2}$$

Thus in a noiseless amplifier $F = 1$.

A noise quantity which is often used is the total equivalent input noise voltage v_{ni} which is defined as output noise voltage divided by the gain. It is also the total noise power in R_s, thus

$$v_{ni} = \frac{v_{no}}{A} = \sqrt{F \cdot 4kTBR_s}$$

Obviously, noise performance of the amplifier is optimum when v_{ni} is minimized.

20.7.2 Noise voltage and current

The actual amplifier is assumed to consist of a noisefree amplifier with two noise generators v_n and i_n connected to the input (Fig. 20.8). The noise voltage v_n is the amplifier noise when R_s is zero, the noise current i_n is the additional noise when R_s is not zero.

These noise sources are not necessarily independent and a possible relationship can be indicated by a correlation coefficient γ $(0 \leqslant \gamma \leqslant 1)$.

Normally the spread in values of v_n and i_n is such that γ can be assumed zero. It follows from Fig. 20.8 that

$$\overline{v_{ni}^2} = 4kTBR_s + \overline{v_n^2} + \overline{i_n^2}R_s^2$$

or

$$\frac{\overline{v_{no}^2}}{A^2} = \overline{v_{ns}^2} + \overline{v_n^2}\,\overline{i_n^2}R_s^2$$

The relationship between v_n, i_n and F is thus

$$F = 1 + \frac{1}{\overline{v_{ns}^2}}(\overline{v_n^2} + \overline{i_n^2}R_s^2)$$

The minimum value of F is

$$F_{min} = 1 + \frac{v_n i_n}{2kTB}$$

which occurs when

$$R_s = R_s^* = \frac{v_n}{i_n}$$

The output signal-to-noise ratio is

$$(S/N)_o = \frac{A^2 v_s^2}{\overline{v_{no}^2}} = \frac{v_s^2}{4kTBR_s + \overline{v_n^2} + \overline{i_n^2}R_s^2}$$

which has a maximum value

$$(S/N)_{o,\,max} = \frac{v_s^2}{\overline{v_n^2}}$$

when $R_s = 0$.

When comparing these results, apparently F is minimum when $R_s = v_n/i_n$ while $(S/N)_o$ is maximum when $R_s = 0$. Therefore a minimum noise figure does not automatically imply minimum amplifier noise. A minimum noise figure indicates that the ratio of amplifier noise to thermal noise is minimum.

This is an example of the limitations in the use of F. Other limitations are:

- If the source is reactive, $F = \infty$.
- If amplifier noise is much less than source noise, F values will be close to unity and differences are hardly noticed.

Maximum signal-to-noise ratio is equivalent to minimum equivalent input noise voltage. When $R_s = 0$, $\overline{v_{ni}^2} = \overline{v_n^2}$. Zero source resistance is un-

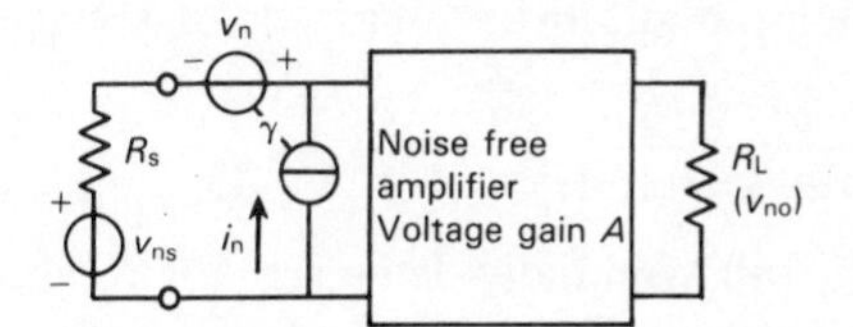

Fig. 20.8 Noise model of amplifier with amplifier noise referred to the input

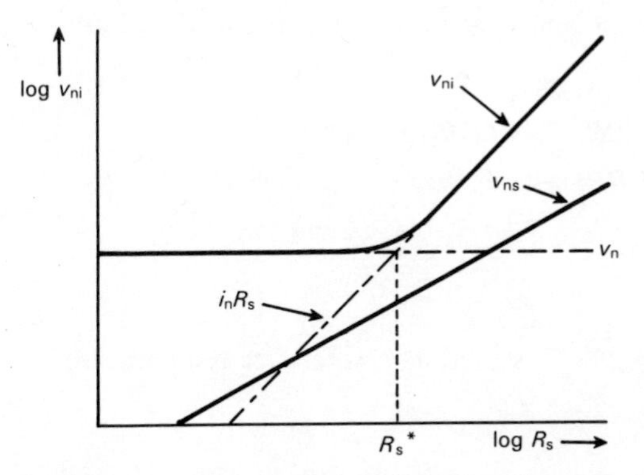

Fig. 20.9 Total input noise voltage as a function of source resistance

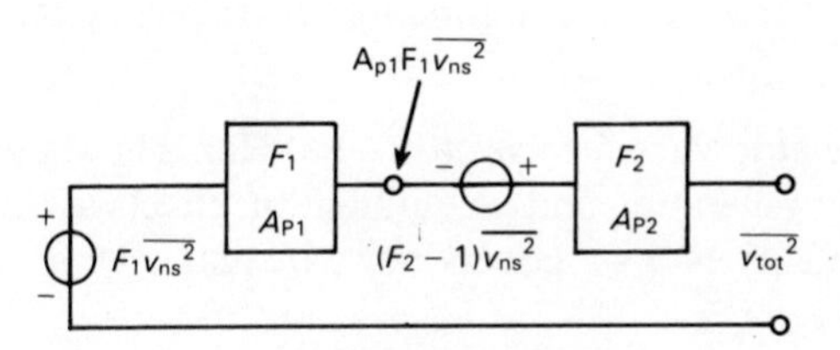

Fig. 20.10 Noise figure calculation of two cascaded stages

realistic. However, a range of R_s values exists where the total noise voltage is constant (Fig. 20.9).

In a **cascade** of two amplifier stages having individual noise factors F_1 and F_2 and power gains A_{P1} and A_{P2} (Fig. 20.10) the total noise power at the output is

$$\overline{v_{tot}^2} = A_{P1}A_{P2}F_1\overline{v_{ns}^2} + A_{P2}\overline{v_{no}^2}\,(F_2 - 1)$$

so that the noise factor of the cascade is

$$F_{tot} = F_1 + \frac{F_2 - 1}{A_{P1}}$$

This formula can be generalized for a cascade of n stages so that

$$F_{tot} = F_1 + \frac{F_2 - 1}{A_{P1}} + \frac{F_3 - 1}{A_{P1}A_{P2}} + \cdots + \frac{F_n - 1}{A_{P1}A_{P2}\ldots A_{P(n-1)}}$$

Thus if $A_{P1} \gg 1$, F_1 will essentially determine the total noise factor.

To distinguish between F values which are close to unity, the **noise temperature** T_n is introduced.

The amplifier noise was expressed as $\overline{v_a^2} = \overline{v_{ns}^2}\,(F - 1)$. This noise can be thought of as being produced by a fictitious resistor equalling R_s but having a different temperature and located at the amplifier input. Thus

$$\begin{aligned}\overline{v_{ni}^2} &= \overline{v_{ns}^2} + \overline{v_a^2} \\ &= 4kTBR_s + 4kTBR_s(F - 1) \\ &= 4kTBR_s + 4kT_nBR_s\end{aligned}$$

so that

$$T_n = T(F - 1)$$

Normally, $T = 290$ K is used as a reference temperature.

20.8 NOISE IN A BIPOLAR TRANSISTOR

In a bipolar transistor, noise as a function of frequency can be divided into three ranges:

(a) Excess noise at low frequencies (0.1–100 Hz). The sources of this noise are probably generation and recombination effects at the crystal surface and leakage current I_{cbo}. Excess noise increases with increasing values of I_B and V_{CE}.

(b) Midfrequency noise which is due to:

- Shot noise, caused by fluctuations in I_B.
- Shot noise, caused by fluctuations in I_C.
- Thermal noise, caused by the base resistance r_b.

(c) High frequency noise which increases at 6 dB octave^{-1} because β decreases as a result of parasitic capacitances.

The noise picture of a bipolar transistor is shown in Fig. 20.11 where noise is expressed in the noise figure *NF*.

The noise model of a bipolar transistor in common-emitter configuration is shown in Fig. 20.12. When assuming $\beta \gg 1$ and I_{cbo} negligible, the

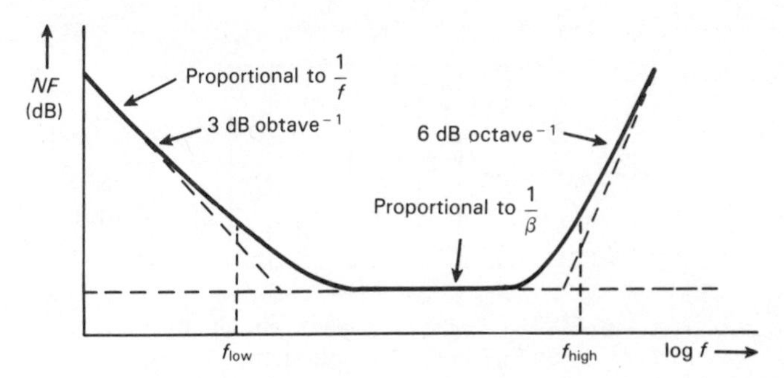

Fig. 20.11 Noise figure of a bipolar transistor as a function of frequency

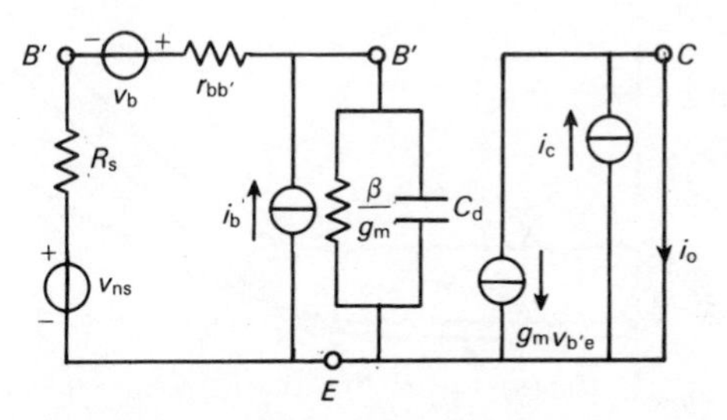

Fig. 20.12 Noise equivalent circuit of bipolar transistor in CE configuration

noise factor at medium and high frequencies is eventually found as

$$F \simeq 1 + \frac{1}{2g_m R_s}\left[1 + 2g_m r_b + \frac{g_m^{\,2}(R_s + r_b)^2}{\beta}\right]$$

while

$$F_{min} \simeq 1 + \frac{g_m R_s^*}{\beta} \quad \text{for } R_s^* = \frac{1}{g_m}\sqrt{\beta(1 + 2g_m r_b)}$$

At low frequencies, where excess noise might be dominant, larger values of R_s^* have to be used.

20.9 NOISE IN AN FET

The principal sources of noise in an FET are:

- Thermal noise in the conducting channel

$$\overline{v_{ch}^{\,2}} = 4kTB\,\frac{1}{g_m}$$

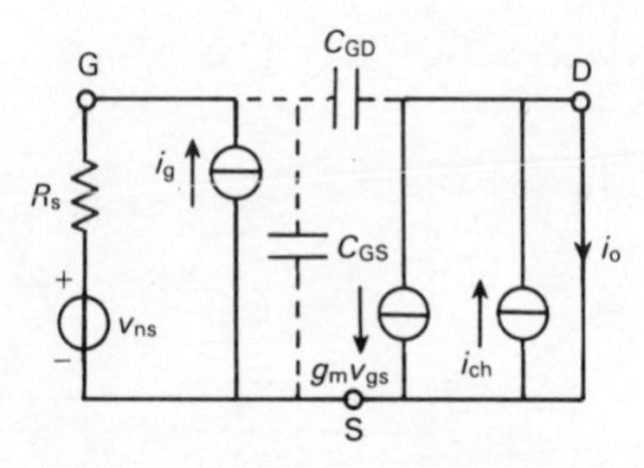

Fig. 20.13 Noise equivalent circuit of an FET in CS configuration

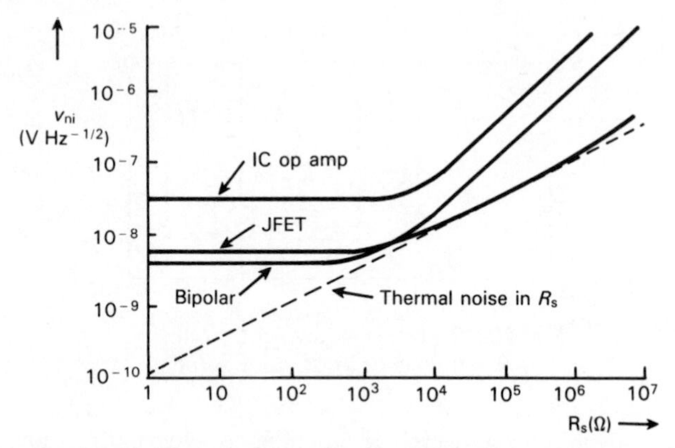

Fig. 20.14 Comparison of noise in bipolar transistor, FET and operational amplifier

- shot noise caused by leakage currents between gate and channel

  $$\overline{i_g^{\,2}} = 2qI_G B$$

  where I_G is the gate leakage current; values of I_G are 10^{-8}–10^{-10} A for a JFET and 10^{-15}–10^{-16} for a MOSFET.

- excess noise due to generation and recombination effects in the depletion layer between gate and channel.

The noise model of an FET is shown in Fig. 20.13. At medium and high frequencies the noise factor is

$$F \simeq 1 + \frac{1}{g_m R_s^*} \quad \text{where } R_s^* = \sqrt{\frac{2kT}{qI_G g_m}}$$

Practical values of F are close to unity. Drain current and drain-source voltage have little effect on F.

In low-frequency applications larger values of R_s^* are required due to the excess noise.

20.10 NOISE IN OTHER COMPONENTS

Operational amplifier input stages often consist of at least two transistors and noise voltage is thus at least $\sqrt{2}$ that of a single transistor stage. With bipolar transistors, β values in integrated circuits are usually not very high which has a negative effect on output noise.

Generally, operational amplifiers produce more noise than discrete amplifiers. A comparison between the three devices is shown in Fig. 20.14.

- Zener diodes produce shot noise when their breakdown voltages are less than about 5 V, namely

  $$\overline{i_z^2} = 2qI_zB$$

  where I_z is the Zener current. When breakdown voltages are 5 V and higher, the avalanche effect increases the noise.

- Tunnel diodes have very low noise; the nature of it is essentially shot noise.

- Capacitors have series resistance and parallel resistance (leakage resistance). These resistances produce some noise.

- Inductors produce some noise due to the loss resistance.

21
Nonlinear Wave Shaping

21.1 LIMITING

Limiting or **clipping** circuits are used to remove unwanted positive or negative extremities of an arbitrary waveform and to retain the other part. For example, in Fig. 21.1 the waveform v_s is clipped when $v_s \geqslant V_r$ where V_r is a reference voltage which establishes the clipping level. The general character of the waveform is thus changed.

Clipping action is achieved by switching devices such as diodes and transistors. In **shunt clipping** the output is in parallel with the clipping device (Fig. 21.2(a)) and clipping occurs when the diode conducts. The clipping level is $V_r + V_D$ where V_D is the forward voltage drop of the diode. Figure 21.2(b) is the voltage transfer characteristic, Fig. 21.2(c) shows the clipping action applied to a sine wave input.

The dotted lines indicate that the discontinuities in the slope of the waveform are not sharp due to the nonideal characteristic of the diode. This implies that clipping deteriorates markedly when V_s is comparable with $V_r + V_D$.

When fast waveforms are applied (Fig. 21.3) the parasitic capacitances in the circuit (diode and wiring) cannot be neglected and a clipped square wave will show exponential rise and decay.

The voltage source V_r is often provided by a Zener diode (Fig. 21.4(a)) where the clipping level is $V_z + V_D$. A practical example is the meter protection circuit of Fig. 21.4(b) where two Zener diodes are used in anti-series connection. The voltage across the meter is limited between the values $|V_z + V_D|$.

Various clipping circuits are shown in Fig. 21.5. In Figs 21.5(a) and 21.5(c) **series clipping** is employed. The clipping device is then in series with the input–output line and clipping occurs when the diode is nonconducting.

When using a transistor as the clipping device, two-level clipping is obtained as shown in Fig. 21.6. If sufficient input signal is applied, the stage is overdriven, clipping the lower part of the output signal at $V_z + V_{sat}$, the upper part at V^+.

When leaving the Zener diode out, the clipping levels of the simple amplifier stage are V_{sat} and V^+. A limiter of this kind is used in FM

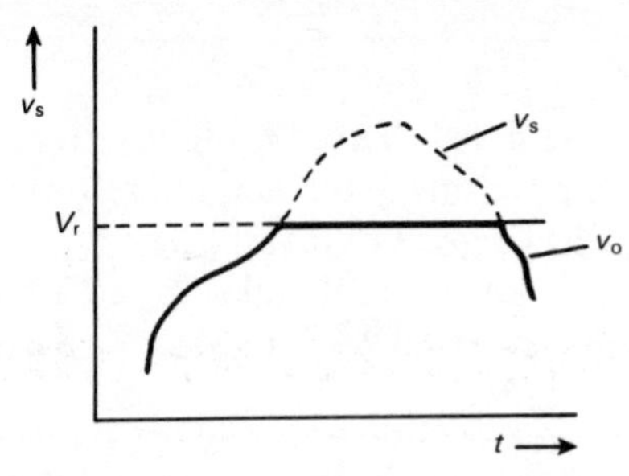

Fig. 21.1 Clipping circuit for voltages $v_s \geqslant V_r$

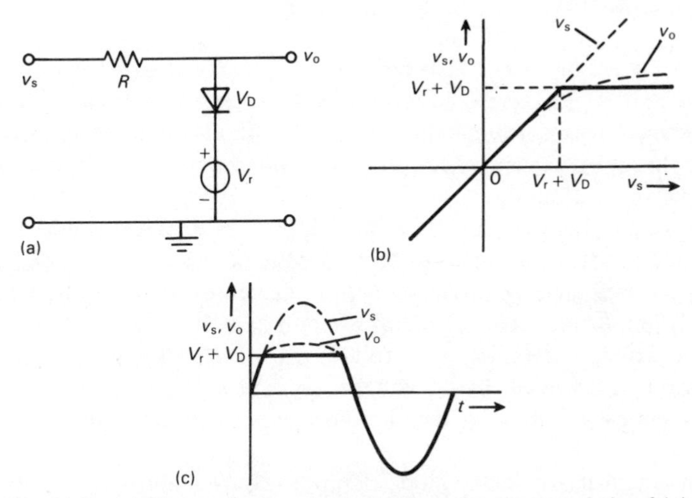

Fig. 21.2 (a) Shunt clipping circuit (b) Transfer characteristic of (a) (c) Clipping of sinusoidal waveform

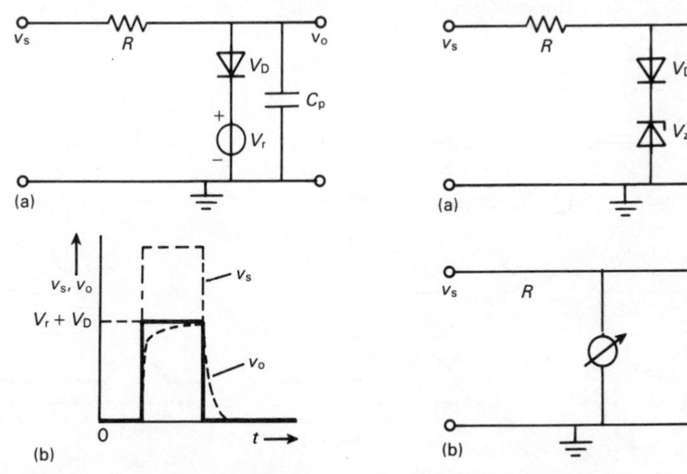

Fig. 21.3 (a) Parallel capacitance in shunt clipper (b) Effect of parallel capacitance on fast waveform

Fig. 21.4 (a) Diodes in clipping circuits (b) Meter protection circuit using anti-series connected Zener diodes

receivers where the FM signal, before being detected, is passed through the limiter in order to remove unwanted amplitude variations.

A limitation of the overdriven transistor is its charge storage effect: the transition from saturation to cutoff takes place after a short delay. This limitation is effectively removed when connecting a Schottky diode between collector and base (Fig. 21.7).

21.2 CLAMPING

A **clamping** circuit or **DC restorer** sets a positive or negative extremity of a waveform to a predetermined level, thereby preserving the character of the waveform. A waveform may have lost its DC level, for example when it has passed through a capacitive coupling. Clamping circuits are commonly used in television and radar.

A simple clamping circuit is shown in Fig. 21.8. When v_s goes positive, the diode conducts, keeping v_o at zero volts (assuming the diode is ideal) while C is being charged. As soon as v_s decreases, v_o will follow and the diode is immediately cut off. Thus as soon as v_s has passed the maximum positive voltage level, v_o will follow v_s, which means that the input waveform is lowered by a voltage equal to its own amplitude. All subsequent excursions of v_o will therefore have their maximum value at zero volts.

In practical circuits the diode clamps the output above zero volts due to the forward voltage drop V_D. In addition, the positive excursions of v_o will extend somewhat above the clamping level due to a slight discharge of C through the reverse biased diode.

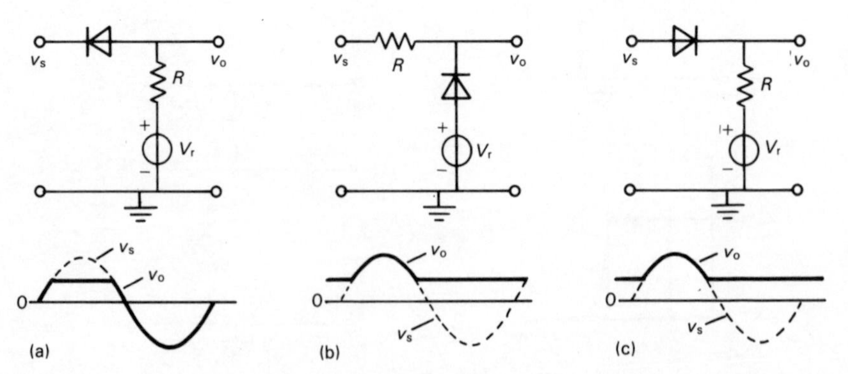

Fig. 21.5 (a) Series clipping circuit and clipped sinusoid (b) Shunt clipping circuit and clipped sinusoid (c) Series clipping circuit and clipped sinusoid

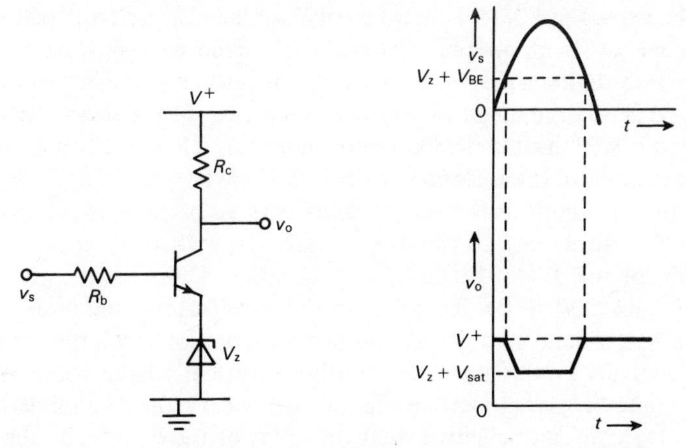

Fig. 21.6 Transistor and Zener diode provide two-level clipping

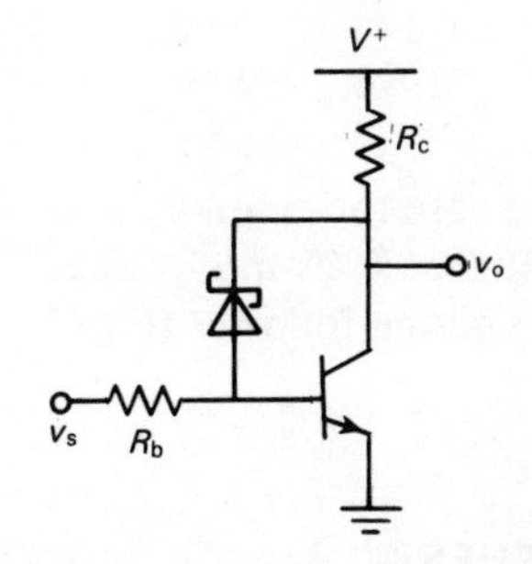

Fig. 21.7 Schottky diode clipping

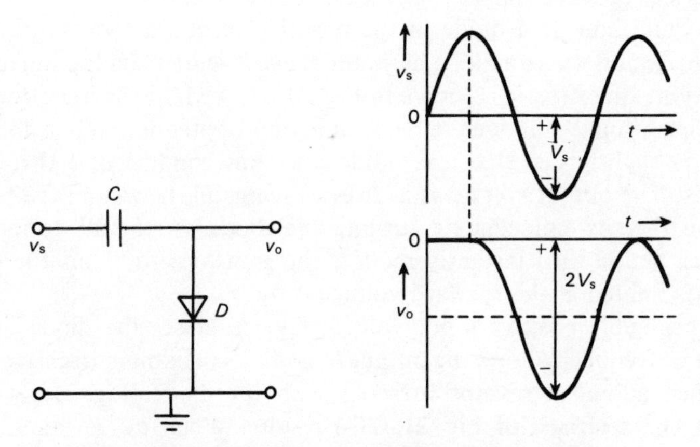

Fig. 21.8 Clamping circuit using a single diode

When the amplitude V_s is suddenly decreased the circuit will show a long recovery time before the reference level is reached since the capacitor is looking into a diode which is cut off. Therefore, a resistor R is often connected across the diode to improve the speed of the clamp circuit. This resistor, as well as the source resistance R_s, affect the waveform. In particular when v_s is a rectangular or square wave signal (Fig. 21.9(a)), the effects of time constants become apparent. A positive excursion of v_s charges C with a time constant $\tau_c = C(R_s + r_d)$ where r_d is the dynamic resistance of the diode. When v_s goes negative, C discharges with a time constant $\tau_d = C(R_s + R)$. To preserve the waveform as much as possible requires $\tau_c \ll T_1$ and $\tau_d \gg T_2$. A compromise solution is a large value of R and a relatively small value of C. With a sufficiently large value of C the output signal is as shown in Fig. 21.9(b) where the clamping level is $V_c = V_r + V_a$. It can be shown that the areas of the waveform above and below V_r are proportional to the relevant time constants in these areas. It follows then that

$$V_a = V_s \frac{1}{1 + \dfrac{T_1}{T_2}\dfrac{R}{r_d}}$$

Thus large values of R decrease the error V_a.

Realization of an adjustable clamp voltage is easy when using an operational amplifier as a voltage follower (Fig. 21.10).

21.3 WAVEFORM SYNTHESIS

An operational amplifier, combined with resistor-diode circuits allows a large variety of wave shaping operations to be performed.

The connection of a diode in the negative feedback loop removes the transition region where a diode hesitates between a little conduction and full conduction. An example is shown in Fig. 21.11(a) where the inverting input of the operational amplifier remains at ground potential. When the input voltage is slightly negative, the diode does not conduct and there is no feedback. The output voltage thus tries to swing all the way in the positive direction thereby immediately turning the diode on to full conduction. Feedback action then instantly reduces the gain to zero. Thus the output voltage is limited at the forward voltage drop V_D.

A negligible positive input voltage reverse biases the diode and the absence of feedback allows the output to go to its maximum negative value.

When adding a resistor in series with the diode (Fig. 21.12(a)) the transfer characteristic of Fig. 21.12(b) results. When v_o' is taken as the

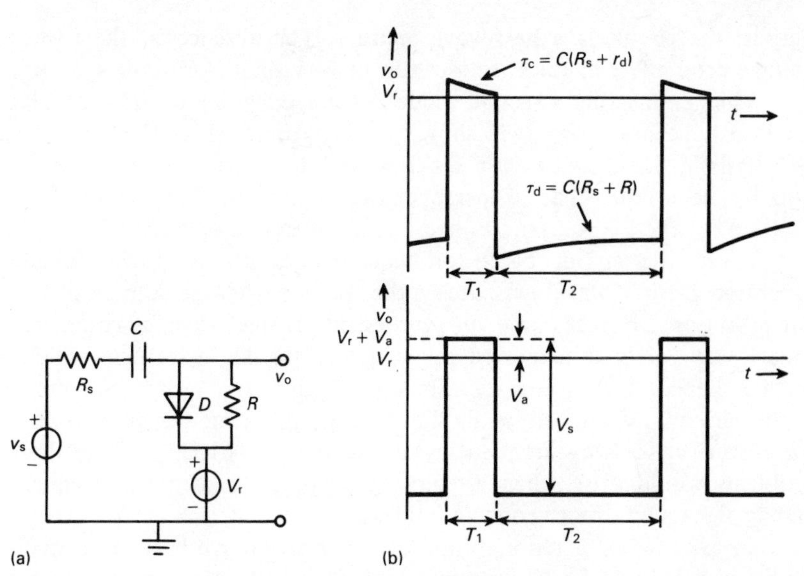

Fig. 21.9 (a) Time constants affect fast waveform clamping (b) Clamped waveforms with two different time constants

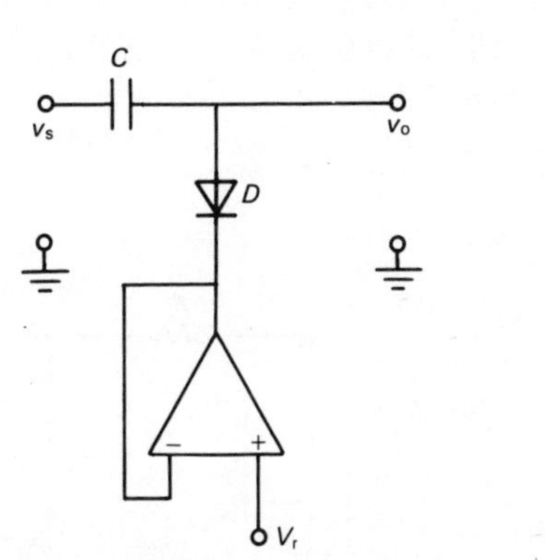

Fig. 21.10 Operational amplifier providing an adjustable clamp voltage

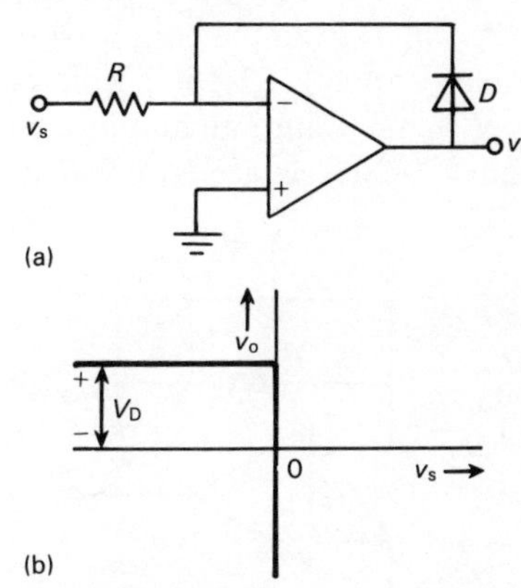

Fig. 21.11 Waveform shaping circuit and transfer characteristic

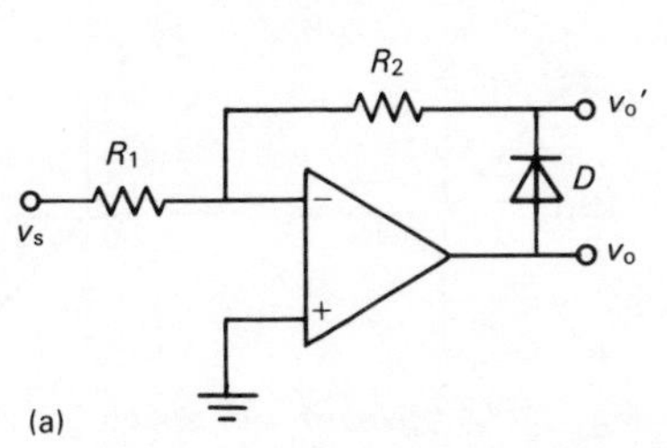

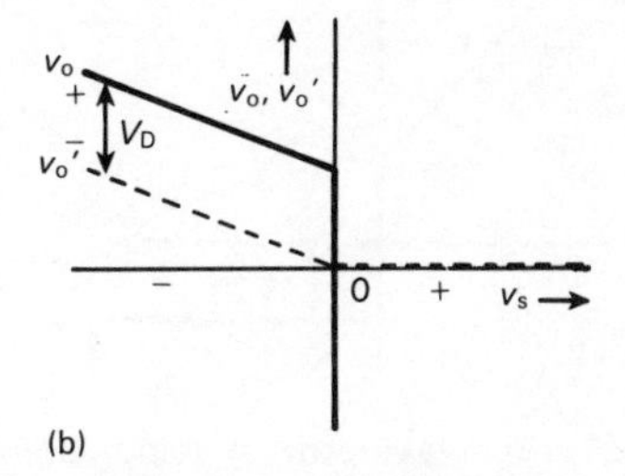

Fig. 21.12 Waveform shaping circuit and transfer characteristics

output, the circuit is a half-wave rectifier. The absence of the forward voltage drop effect enables rectification of very small AC signals.

When connecting a second diode in the feedback loop (Fig. 21.13(a)) the transfer characteristic will show two breakpoints (Fig. 21.13(b)).

Resistor R in series with D_2 (Fig. 21.14(a)) turns the circuit into a **polarity separator**. The transfer characteristics of Figs. 21.14(b) and 21.14(c) for the two outputs v_o and v_o' illustrate the operation.

A versatile circuit based on waveform synthesis is the **function generator**. Desired waveforms are synthesized from a large number of small linear sections. For example, in synthesizing a sine wave, a symmetrical square wave is first integrated to yield a triangular waveform which is further shaped into a sine wave. The advantage of synthesizing waveforms is the absence of capacitors in the process, allowing the generation of waveforms of very low frequencies such as 0.01–0.001 Hz. Serious design problems would arise when attempting to generate such low frequency analog signals in a more conventional way.

An example of a more complicated waveform synthesizing circuit is shown in Fig. 21.15(a). The circuit contains four diodes, ten resistors and two bias voltages. The resulting transfer characteristic, having four breakpoints, is shown in Fig. 21.15(b).

A final example of waveform synthesis is the pulse-amplitude modulator of Fig. 21.16(a) which combines a modulation signal v_m and a carrier signal v_c into an amplitude modulated signal v_o. Equal positive and negative DC levels are established by the two amplifiers. The modulation

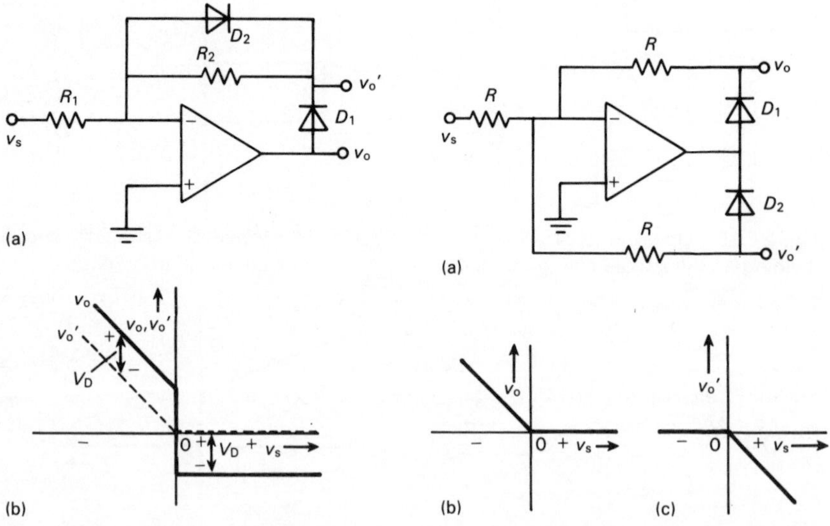

Fig. 21.13 Waveform shaping circuit and transfer characteristics

Fig. 21.14 Waveform shaping circuit and transfer characteristics

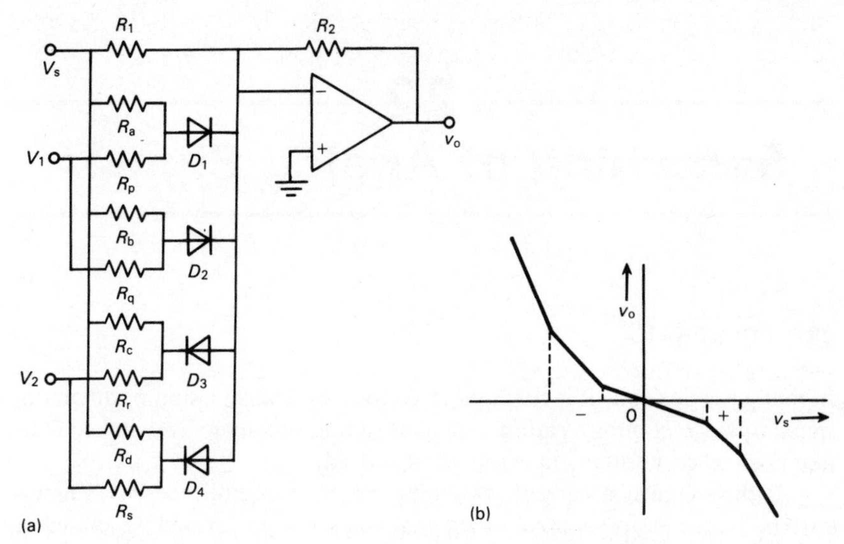

Fig. 21.15 More complex waveshaping circuit and transfer characteristic

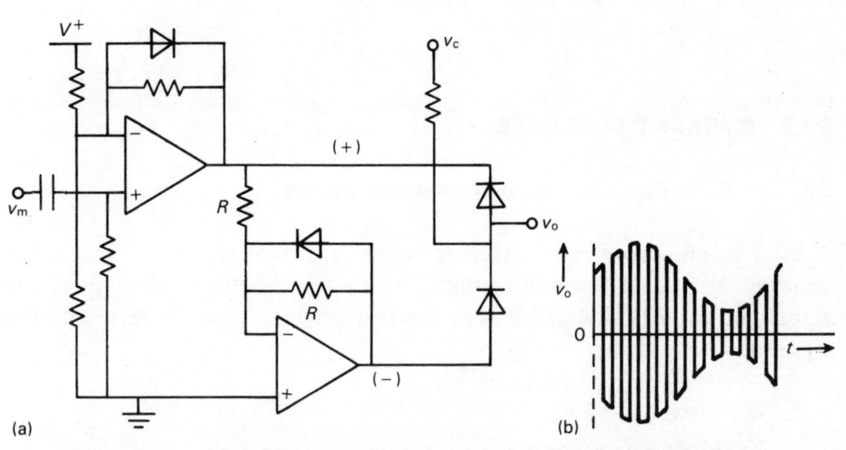

Fig. 21.16 Example of amplitude modulation circuit and waveform

signal v_m is superimposed on these levels. The instantaneous output voltages of the amplifiers are the limiting levels for the carrier signal v_c which is assumed to consist of large rectangular pulses. The diodes in the amplifier feedback loops prevent the outputs from reversing polarity when large modulation signals are applied. Figure 21.16(b) is the resulting amplitude modulated waveform.

22
Switching of Analog Signals

22.1 GENERAL

In many applications it is required to control the transmission of an analog signal through communications channels so that this signal is switched from one channel to another, interrupted or passed.

If the signal is a current, switching can be accomplished by a **current switch.** Transmission control of a signal voltage is performed by an **analog gate** or **transmission gate**. A digital waveform (called gating signal or control signal) controls the switching action.

22.2 CURRENT SWITCHES

22.2.1 Current switch with bipolar transistors

A differential stage (Fig. 22.1(a)) is designed so that transistors can never saturate. When a difference voltage ΔV is applied between the inputs, the signal current I is shared between the two transistors as expressed by the equations:

$$\frac{I_{C1}}{I} = \frac{\exp\left(\frac{q}{kT}\,\Delta V\right)}{1 + \exp\left(\frac{q}{kT}\,\Delta V\right)}$$

$$\frac{I_{C2}}{I} = \frac{1}{1 + \exp\left(\frac{q}{kT}\,\Delta V\right)}$$

Figure 22.1(b) shows I_{C1}/I and I_{C2}/I as a function of ΔV. Switching back and forth between the two channels thus requires approximately $|\Delta V| = 340$ mV.

An application is the modulation circuit (Fig. 22.2). The signal current is derived from the modulation voltage while the carrier switches the current between the two transistors (Chapter 23).

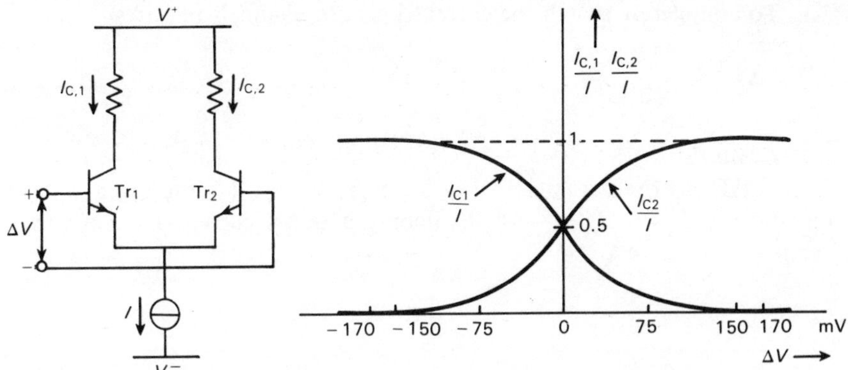

(a) (b)

Fig. 22.1 (a) Differential stage current switch with bipolars (b) Transfer characteristics of (a)

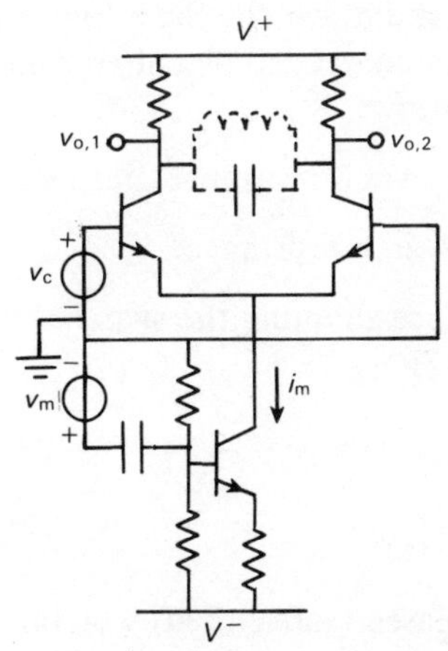

Fig. 22.2 Application of current switch in modulator circuit

22.2.2 Current switch with FETs

If the drain current of the identical FETs is expressed as

$$I_D = a(V_{GS} \pm b)^2$$

(Fig. 22.3(a)) the current equations are

$$\frac{I_{D1}}{I} = \frac{1}{2}\left[1 + \Delta V \sqrt{\frac{a}{I}} \sqrt{2 - (\Delta V)^2 \frac{a}{I}}\right]$$

$$\frac{I_{D2}}{I} = \frac{1}{2}\left[1 - \Delta V \sqrt{\frac{a}{I}} \sqrt{2 - (\Delta V)^2 \frac{a}{I}}\right]$$

These equations are shown in Fig. 22.3(b).

To completely switch the current between channels requires

$$|\Delta V| = 2\sqrt{\frac{I}{a}}$$

Example

If a JFET has the parameters $I_{DSS} = 8$ mA, $V_p = -4$ V the drain current equation is $I_D = 0.5(V_{GS} + 4)^2$. Switching a signal current $I = 2$ mA then requires $|\Delta V| = 4$ V.

22.3 ANALOG SWITCHES

The three basic circuits are shown in Figs. 22.4(a), (b) and (c): series switch, shunt switch and series-shunt switch. Since these switches are (nonideal) electronic devices, some unintended signal transmission may occur. It is therefore important to know:

- When S is closed: the switch resistance and its characteristic.
- When S is open: the switch resistance.
- The parasitic capacitance shunting the switch.

22.3.1 Diode gates

Two-diode gate (Fig.22.5(a))

The diodes are reverse biased (gate closed) when the control signals are V_2 and $-V_2$ respectively.

For reasons of balance, the control voltages should be of identical shape (including rise and fall times). If not, spikes or **glitches** will appear during opening or closing.

For the same reason the diodes should be identical (matched).

Four-diode gate (Fig. 22.5(b))

The gate is open when $V_c = V_1$ and closed when $V_c = -V_2$. Diode gates should be so designed that during transmission the diodes remain conducting over the full range of the analog signal and remain cut off during nontransmission.

Often Schottky diodes are used for higher operating speed. Matched diodes are available as integrated arrays.

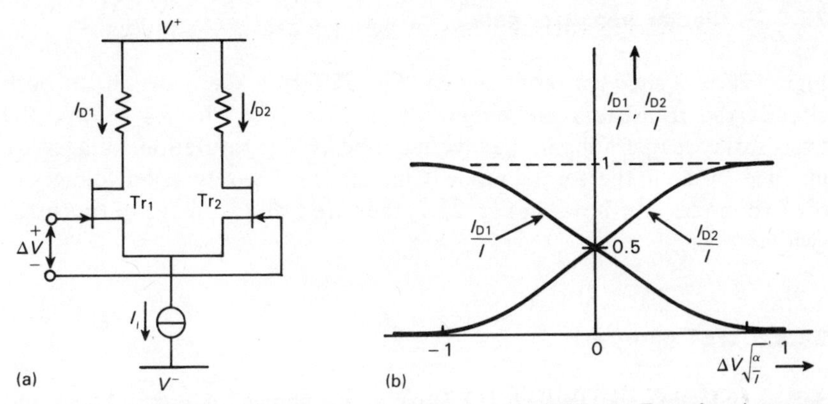

Fig. 22.3 (a) Differential stage current switch with JFETs (b) Transfer characteristics of (a)

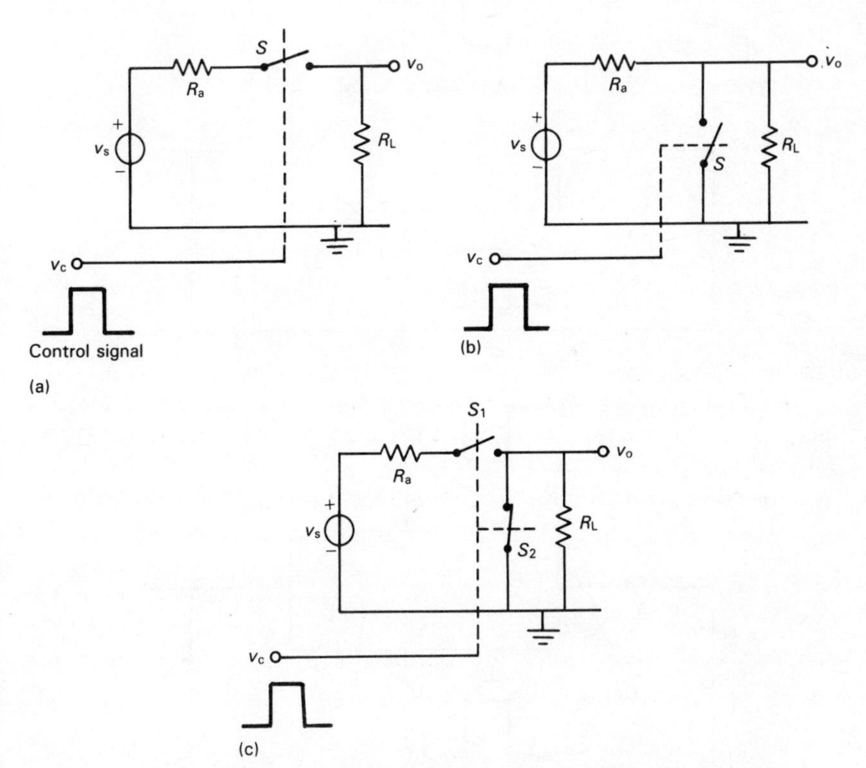

Fig. 22.4 (a) Series switch (b) Shunt switch (c) Series-shunt switch

22.3.2 Bipolar transistor gate

Figure 22.6(a) shows a series switch, Fig. 22.6(b) a shunt switch. In both circuits the transistors are either saturated or cut off. As shown, the transistors are operating in the inverted mode where saturation voltages are much less than in the normal mode (Chapter 10). The saturation voltages in the two modes are shown in Fig. 22.7; they are a disadvantage of transistor switches.

22.3.3 FET gate

Series and shunt controlled JFET gates are shown in Figs. 22.8(a) and 22.8(b) respectively. The advantage of using FETs is the absence of offset voltages. Moreover, FET characteristics extend through the origin

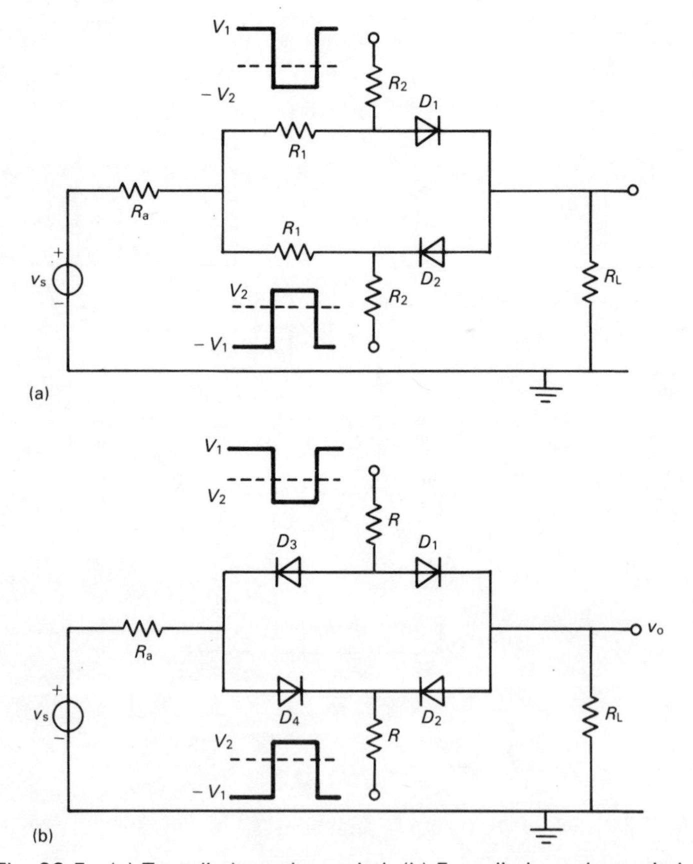

Fig. 22.5 (a) Two-diode analog switch (b) Four-diode analog switch

(Fig. 22.9) enabling operation in the forward and inverse mode. A disadvantage is the relatively large voltage swing to turn the FET on and off (Section 22.2.2).

Driver circuits

The JFET of Fig. 22.8(a) is turned on and off when V_{GS} swings between 0 V and V_P (pinch-off voltage). If R_a is small and $v_c = 0$ V (FET on), the analog

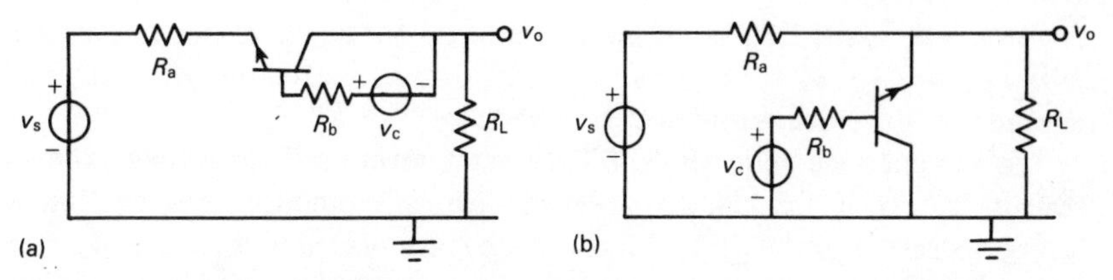

Fig. 22.6 (a) Series transistor analog switch (b) Shunt transistor analog switch

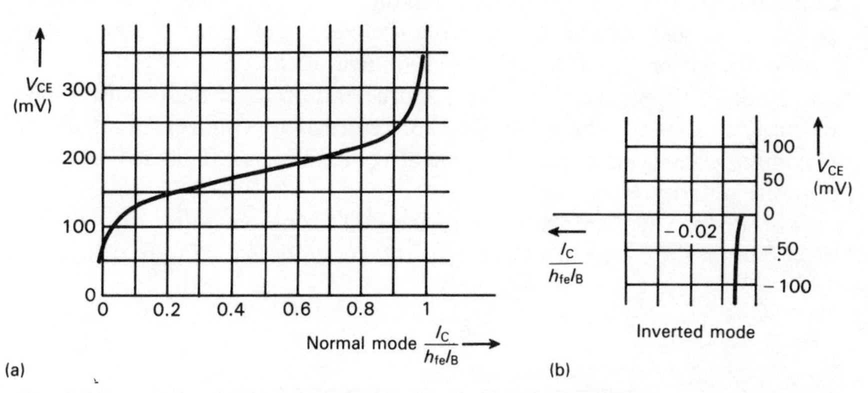

Fig. 22.7 (a) Saturation characteristic of bipolar transistor in normal mode (b) Saturation characteristic of bipolar transistor in inverted mode

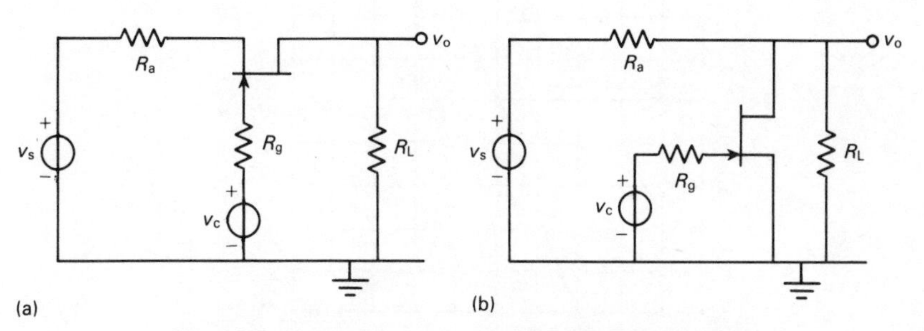

Fig. 22.8 (a) JFET series gate (b) JFET shunt gate

signal v_s cannot be more negative than about 0.6 V to prevent foward biasing of the gate–source junction. When v_s goes positive, V_{GS} goes negative and the FET will begin to turn off before V_{GS} has reached the pinch-off voltage. Hence the allowable range of v_s is rather limited.

To allow larger excursions of v_s the gate is made to follow the source voltage as shown in Fig. 22.10. If $v_c = V^+$, D is reverse biased and the FET is on. $V_{GS} = 0$ V for negative excursions of v_s. If v_s goes positive, source and drain reverse and $V_{GS} > 0$ V. The FET remains on until $v_s \simeq V^+ + 0.6$ volts. A further increase of v_s turns D on and cutoff of the FET occurs when $v_s \simeq V^+ + V_p$.

If $v_c = V^-$, the FET is off and remains off for negative excursions of v_s until $V_{GS} \simeq V^- - v_s - 0.6$ volts. During positive excursions of v_s the FET remains off since drain and source reverse.

In the enhancement MOSFET an extra terminal is available, namely the substrate (Fig. 22.11). In order to keep all junctions reverse biased during operation of the FET, the substrate is connected to a voltage which is at least as negative as the most negative voltage to be expected in the circuit. When using this FET as a series gate (Fig. 22.12) with the substrate connected to a negative voltage equal to the maximum negative excursion of v_s, the control voltage must swing between the value $V_s + V_{th}$ and $-V_s + V_{th}$ where V_{th} is the threshold voltage of the FET.

In the CMOS gate (Fig. 22.13) source and drain of the two FETs are reversed as shown. The substrates are returned to voltages equal to the maximum excursions of v_s. The control voltages are well above V_{th} and of opposite polarity.

In the 'on' mode at least one of the FETs is on regardless of the value of v_s (within the range of v_c). In the 'off' mode both FETs are off.

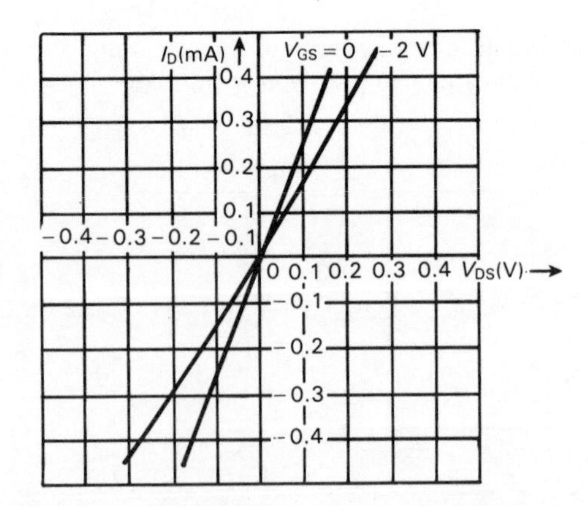

Fig. 22.9 Expanded FET-characteristics near the origin

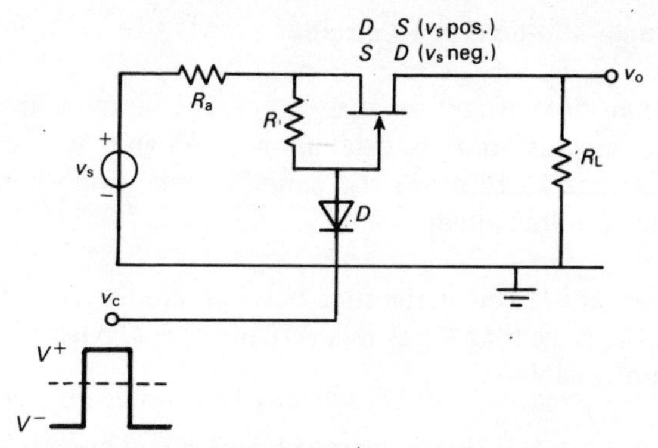

Fig. 22.10 JFET gate with driver circuit

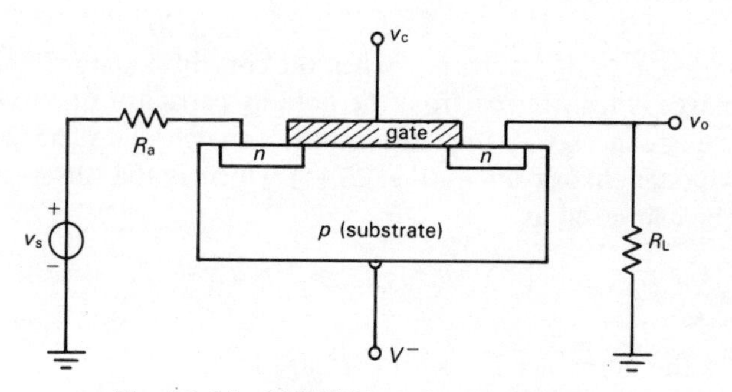

Fig. 22.11 MOSFET used as series gate

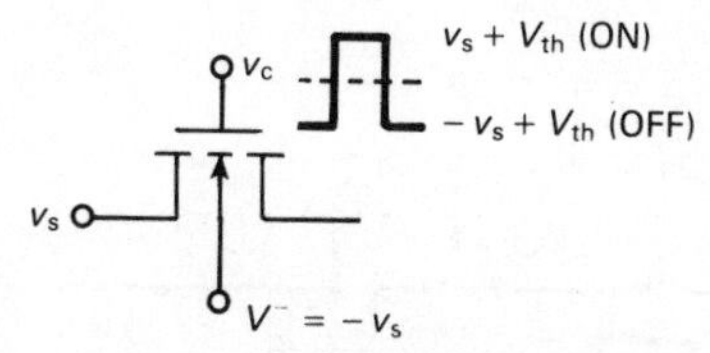

Fig. 22.12 MOSFET series gate and control signal

22.4 APPLICATIONS

22.4.1 Multiplexer

In a **multiplexer** a number of analog signals, one at a time, is connected to a common load as shown in the basic circuit of Fig. 22.14(a). A three-stage multiplexer using MOSFETs is shown in Fig. 22.14(b).

22.4.2 Sample-and-hold (S/H) circuit

In a **sample-and-hold circuit** an analog signal is sampled at regular times and its instantaneous value is held upon command by a control signal (Fig. 22.15). Figure 22.16 shows the sampling operation where the following definitions are illustrated:

Aperture time. This is the time between an $S \rightarrow H$ transition of the control signal and the time the switch actually opens. Aperture times of less than 10 ns are feasible.

Acquisition time. This is the time between an $H \rightarrow S$ transition of the control signal and the time the output has acquired the new signal voltage (to within a stated accuracy).

Sample-to-hold offset error. When the control signal switches from S to H, a charge is transferred from the holding capacitor due to stray and parasitic capacitances C_d associated with the switch. This charge transfer causes a voltage change on C (Fig. 22.15) which is the offset error. The error can be expressed as

$$\Delta V = \frac{C_d}{C} V_g$$

where V_g is the difference in control voltage levels.

Drift rate. In the hold mode, leakage currents cause the output voltage to drift. These leakage currents are caused by the bias current of the

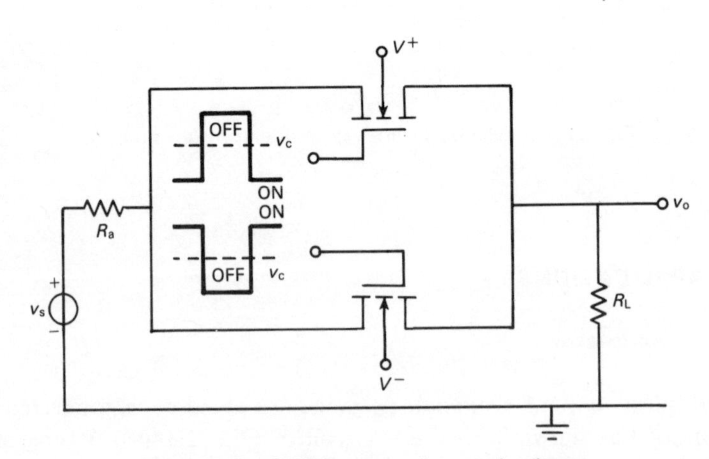

Fig. 22.13 CMOS gate and control voltages

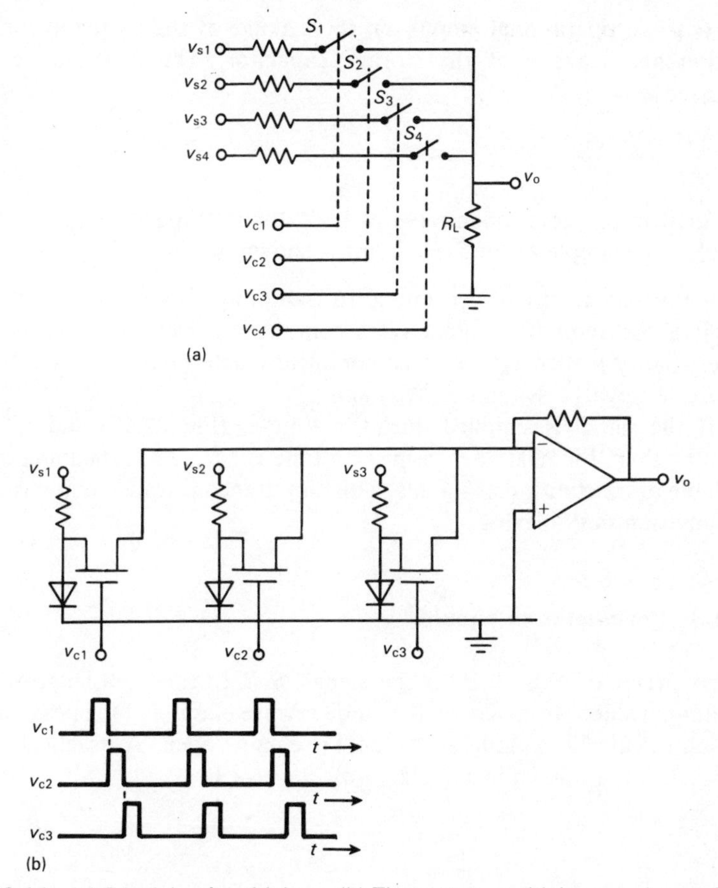

Fig. 22.14 (a) Principle of multiplexer (b) Three-stage multiplexer and waveforms

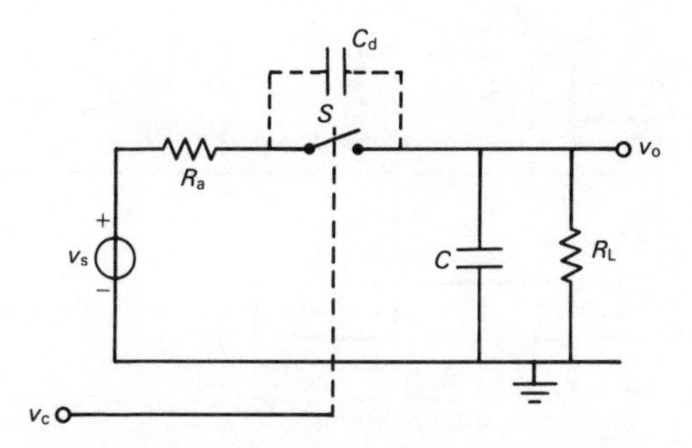

Fig. 22.15 Principle of sample-and-hold circuit

load (e.g. an operational amplifier), the leakage of the switch in the 'off' position and leakage of the storage capacitor. The drift rate can be expressed as

$$\frac{\Delta V_o}{\Delta t} = \frac{I_{leakage}}{C}$$

Low drift in the hold mode requires high quality capacitors (polystyrene, teflon). An example of an S/H circuit is shown in Fig. 22.17.

An important theorem pertaining to sampling theory is the **Nyquist sampling theorem:** if a signal $v_s(t)$ contains no frequency components higher than f_m, then $v_s(t)$ can be completely determined by its values at uniform intervals less than $1/(2f_m)$ apart.

If the signal is sampled during a time t_c (Fig. 22.18) and t_s is the sampling time, the relative transmission time is t_c/t_s. The remaining time is available to transmit other signals. This simultaneous transmission is called **time-division multiplexing**.

22.4.3 Peak detector circuit

In the circuit of Fig. 22.19(a) the signal v_s is tracked until it reaches a maximum value. From here on the diode reverse biases and the peak value is held automatically as long as switch S_2 remains open. A practical circuit configuration is shown in Fig. 22.19(b).

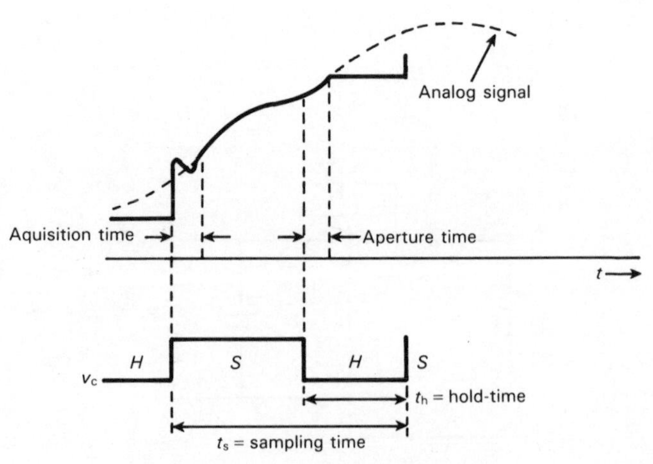

Fig. 22.16 Sampled waveform

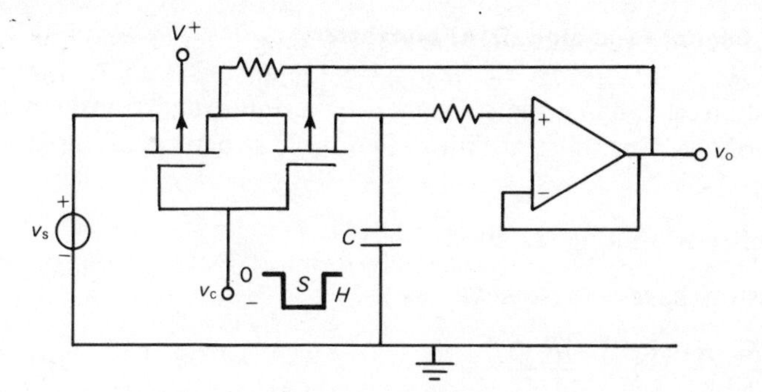

Fig. 22.17 Example of sample-and-hold circuit

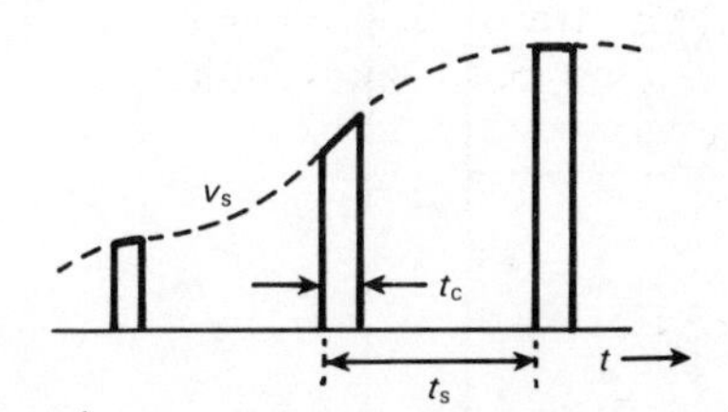

Fig. 22.18 Sampling of waveform v_s

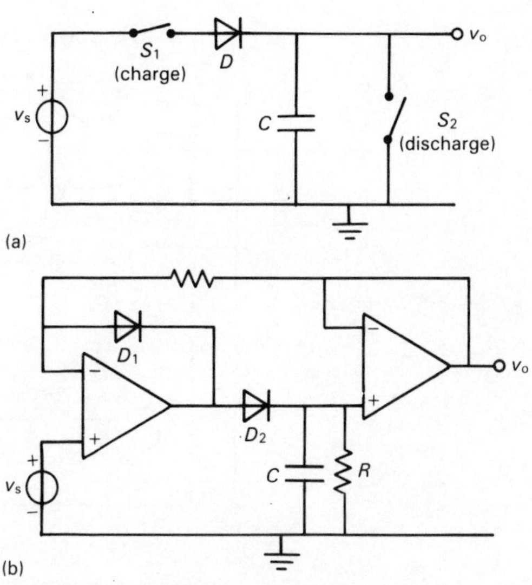

Fig. 22.19 (a) Basic peak detector circuit (b) Peak detector circuit with two operational amplifiers

22.4.4 Digital-to-analog (D/A) converter

A D/A converter has a digital signal as its input and translates it to an analog voltage. For this translation two basic structures are used.

Weighted resistors (Fig. 22.20(a))

The resistors have successive values

$$R, \frac{R}{2}, \frac{R}{4}, \frac{R}{8}, \ldots, \frac{R}{2^{n-1}}$$

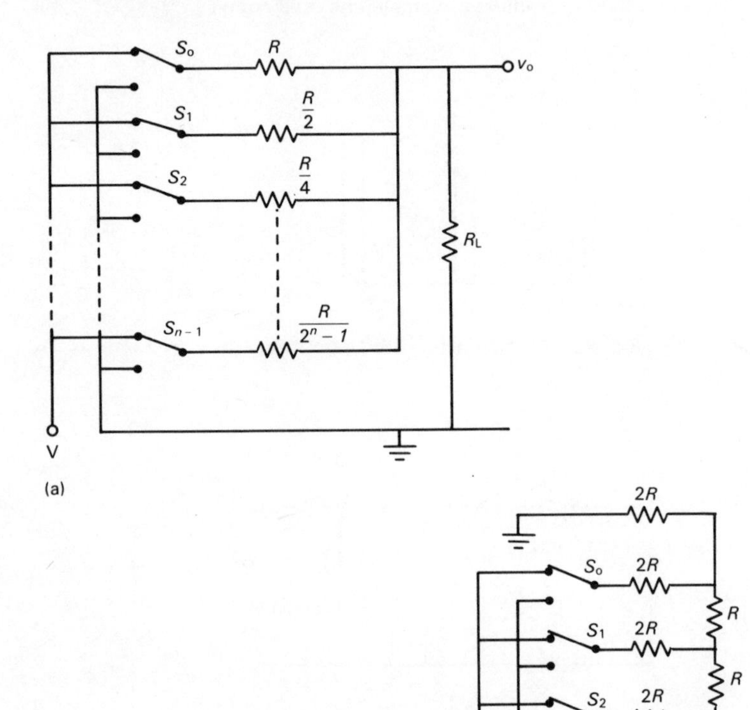

Fig. 22.20 (a) Principle of D/A converter (weighted resistors) (b) Principle of D/A converter (R–$2R$ ladder)

The output voltage is

$$v_o = \frac{VR_L}{R + R_L(2^{n-1})} \sum_{k=0}^{n-1} S_k 2^k$$

where $S_k = 1$ if the switch connects to V and $S_k = 0$ if the switch connects to ground. A disadvantage of this circuit is the requirement of a wide range of (precision) resistors.

$R - 2R$ ladder (Fig. 22.20(b))

The resistors have only two values: R and $2R$. Under the same assumption as for weighted resistors the output voltage is

$$v_o = \frac{R_L}{R + R_L} \frac{V}{2^n} \sum_{k=0}^{n-1} S_k 2^k$$

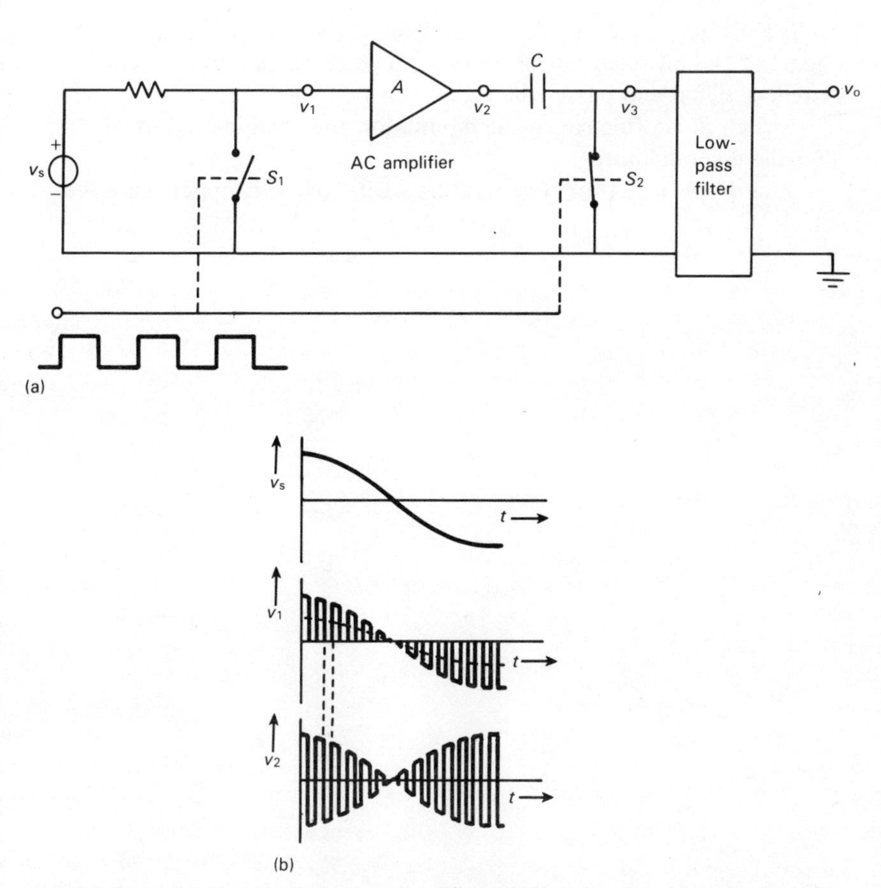

Fig. 22.21 (a) Principle of chopper amplifier (b) Waveforms in chopper amplifier

22.4.5 Chopper amplifier

A very small and very slow change in amplitude of a signal may, when DC amplified, be impossible to distinguish from drift in the amplifier (e.g. thermocouple output signal).

A **chopper amplifier** eliminates this drift problem substantially. In Fig. 22.21 switch S_1 is alternately opened and closed. The chopped waveform v_1 can be expressed as

$$v_1(t) = \tfrac{1}{2} v_s(t) + v_s(t) \sum_{n=1}^{\infty} \frac{\sin(n\pi/2)}{n\pi/2} \cos \frac{2\pi n t}{T}$$

where T is the period of the switching waveform.

The AC amplifier cuts off the low-frequency signal $\tfrac{1}{2} v_s(t)$ so that

$$v_2(t) = A v_s(t) \sum_{n=1}^{\infty} \frac{\sin(n\pi/2)}{n\pi/2} \cos \frac{2\pi n t}{T}$$

As S_1 and S_2 operate in antisynchronism, the signal v_3 is a replica of the signal v_s. The low-pass filter rejects the high-frequency components of the switching signal and passes the signal v_3.

Switch S_1 is often called the **modulator,** the combination of S_2, C and filter the **demodulator**.

A typical value of temperature drift of a chopper amplifier is $0.1\ \mu\text{V}\ ^\circ\text{C}^{-1}$.

23 Modulation

23.1 AMPLITUDE MODULATION

23.1.1 General

Modulation is the process by which one of the characteristics (amplitude, angle) of a carrier is changed as a function of the instantaneous value of an intelligence signal. This signal is called the **modulation signal**.

When the carrier frequency is sufficiently high, transmission or radiation into space is possible. Since modulation is a nonlinear process involving multiplication of electrical signals, nonlinear electronic devices are required. If the carrier is represented as a sine wave

$$v_c = V_c \cos \omega_c t$$

and the modulation signal as a single sine wave

$$v_m = V_m \cos \omega_m t$$

the amplitude of the modulated carrier is $V_c + k_a v_m$ where k_a is a system constant. The amplitude modulated carrier is thus expressed as

$$v_c = V_c\left(1 + k_a \frac{V_m}{V_c} \cos \omega_m t\right) \cos \omega_c t$$

(Fig. 23.1a). This multiplication can be translated into the functional block diagram of Fig. 23.1(b). The constant

$$k_a \frac{V_m}{V_c} = m_a$$

is the degree of modulation or **modulation factor** and it is commonly expressed as a percentage.

The modulated carrier can be broken down in its frequency components (the spectrum) by trigonometric methods:

$$v_c = V_c[\cos \omega_c t + \tfrac{1}{2} m_a \cos(\omega_c + \omega_m)t + \tfrac{1}{2} m_a \cos(\omega_c - \omega_m)t]$$

The first term is the unmodulated carrier which contains no information. The other two terms are sum and difference frequency terms, called

sideband frequencies. Both contain the information ω_m in equal amounts. The spectrum is shown in Fig. 23.1(c).

A phasor representation of the modulated carrier (Fig. 23.1(d)) consists of a pair of rotating conjugate phasors and a fixed carrier phasor. The resultant phasor is the modulated carrier.

Generally, the modulation covers a band of frequencies. The corresponding spectrum then contains **sidebands** (Fig. 23.1(e)).

The ratio of carrier power to sideband power is $1 : \frac{1}{2} m_a^2$. Thus in case of complete modulation the sideband power is 50 percent of the carrier power but normally much less. The efficiency of amplitude modulation is thus rather poor. A saving in power is effected when suppressing the carrier. This modulation is called **double sideband suppressed carrier (DSB) modulation.**

Further enhancement of efficiency, a saving in bandwidth and a reduction in noise, is achieved in **single sideband (SSB) modulation** where in addition to the carrier, one of the two sidebands is suppressed.

23.1.2 Generation of AM signals

AM signals can be generated by employing various principles.

A nonlinear input characteristic (*square-law modulation*)

The sum of carrier and modulation signals forms the input signal of a device having a nonlinear transfer characteristic. This characteristic can be represented by the series expansion

$$i_o = I_o + a_1 v_s + a_2 v_s^2 + a_3 v_s^3 + \cdots$$

When substituting $v_s = V_c \cos \omega_c t + V_m \cos \omega_m t$ and using trigonometric identities, it follows that

$$i_o = \underbrace{I_o + \tfrac{1}{2} a_2 (V_c^2 + V_m)^2}_{A} \underbrace{+ a_1 V_m \cos \omega_m t + \tfrac{1}{2} a_2 V_m^2 \cos 2\omega_m t}_{B}$$

$$\underbrace{+ a_1 V_c \cos \omega_c t + a_2 V_c V_m [\cos(\omega_c - \omega_m)t + \cos(\omega_c + \omega_m)t]}_{C}$$

$$\underbrace{+ \tfrac{1}{2} a_2 V_c^2 \cos 2\omega_c t + \ldots}_{D}$$

The terms A represent a DC current, the terms B the modulation frequency and its second harmonic, the terms C the carrier and sidebands and the

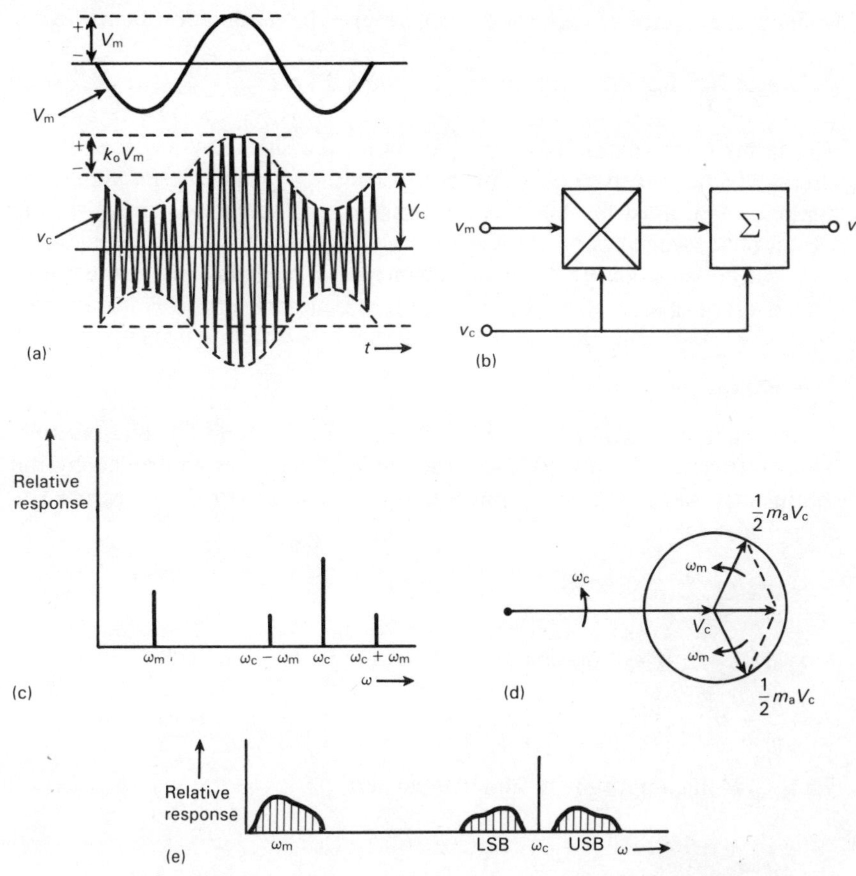

Fig. 23.1 (a) Amplitude modulated sinusoidal carrier (b) A possible block diagram for amplitude modulation (c) Spectrum of amplitude modulated carrier (d) Phasor representation of amplitude modulated carrier (e) Spectrum of amplitude modulated carrier

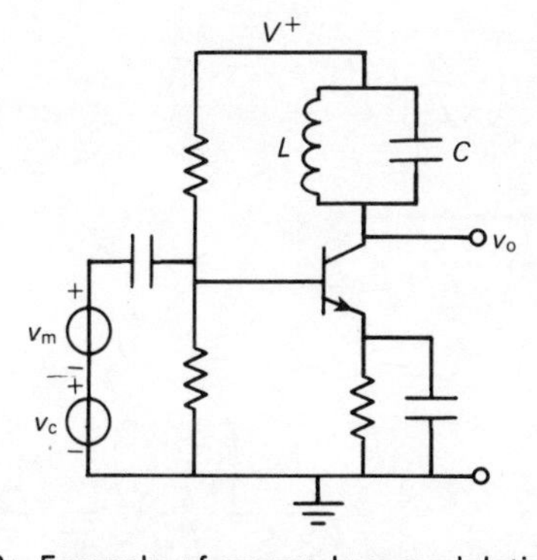

Fig. 23.2 Example of square law modulation circuit

terms D the higher harmonics of the carrier. The modulation factor is

$$m_a = \frac{2a_2}{a_1} V_m$$

The terms C are extracted from the composite signal by a resonant LC circuit which is tuned to ω_c. In order to keep the unwanted terms small, the applied signal must be small. An example of this method is shown in the CE-amplifier stage of Fig. 23.2.

When using an FET, unwanted terms are smaller since the series expansion of the transfer characteristic terminates with the quadratic term.

A nonlinear resistance

In the single stage amplifier of Fig. 23.3 the JFET has no DC bias current. Thus it functions as a resistance, the value of which is controlled by the modulation signal v_m. The input signal of the stage is the carrier v_c. The stage gain is thus

$$\frac{v_o}{v_c} \simeq \frac{R_c}{R_e // r_{DS}}$$

If $r_{DS}/R_e \ll 1$, it follows that

$$\frac{v_o}{v_c} \simeq R_c g_{mo}\left(1 - \frac{v_m}{V_p}\right)$$

Thus v_o is the amplitude modulated carrier.

A current switch

Current switching in differential stages was described in Sections 22.2.1 and 22.2.2. In the bipolar stage of Fig. 23.4 the stage current is modulated by v_m, thus $I = I_o + i_m$. The currents in the two branches are:

$$\frac{I_{C1} + i_{c1}}{I_o + i_m} = \frac{1}{1 + \exp\left(\frac{-q}{kT} v_c\right)}$$

$$\frac{I_{C2} + i_{c2}}{I_o + i_m} = \frac{1}{1 + \exp\left(\frac{q}{kT} v_c\right)}$$

The difference is:

$$\frac{(I_{C1} - I_{C2}) + (i_{c1} - i_{c2})}{I_o + i_m} = \frac{\left[\exp\left(\frac{q}{kT} v_c\right) - \exp\left(\frac{-q}{kT} v_c\right)\right]}{\left[2 + \exp\left(\frac{q}{kT} v_c\right) + \exp\left(\frac{-q}{kT} v_c\right)\right]}$$

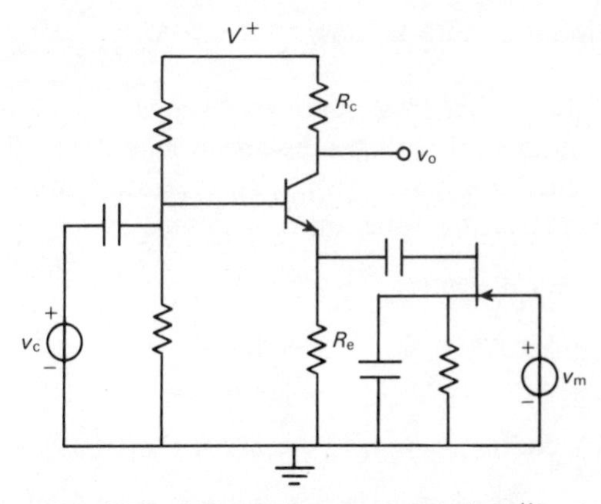

Fig. 23.3 Modulation circuit with JFET

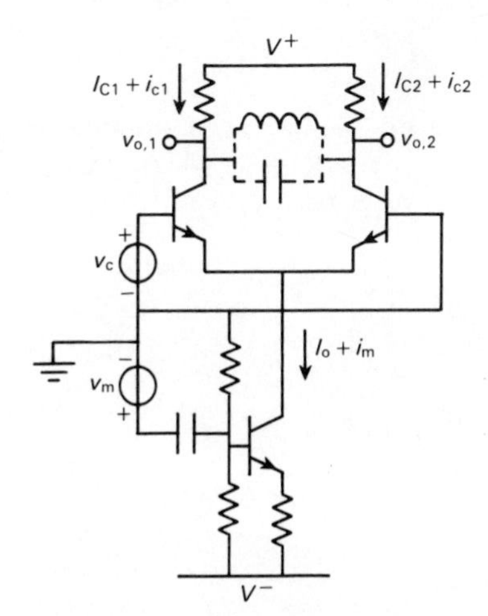

Fig. 23.4 Modulation circuit with current switch

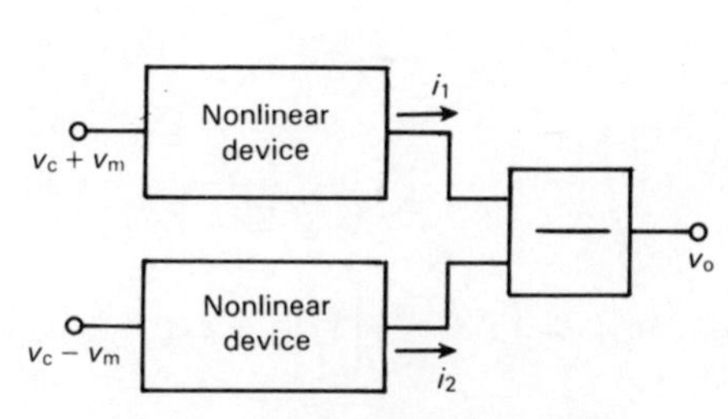

Fig. 23.5 Principle of balanced modulator

When using series expansions for the exponentials and separating DC signals and AC signals, the result is

$$i_{c1} - i_{c2} \simeq (I_o + i_m)\,\frac{1}{2}\,\frac{q}{kT}\,v_c$$

The parallel resonant circuit between the collectors (tuned to f_c) converts the differential current to a differential voltage while suppressing higher frequency components. As a result, the differential voltage is an amplitude modulated signal.

A similar derivation applies to an FET-stage.

23.1.3 Generation of DSB signals

Suppression of the carrier in a modulated signal is achieved by **compensation** in a circuit which is called a **balanced modulator**. The principle of operation is sketched in Fig. 23.5 where two identical nonlinear devices are used. If the device transfer function is expressed by

$$i = I_o + a_1 v_s + a_2 {v_s}^2 + a_3 {v_s}^3 + \cdots$$

and $v_{s1} = v_c + v_m$, $v_{s2} = v_c - v_m$, it follows that

$$\begin{aligned} v_o &= k(i_1 - i_2) \\ &= 2ka_1 V_m \cos \omega_m t + 4ka_2 V_c \cos \omega_c t \times V_m \cos \omega_m t + \cdots \end{aligned}$$

A resonant circuit turned to ω_c leaves only the second term so that

$$\begin{aligned} v_o &= 4ka_2 V_c \cos \omega_c t \times V_m \cos \omega_m t \\ &= 2ka_2 V_c V_m[\cos(\omega_c + \omega_m)t + \cos(\omega_c - \omega_m)t] \end{aligned}$$

The modulation factor here is $m_a = 4ka_2 V_m$.

This process can be implemented in the **double-balanced modulator** (Fig. 23.6(a)). The signal v_m causes currents $I + (v_m/\mathrm{R})$ and $I - (v_m/R)$ to flow in Tr_5 and Tr_6. The AC currents in Tr_1, Tr_2, Tr_3 and Tr_4 are respectively

$$i_{c1} = \left(I + \frac{v_m}{R}\right)\left\{\exp\left(\frac{q}{kT} v_c\right) \Big/ \left[1 + \exp\left(\frac{q}{kT} v_c\right)\right]\right\}$$

$$i_{c2} = \left(I - \frac{v_m}{R}\right)\left\{\exp\left(\frac{q}{kT} v_c\right) \Big/ \left[1 + \exp\left(\frac{q}{kT} v_c\right)\right]\right\}$$

$$i_{c3} = \left(I + \frac{v_m}{R}\right)\left\{1 \Big/ \left[1 + \exp\left(\frac{q}{kT} v_c\right)\right]\right\}$$

$$i_{c4} = \left(I - \frac{v_m}{R}\right)\left\{1 \Big/ \left[1 + \exp\left(\frac{q}{kT} v_c\right)\right]\right\}$$

Also $i_{o1} = i_{c1} + i_{c4}$ and $i_{o2} = i_{c2} + i_{c3}$ from which it follows that

$$v_{o1} - v_{o2} = \frac{2R_c}{R} v_m \left[\frac{qv_c}{kT} - \frac{1}{12}\left(\frac{qv_c}{kT}\right)^3 + \cdots\right]$$

Suppression of the higher harmonics yields

$$v_{o1} - v_{o2} = \frac{2R_c}{R}\frac{q}{kT} V_m V_c \cos \omega_m t \cos \omega_c t$$

which represents a DSB signal (Fig. 23.6(b)).

It should be noticed that the phase changes 180° in the DSB signal when the modulation passes through zero.

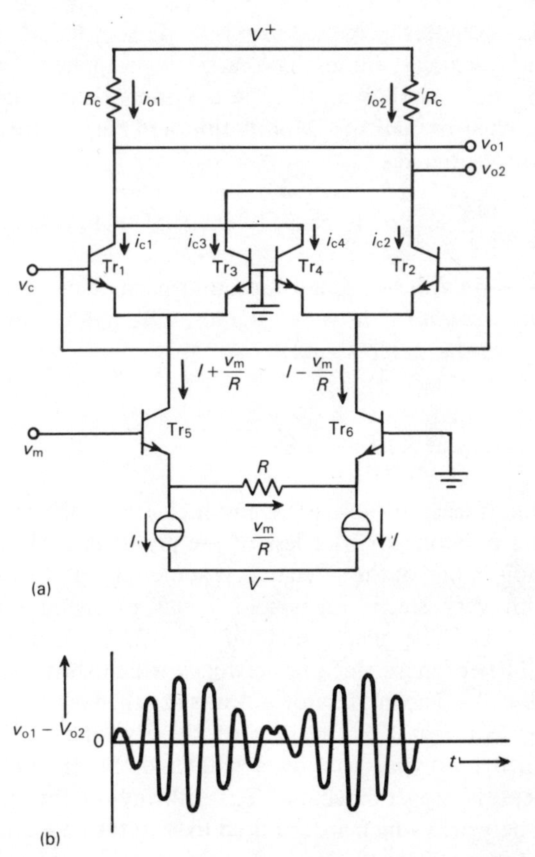

Fig. 23.6 (a) Basic circuit of double-balanced modulator (b) DSB waveform

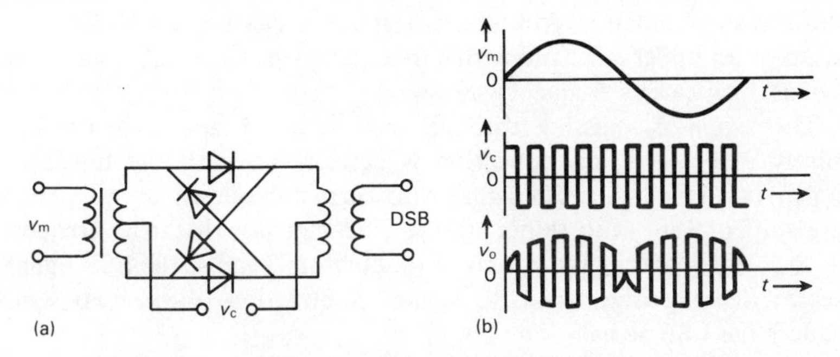

Fig. 23.7 (a) Ring modulator circuit (b) Waveforms of (a)

The **ring modulator** (Fig. 23.7(a)) is a passive balanced modulator containing four (identical) diodes. The carrier v_c is applied at the input and output transformers as shown. If v_c is a symmetrical square wave, the diodes are switched on and off. Modulation and carrier are multiplied to produce the output signal

$$v_o = kv_m v_c = \frac{4k}{\pi} V_c[\tfrac{1}{2} m_a V_c \cos(\omega_c - \omega_m)t - \tfrac{1}{2} m_a V_c \cos(\omega_c + \omega_m)t + \cdots]$$

Thus v_o contains only the sidebands and odd harmonics which are filtered out. In the transmission of FM stereo signals, DSB modulation is employed in the synthesis of the stereo signal.

23.1.4 Generation of SSB signals

SSB modulation is used in short-wave and high-frequency communications since it requires only one-sixth or less of the power of AM modulation.

Suppression of one of the sidebands is sometimes performed by precise **SSB-filters** with very steep slopes and a flat passband (crystal filters, mechanical filters). The **phase method** (Fig. 23.8) uses two balanced modulators and two phase shifting networks which shift the phase of all input signals by 90°. The modulator outputs are then $V_m V_c \sin \omega_m t \sin \omega_c t$ and $V_m V_c \cos \omega_m t \cos \omega_c t$. The sum of these signals is $v_o = V_m V_c \cos (\omega_m - \omega_c)t$ which represents the lower sideband. Subtraction of the two signals produces the upper sideband. The difficulty of the circuit is in the phase shifting networks which are required to shift the phase of all signals in a frequency band by 90°.

A more complex circuit is the **Weaver method** (Fig. 23.9(a)). The modulation signal, covering the frequency band ω_1 to ω_2 and an oscillator signal are the inputs of balanced modulator 3 (Fig. 23.9(b)). The oscillator frequency is exactly in the center of the frequency range of v_m. The output signal of modulator 3 is a DSB signal having a folded lower sideband from 0 to $\frac{1}{2}(\omega_2 - \omega_1)$. The upper sideband is filtered out. The lower sideband is multiplied in modulator 4 by the carrier v_c; it can be considered as the overlap of an upper sideband with carrier frequency $\omega_c - \omega_{osc}$ and a lower sideband with carrier frequency $\omega_c + \omega_{osc}$.

The signal v_m passing through modulators 1 and 2 is processed similarly except that multiplication is carried out with the quadrature signals. As a result the LSB signal with carrier frequency $\omega_c + \omega_{osc}$ has a phase shift of 180° with respect to the USB signal with carrier frequency $\omega_c - \omega_{osc}$. When adding the signals of modulators 2 and 4, the LSB signal is canceled, leaving only the USB signal. Subtraction of the two signals produces the LSB signal.

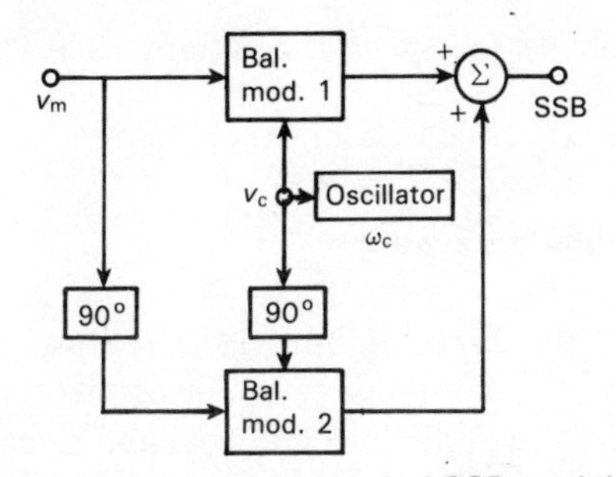

Fig. 23.8 The phase method of SSB modulation

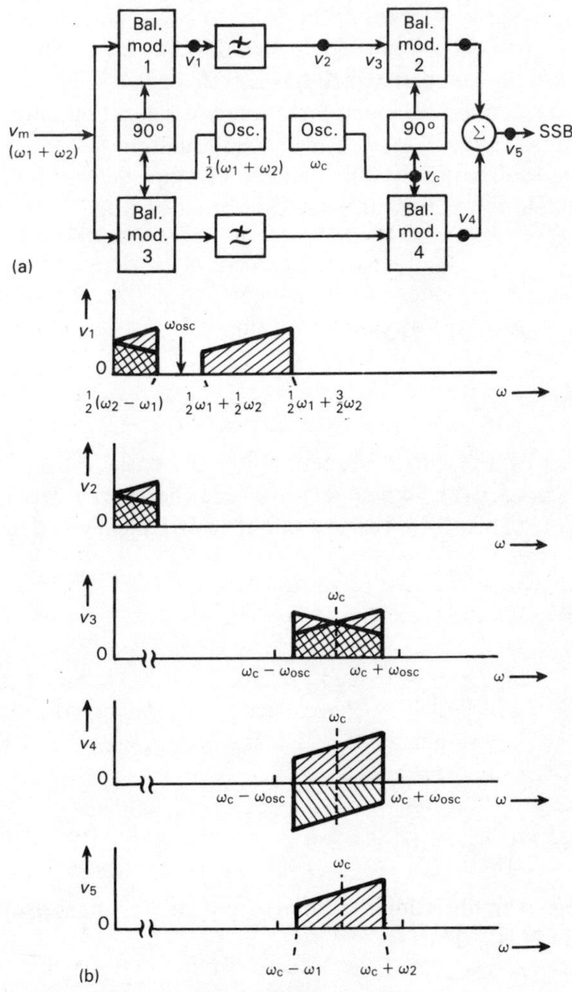

Fig. 23.9 (a) The Weaver method of SSB modulation (b) Spectral components at different points in (a)

Figure 23.9(b) shows the spectral components at various points in the circuit.

23.1.5 Other modulation systems

Independent sideband modulation (ISB). This is a variation of SSB modulation. The carrier contains two independent modulation signals which are converted to two independent sideband signals.

Vestigial sideband modulation (VSB). Here one of the sidebands of an AM signal is partly suppressed in order to save bandwidth (Fig. 23.10).

Range A represents a normal AM signal, range B an SSB signal. Range A is filtered in a linear filter which has the effect that the resultant of two corresponding sideband frequencies in range A is constant and equal to the SSB amplitude in range B. As a result, the modulation factor is independent of the modulation frequency which allows envelope detection of the signal.

VSB modulation is used in television broadcasting.

23.2 ANGLE MODULATION

23.2.1 General

The argument of the carrier function $v_c = V_c \cos(\omega_c t + \phi)$ is the angle function $\Phi = \omega_c t + \phi(t)$. When Φ is varied, the carrier frequency is no longer constant. Therefore, the instantaneous frequency is defined as

$$\omega(t) = \frac{d\Phi}{dt} = \omega_c + \frac{d\phi}{dt}$$

Conversely,

$$\Phi(t) = \int_0^t \omega(t)\, dt$$

so that

$$v_c = V_c \cos \int_0^t \omega(t)\, dt$$

In **frequency modulation** the carrier frequency changes proportional to the amplitude of v_m. Thus

$$\frac{d\phi}{dt} = k_f V_m \cos \omega_m t$$

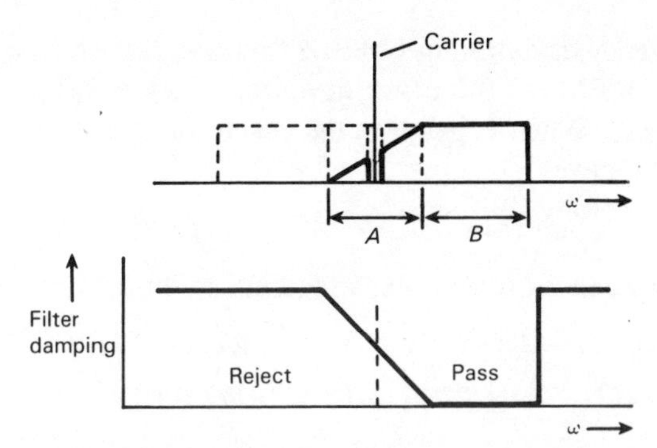

Fig. 23.10 Vestigial sideband modulation

so that

$$\Phi_f = \omega_c t + \frac{k_f V_m}{\omega_m} \sin \omega_m t$$

and

$$\omega(t) = \frac{d\Phi_f}{dt} = \omega_c + k_f V_m \cos \omega_m t = \omega_c + \Delta\omega \cos \omega_m t$$

Hence the FM signal is described as

$$v_c = V_c \cos\left(\omega_c t + \frac{k_f V_m}{\omega_m} \sin \omega_m t\right)$$
$$= V_c \cos(\omega_c t + m_f \sin \omega_m t)$$

The factor $k_f V_m/2\pi = \Delta f_c$ is called the **frequency deviation** and $m_f = k_f V_m/\omega_m = \Delta f_c/f_m$ is called the **modulation index** or **deviation ratio.**

In **phase modulation** the carrier phase changes proportional to the amplitude of v_m thus $\phi = k_p V_m \cos \omega_m t$. Then

$$\Phi_p = \omega_c t + k_p V_m \cos \omega_m t$$

and

$$\omega(t) = \frac{d\Phi_p}{dt} = \omega_c + k_p V_m \omega_m \sin \omega_m t$$

The phase modulated signal is thus described by

$$v_c = V_c \cos(\omega_c t + k_p V_m \cos \omega_m t)$$
$$= V_c \cos(\omega_c t + m_p \cos \omega_m t)$$

where m_p is the modulation index.

Obviously, FM and PM are closely related. The difference is a 90° phase shift of the modulation signal as illustrated in Fig. 23.11.

The frequency deviation in FM is $\Delta f_c = k_f V_m/2\pi$ which is independent of ω_m whereas in PM the frequency deviation is $\Delta f_c = k_p V_m \omega_m/2\pi$ which is dependent on ω_m. When expanding the cosine functions of the carriers of FM and PM we have

$$v_c = V_c \cos \omega_c t \cos(m \sin \omega_m t) - V_c \sin \omega_c t \sin(m \sin \omega_m t)$$

These functions can be further expanded in the series

$$\cos(m \sin \omega_m t) = J_o(m) + 2J_2(m) \cos 2\omega_m t + 2J_4(m) \cos 4\omega_m t + \cdots$$

$$\sin(m \sin \omega_m t) = 2J_1(m) \sin \omega_m t + 2J_3(m) \sin 3\omega_m t + 2J_5(m) \sin 5\omega_m t + \cdots$$

where the coefficients $J_n(m)$ are the Bessel functions of the first kind and order n. These functions are shown in Fig. 23.12.

The spectral components of the carrier can then be written as

$$\begin{aligned} v_c = {} & J_o(m) V_c \sin \omega_c t \\ & + J_1(m) V_c[\sin(\omega_c + \omega_m)t - \sin(\omega_c - \omega_m)t] \\ & + J_2(m) V_c[\sin(\omega_c + 2\omega_m)t + \sin(\omega_c - 2\omega_m)t] \\ & + J_3(m) V_c[\sin(\omega_c + 3\omega_m)t - \sin(\omega_c - 3\omega_m)t] + \cdots \end{aligned}$$

It should be noticed that the carrier, which is represented by $J_o(m)$, vanishes when $m = 2.4$. This property is used as a means for measuring the frequency deviation.

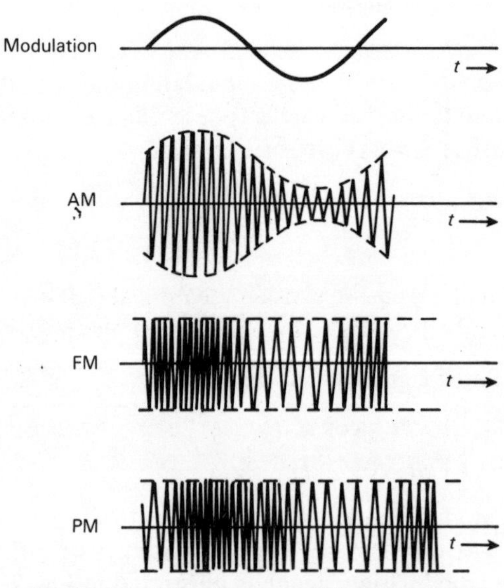

Fig. 23.11 Waveforms in AM, FM and PM

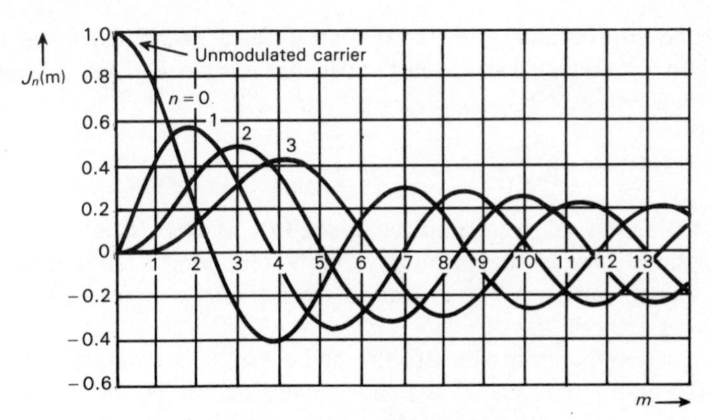

Fig. 23.12 A number of Bessel functions

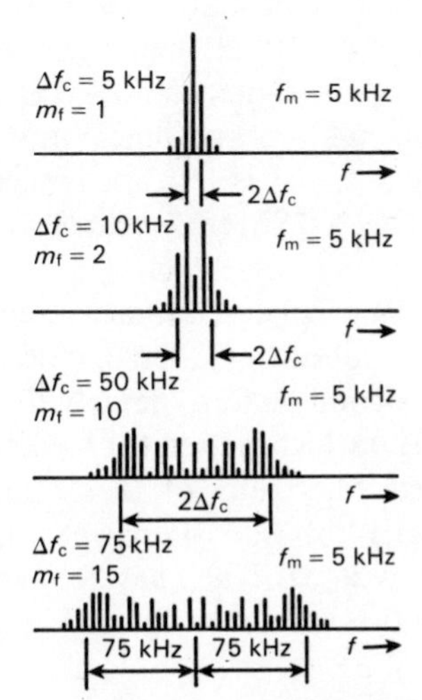

Fig. 23.13 Spectra of frequency modulated signals

The spectrum of a frequency modulated signal thus contains an infinite number of sideband frequencies. Figure 23.13 shows a few spectra of FM signals at different frequency deviations and modulation indexes. Apparently, the sideband amplitudes diminish rapidly outside a region $2\Delta f_c$ around the carrier. In practice the contribution of harmonics higher than $m_f + 3$ is negligible and a total bandwidth $B = 2(m_f + 3)f_m = 2\Delta f_c + 6f_m$ is considered adequate.

If, for example, $\Delta f_c = 75$ kHz and $f_m = 20–15\,000$ Hz, $B \simeq 240$ kHz.

The energy of each sideband is proportional to the square of the corresponding Bessel coefficient and it can be shown that the total energy equals the sum of the energies of carrier and sidebands. Thus angle modulation removes energy from the carrier and puts it in the sidebands.

23.2.2 Synthesis of the stereo signal

The FM stereo signal is standardized by the FCC (Federal Communications Commission). One of these standards is the requirement of compatibility with mono signals such that monaural receivers are able to receive stereo broadcasts without loss of quality.

For stereophonic broadcasts the audio signal is separated in a left signal (L) and a right signal (R). Both signals cover the audio range of 30–15 000 Hz. First these signals are passed through a **pre-emphasis** network which is a high-pass network having a time constant of 75 μs (USA) or 50 μs (Europe). The higher audio freqencies (above 2 kHz in the USA) are then transmitted at a higher level than the lower frequencies which results in an improvement of the signal-to-noise ratio.

The left and right signals are combined into a sum L + R and a difference L – R. The L + R signal is left unchanged, the L – R signal is used to amplitude modulate a 38 kHz subcarrier, thus forming a DSB signal.

In addition, a small 19 kHz pilot signal is generated as a reference signal to recover the 38 kHz subcarrier when demodulating. In the USA an extra signal is added to this combination, namely the SCA signal. This is a monaural signal (60–74 kHz) which is used for background music in stores, offices, etc. The complete spectrum of a stereo signal is shown in Fig. 23.14(a) while Fig. 23.14(b) is a simple example of the composite waveform, called a **multiplex signal**. The final step is to frequency modulate the FM carrier with the entire multiplex signal (Fig. 23.14(c)).

23.2.3 Generation of FM signals

The generation of an FM signal requires an oscillator signal to be changed in frequency in accordance with the modulating signal.

In Fig. 23.15 a **capacitance diode** (varicap) is part of an oscillator circuit. The diode is reverse biased and the modulation v_m is superimposed on this reverse bias. The components L, C and C_d determine the oscillation frequency. The frequency deviation can be adjusted with resistor R.

A limitation of this method is the nonlinear relation between reverse bias voltage and capacitance value which may cause distortion when large modulation signals are applied.

A FET circuit can be used to function as a **capacitive reactance**, the value of which is controlled by the gate signal v_m. Figure 23.16 shows the AC circuit as well as the equivalent circuit.

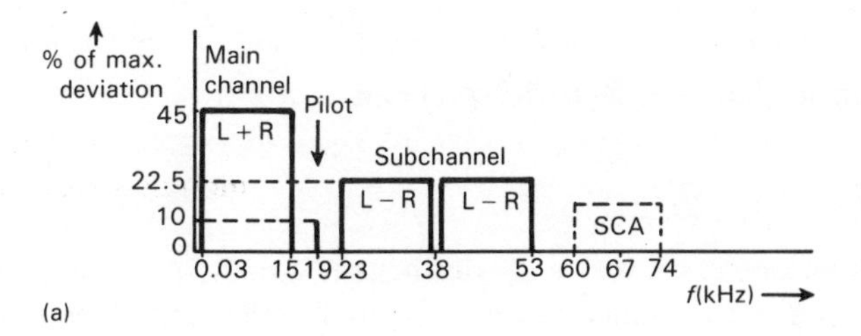

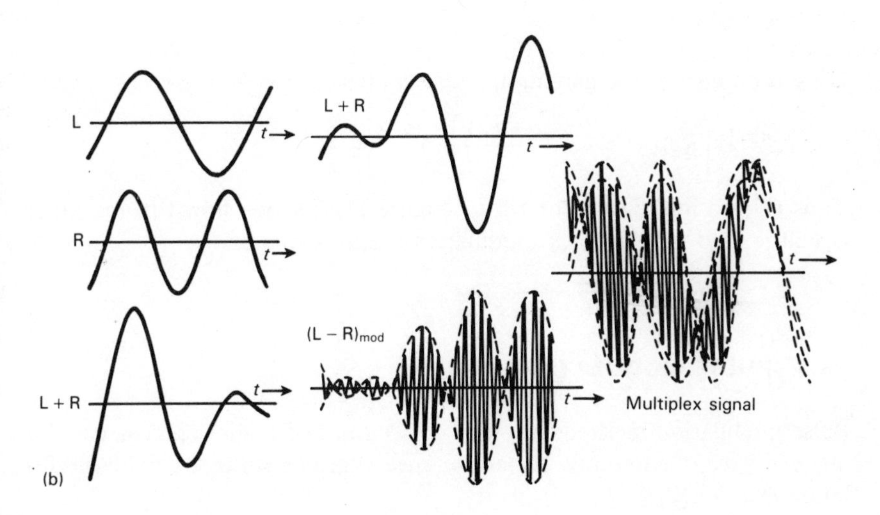

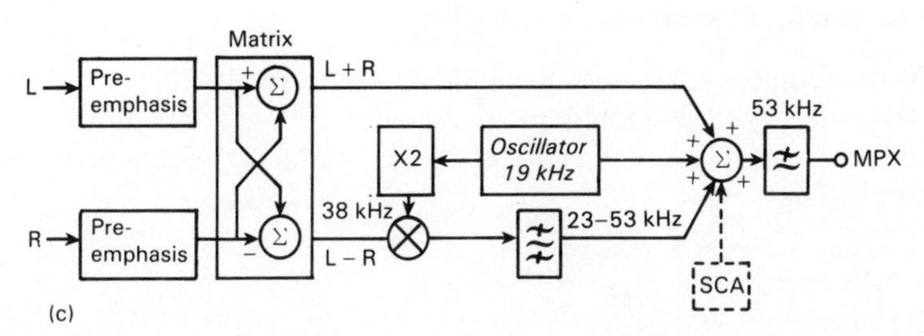

Fig. 23.14 (a) Spectrum of the stereo signal (b) Synthesis of a multiplex signal (c) Block diagram showing synthesis of multiplex signal

It follows that the impedance between drain and source is

$$Z = \frac{R + Z_C}{1 + g_m R}$$

If R is chosen so that $R \ll Z_C (\omega RC \ll 1)$ and $g_m R \gg 1$

$$Z \simeq \frac{Z_C}{g_m R} = \frac{1}{j\omega(g_m RC)} = \frac{1}{j\omega C^*}$$

Apparently, between source and drain the circuit is a capacitance $C^* = g_m RC$. The transconductance g_m is related to the gate–source voltage as

$$g_m = g_{max}\left(1 - \frac{V_{GS}}{V_P}\right)$$

If v_m is applied at the gate input, $v_{gs} \simeq v_m$ (since $R \ll Z_C$) so that

$$C^* \simeq C\left[Rg_{max}\left(1 - \frac{V_{GS} + v_m}{V_P}\right)\right]$$

Thus C^* is a linear function of v_m. Figure 23.17 shows how this reactance circuit is used to frequency modulate a Clapp oscillator.

23.3 PULSE MODULATION

Pulse modulation methods are used for reasons of higher efficiency and a larger degree of immunity to interference. Various forms of pulse modulation exist.

Pulse amplitude modulation (PAM)

The instantaneous values of v_m modulate a pulse carrier (Fig. 23.18(a)). If T is the sampling period and τ the pulse width, the duty cycle is $\delta = \tau/T$ and the sampling frequency $\omega_o = 2\pi/T$. The sampling signal can then be expressed as

$$p(t) = \delta \sum_{n=-\infty}^{+\infty} \frac{\sin n\pi\delta}{n\pi\delta} \exp(jn\omega_o t)$$

When sampling the modulation $v_m = V_m \cos \omega_m t$, the spectrum can be written as

$$v_c = \delta V_m \sum_{n=-\infty}^{+\infty} \frac{\sin \pi\delta[n + (\omega_m/\omega_o)]}{\pi\delta[n + (\omega_m/\omega_o)]} \cos \omega_o t[n + (\omega_m/\omega_o)](1 - \tfrac{1}{2}\delta)$$

The spectrum thus contains an infinite number of sideband frequencies around ω_o.

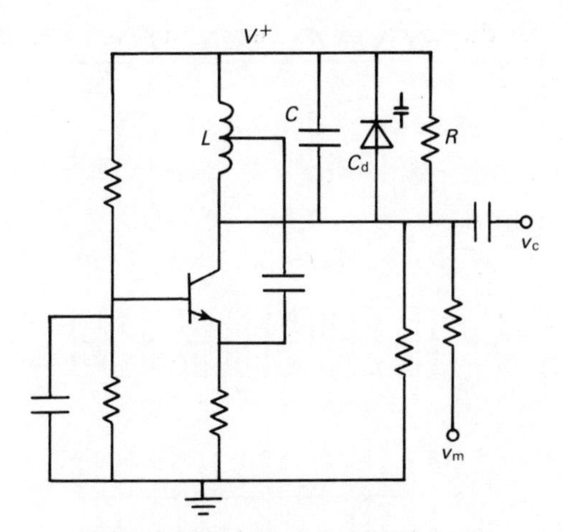

Fig. 23.15 Frequency modulation by capacitance diode

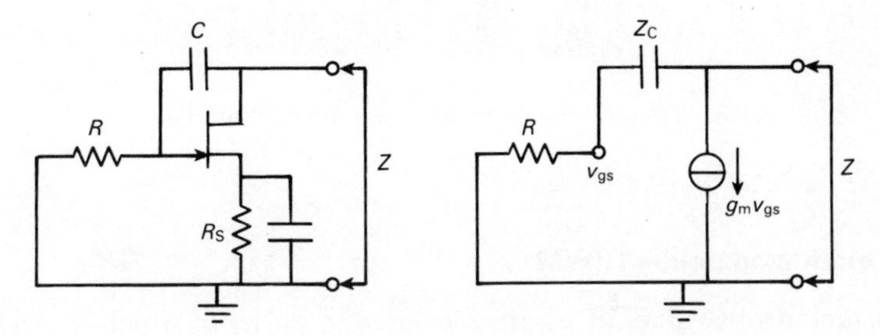

Fig. 23.16 Reactance circuit and equivalent circuit

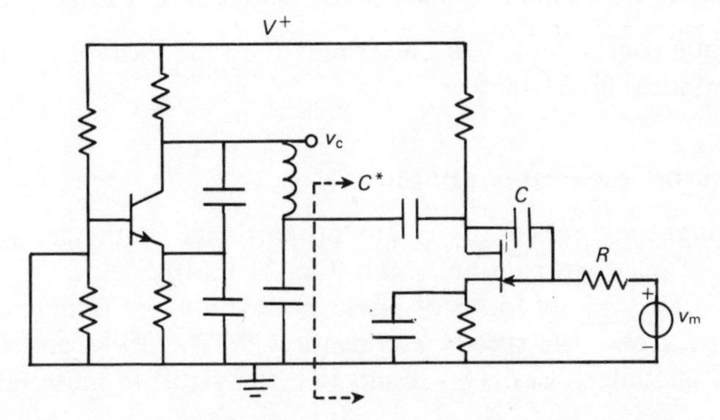

Fig. 23.17 The reactance circuit as part of a frequency modulator circuit

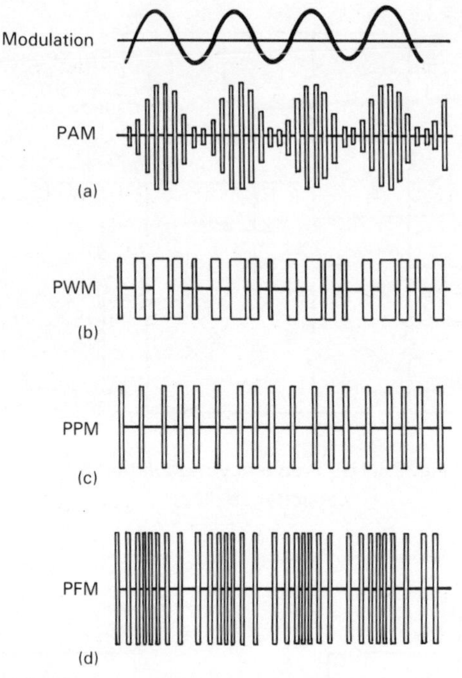

Fig. 23.18 Waveforms in various pulse modulation methods

Pulse width modulation (PWM)

The instantaneous values of v_m are converted to pulses of which the widths are proportional to the amplitude (Fig. 23.18(b)). Recovery of the modulation can be accomplished by a low-pass filter.

Pulse position modulation or pulse phase modulation (PPM)

The instantaneous values of v_m determine the time location of pulses of constant width (Fig. 23.18(c)).

Pulse frequency modulation (PFM)

The instantaneous values of v_m are proportional to the instantaneous frequency of pulses of constant width (Fig. 23.18(d)).

PPM and PFM are forms of phase modulation and frequency modulation respectively. The spectra are identical. PWM, PPM and PFM use constant amplitude pulses. This results in better signal-to-noise ratios than in PAM.

Pulse code modulation (PCM)

The amplitudes of the modulation signals are divided into a number of levels. Each level corresponds with a number in digital notation as shown in the example of Fig. 23.19(a).

PCM is used in telephone systems where the frequency of voice signals is limited to about 3400 Hz. The signals are sampled at a rate of 8000 times a second, in agreement with the Nyquist sampling theorem. The sampling pulses contain a large number of harmonics. Only four of these are shown in Fig. 23.19(b), where $f_o = 1/T$ is the sampling frequency. When the modulation signal is sampled, these carriers are amplitude modulated. The spectrum of each carrier has two sidebands and a bandwidth equal to the highest modulation frequency f_{max}. Recovery of the signal is accomplished by filtering all carriers and sidebands. Thus the filter cutoff frequency must equal f_{max} and the sampling frequency $f_o > 2f_{max}$.

The number of levels is usually limited to $2^7(128)$. Quantization approximates the voice signal when the PCM signal is decoded and filtered. The difference in the two signals is called **quantization noise**. Increasing the number of levels decreases the quantization noise but more levels increase the complexity and the signal bandwidth.

A sampled waveform and its quantization noise are shown in Fig. 23.19(c). When v_q is the quantization level, the error signals are limited to between $-\frac{1}{2}v_q$ and $+\frac{1}{2}v_q$. Figure 23.19(d) shows the noise power as a function of error voltage. Since all values of $e(t)$ have the same probability of occurence, the average noise power is the average value of the parabola:

$$\overline{e^2(t)} = \frac{1}{12} {v_q}^2$$

With seven bits, $v_q = 2V_m/128$ where V_m is the signal amplitude.

The quantization noise power is

$$\frac{1}{12} {v_q}^2 = \frac{1}{12}\frac{4{V_m}^2}{128^2} = \frac{{V_m}^2}{49\,152}$$

The signal power is $\frac{1}{2}{V_m}^2$ so that the signal-to-noise ratio is

$$S/N = \tfrac{3}{2}(128)^2 \simeq 25\,000 \ (\simeq 44 \text{ dB})$$

In telephone systems, 24 audio signals are usually multiplexed at a time by sequentially taking a sample from each of the 24 signals (time-division multiplex). Thus there are a total of 192 000 samples per second. A seven-bit code thus requires 1.344×10^6 bits per second. An extra bit is added for synchronization purposes which raises the number to 1.576×10^6 bits per second. An example of time division multiplex with three signals is shown in Fig. 23.19(e).

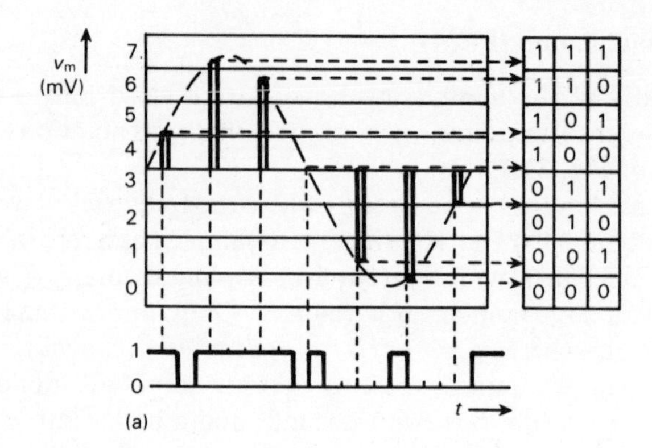

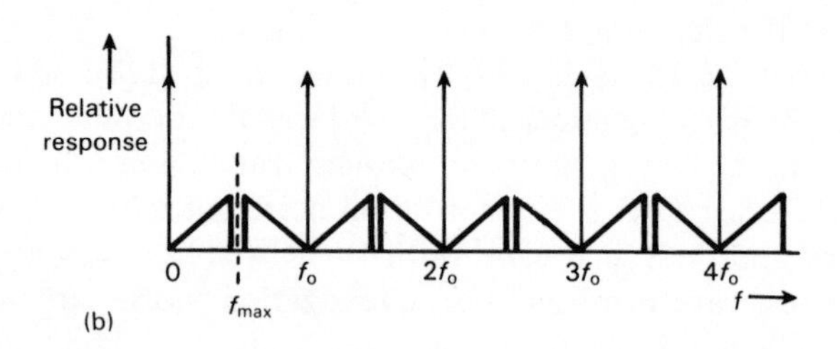

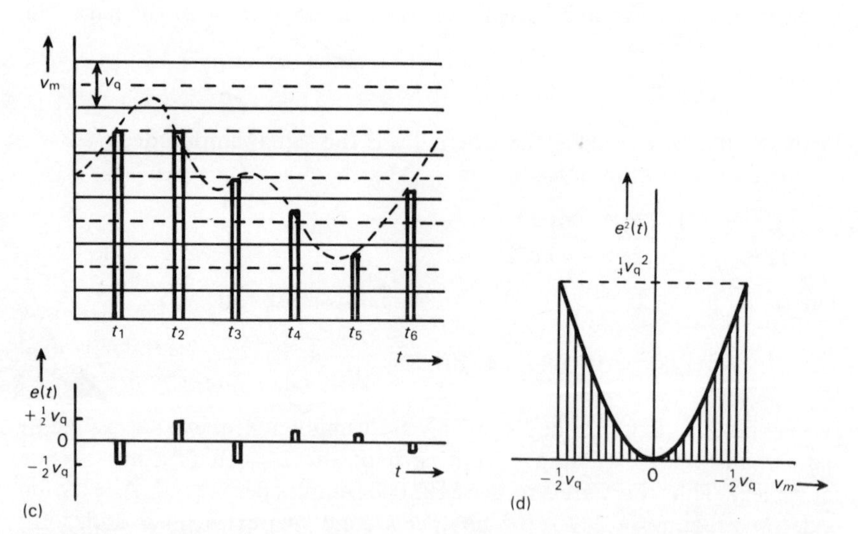

Fig. 23.19 (a) Pulse code modulation of a sinusoid (b) Illustration of the Nyquist sampling theorem (c) Sampled waveform and quantization noise (d) Average noise power (e) Example of time-division multiplex with three signals (f) Compression in pulse code modulation

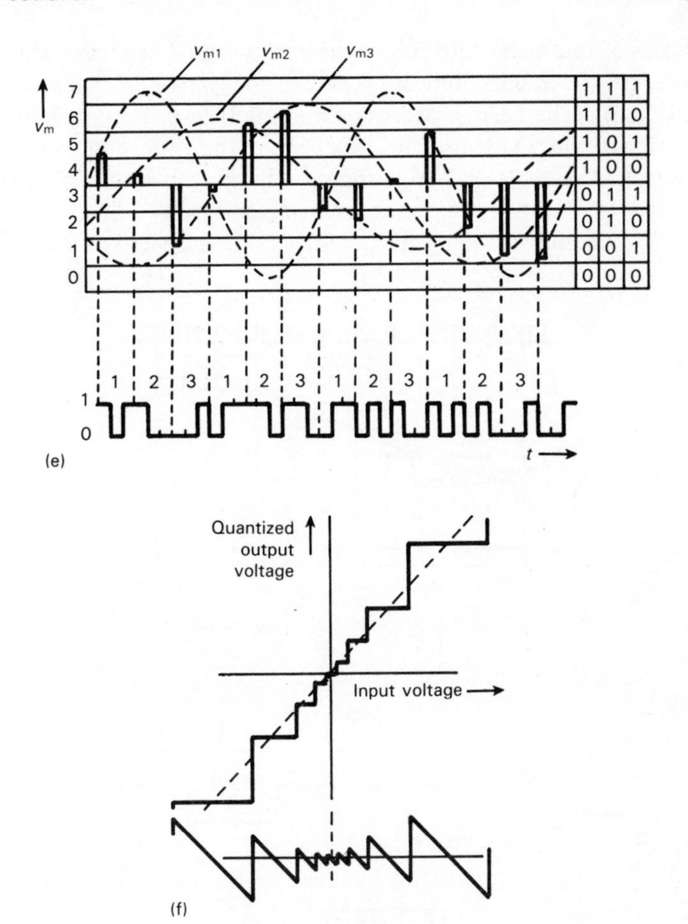

Fig. 23.19 *(Continued)*

A high voice level would fall into all 128 voltage steps whereas low-level signals will pass through the lower voltage steps only. Since low-level signals tolerate smaller signal-to-noise ratios, a technique called **compression** is applied: low-level signals are sampled with smaller voltage steps than stronger signals (Fig. 23.19(f)). This way the signal-to-noise ratio is fairly constant.

At the receiving end the inverse operation, called **expansion** is used to convert the signal back to its original form.

Delta modulation (DM)

Digital voice systems often use **delta modulation** because lower bit rates (smaller bandwidths) are required than in PCM.

Pulses of uniform width follow the analog signal in a way such that the polarities of the pulses indicate whether the signal is rising or falling (Fig. 23.20(a)). The basic processing circuit is shown in Fig. 23.20(b). The analog signal and the pulses are applied to the inputs of a differential comparator. A sample-and-hold circuit converts the output signal of the comparator to '1' when the pulses are rising and to '0' when the pulses are falling. The analog signal is thus quantized. At the receiver the pulses are

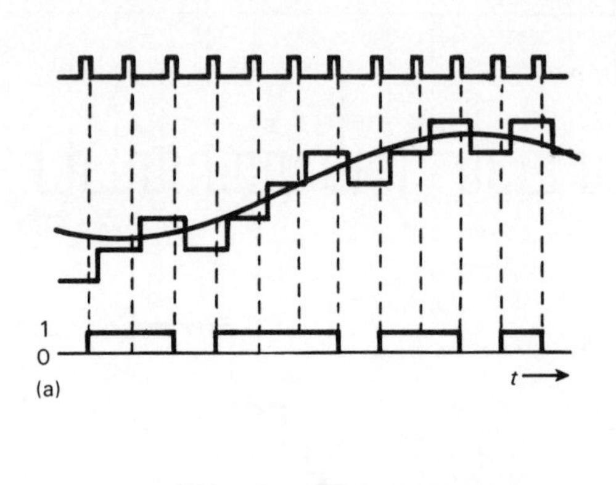

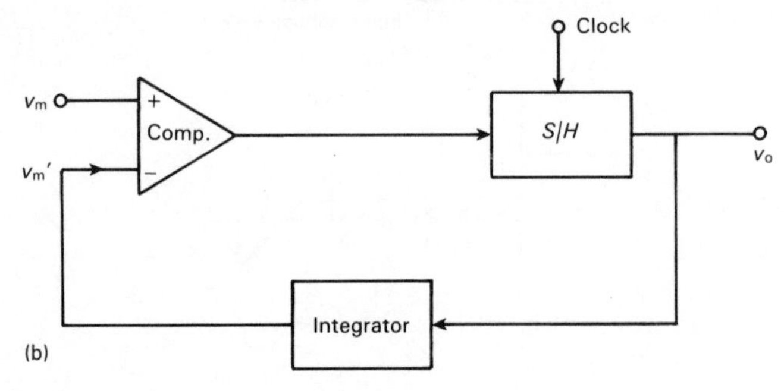

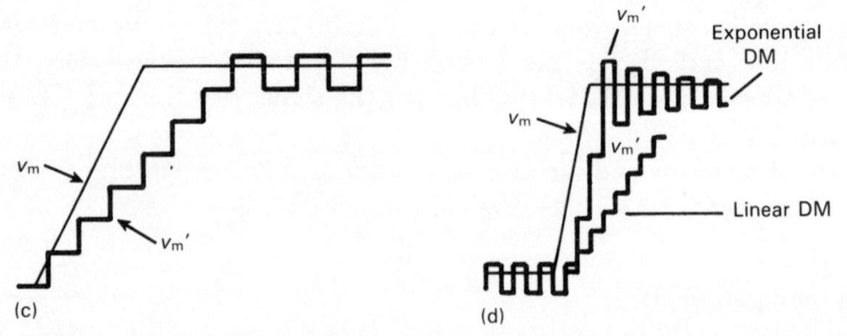

Fig. 23.20 (a) Principle of delta modulation (b) Basic circuit for delta modulation (c) Slope overload in delta modulation (d) Principle of adaptive delta modulation

reshaped and applied to a low-pass filter to remove the quantization noise. The original signal is then recovered.

If the rate of change of the analog signal is too high, the staircase pulses are unable to follow this fast rise (Fig. 23.20(c)) causing an error in the modulation. This effect is called **slope overload**. To decrease errors due to slope overload, a variation of DM called **adaptive delta modulation** (ADM) was developed. An amplifier precedes the integrator. The amplifier gain is controlled by the average pulse rate. If more positive pulses are transmitted than negative pulses, the gain increases. If more negative pulses are transmitted than positive pulses, the gain increases also.

The gain is minimum when positive and negative pulses are equal in number. In this way the height of the pulses is larger when the signal rises faster. Adaptive delta modulation produces a larger quantization noise than delta modulation. An illustration of exponential delta modulation is shown in the waveform of Fig. 23.20(d).

24
Frequency Conversion and Demodulation

24.1 GENERAL

To recover information from modulated carriers, the receiving system of Fig. 24.1 is generally employed. The system is required to select, from a vast number of carriers, a desired carrier and its sidebands and to reject all others. The selected signal is usually very small and needs to be amplified to a suitable level. The amplified signal is often converted or shifted to a more convenient frequency in a **mixer** or **converter** (Section 24.2). The **detector** or **demodulator** recovers the information which is amplified to operate the output device. A few of the various methods of amplitude demodulation are given below.

Tuned radio frequency receiver (TRF) (Fig. 24.2)

A number of RF amplifiers is cascaded and followed by adjustable resonant circuits which are tuned to the carrier frequency f_c. Restrictions of this system are as follows:

(a) All amplifiers operate at the same frequency. Since the overall gain is high (10^6–10^7), oscillations are likely to occur.

(b) Sensitivity and selectivity depend upon the tuning frequency.

(c) Tracking of a number of tuned circuits is difficult.

Synchrodyne receiver (Fig. 24.3)

This is a direct conversion receiver. A local oscillator is tuned to the carrier frequency f_c. One of the demodulation products is the difference frequency which is the low frequency modulation.

The sensitivity of this type is generally low and tuning causes whistles.

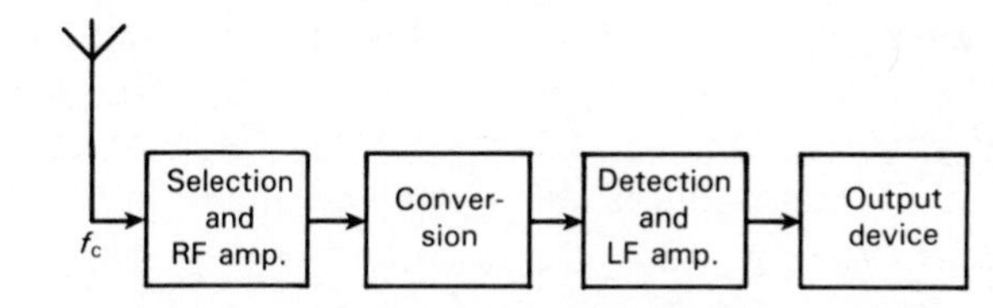

Fig. 24.1 Block diagram of demodulator circuit

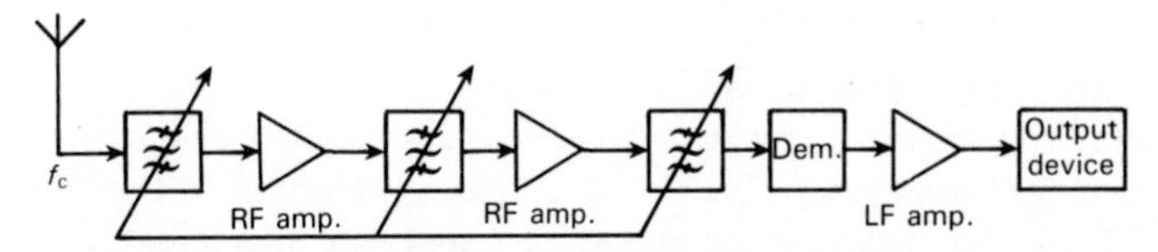

Fig. 24.2 Block diagram of TRF receiver

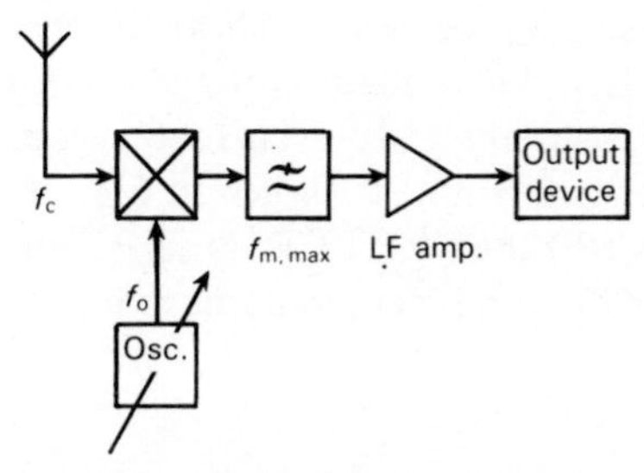

Fig. 24.3 Block diagram of synchrodyne receiver

Superheterodyne system

The most common demodulation system is the **superheterodyne system** (Fig. 24.4). The modulated carrier is multiplied by a signal with frequency f_o from a local oscillator. Oscillator frequency and carrier frequency track in such a way that the difference frequency $|f_o - f_c| = f_i$ remains constant. All incoming carriers and sidebands are thus translated to a carrier with constant frequency f_i which is called the **intermediate frequency**. Most of the amplification and selectivity is carried out at this frequency.

The process of obtaining the intermediate frequency is called heterodyne action. The multiplying circuit which produces the intermediate frequency is called a **converter** or **mixer**. The mixer thus produces the frequency $f_i = |f_o - f_c|$ but also $f_o + f_c$, f_o, f_c and harmonics of these frequencies. To remove unwanted frequencies, the mixer is followed by an IF filter which passes only the intermediate frequency and the sidebands.

If $f_i > f_c$, the conversion is called 'up-conversion', if $f_i < f_c$ the conversion is called 'down-conversion' (Fig. 24.5).

The superheterodyne method is applied both to AM and FM demodulation. AM broadcasts range from 150 kHz to 30 MHz and have a bandwidth of about 6 kHz. The IF is 455 kHz. The FM range is from 87.5–104 MHz, the bandwidth is about 240 kHz, the IF is 10.7 MHz.

24.2 FREQUENCY CONVERSION

Square law conversion or **additive mixing** is the method whereby the modulated carrier and a local oscillator signal are added and applied to the input of a nonlinear device (Section 23.1.2). A limitation of this method is the superposition of the two signals which may cause interference. Addition of the two signals without interference is possible by using a balanced mixer circuit as in Fig. 24.6.

In **multiplicative mixing** the two signals are applied to isolated inputs of a nonlinear device so that the oscillator voltage varies the transconductance of the mixer. As a result, the output current becomes a function of the product of v_s and v_o. An example is shown in Fig. 24.7(a) with a dual-gate MOSFET. In Fig. 24.7(b) the MOSFET is the RF amplifier, the output of which is coupled to the ring demodulator consisting of four matched Schottky diodes.

When converting an AM signal

$$v_c = V_c(1 + m_a \cos \omega_m t)$$

to its intermediate frequency by an oscillator signal

$$v_o = V_o \cos \omega_o t$$

the multiplication process results in an output signal

$$v_o = kV_oV_c(1 + m_a \cos \omega_m t)\cos(\omega_o - \omega_c)t + kV_oV_c(1 + m_a \cos \omega_m t)\cos(\omega_o + \omega_c)t$$

The IF filter passes the constant frequency term $\omega_i = \omega_o - \omega_c$.

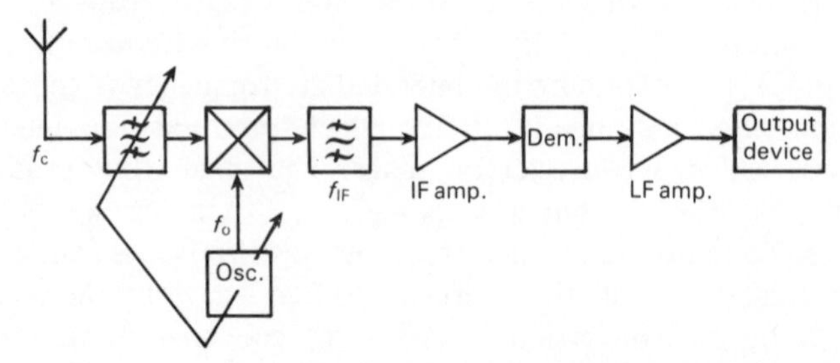

Fig. 24.4 Block diagram of superheterodyne receiver

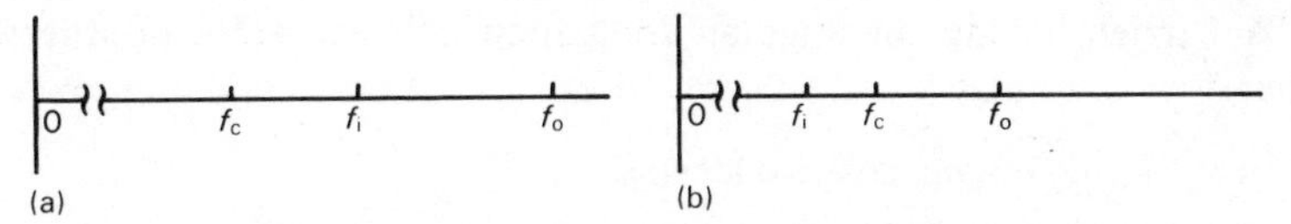

Fig. 24.5 (a) Spectral components in up-conversion (b) Spectral components in down-conversion

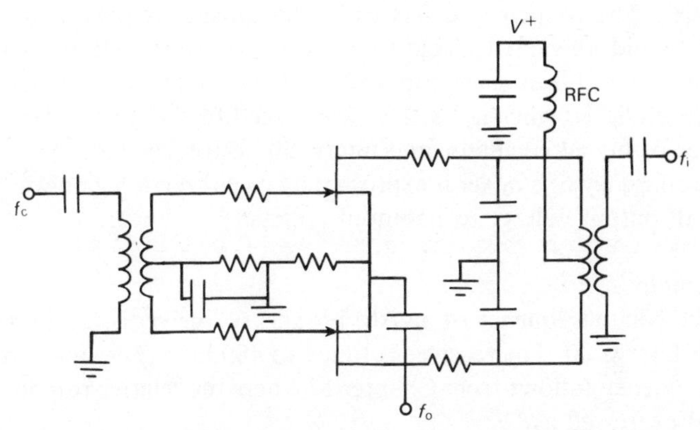

Fig. 24.6 Basic circuit of balanced mixer

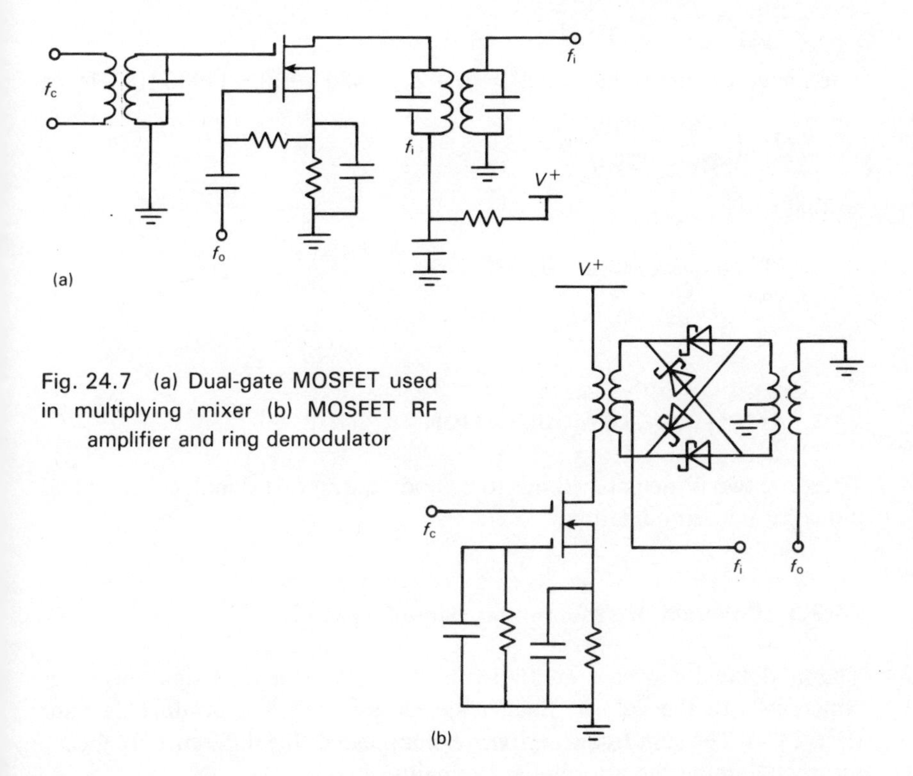

Fig. 24.7 (a) Dual-gate MOSFET used in multiplying mixer (b) MOSFET RF amplifier and ring demodulator

A carrier having an angular frequency $\omega_c^* = \omega_c + 2\omega_i$ produces an output

$$v_o = kV_oV_c(1 + m_a \cos \omega_m t)\cos(\omega_o - \omega_c^*)$$
$$= kV_oV_c(1 + m_a \cos \omega_m t)\cos \omega_i t$$

The intermediate frequency can thus be produced by two different carrier frequencies. The frequency ω_i^* is called the **image frequency** of ω_c (Fig. 24.8). To avoid unwanted image frequencies, sufficient selectivity prior to mixing must have taken place. Figure 24.8 shows that this selectivity is more easily accomplished when ω_c^* and ω_c are spaced further apart. This implies the choice of higher oscillator and intermediate frequencies.

The efficiency of a mixer is expressed by its **conversion gain** which is the ratio of IF output voltage to RF input voltage.

Example

The intermediate frequency of an AM broadcast receiver is 455 kHz. The Q of the IF filter is 60. The receiver is tuned to 480 kHz. The image rejection at this frequency follows from Chapter 17 where the relative response of the filter was expressed as

$$\frac{|Z_p|}{Z_{p,\max}} = \frac{1}{\sqrt{1 + \gamma^2 Q^2}}$$

The image frequency here is $f_i^* = f_s + 2f_i = 480 + 910 = 1390$ kHz. Hence

$$\gamma Q = \left(\frac{1390}{480} - \frac{480}{1390}\right)60 = 153 \quad (\gg 1)$$

so that

$$\frac{|Z_p|}{Z_{p,\max}} \simeq \frac{1}{\gamma Q} = \frac{1}{153} = -44 \text{ dB}$$

24.3 AMPLITUDE DEMODULATION

There are two principal methods to demodulate an AM signal: coherent and noncoherent demodulation.

24.3.1 Coherent or synchronous demodulation

The modulated carrier is multiplied by a locally generated signal which is coherent with the carrier. Such a detector is called a **product detector** (Fig. 24.9). The sum frequency term is suppressed, the difference frequency term, containing the information, remains.

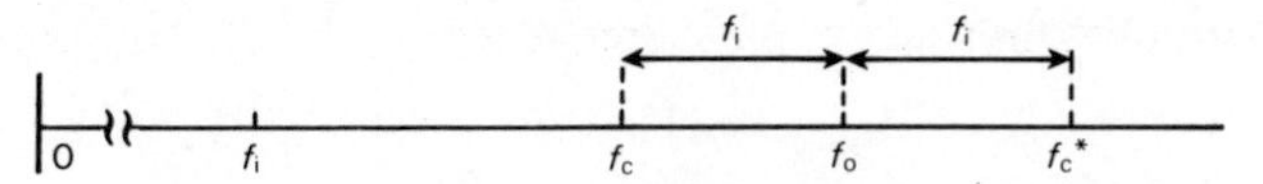

Fig. 24.8 Spectral components showing image frequency

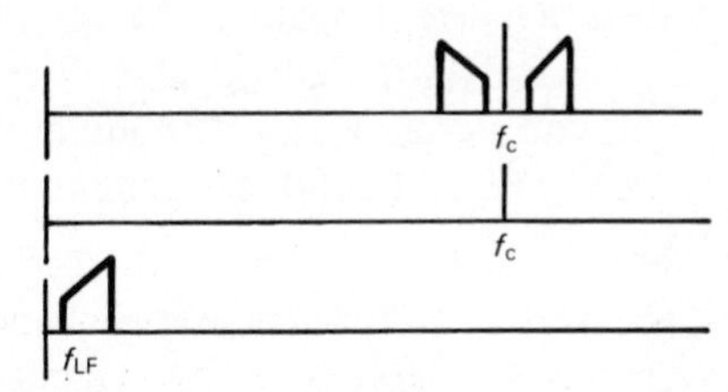

Fig. 24.9 Spectral components in product detector

- If the AM signal is

$$v_c = V_c(1 + m_a \cos \omega_m t)\cos(\omega_c t + \phi_c)$$

and the oscillator signal

$$v_o = V_o \cos(\omega_o t + \phi_o)$$

the low frequency output signal is

$$v_{LF} = kV_oV_c(1 + m_a \cos \omega_m t)\cos[(\omega_o - \omega_c)t + (\phi_o - \phi_c)]$$

When the signals are coherent ($\omega_o = \omega_c$, $\phi_o = \phi_c$)

$$v_{LF} = kV_oV_c(1 + m_a \cos \omega_m t)$$

The amplitude of the low frequency signal is smaller when phase coherence is lacking. In the case where $\phi_o - \phi_c$ is not constant, the signal is **fading**. When $\omega_o \neq \omega_c$, signal distortion will occur.

- A DSB signal is expressed as

$$v_c = kv_m \cos(\omega_c t + \phi_c)$$

The low frequency output is

$$v_{LF} = kv_m V_o \cos[(\omega_o - \omega_c)t + (\phi_o - \phi_c)]$$

In the case of coherence

$$v_{LF} = kv_m V_o$$

- An SSB signal is

$$v_c = kV_m \cos[(\omega_c \pm \omega_m)t + \phi_c]$$

and the low frequency signal

$$v_{LF} = kV_m V_o \cos[(\omega_o - \omega_c \pm \omega_m)t + (\phi_o - \phi_c)]$$

In case of coherence

$$v_{LF} = kV_m V_o \cos \omega_m t = kv_m V_o$$

Many types of synchronous detector circuits exist. In most cases the oscillator signal is a square wave.

Figure 24.10(a) shows a current switch, the relevant waveforms of which are shown in Fig. 24.10(b). Obviously, the double balanced modulator of Fig. 23.6(a) can be used as a demodulator. A phase locked loop demodulator is shown in Fig. 24.10(c). This circuit is described in more detail in Section 24.4.6.

24.3.2 Noncoherent or nonsynchronous demodulation

Noncoherent demodulation is performed by using some form of rectifying circuit followed by a low-pass filter. If the rectifier has a linear characteristic, the output signal is $v_2 = a|v_1|$. If v_1 is the AM signal

$$v_1 = v_c = V_c(1 + m_a \cos \omega_m t)\cos(\omega_c t + \phi_c)$$

then

$$v_2 = aV_c(1 + m_a \cos \omega_m t)|\cos(\omega_c t + \phi_c)|$$

When expanding the carrier term in a series, the low frequency terms are

$$v_{2,\,LF} = \frac{2}{\pi} aV_c(1 + m_a \cos \omega_m t)$$
$$= \frac{2}{\pi} aV_c + \frac{2}{\pi} ak_a v_m$$

The last term is the recovered modulation; the result is independent of the phase term ϕ_c.

If the rectifier has the square law characteristic $v_2 = av_1^2$, the low frequency term is

$$v_{2,\,LF} = aV_c^2 m_a(\cos \omega_m t + \tfrac{1}{2} m_a \cos^2 \omega_m t)$$

The square law detector thus produces second harmonic distortion of the modulation.

A series diode **envelope detector** is shown in Fig. 24.11(a), and the demodulated waveform in Fig. 24.11(b). It can be derived that the envelope follows the modulation closely when the time constant of the detector is chosen so that

$$\frac{1}{f_c} \ll RC \leqslant \frac{1}{m_a \omega_{m,\,max}}$$

Weak signals or large values of m_a result in distortion of the modulation due to the curvature of the diode characteristic.

The previous method cannot be applied to DSB and SSB signals. Envelope detection of these signals is only possible when the carrier is reinserted. If the reinserted carrier is expressed by

$$v_o = V_o \cos(\omega_o t + \phi_o)$$

a DSB signal gives an input to the envelope detector

$$v_1 = kv_m \cos \omega_c t + V_o \cos(\omega_o t + \phi_c - \phi_o)$$

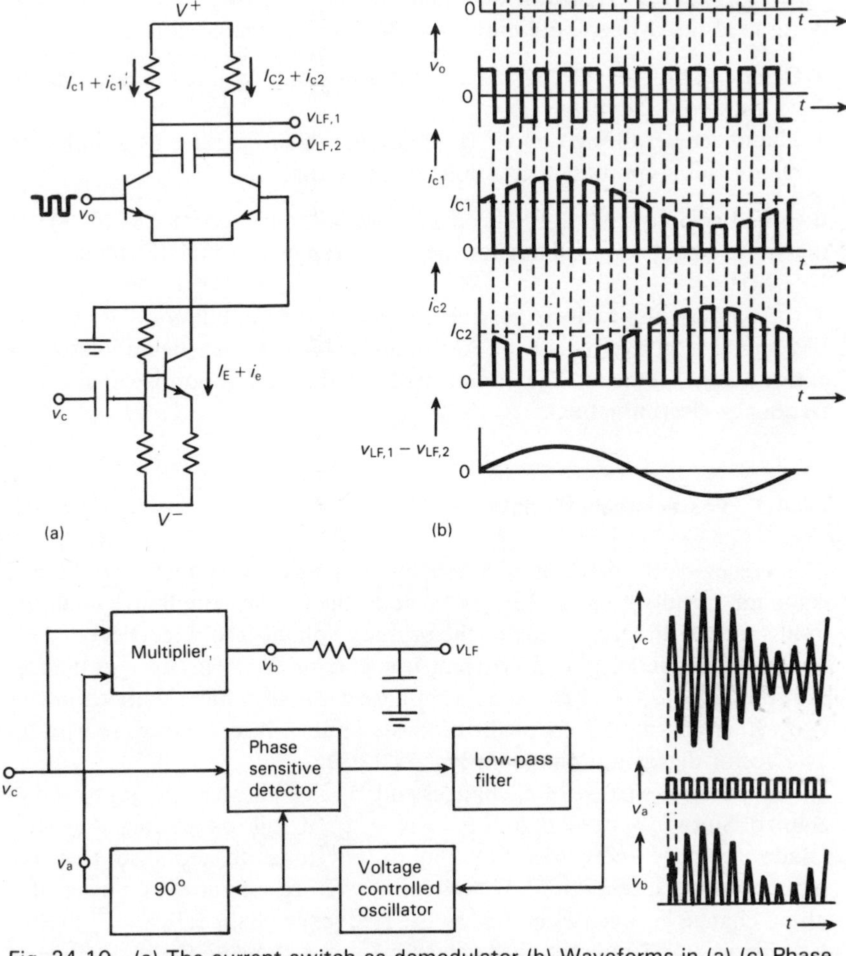

Fig. 24.10 (a) The current switch as demodulator (b) Waveforms in (a) (c) Phase locked loop AM demodulator circuit and waveforms

If $\omega_o = \omega_c$

$$v_i = [kv_m + V_o \cos(\phi_c - \phi_o)]\cos \omega_c t - [V_o \sin(\phi_c - \phi_o)]\sin \omega_c t$$

Evidently, when $\phi_c - \phi_o = 0$, the signal is a normal AM signal with $m_a = k(V_m/V_o)$. If frequency and phase of the reinserted carrier are not correct, DSB and SSB signals will be shifted up or down from their original location in the spectrum. Generation of such a stable and exact frequency is the main problem in reception of these signals.

24.4 FREQUENCY DEMODULATION

The principles underlying the demodulation of an FM signal are the same as those used in AM demodulation (Fig. 24.12). The major differences are:

- Demodulation of FM broadcast transmissions requires a large bandwidth: about 240 kHz.
- The mixer is followed by an IF amplifier which operates as a limiter to remove all amplitude fluctuations of the signal.

It should be noted that in FM broadcasts which range from 87.5–104 MHz, the image frequencies fall outside the tuning range, since the IF frequency is 10.7 MHz.

A variety of circuits is possible to recover the modulation. Usually the FM signal is converted to an AM signal which is then demodulated by conventional methods. The circuit used for this conversion is often called **frequency discriminator**.

24.4.1 Foster–Seely detector

The Foster–Seely detector is based on properties of inductively coupled resonant circuits (Fig. 24.13(a)). In addition to the coupling by mutual inductance, a direct coupling of the primary voltage (the IF carrier) to L_2 is made. At resonance ($f = f_c$) current i_1 lags v_c by 90°. This current induces the voltage $-j\omega M i_2$ in L_2. As i_2 is in phase with the induced voltage, v_1 lags i_2 by 90° (ignoring effects of the diodes). Thus v_1 is in quadrature with v_c. The vector diagram is shown in Fig. 24.13(b).

If $f \neq f_c$ the phase of v_1 changes and, as a result, the voltages v_2 and v_3 change. Since $v_2 = v_c + \frac{1}{2}v_1$ and $v_3 = v_c - \frac{1}{2}v_1$, it follows that $v_{LF} = v_2 - v_3$. Diodes and RC networks function as envelope detectors so that v_{LF} represents the modulation. The circuit therefore operates by virtue of a phase change between v_1 and v_c as the frequency changes.

Figure 24.13(c) is the characteristic S-curve showing the magnitude of v_{LF} as a function of frequency deviation.

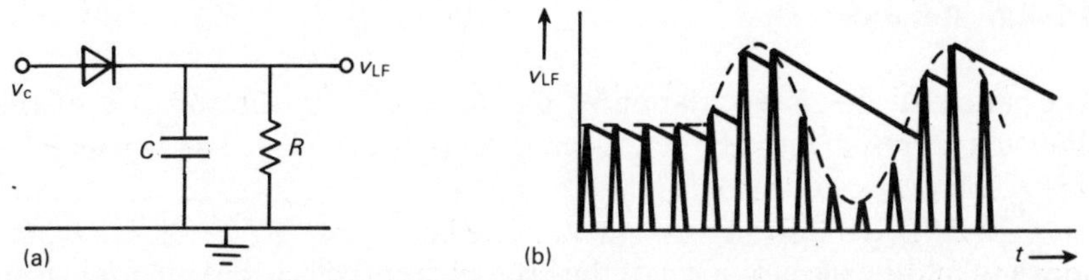

Fig. 24.11 (a) Envelope detector with series diode (b) Waveforms in envelope detection

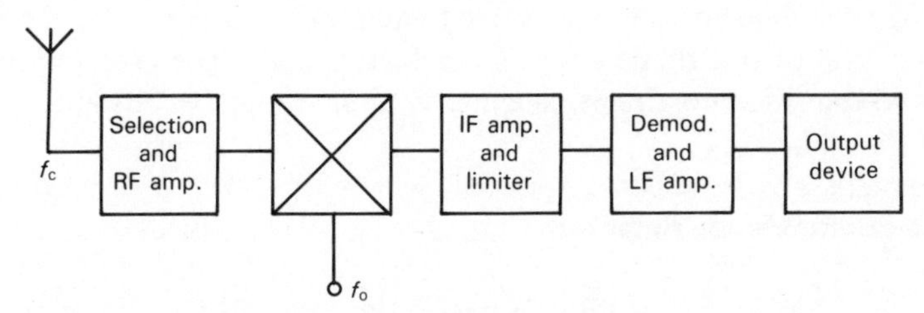

Fig. 24.12 Basic FM demodulator circuit

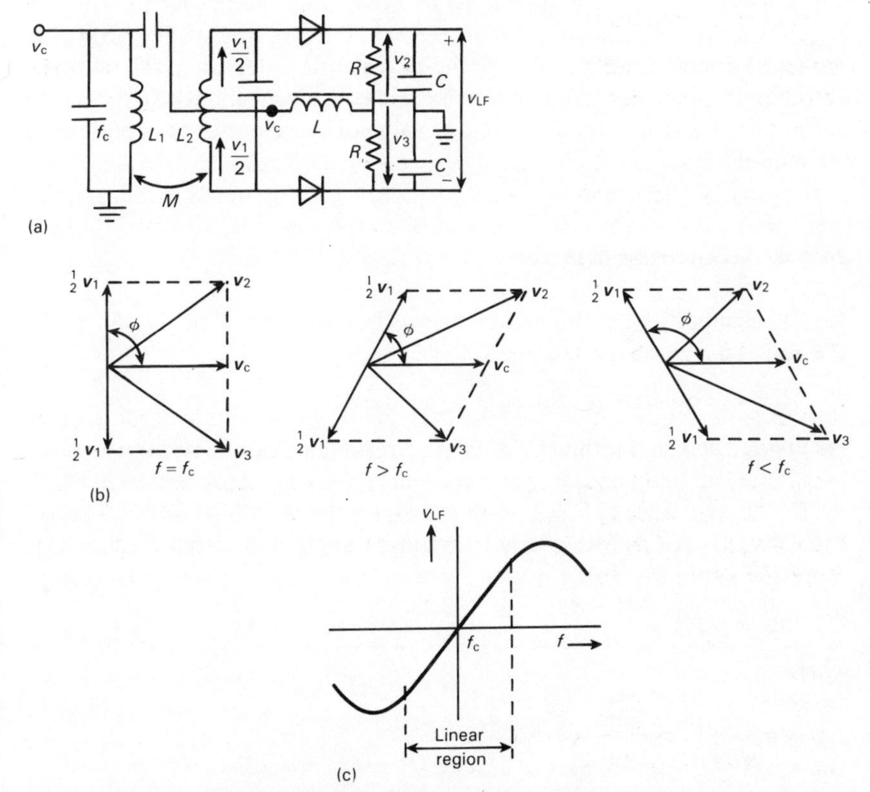

Fig. 24.13 (a) The Foster–Seely detector (b) Phasor diagram of the Foster–Seely detector (c) Characteriostic response of Foster–Seely detector

24.4.2 Ratio detector

The ratio detector is a variation of the Foster–Seely detector: one of the diodes is reversed and a large capacitor C_2 is added (Fig. 24.14). Here

$$v_{LF} = \tfrac{1}{2}(v_2 - v_3)$$

which is half the output signal of the Foster–Seely detector. As before, as v_2 increases, v_3 decreases keeping $v_2 + v_3$ approximately constant. This allows the connection of a large capacitor C_2 which improves the constancy.

The ratio detector is less sensitive to amplitude variations of the signal since reversal of one diode opens a conductive path. The detector circuits thus load the resonant circuit, keeping v_1 approximately constant.

24.4.3 Coincidence detector

The limited signal (Fig. 24.15(a)) is passed through an LCR circuit of which the phase characteristic is shown in Fig. 24.15(b). The input and output signals of this circuit are applied to an AND gate the output of which is integrated. At $f = f_c$ the phase shift is 90° and the output of the AND is as shown in Fig. 24.15(d). If $f > f_c$ the phase shift is less than 90° and the output pulses of the AND gate become wider. Conversely, when $f < f_c$ the pulses become narrower. The average value of these pulses thus represents the modulation.

24.4.4 Quadrature detector

An FM signal, v_c, is multiplied by its quadrature signal (Fig. 24.16). If the FM signal is

$$v_c = V_c \cos(\omega_c t + m_f \sin \omega_m t)$$

the voltage across the tuned circuit near resonance can be written as

$$v_z = V_z \sin(\omega_c t - \phi)$$

It follows from Chapter 17 that the phase angle ϕ between v_c and v_z is expressed as

$$\tan \phi = \gamma Q$$

where

$$\gamma = \frac{\omega}{\omega_o} - \frac{\omega_o}{\omega} \simeq \frac{2\Delta\omega}{\omega_o} = k_1 v_m$$

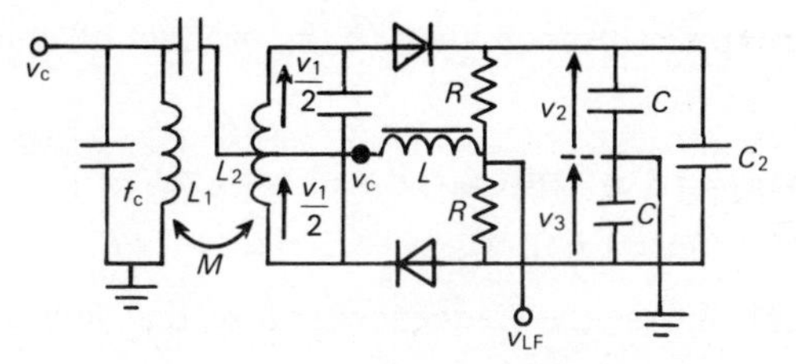

Fig. 24.14 The ratio detector

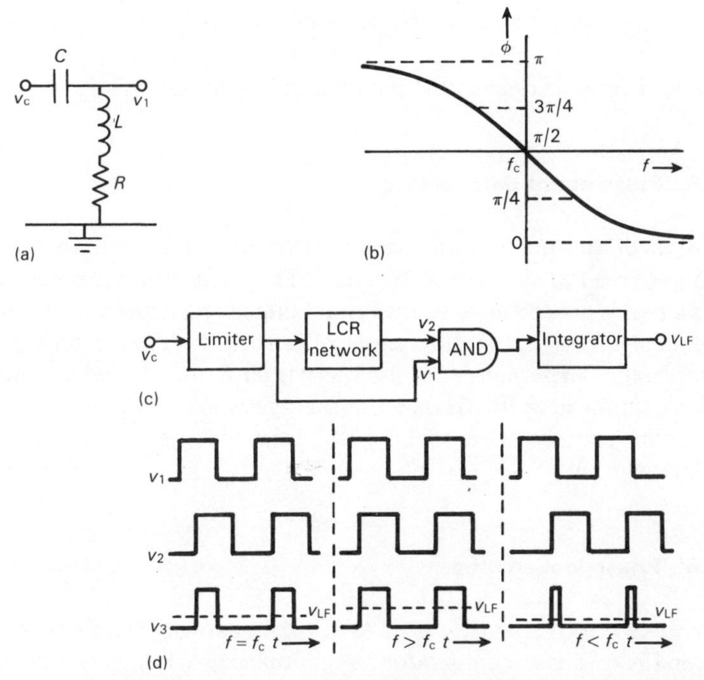

Fig. 24.15 (a) Network used in coincidence detector (b) Phase characteristic of (a) (c) Coincidence detector circuit (d) Waveforms in coincidence detector

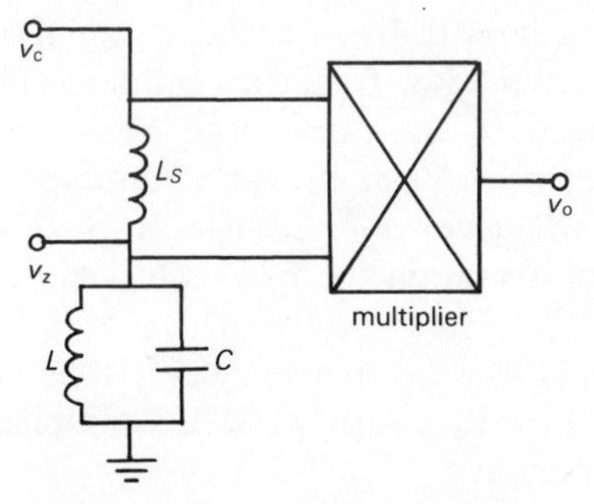

Fig. 24.16 Quadrature detector

The multiplier output is proportional to the product of v_c and v_z so that

$$v_o = k_2 v_c v_z$$
$$= k_2 V_c \cos(\omega_c t + m_f \sin \omega_m t) V_z \sin(\omega_c t - \phi)$$
$$= k_3 [\sin(2\omega_c t + \phi) + \sin \phi]$$

A low pass filter removes the frequency term so that finally

$$v_o = k_3 \sin \phi \simeq k_3 \tan \phi$$
$$= k_3 \gamma Q$$
$$= k_3 k_1 Q v_m$$

Hence v_o is proportional to the modulation signal v_m.

24.4.5 Pulse count demodulator

Operation of the pulse count demodulator (PCD) is based on the zero crossings of the FM signal (Fig. 24.17(a)). The number of signal periods in a time interval is proportional to the instantaneous frequency in that interval. The limited FM signal is differentiated in such a way that only positive pulses remain. These pulses are used to trigger a one-shot which produces pulses of uniform width. Hence the average value of these pulses is the modulation. The advantage of this circuit is the absence of tuning circuits or other adjustments.

24.4.6 Phase locked loop

The basic circuit of a phase locked loop (PLL) consists of a phase sensitive detector (PD) (phase comparator or multiplier), a low pass filter and a voltage controlled oscillator (VCO), connected as a feedback system (Fig. 24.18). The PD multiplies the signals v_c and v_o so that the output contains the signals with frequencies $|\omega_c - \omega_o|$ and $\omega_c + \omega_o$. The low-pass filter suppresses the signal with frequency $\omega_c + \omega_o$ and other high-frequency components and passes the low-frequency components. The output signal of the filter is v_f.

The VCO frequency ω_o can be varied around a fixed free-running frequency or **center frequency** ω_o^*. Changes in ω_o result from changes in control voltage v_f. In operation the PLL is locked or synchronized, thus $\omega_o = \omega_c$.

If the input signal has the frequency ω_o^*, thus $v_c = V_c \sin \omega_o^* t$, the oscillator output is $v_o = V_o \cos(\omega_o^* t - \phi_o)$. Multiplication of these two signals in the PD results in

$$v_d = \tfrac{1}{2} k V_c V_o [\sin \phi_o + \sin(2\omega_o^* t - \phi_o)]$$

and

$$v_f = \tfrac{1}{2} k V_c V_o \sin \phi_o$$

The output of the PD thus is a function of ϕ_o. The factor $\tfrac{1}{2} k V_c V_o$ is called the **gain factor** of the PD.

The VCO transfer characteristic is expressed as

$$\omega_o = \omega_o^* + k_o v_f$$

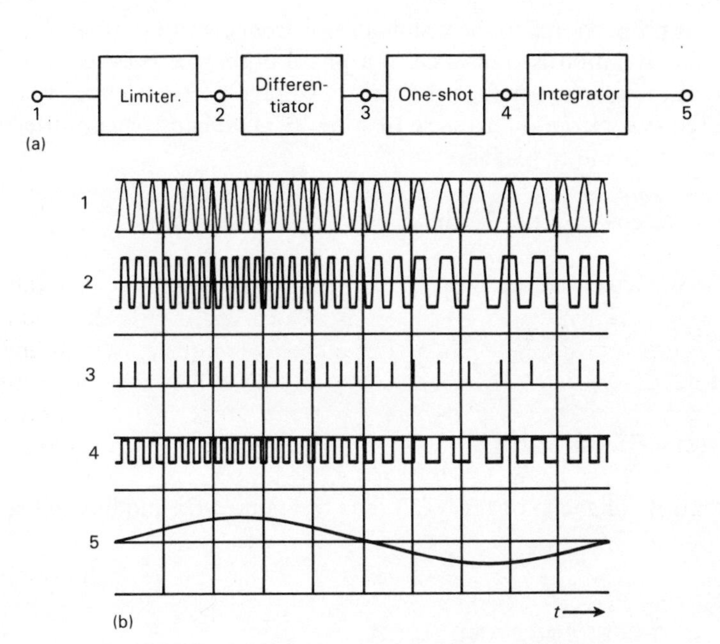

Fig. 24.17 (a) Pulse count demodulator (b) Waveforms in pulse count demodulator

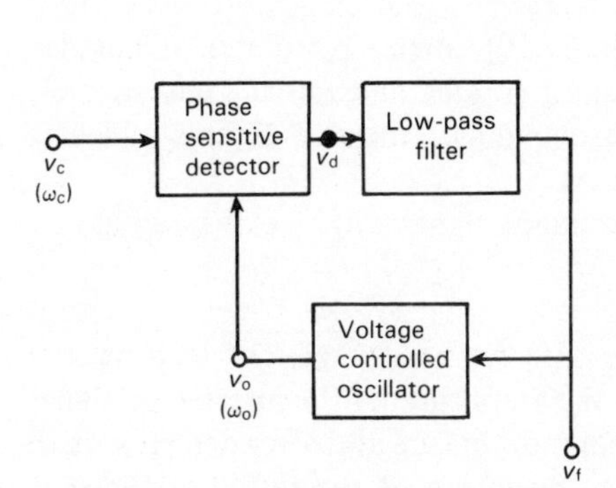

Fig. 24.18 Block diagram of phase locked loop

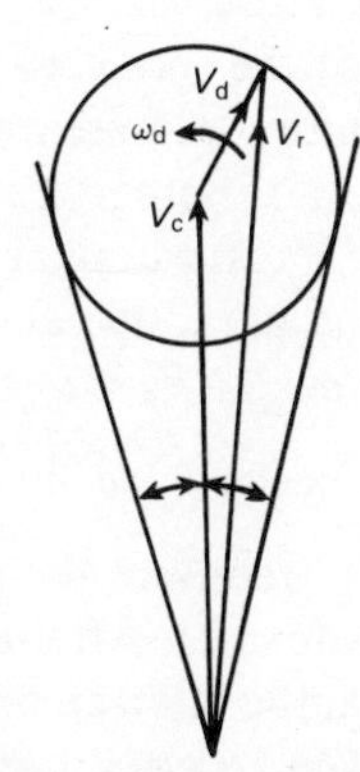

Fig. 24.19 Phasor representation of carrier and interference signal

where k_o is the **conversion gain** of the VCO. If now $\omega_c = \omega_o = \omega_o^*$, then $v_f = 0$ and $\phi_o = 0$. This implies that v_c and v_o are in quadrature when $\omega_c = \omega_o^*$ since v_c was assumed as a sine function, v_o as a cosine function.

When changing ω_c so that $\omega_c = \omega_o^* + \Delta\omega$, the VCO frequency is $\omega_o = \omega_c = \omega_o^* + k_o v_f = \omega_o^* + \Delta\omega$ so that

$$v_f = \frac{\omega_s - \omega_o^*}{k_o} = \frac{\Delta\omega}{k_o}$$

Thus v_f is proportional to the instantaneous frequency deviation of v_c. This frequency deviation is translated in a phase difference between v_c and v_o which is necessary to generate the control voltage v_f. This voltage shifts the VCO frequency from ω_o^* to ω_c so that the VCO remains synchronized with v_c. When v_c is the FM signal

$$v_c = V_c \cos\left(\omega_o^* t + \frac{\Delta\omega}{\omega_m} \sin \omega_m t\right)$$

the instantaneous frequency is

$$\omega_c(t) = \omega_o^* + \Delta\omega \cos \omega_m t$$

so that

$$v_f(t) = \frac{\Delta\omega}{k_o} \cos \omega_m t$$

The control voltage v_f of the VCO thus reproduces the modulation signal.

24.5 INTERFERENCE AND NOISE

Suppose an unmodulated carrier $v_c = V_c \cos \omega_c t$ is disturbed by an interference signal $v_d = V_d \cos \omega_d t$ so that both carriers fall in the IF bandwidth of the receiver.

In a phasor representation (Fig. 24.19) carrier $\boldsymbol{v}_c$ rotates at angular frequency ω_c, the interference signal $\boldsymbol{v}_d$ rotates at angular frequency ω_d around $\boldsymbol{v}_c$. The resultant is the phasor sum of the two signals which is modulated in amplitude and frequency.

Assuming $V_d / V_c = a \ll 1$, the resultant signal can be expressed as

$$v_r \simeq V_c \cos(\omega_c t + a \sin \omega_d t)$$

In an **AM receiver** the envelope of v_r is $\sqrt{V_c^2(1 + a^2 + 2a \sin \omega_d t)} \simeq V_c(1 + a \sin \omega_d t)$. After detection the amplitude of the interfering signal is aV_c so that the interference-to-signal ratio before and after detection is a.

In an **FM receiver** the frequency deviation of the resultant signal is $\Delta f = af_d$. If $(\Delta f)_{max}$ is the maximum frequency deviation, the relative signal strength is $af_d/(\Delta f)_{max}$. This is the interference-to-signal ratio at the

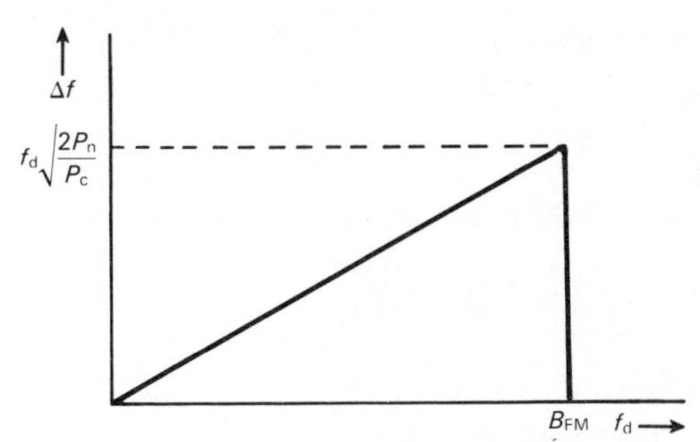

Fig. 24.20 Noise spectrum of FM receiver

detector output. These results show the following:

(a) The improvement ratio of FM with respect to AM is $f_d/(\Delta f)_{max}$.
(b) In an FM receiver the effect of interference can be made arbitrarily small by increasing the maximum frequency deviation.
(c) The effect of interference in an FM receiver is less when f_d is lower. It is for this reason that pre-emphasis is employed in FM broadcast transmissions.

As long as $a < 1$, the phasor phase variations are less than 90° and the resultant phasor follows the motions of v_c while the effect of v_d is suppressed. If $a \geqslant 1$, the situation suddenly reverses and the combined signal follows the phase of the interference signal. This effect is called **capture**. When the interfering signal consists of band-limited white noise surrounding the carrier, the noise spectrum can be divided into a large number of narrow band elements. Each element can be treated as a separate interfering signal. If the bandwidth of an element is 1 Hz and the noise power is P_n W Hz^{-1}, this noise component can be considered as a sinusoidal signal with rms value $\sqrt{P_n}$ and peak value $\sqrt{2P_n}$. If the carrier power is P_c, its peak value is $\sqrt{2P_c}$. The relative strength of the interfering signal is

$$a = \sqrt{\frac{P_n}{P_c}}$$

In an **FM receiver** the frequency deviation

$$\Delta f = af_d = f_d\sqrt{\frac{P_n}{P_c}}$$

The net deviation is thus the rms addition of these individual deviations

$$\Delta f = f_d\sqrt{\frac{2P_n}{P_c}}$$

This noise spectrum is of triangular shape (Fig. 24.20). When B_{FM} is the bandwidth of the audio filter, the spectrum ends at $f_d = B_{FM}$. The noise power as a function of f_d is

$$P_n = f_d^2 \frac{2P_n}{P_c}$$

and the total noise power in the bandwidth B_{FM} is

$$P_{n,t} = \int_0^{B_{FM}} P_n \, df_d = \frac{1}{3} B_{FM}^3 \frac{2P_n}{P_c}$$

The signal-to-noise ratio is thus

$$\left(\frac{S}{N}\right)_{FM} = \frac{(\Delta f)_{max}^2}{P_{n,t}} = \frac{3(\Delta f)_{max}^2}{B_{FM}^3} \frac{P_c}{2P_n}$$

In an **AM receiver** with audio bandwidth B_{AM} the total detected noise power is

$$P_{n,t} = \int_0^{B_{AM}} 2P_n \, df_d = 2P_n B_{AM}$$

If P_c is again the carrier power, the signal-to-noise ratio is

$$\left(\frac{S}{N}\right)_{AM} = \frac{P_c}{2P_n B_{AM}}$$

The improvement ratio is the ratio of the signal-to-noise levels:

$$\frac{(S/N)_{FM}}{(S/N)_{AM}} = \frac{3(\Delta f)_{max}^2}{B_{FM}^3} B_{AM}$$

In broadcast transmissions $(\Delta f)_{max} = 75$ kHz, $B_{FM} = 15$ kHz, $B_{AM} \simeq 6$ kHz so that the improvement ratio is about 30.

Figure 24.21 shows the signal-to-noise ratio at the detector output for AM and FM as a function of carrier signal-to-noise level with the modulation index as the parameter.

Evidently, the superiority of FM over AM holds only at higher values of carrier signal-to-noise ratio. When this ratio drops below a certain threshold value, the output signal-to-noise ratio also drops rapidly. This threshold signal value increases with the modulation index.

The behavior of a receiver to a single interfering signal and noise is similar. When the noise exceeds the signal in an FM receiver, the noise tends to wipe out the signal and FM becomes inferior to AM.

24.6 STEREO DECODER

One of the circuits to decode a stereo signal is the **matrix decoder** (Fig. 24.22). The multiplex signal is applied to the input of a phase locked

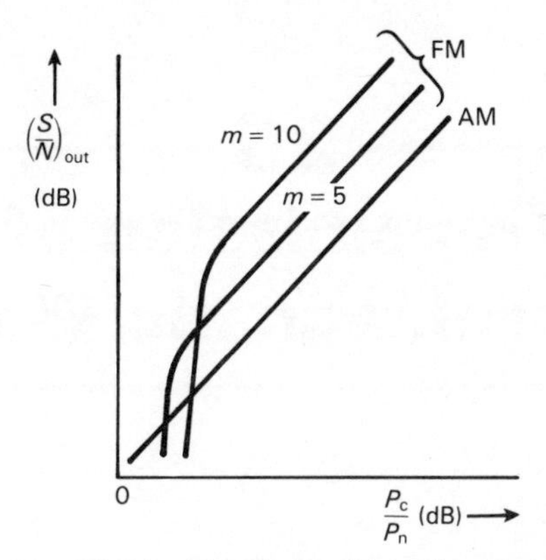

Fig. 24.21 Signal-to-noise characteristics of AM and FM

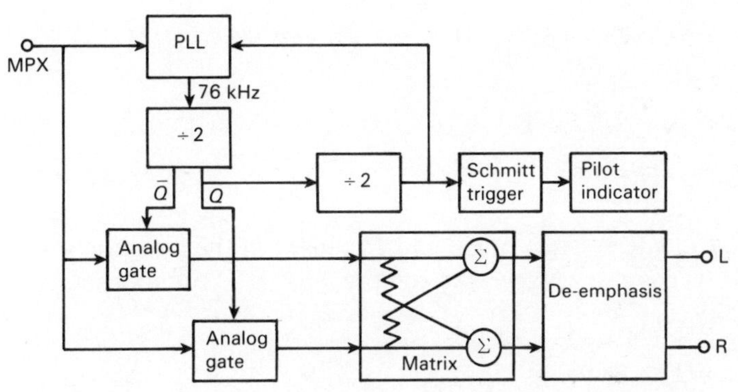

Fig. 24.22 Matrix decoder for decoding stereo signals

loop with a VCO oscillating at 76 kHz. A divide-by-two circuit produces signals Q and $\bar{Q}$ of 38 kHz. These signals are used to turn two analog gates on and off thus synchronously demodulating the L – R signal. Both outputs of the analog gates contain L and R components. The gates are followed by a matrix circuit which combines the composite signals so that at the outputs of the matrix the L and R signals are separated.

To offset the pre-emphasis which was introduced at the transmitting end of the signal, a de-emphasis network is added as the final stage. On reception of a monaural signal the gates are disabled so that the L + R signal is passed to the matrix.

A 19 kHz signal from a second divide-by-two circuit is used to close the loop of the PLL and to drive a Schmitt trigger which controls a pilot tone indicator.

25
Transmission of Electromagnetic Waves

25.1 TRANSMISSION IN A MEDIUM

An electromagnetic wave traveling in a medium obeys the Maxwell equations:

$$\nabla \times E = -\mu \frac{\partial H}{\partial t}$$

$$\nabla \times H = \varepsilon \frac{\partial E}{\partial t}$$

In a coordinate system with E and H chosen in the direction of the x-axis and the y-axis respectively, the solutions of these equations are

$$E_x = E \exp[j\omega(t - z\sqrt{\varepsilon\mu})]$$
$$E_y = E_z = 0$$
$$H_y = \pm H \exp[j\omega(t - z\sqrt{\varepsilon\mu})]$$

Apparently the wave propagates along the z-axis with phase velocity $v_p = 1/\sqrt{\varepsilon\mu}$.

Figure 25.1 illustrates the wave at $t = 0$. As the solutions indicate, the electric and magnetic fields are always perpendicular to each other. Such a wave can be generated by a resonating dipole (Hertzian dipole), the electric and magnetic fields of which are shown in Figs. 25.2(a) and 25.2(b).

In vacuum the wave travels at the velocity of light so that $c = 1/\sqrt{\varepsilon_v\mu_v}$ which allows v_p to be expressed as $v_p = c/\sqrt{\varepsilon_r\mu_r}$ (m s^{-1}).

The **wave impedance** in free space is $E/H = \sqrt{\mu/\varepsilon} = 120\pi = 377\ \Omega$.

According to Chapter 3 the energy density of the electric component is

$$w_e = \tfrac{1}{2}\varepsilon E^2 \text{ (W s m}^{-3}\text{)}$$

and of the magnetic component

$$w_m = \tfrac{1}{2}\mu H^2 \text{ (W s m}^{-3}\text{)}$$

The average power density can be expressed as the vector product

$$w = E_{\text{rms}} \times H_{\text{rms}}$$

This is the **Poynting vector** and its direction indicates the way in which the energy travels.

When the electric and magnetic fields are both perpendicular to the direction of propagation, the wave is called the **transverse electromagnetic wave** (TEM). A wave in free space is a TEM wave as well as waves in transmission lines.

25.2 TRANSMISSION LINES

If a signal is applied to a two-wire closed line and the signal frequency is low, the current through any cross section of the wire will be the same. However, if the frequency is so high that the length of the line is no longer negligible with respect to the wavelength, currents at different cross sections

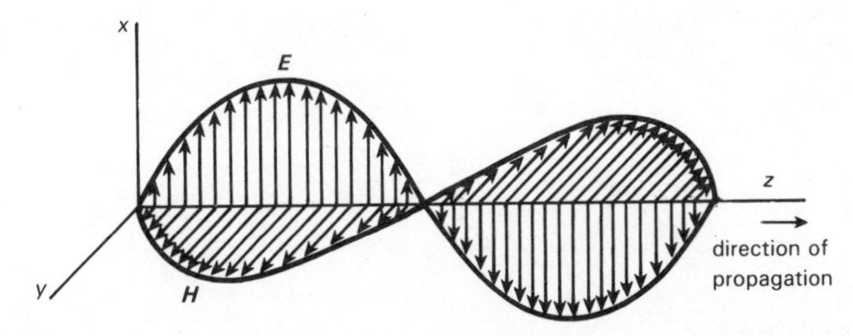

Fig. 25.1 Field components of an electromagnetic wave in free space

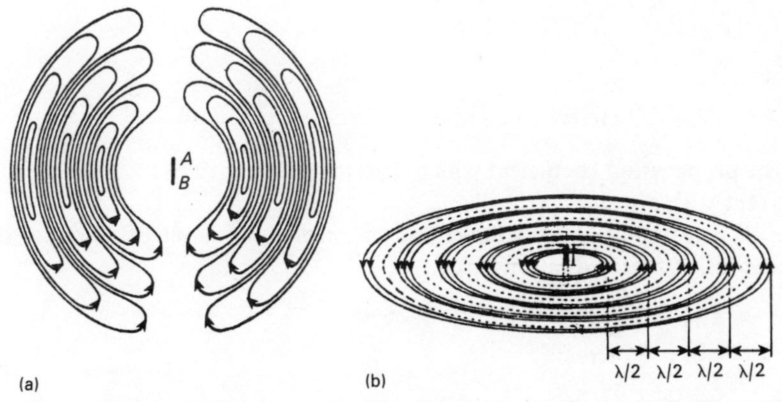

Fig. 25.2 (a) Electric field lines of a Hertzian dipole (b) Magnetic field lines of a Hertzian dipole

will be different such that currents flow in opposite directions in a section of the wire. This obviates the need for a return wire. Once initiated, the current pattern becomes independent of the source.

In most cases a single wire is inadequate for transmission of energy since most of it will be lost by radiation. An exception is the antenna which is based on transmission by radiation into space.

A transmission line connects a signal source v_s to a load impedance Z_L. The line can be balanced such as the two-wire line of Fig. 25.3(a) or unbalanced like the coaxial line of Fig. 25.3(b), the strip line (Fig. 25.3(c)) and the strip transmission line (Fig. 25.3(d)).

Balanced transmission lines are used in telephone circuits, coaxial lines in high-frequency applications. When a balanced radio antenna is connected to a coaxial line, a **balun** (balanced to unbalanced) transmission line transformer is used. A small element dx of the line can be represented as shown in Fig. 25.4. The line has series resistance R ($\Omega\,m^{-1}$), series inductance L ($H\,m^{-1}$), shunt conductivity G ($S\,m^{-1}$) and shunt capacitance C ($F\,m^{-1}$). These four parameters are the **primary line constants**.

It follows from Fig. 25.4 that

$$\frac{-dv_x}{dx} = i_x(R + j\omega L)$$

$$\frac{-di_x}{dx} = v_x(G + j\omega C)$$

Differentiation leads to the **telegraph equations** (developed by Kirchhoff in 1860):

$$\frac{d^2v_x}{dx^2} = \gamma^2 v_x$$

$$\frac{d^2i_x}{dx^2} = \gamma^2 i_x$$

where

$$\gamma = \sqrt{(R + j\omega L)(G + j\omega C)}$$

is the **propagation coefficient** which determines the variation of voltage and current with distance x.

When v_s and i_s are voltage and current at the input of the line, the solutions of these equations are

$$\begin{aligned} v_x &= \tfrac{1}{2}(v_s + i_s Z_o)e^{-\gamma x} + \tfrac{1}{2}(v_s - i_s Z_o)e^{\gamma x} \\ &= v_i\, e^{-\gamma x} + v_r\, e^{\gamma x} \end{aligned}$$

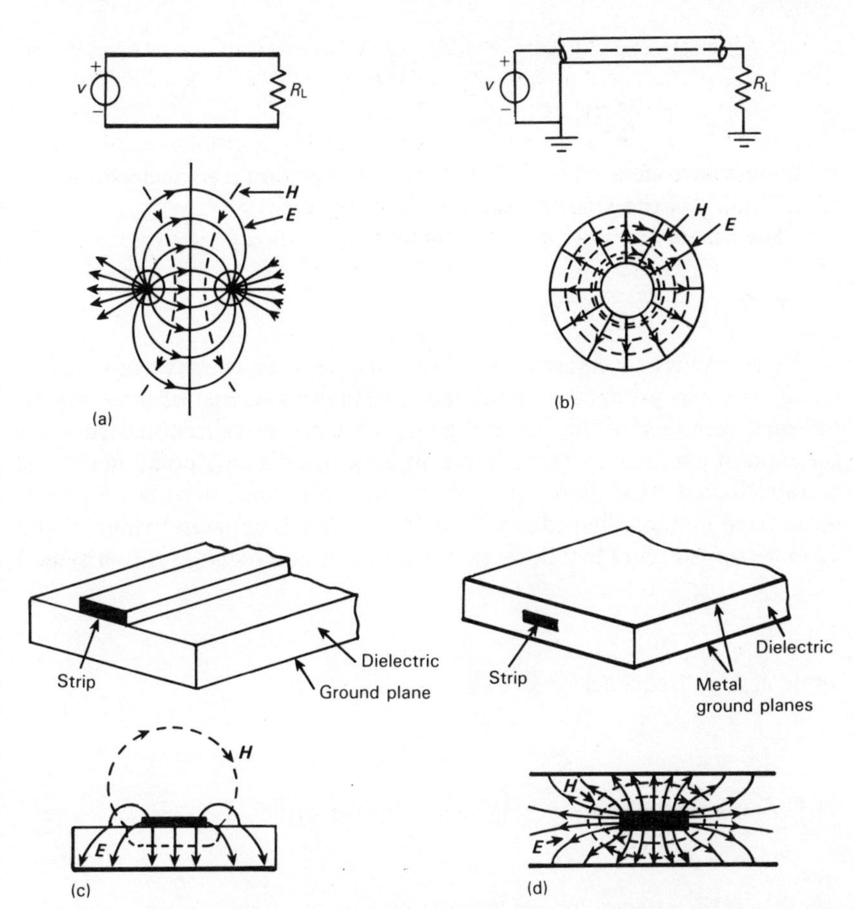

Fig. 25.3 (a) Field lines of two-wire line (b) Field lines of coaxial line (c) Strip line and field lines (d) Strip transmission line and field lines

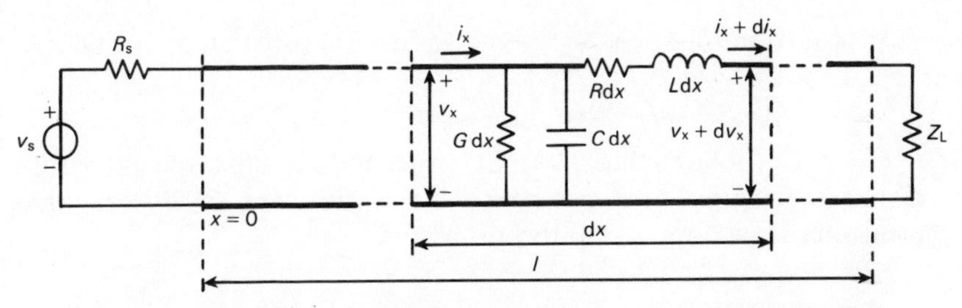

Fig. 25.4 Electrical components of an element of a transmission line

$$i_x = \tfrac{1}{2}\,\frac{v_s + i_s Z_o}{Z_o}\,e^{-\gamma x} - \tfrac{1}{2}\,\frac{v_s - i_s Z_o}{Z_o}\,e^{\gamma x}$$

$$= i_i\,e^{-\gamma x} - i_r\,e^{\gamma x}$$

In these expressions $v_i\,e^{-\gamma x}$ and $i_i\,e^{-\gamma x}$ represent the incident wave, $v_r\,e^{\gamma x}$ and $i_r\,e^{\gamma x}$ the reflected wave.

The parameter

$$Z_o = \sqrt{\frac{R + j\omega L}{G + j\omega C}} = \frac{\gamma}{G + j\omega C}$$

is the **characteristic impedance** of the line. The forward wave contains energy which is propagated along the line. In general, part of the energy is reflected at the end of the line. If the line is infinite, no reflection occurs and the ratio of maximum voltage to maximum current at any point on the line is the characteristic impedance Z_o. Thus if a finite length of line is terminated in a load impedance $Z_L = Z_o$, the line is apparently infinite and no reflections occur. It follows that at low frequencies ($R \gg \omega L$, $G \gg \omega C$)

$$Z_o \simeq \sqrt{\frac{R}{G}}$$

while at high frequencies ($R \ll \omega L$, $G \ll \omega C$)

$$Z_o \simeq \sqrt{\frac{L}{C}}$$

The complex propagation coefficient γ can be written as

$$\gamma = \alpha + j\beta$$

where α is the **attenuation coefficient**. This is the attenuation of voltage or current amplitude per unit length. It is usually expressed in dB m^{-1}. Here β is the **phase shift coefficient** which is equal to the phase shift per unit length.

When expanding γ in a series, the following approximations are obtained:

$$\alpha \simeq \frac{R}{2Z_o} + \frac{GZ_o}{2}$$

$$\beta \simeq \omega\sqrt{LC}\left[1 + \frac{1}{8}\left(\frac{R}{\omega L} - \frac{G}{\omega C}\right)^2\right]$$

If $R/G = L/C$ it follows that $\beta = \omega\sqrt{LC}$ which indicates that the phase shift is a linear function of frequency. This is the case of **distortionless transmission** since no phase distortion occurs.

25.3 TRAVELING WAVES

It follows from the basic equation $\lambda f = v_p$ that

$$\frac{\omega}{\beta} = v_p = \frac{1}{\sqrt{LC}}$$

which is the propagation velocity of the signal. If $R/G \neq L/C$ the envelope of the signal, as well as the energy, travel with the group velocity

$$v_g = \frac{\partial \omega}{\partial \beta}$$

If a line of length l is terminated with a load impedance Z_L, it can be derived that the input impedance of the line is

$$Z_i = Z_o \frac{Z_L \cosh \gamma l + Z_o \sinh \gamma l}{Z_o \cosh \gamma l + Z_L \sinh \gamma l}$$

Then if

$$l = \infty: \quad Z_i = Z_o$$

$$Z_L = Z_o: \quad Z_i = Z_o$$

$$Z_L = 0: \quad Z_i = Z_o \tanh \gamma l$$

$$Z_L = \infty: \quad Z_i = Z_o \operatorname{cotanh} \gamma l$$

Figure 25.5(a) shows Z_i as a function of l when $Z_L = 0$. Thus if

$$l = \tfrac{1}{4}\lambda + \tfrac{1}{2}k\lambda \quad (k = 0, 1, 2, \ldots)$$

the line impedance is infinite and the line behaves as a parallel resonant circuit. If

$$l = \tfrac{1}{2}k\lambda \quad (k = 0, 1, 2, \ldots)$$

$Z_i = 0$ and the line behaves as a series resonant circuit.

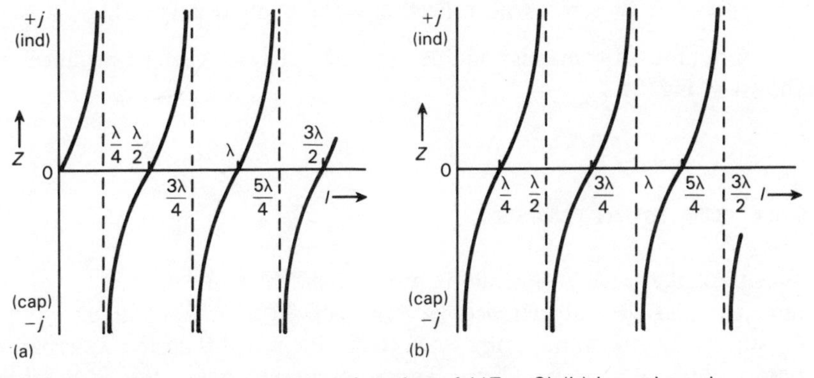

Fig. 25.5 (a) Input impedance as a function of l ($Z_L = 0$) (b) Input impedance as a function of l ($Z_L = \infty$)

Similarly, when $Z_L = \infty$ (Fig. 25.5(b)) and

$$l = \tfrac{1}{4}\lambda + \tfrac{1}{2}k\lambda \quad (k = 0, 1, 2, 3, \ldots)$$

$Z_i = 0$, while for

$$l = \tfrac{1}{2}k\lambda \quad (k = 0, 1, 2, 3, \ldots)$$

$Z_i = \infty$.

Figures 25.6(a) and 25.6(b) show the voltage and current distribution in open and shorted $\frac{1}{4}\lambda$ and $\frac{1}{2}\lambda$ lines.

Reflections occur when $Z_L \neq Z_o$. When the wave travels along the line, the voltage-to-current ratio is Z_o. At $x = l$, the impedance is Z_L and it is required that $v_i/i_i = Z_L$. A potential inconsistency thus suddenly develops at the termination. This results in a reflection of voltage and current so that the ratio

$$\frac{\text{incident + reflected voltage}}{\text{incident + reflected current}} = Z_L$$

is maintained. When the reflected voltage v_r is ρv_i and the reflected current $-\rho v_i/Z_o$ then at the termination:

$$\frac{v_i + \rho v_i}{\dfrac{v_i}{Z_o} - \dfrac{\rho v_i}{Z_o}} = Z_L$$

from which it follows that

$$\rho = \frac{Z_L - Z_o}{Z_L + Z_o} \quad (-1 \leqslant \rho \leqslant +1)$$

The parameter ρ is called the **reflection coefficient**. Thus

$\rho = 0$ if $Z_L = Z_o$

$\rho = -1$ if $Z_L = 0$ (full reflection and phase reversal)

$\rho = +1$ if $Z_L = \infty$ (full reflection without phase reversal)

A number of transmission lines and their characteristic impedances are shown in Fig. 25.7.

25.4 STANDING WAVES

When voltage waves travel along a line, incident and reflected waves will interfere and the interference pattern will show maxima and minima. Obviously, a maximum occurs when the incident and reflected waves are in phase, thus $V_{max} = |v_i| + |v_r|$, and a minimum when the waves have

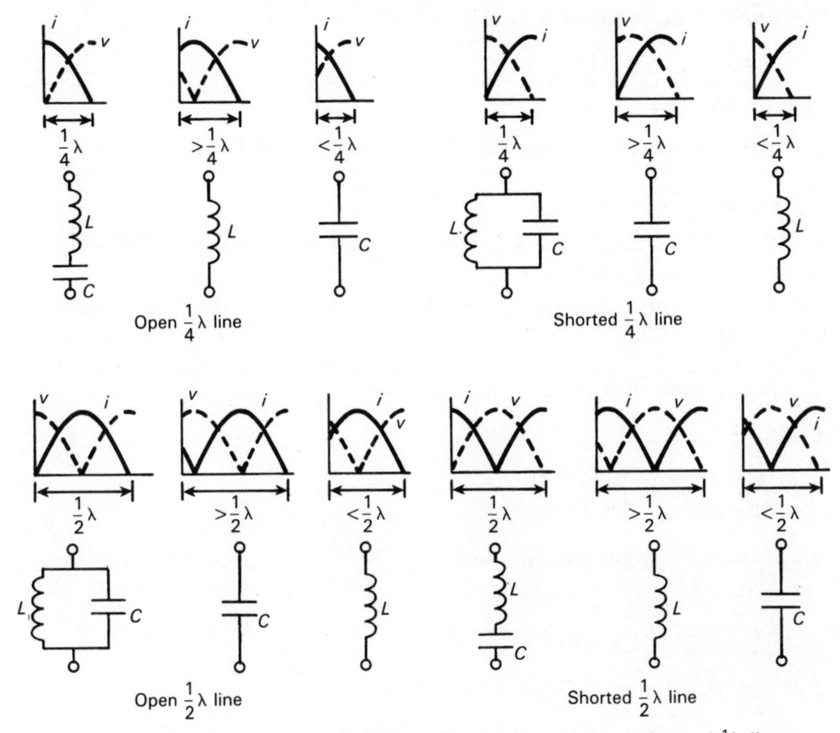

Fig. 25.6 (a) Voltage and current distribution in open and shorted $\frac{1}{4}\lambda$ line
(b) Voltage and current distribution in open and shorted $\frac{1}{2}\lambda$ line

opposite phase: $V_{\min} = |v_i| - |v_r|$. The ratio of these two values is called the **voltage standing wave ratio** (VSWR). Since $v_r = \rho v_i$, it follows that

$$\text{VSWR} = \frac{1 + |\rho|}{1 - |\rho|} \quad (1 \leqslant \text{VSWR} \leqslant \infty).$$

The pattern of standing waves along the length of the line is stationary as regards position but time varying since the value of v_i is a function of time. Figure 25.8 illustrates such a standing wave pattern where $|\rho| = r/s$ and VSWR $= p/q$.

Example 1

The primary constants of a line are $L = 250\ \text{nH m}^{-1}$, $C = 100\ \text{pF m}^{-1}$, $R = 0.5\ \Omega\,\text{m}^{-1}$, and $G = 0$. The signal frequency $f = 20$ MHz.

Since $R \ll \omega L$, the characteristic impedance can be written as

$$Z_o \simeq \sqrt{\frac{L}{C}} = \sqrt{\frac{250 \times 10^{-9}}{100 \times 10^{-12}}} = 50\ \Omega$$

The attenuation coefficient is

$$\alpha \simeq \frac{R}{2Z_o} = \frac{0.5}{100} = 0.005$$

This corresponds to an attenuation of (20 log e)0.005 $\simeq$ 0.043 dB m^{-1}.

Example 2

A shorted 50 Ω line is 0.2λ in length. The signal frequency is 500 MHz, the attenuation coefficient α is negligible.

The line impedance is inductive (Fig. 25.5(a)) and $Z_i = Z_o \tanh j\beta l = jZ_o \tan \beta l$. Thus

$$j\omega L_{eq} = jZ_o \tan \beta l$$

so that

$$L_{eq} = \frac{Z_o \tan \beta l}{\omega}$$

Since $\beta l = 2\pi \times 0.2$ it follows that

$$L_{eq} = \frac{50 \tan 2\pi \times 0.2}{2\pi \times 5 \times 10^8} = 0.35 \text{ nH}$$

If $\beta l = 0.4\lambda$, the line is capacitive and

$$\frac{1}{j\omega C_{eq}} = -jZ_o \tan \beta l$$

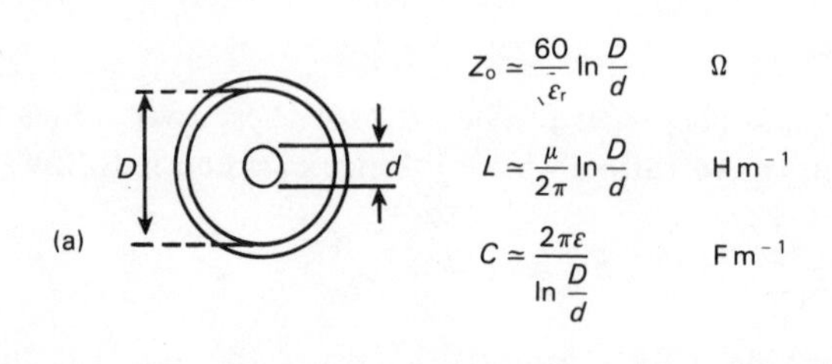

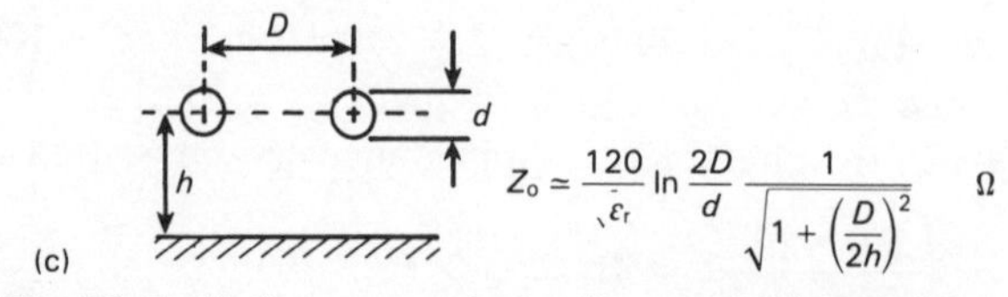

Fig. 25.7 Various transmission lines and relevant formulas

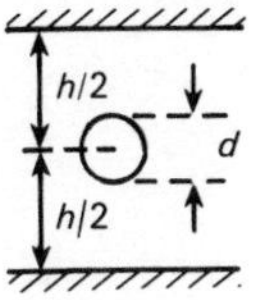

(d)

$$Z_o \simeq \frac{60}{\sqrt{\varepsilon_r}} \ln \frac{4h}{d} \quad \Omega \quad \left(\frac{h}{d} \geqslant 1.5\right)$$

(e)

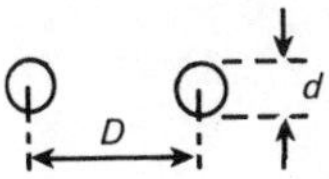

$$Z_o \simeq \frac{120}{\sqrt{\varepsilon_r}} \ln \frac{2D}{d} \quad \Omega \quad \left(\frac{D}{d} \geqslant 3\right)$$

$$L \simeq \frac{\mu}{\pi} \ln \frac{D}{d} \quad \mathrm{H\,m^{-1}}$$

$$C \simeq \frac{\pi\varepsilon}{\ln \dfrac{2D}{d}} \quad \mathrm{F\,m^{-1}}$$

(f)

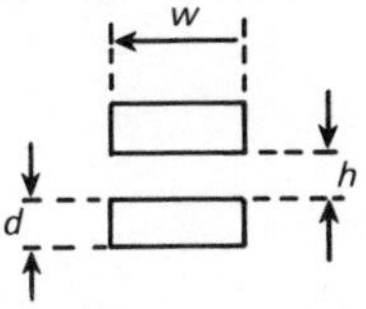

$$Z_o \simeq \frac{377}{\sqrt{\varepsilon_r}} \frac{h}{w} \quad \Omega \quad (w \gg h \gg d)$$

(g)

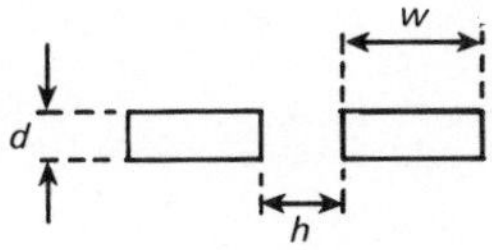

$$Z_o \simeq \frac{120}{\sqrt{\varepsilon_r}} \ln \frac{\pi h}{w+d} \quad \Omega \quad (w \gg d)$$

(h)

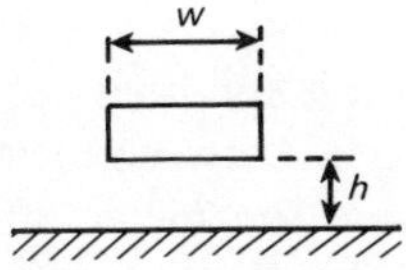

$$Z_o \simeq \frac{377}{\sqrt{\varepsilon_r}} \frac{h}{w} \quad \Omega \quad (w \gg h)$$

(i)

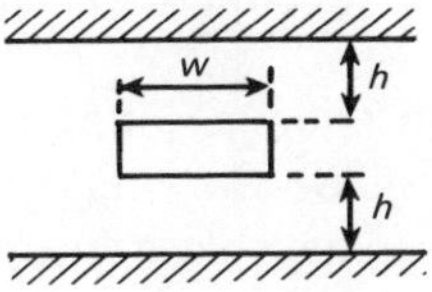

$$Z_o \simeq \frac{188}{\sqrt{\varepsilon_r}} \frac{h}{w} \quad \Omega \quad (w \gg h)$$

Fig. 25.7 (*Continued*)

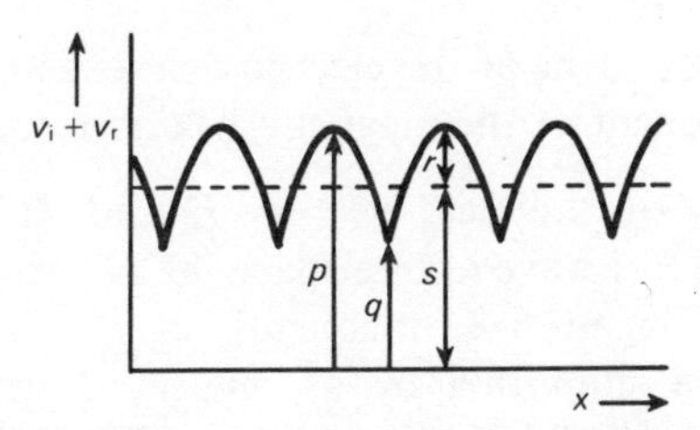

Fig. 25.8 Standing-wave voltage pattern

so that

$$C_{eq} = \frac{1}{\omega Z_o \tan \beta l}$$

Here we find

$$C_{eq} = \frac{1}{2\pi \times 5 \times 10^8 \times 50 \tan 0.8\pi} = 145 \text{ pF}$$

Example 3

An example of multiple reflections is shown in Fig. 25.9 where $Z_L \neq Z_o$ and $R_s \neq Z_o$. At $t = 0$ a step voltage V is applied. At the input of the line this step is reduced to

$$V_i = V \frac{Z_o}{R_s + Z_o}.$$

At the end of the line an attenuated reflection develops which returns to the input of the line and so on.

25.5 WAVEGUIDES

At frequencies above 3 GHz the use of transmission lines becomes impractical due to losses in the wires and the dielectric between them. In this part of the microwave region waveguides provide the means of transmitting electromagnetic energy. Normally, waveguides have a rectangular shape with the inner walls plated to reduce the resistance to induced currents. The electric characteristics are determined by the inner dimensions of the waveguide.

Actually, the space between the earth and the ionosphere (at about 80 km) can be viewed as a large scale waveguide allowing the transmission of radio waves. At the other end of the scale, glass fibers are used to transmit optical signals. Electromagnetic waves traveling through a waveguide obey two boundary conditions:

1. The tangential component of the electric field is zero.
2. The normal component of the magnetic field is zero.

Hence, the direction of the field vectors $\boldsymbol{E}$ and $\boldsymbol{H}$ is as sketched in Fig. 25.10(a). In the TEM wave in free space the $\boldsymbol{H}$ vector is at right angles to the direction of propagation. Consequently, according to Fig. 25.10(a) a TEM-wave cannot be supported by a waveguide and other modes of propagation will result. A solution is shown in Fig. 25.10(b). The $\boldsymbol{E}$ field

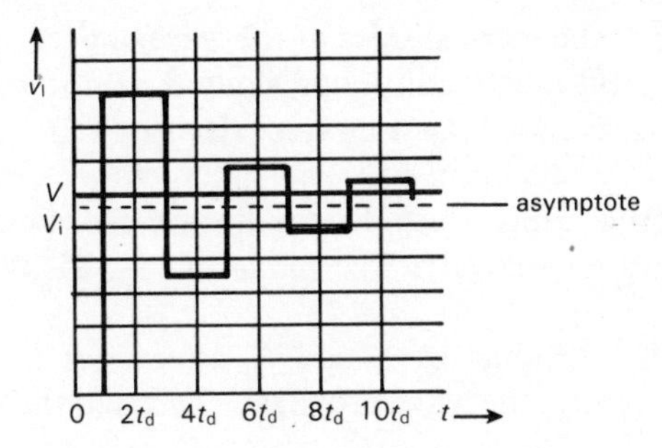

Fig. 25.9 Multiple reflections in case $Z_L \neq Z_o$

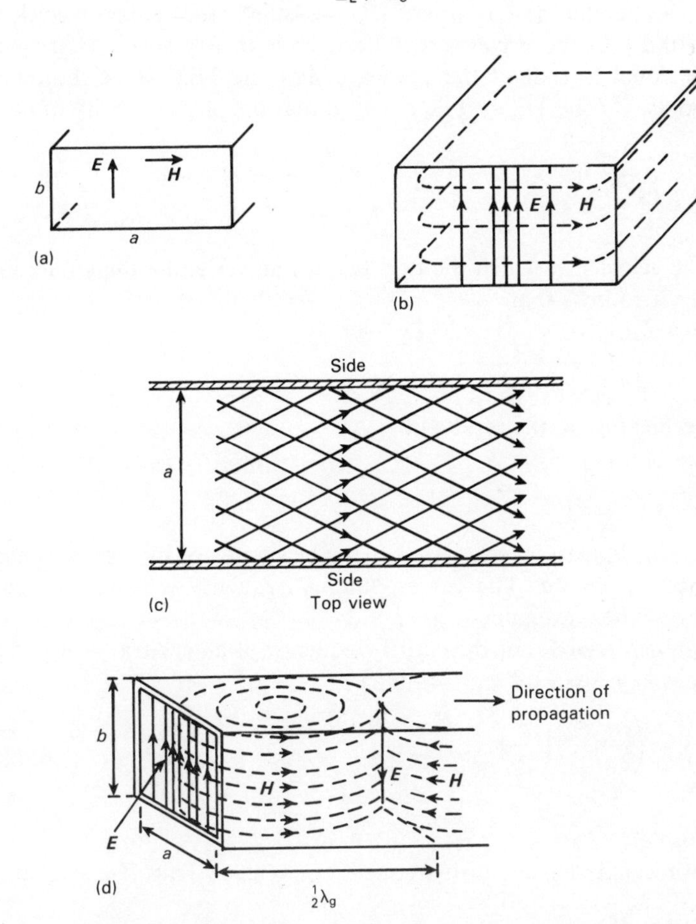

Fig. 25.10 (a) Waveguide with electric and magnetic field vectors (b) Possible ***E*** and ***H*** lines in a waveguide (TE_{10} mode) (c) Propagation of an electromagnetic wave through a waveguide (d) Guide wavelength in a waveguide

must be zero at the vertical sides of the waveguide, requiring a nonuniform field. The H field must form loops along the direction of propagation.

This mode is called the **transverse electric** (TE) mode since the E field is transverse to the direction of propagation. More details of the mode are indicated by subscripts which denote the number of maxima along the short and long sides respectively. The mode of Fig. 25.10(b) is called the TE_{10} mode.

The vector product of E and H determines the direction of propagation. Apparently, the wave propagates through the waveguide in a zigzag fashion (Fig. 25.10(c)).

The loops of the H-field alternate so that one loop covers a distance of a half wavelength ($\frac{1}{2}\lambda_g$) where λ_g is called the **guide wavelength** (Fig. 25.10(d)). If the wave were a TEM wave in free space its frequency would be equal to that of the TE wave. For the TEM wave the relation $\lambda f = c$ holds, for the TE wave $\lambda_g f = v_p$ so that the phase velocity of the TE wave is

$$v_p = c\frac{\lambda_g}{\lambda} \quad (v_p > c)$$

This is the velocity at which the field pattern moves along the guide. From Fig. 25.11 it follows that

$$\cos\alpha = \frac{\lambda}{2a} = \frac{\frac{1}{2}\lambda_g}{\sqrt{\frac{1}{4}\lambda_g^2 + a^2}}$$

Rearrangement of terms results in

$$\frac{1}{\lambda_g^2} = \frac{1}{\lambda^2} - \frac{1}{(2a)^2}$$

Evidently, the longest wavelength that can be supported by the waveguide is determined by $\lambda = 2a$. The corresponding frequency is called the **cutoff frequency** of the waveguide.

It can be derived that the cutoff frequency of an arbitrary mode TE_{mn} or TM_{mn} is expressed by the formula

$$\left(\frac{f_c}{c}\right)^2 = \left(\frac{m}{2a}\right)^2 + \left(\frac{n}{2b}\right)^2$$

Example

A TE_{10} waveguide has a cutoff frequency $f_c = c/2a$. Thus if $a = 2$ cm,

$$f_c = \frac{3\times10^8}{4\times10^{-2}} = 7.5 \text{ GHz}$$

The TE_{10} mode which is most frequently used, has no changing field in the x-direction. A mode which has a changing field in the x-direction is the TE_{11}

mode (Fig. 25.12(a)) or the TE_{20} mode (Fig. 25.12(b)). The transverse magnetic mode TM_{11} is shown in Fig. 25.12(c).

When terminating a waveguide by a conducting sheet, a situation similar to that of transmission lines is obtained. Incident waves and reflected waves may interfere so that standing waves result. Quarter wavelength and half wavelength sections of the guide have similar properties to those of transmission lines.

A section of a waveguide which is terminated at two sides can produce standing waves if the length of the section equals $\frac{1}{2}k\lambda$ ($k = 1, 2, 3, \ldots$). Such a section is a **resonant cavity** or **cavity resonator**. These cavities are used to adjust the frequency of microwave oscillators or as load impedance of microwave amplifiers. They are the equivalent of LC-resonant circuits. The principle is shown in Fig. 25.13(a) where the cavity is obtained by rotating a

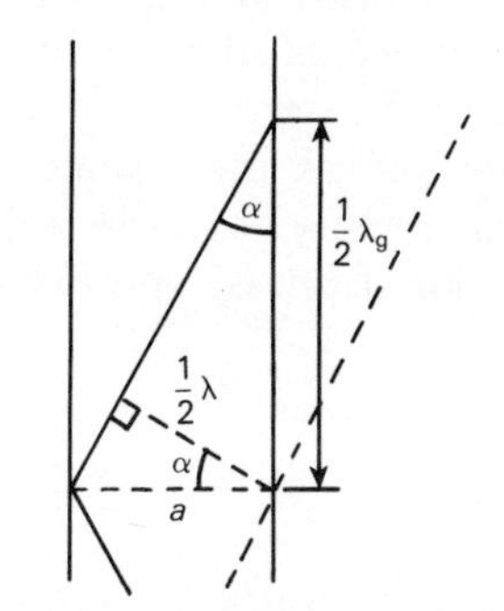

Fig. 25.11 Deriving the relation between λ_g, λ and a

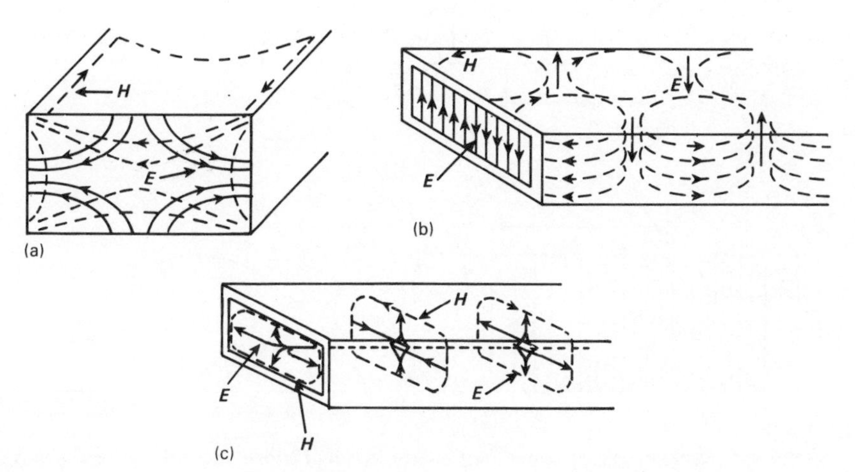

Fig. 25.12 (a) The TE_{11} mode (b) The TE_{20} mode (c) The TM_{11} mode

$\frac{1}{4}\lambda$ line along the axis AB. Electric field lines and field strength are shown in Fig. 25.13(b). Magnetic field lines form circular loops around them. It can be derived that the impedance of such a cavity is

$$Z_o \simeq 100 \frac{h}{\delta(r+h)}$$

where δ is the **skin depth** (m).

25.6 SKIN EFFECT

It can be derived from the laws of electromagnetism that in a conductor which carries alternating current, the current density in the center of the conductor will be smaller than in the outer circumference. Thus current tends to flow near the surface of the conductor so that the effective resistance of the conductor increases.

The **skin depth** is defined as the distance below the surface of a conductor where the current density has decreased to $1/e$ (37 percent) of the value at the surface. The skin depth is expressed by the formula

$$\delta = \frac{1}{\sqrt{\pi f \mu \gamma}} \text{ (m)}$$

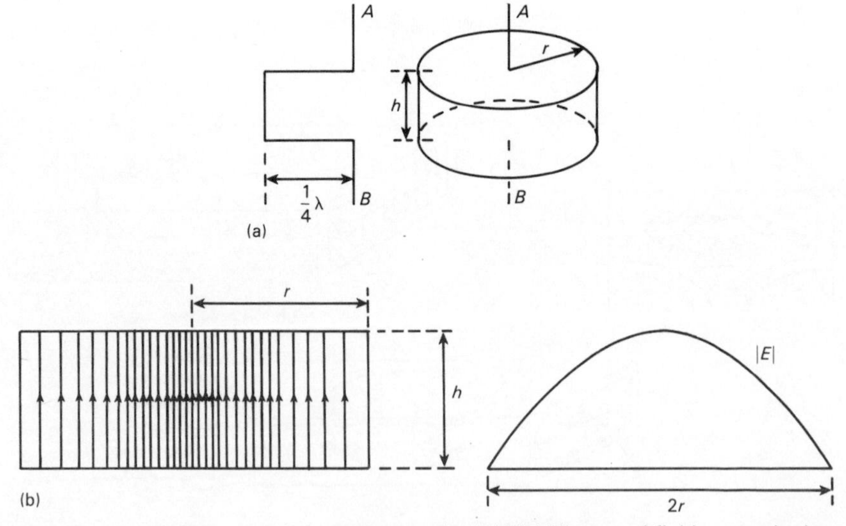

Fig. 25.13 (a) Example of a resonant cavity (b) Field lines and field strength along the diameter of the cavity

where

f is the frequency (Hz)
μ is the permeability (in air: $\mu \simeq 4\pi \times 10^{-7}$ H m^{-1})
γ is the specific conductivity (S m^{-1})

Example
Copper has a specific conductivity of 5.8×10^7 S m^{-1}. At $f = 20$ GHz it follows that $\delta \simeq 500$ nm.

For copper the formula can be rewritten as

$$\delta \simeq \frac{2100}{\sqrt{f}} \text{ (nm, GHz)}$$

26 Propagation of Electromagnetic Waves; Antennas

26.1 PROPAGATION PATHS

The Earth's atmosphere is the propagation medium for the transmission of electromagnetic waves from transmitter to receiver. It is divided into two main parts: the **troposphere** extending from the Earth's surface to about 50 km and the **ionosphere** from 50 km to approximately 300 km. The ionosphere results from ultraviolet radiation which ionizes the rarefied gases. Ionization occurs in layers where higher electron densities exist. These electrons reduce the value of the permittivity ε and, since the propagation velocity is $v = 1/\sqrt{\varepsilon\mu}$, the phase velocity of the wave increases to values greater than the velocity of light. This has the effect of refraction of the wave. Depending upon the frequency and the angle of the wave with respect to the earth's surface, refraction will cause the wave either to return to the surface or to vanish into space (Fig. 26.1).

There are two causes of attenuation of propagating waves:

- Due to the law of conservation of energy, attenuation is proportional to the distance to the source. This is called **space attenuation**.
- The medium in which the wave travels causes **absorption**.

Many waves are polarized. When the E vector has a fixed direction with respect to the direction of propagation, the wave is **linearly polarized.** If the E vector rotates about the direction of propagation in a circular fashion, the wave is **circularly polarized.**

In actual space a wave can travel along different paths (Fig. 26.1).

Surface wave

The wave follows the curvature of the Earth as a result of refraction. In calculations this effect is usually allowed for by increasing the Earth's radius by a factor of 1.33.

Conduction and displacement currents below the surface cause losses and attenuation of the wave. Frequencies are limited to about 2 MHz.

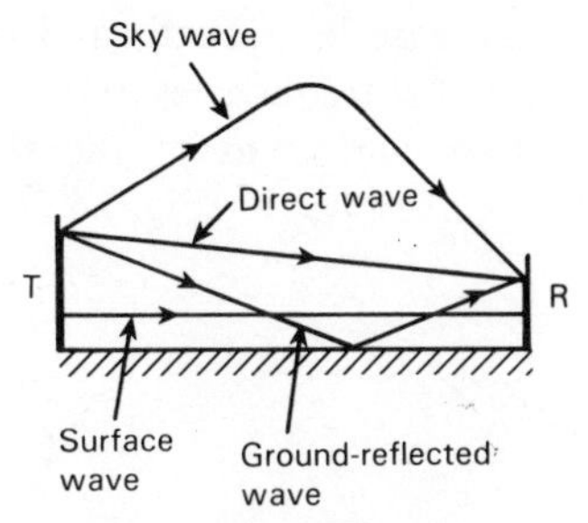

Fig. 26.1 Nomenclature of wave propagation

Direct wave

When transmitter and receiver have a direct line-of-sight, direct transmission is possible. The maximum distance is called the **optical range** which is

$$d_{max} \simeq 3.6(\sqrt{h_t} + \sqrt{h_r})$$

where h_t and h_r are the heights of transmitter and receiver antennas (in m) and d_{max} is the range in km.

When taking into account the refraction factor of 1.33, the range of reception is increased. This range is called the **radio horizon:**

$$d_{max} \simeq 4.2(\sqrt{h_t} + \sqrt{h_r})$$

Ground-reflected wave

At the receiver, direct wave and ground-reflected wave interfere. When the waves are in opposite phase, attenuation occurs. Figure 26.2(a) shows how the field strength varies as a function of distance in the case where the earth's curvature can be neglected.

Sky wave

The sky wave reflects against one of the ionized layers in the ionosphere. In the nomenclature of wave propagation the **space wave** is the combination of ground-reflected wave and direct wave while the **ground wave** is the combination of surface wave and space wave.

Reflection of the sky wave occurs under certain conditions of frequency and angle of elevation and is expressed as **maximum usable frequency** (MUF):

$$\text{MUF} = f \sec \phi$$

The value of MUF depends on the density of the ionized layers, time of day, season, geographical latitude, etc. Bending of sky waves in the ionosphere allows communication over great distances.

At night, sky waves increase in strength and become comparable in magnitude to the surface waves at frequencies of 0.5–1.5 MHz. Ionospheric fluctuations then cause phase variations of the sky waves, resulting in **fading** (distortion) of the received waves (Fig. 26.2(b)).

26.2 IONIZED LAYERS

The ionized layers in the ionosphere are classified according to the electron density (Fig. 26.3):

- D-layer: height 50–90 km. It is the most unstable layer that disappears at night since the ionized particles recombine. Low frequencies are reflected, medium frequencies are absorbed and high frequencies are attenuated. The usable range is about 150 km.
- E-layer: height about 110 km. The layer disappears at night. The usable range during the daytime is about 1600 km, the frequency limit about 4 MHz.
- F_1-layer: height about 175–250 km. This layer combines at night at about 300 km with the F_2-layer. The range is about 3000 km, the frequency limit about 5 MHz.
- F_2-layer: height about 250–400 km. At night the reflection decreases due to the absence of the F_1-layer which causes an increase of the noise. The layer is particularly useful for high frequency waves over long distances. The range is about 4000 km, the frequency limit about 8 MHz.

Propagation can also be classified in terms of wavelength or frequency:

- Long waves (15–500 kHz): ground waves suffer minor attenuation. Earth and E-layer act as conducting surfaces which produce reflection. Transmission over great distances is possible.
- Medium waves (0.5–2 MHz): the losses in the ionosphere are relatively large and returning waves are attenuated. Communication over a limited range is possible. At night, fading usually occurs.
- Short waves (3–30 MHz): the ground wave is highly attenuated so that transmission over larger distances is by sky waves. They penetrate the E-layer and are refracted by the F-layers. The **skip distance** is the distance that can be covered by a single reflection (Fig. 26.4). The skip distance depends on frequency. In Fig. 26.4 point A is the closest point which can be reached by a 20 MHz wave, B is the farthest point and AB is the skip distance or skip zone. Within the skip distance no signal can be received. When transmission occurs by a single reflection, it is called **single hop**. Many transmissions are accomplished by **multihop** (Fig. 26.5).

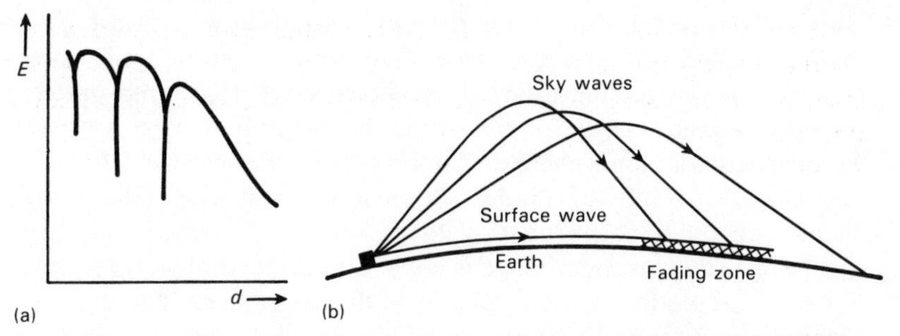

Fig. 26.2 (a) Field strength modulus versus distance (b) Illustration of fading

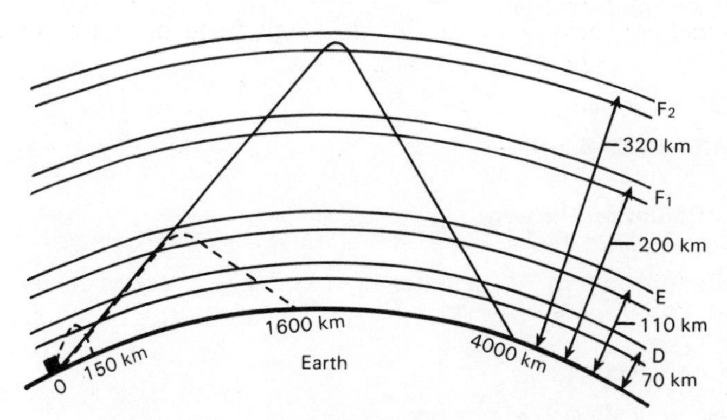

Fig. 26.3 The ionized layers and reflections

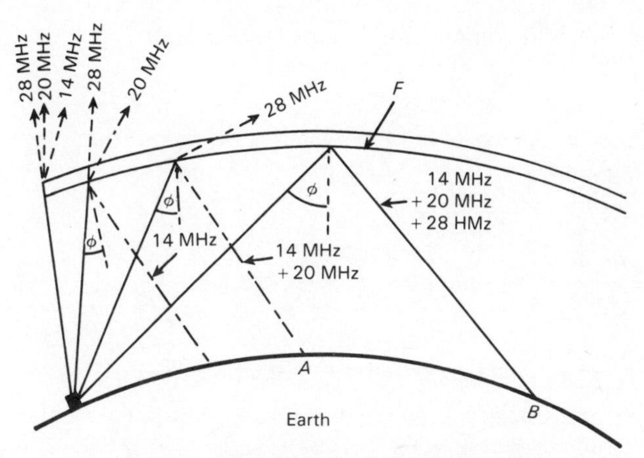

Fig. 26.4 Example showing skip distance

- Ultrashort waves (>30 MHz): the sky wave is not returned by the ionosphere and vanishes into space. The surface wave is rapidly attenuated so that communication is by direct wave from transmitter to receiver. Longer distances can often be covered due to **scattering**. Troposphere and ionosphere are in a state of turbulence which changes the values of the refractive index. At the lower frequencies (30–70 MHz) changes in the ionospheric density of the E-layer cause ionospheric scattering. The maximum range is about 2000 km. At higher frequencies (0.4–5 GHz) scattering occurs mainly in the troposphere (Fig. 26.6). The range is about 1000–2000 km.

  Selective fading occurs with amplitude modulated waves which have covered a large distance. Due to several paths, carrier and sideband frequencies will fade in and out in a more or less random way. This causes severe distortion.

26.3 ANTENNAS

26.3.1 Radiation Pattern

The radiation pattern of an antenna is commonly described by using a spherical coordinate system (Fig. 26.7). The coordinates of a point P are

$$x = r \sin\theta \cos\phi$$

$$y = r \sin\theta \sin\phi$$

$$z = r \cos\theta$$

where θ is the elevation angle and ϕ is the azimuth angle.

If an elementary dipole (Hertzian dipole) of length l ($l \ll 0.1\,\lambda$) is placed at the origin in the direction of the z-axis and the current is $i = I \sin\omega t$, the field components at a distance r are

$$H_r = 0$$

$$H_\theta = 0$$

$$H_\phi = \frac{Il \sin\theta}{4\pi r}\left[\frac{\omega}{c}\cos\omega\left(t-\frac{r}{c}\right)+\frac{1}{r}\sin\omega\left(t-\frac{r}{c}\right)\right]$$

$$E_r = \frac{2Il\cos\theta}{4\pi r\varepsilon}\left[\frac{1}{rc}\sin\omega\left(t-\frac{r}{c}\right)-\frac{1}{\omega r^2}\cos\omega\left(t-\frac{r}{c}\right)\right]$$

$$E_\theta = \frac{Il\sin\theta}{4\pi r\varepsilon}\left[\frac{\omega}{c^2}\cos\omega\left(t-\frac{r}{c}\right)+\frac{1}{rc}\sin\omega\left(t-\frac{r}{c}\right)-\frac{1}{\omega r^2}\cos\omega\left(t-\frac{r}{c}\right)\right]$$

$$E_\phi = 0$$

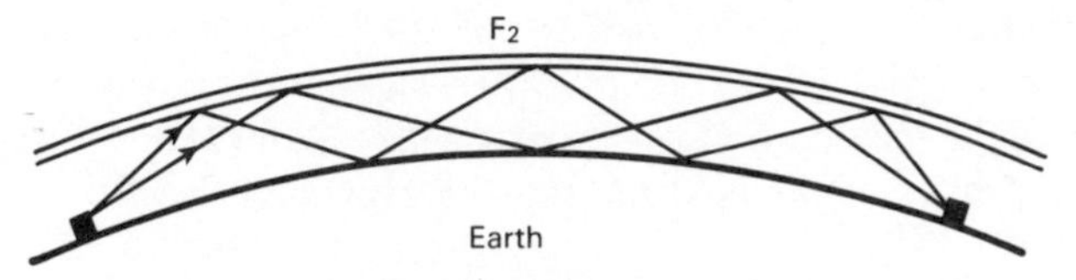

Fig. 26.5 Illustration of multihop

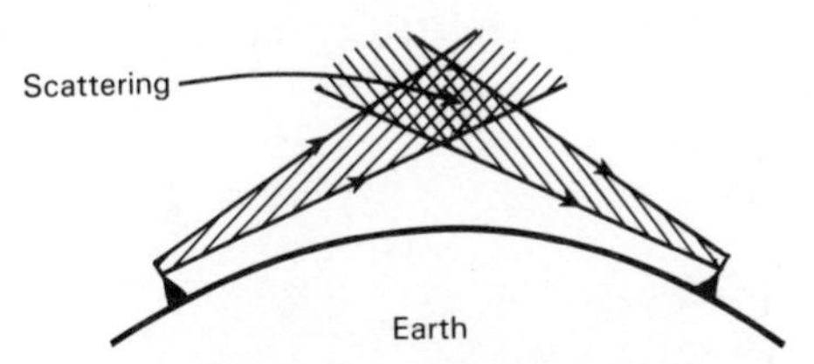

Fig. 26.6 Tropospheric scattering

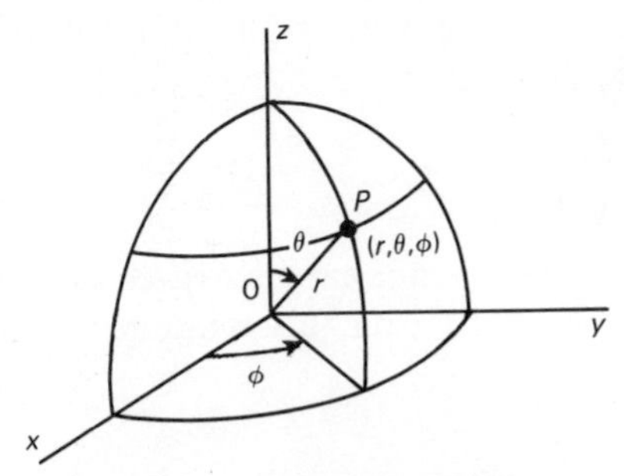

Fig. 26.7 The spherical coordinate system

Close to the dipole (the near-field region) the terms with $1/r^3$ and $1/r^2$ are predominant while at greater distances (the far-field region) the $1/r$ terms become significant. This far field is called the **radiation field.**

The radiation field is defined by its minimum distance

$$r_{\min} = \frac{2D^2}{\lambda}$$

where D is the largest dimension of the antenna and λ is the wavelength. The field components are then approximately

$$H_\phi \simeq \frac{Il \sin\theta}{4\pi r} \frac{\omega}{c} \cos\omega\left(t - \frac{r}{c}\right) = \frac{Il \sin\theta}{2\lambda r} \cos\omega\left(t - \frac{r}{c}\right)$$

$$E_\theta \simeq \frac{Il \sin\theta}{4\pi r\varepsilon} \frac{\omega}{c^2} \cos\omega\left(t - \frac{r}{c}\right) = \frac{Il \sin\theta}{2\lambda r} \sqrt{\frac{\mu}{\varepsilon}} \cos\omega\left(t - \frac{r}{c}\right)$$

Figure 26.8 shows the relative field components of E_θ as a function of r.

26.3.2 Power

In free space the relation $E = 377H$ exists where $\boldsymbol{E}$ and $\boldsymbol{H}$ are at right angles. The power density of the field can thus be expressed as

$$N = \frac{E^2}{377} \ (\mathrm{W\ m^{-2}})$$

The total radiated power can be calculated by integration over a sphere with radius r. The result is

$$P_t = I^2 80\pi^2 \left(\frac{l}{\lambda}\right)^2 \ (\mathrm{W})$$

26.3.3 Radiation resistance

This is the fictitious resistance between the terminals of the antenna which radiates a power P_t. Thus

$$R_r = \frac{P_t}{I^2} = 80\pi^2 \left(\frac{l}{\lambda}\right)^2$$

The total resistance of the antenna can be written as $R_a = R_o + R_r$ where R_o is the ohmic resistance. Based on this relation the antenna efficiency is defined as

$$\eta_a = \frac{R_r}{R_a}$$

26.3.4 Antenna gain

To define antenna gain, use is made of a reference called an **isotropic radiator**. This is a source which radiates power at a constant rate uniformly in all directions (spherical pattern).

If the isotropic radiator radiates P_t watts, the power density over a sphere of radius r is

$$P_{is} = \frac{P_t}{4\pi r^2} \ (W\mathrm{m^{-2}})$$

The **directive gain** G is defined as the ratio of power density in a particular direction to the power density radiated by an isotropic radiator at the same distance. Thus

$$G = \frac{P_r}{P_{is}}$$

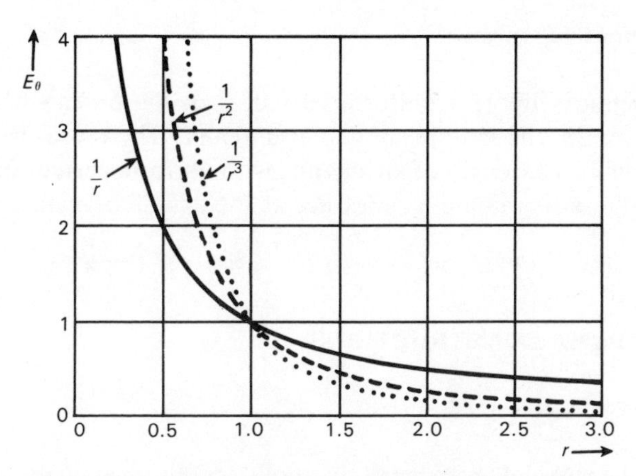

Fig. 26.8 Relative variation of electric field strength components for a short dipole

If a receiving antenna at distance r collects this radiation, the power density at this distance will be P_r. If the wave direction with respect to the receiving antenna is determined by the angular coordinates θ and ϕ, the power P_L delivered by the receiving antenna to a matched load will be less than P_r. If the receiving antenna is oriented so that maximum power is collected, the maximum power depends on the **effective area** S_{eff} of the antenna, thus

$$P_L = P_r S_{eff}$$

It follows that the value of S_{eff} is proportional to the directive gain G, so that

$$S_{eff} = \alpha G$$

It can be shown that the constant of proportionality α is the same for all antennas, namely

$$\alpha = \frac{\lambda^2}{4\pi}$$

and the effective area is thus expressed as

$$S_{eff} = \frac{\lambda^2}{4\pi} G$$

26.3.5 Beamwidth

Beamwidth is defined as the angle between points of the radiation pattern where the power density has decreased to one-half of its maximum value. An example of a radiation pattern is shown in Fig. 26.9.

26.3.6 Effective length

In an elementary dipole a uniform current distribution can be assumed. This is no longer the case in an actual antenna. Therefore, the **effective length** l_e is defined as length of an antenna with uniform current distribution which has the same radiation resistance as the actual antenna.

26.4 ANTENNA CONSTRUCTIONS

26.4.1 Short dipole

A dipole is considered short when the actual length is small compared with $\frac{1}{2}\lambda$. When the current distribution is assumed to be a cosine function (Fig. 26.10) the radiated power is, according to Section 26.3.2,

$$P = I^2 80\pi^2 \left(\frac{l_e}{\lambda}\right)^2$$

where l_e is the effective length.

26.4.2 Half-wave dipole

This is a resonant antenna so that standing voltage and current waves exist along the antenna.

$$P = I^2 \times 80 \text{ (W)}$$
$$R_r = 80 \text{ } (\Omega)$$

These values are not exact since phase differences between waves reaching a point in space from different parts of the dipole are ignored.

When these phase differences are taken into account by dividing the dipole into a number of elementary dipoles and integrating over the length of the dipole, a more exact value is obtained:

$$P = \mathrm{I}^2 \times 73 \text{ (W)}$$
$$R_r = 73 \text{ } (\Omega)$$

The radiation pattern is not very different from that of a Hertzian dipole (Fig. 26.11)

When used at wavelengths $n \cdot \frac{1}{2}\lambda$, different patterns result.

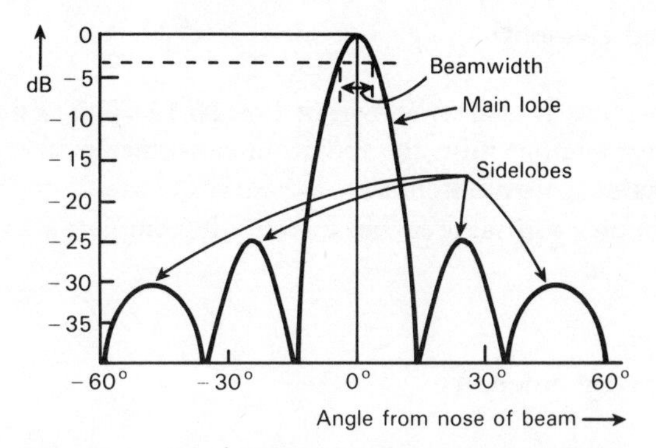

Fig. 26.9 Example of a radiation pattern showing 3-dB points

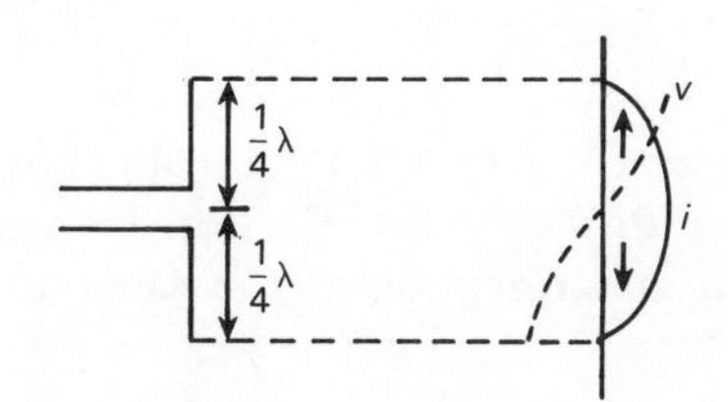

Fig. 26.10 Current and voltage distribution in a $\frac{1}{2}\lambda$ dipole

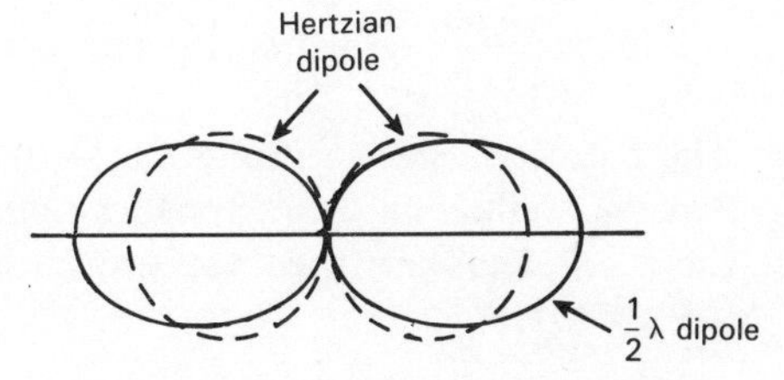

Fig. 26.11 Comparison of radiation patterns of Hertzian and $\frac{1}{2}\lambda$ dipole

26.4.3 Short vertical antenna

The radiated power can be derived as

$$P = I^2 160\pi^2 \left(\frac{l_e}{\lambda}\right)^2 \text{ (W)}$$

so that

$$R_r = 160\pi^2 \left(\frac{l_e}{\lambda}\right)^2 \ (\Omega)$$

If, for example, $l = \frac{1}{4}\lambda$, then $l_e = \lambda/2\pi$ so that $R_r = 40\ \Omega$, or more exactly

$$R_r \simeq 36.5\ \Omega$$

Figure 26.12(a) shows the current and voltage distribution of a $\frac{1}{4}\lambda$ antenna. The antenna can be tuned to a longer wavelength (lower frequency) by connecting a series inductance L as shown in Fig. 26.12(b). Similarly, the antenna can be tuned to shorter wavelengths when adding a capacitor C (Fig. 26.12(c)). These antennas are used for MF broadcast and at VHF for mobile service.

26.4.4 Folded elements

Often, antennas are folded as shown in Fig. 26.13. The radiation pattern does not change significantly; the radiation resistance is about four times that of the single element antenna.

Folded dipoles are favored because of the simple and sturdy mechanical construction.

26.4.5 Long wire antenna

When a single straight wire has a length which is substantially greater than $\frac{1}{2}\lambda$ ($l = n \cdot \frac{1}{2}\lambda$) (Fig. 26.14(a)), the radiation pattern for a traveling wave current is described by

$$E_\theta = \left| \frac{60I \sin\theta}{r(1-\cos\theta)} \sin\left\{\frac{\pi l}{\lambda}(1-\cos\theta)\right\} \right|$$

The pattern is shown in Fig. 26.14(b) for $n = 3$. As there are no standing waves the antenna characteristics remain rather constant. For this reason these antennas are used for transmission and reception in the range of 0.5–30 MHz.

26.4.6 Loop antenna

Loop antennas can have various forms and they can be made up of more turns (Fig. 26.15(a)). If the dimensions of the loop are small with respect to the wavelength, the radiation resistance is approximately

$$R_r \simeq 31\,200 \frac{n^2 S^2}{\lambda^4} \; (\Omega)$$

where n is the number of turns and S the loop area. The directional pattern is shown in Fig. 26.15(b) and applies to loops of all shapes.

These are used in portable receivers and for direction finding purposes. A variation of the loop antenna is the ferrite rod antenna.

26.4.7 Rhombic antenna

The rhombic antenna (Fig. 26.16(a)) consists of four interconnected long wire antennas, terminated in a resistor. Rhombic antennas have a very directive pattern which does not change appreciably with frequency (Fig. 26.16(b)). They are used in the HF bands for skywave propagation.

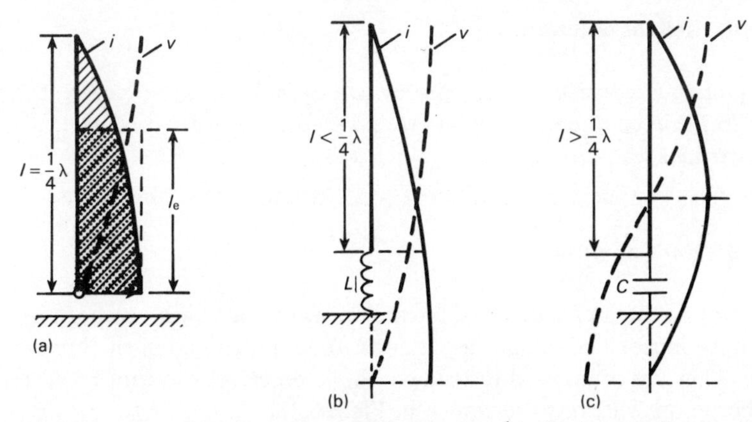

Fig. 26.12 (a) Current and voltage distribution in a $\frac{1}{4}\lambda$ dipole and effective length (b) Tuning of a $\frac{1}{4}\lambda$ dipole to a longer wavelength (c) Tuning of a $\frac{1}{4}\lambda$ dipole to a shorter wavelength

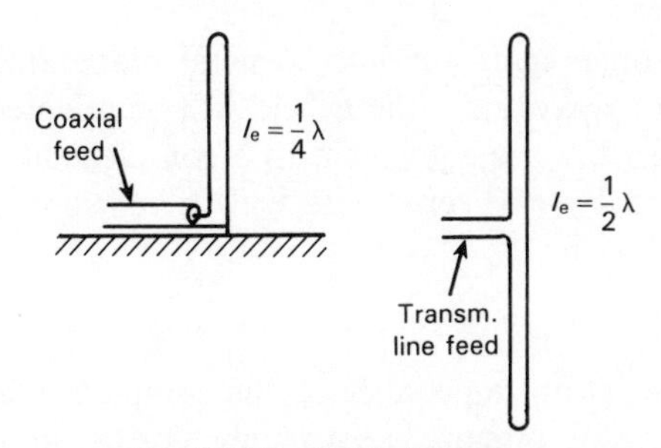

Fig. 26.13 Folded dipole

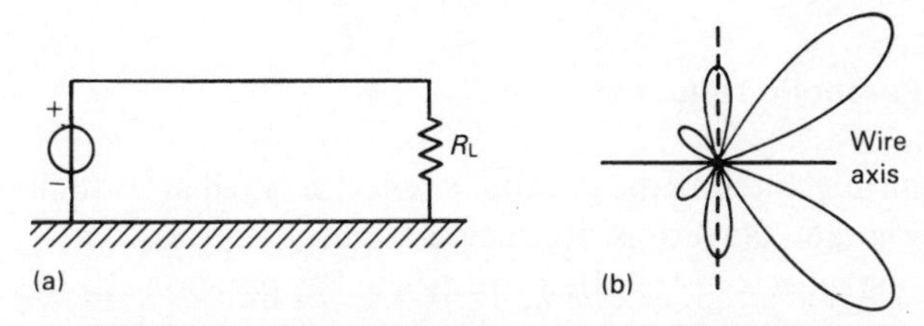

Fig. 26.14 Long wire antenna and radiation pattern

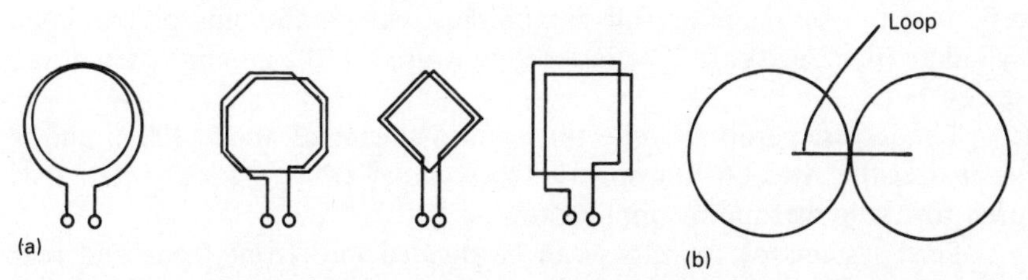

Fig. 26.15 (a) Examples of loop antennas (b) Radiation pattern of loop antennas

26.4.8 Helical antenna

This antenna consists of a wire which is wound as a screw thread (Fig. 26.17) in conjunction with a ground plane. These are used at relatively high frequencies.

26.4.9 Horn antenna

The horn antenna can be considered as a waveguide with the horn as the load impedance. The radiation pattern is omnidirectional in the horizontal plane. They are often used to feed parabolic reflectors (Section 26.4.11). A few horn constructions are shown in Fig. 26.18.

26.4.10 Slot antenna

When a slot-like opening is cut into a metal plane and connected to a transmission line or waveguide, the *E* field will be oriented along the short side. The characteristics are equivalent to a complementary antenna when the slot is replaced by a sheet dipole with a reflector (Fig. 26.19). The impedance of the slot is

$$Z_{s1} Z_c = (60\pi)^2$$

where Z_c is the radiation impedance of the complementary antenna. The construction of the slot antenna is convenient for use in airplanes since no parts project from the surface.

26.4.11 Parabolic reflector

As a transmitting antenna the parabolic reflector is fed at its focus by a horn or dipole (Fig. 26.20(a)) thus producing a parallel beam.

Parabolic reflectors are often constructed as parabolic dishes for use in radar systems, radiotelescopes, etc. They can often be rotated horizontally and vertically to allow tracking of satellites, aircraft or missiles. The reflectors can be made as full paraboloids, cut paraboloids or parabolic cylinders (Fig. 26.20(b)). The latter type is used in the antenna systems of search radar.

The largest parabolic reflector has a diameter of about 300 m and is located at the Arecibo Ionospheric Observatory (Puerto Rico) where it is used for radio astronomy applications.

Feed systems of reflectors can be divided into front feeds and rear feeds. In a rear feed, the transmission line or waveguide is passed through

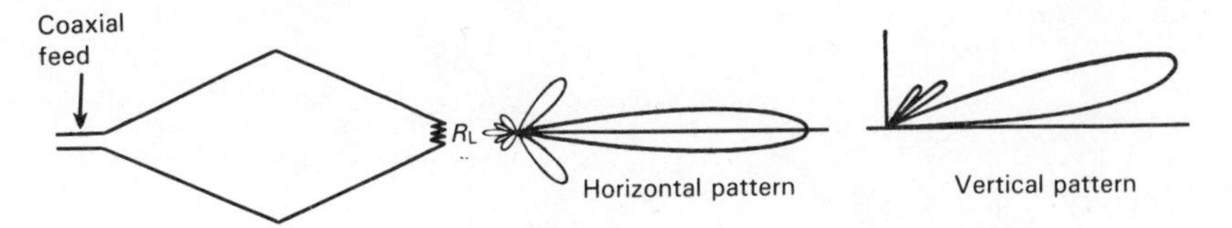

Fig. 26.16 Rhombic antenna and radiation patterns

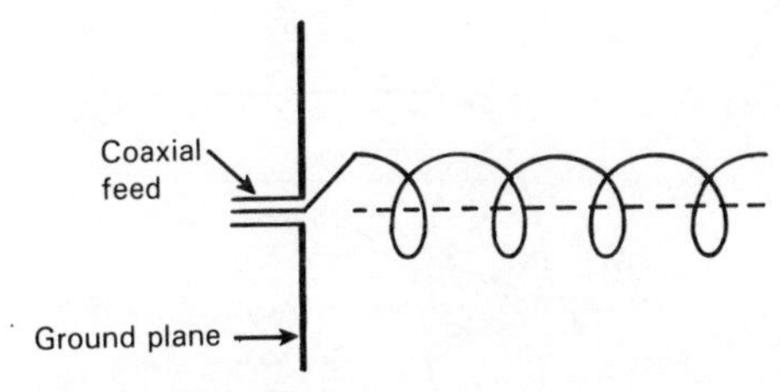

Fig. 26.17 Helical antenna

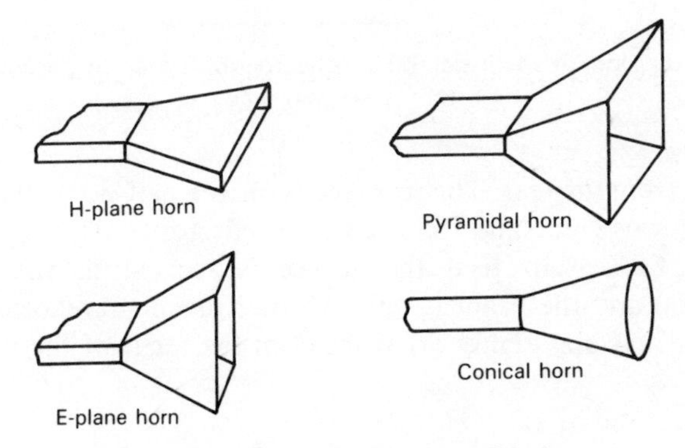

Fig. 26.18 Examples of waveguide horns

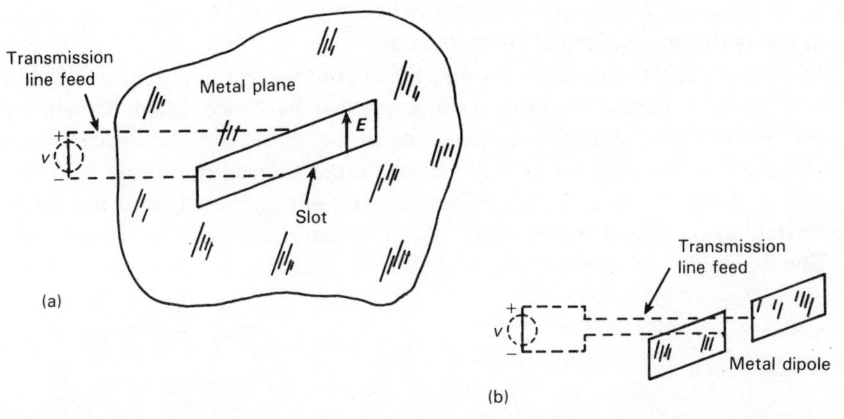

Fig. 26.19 (a) Radiating slot in an infinite plane (b) Complementary antenna

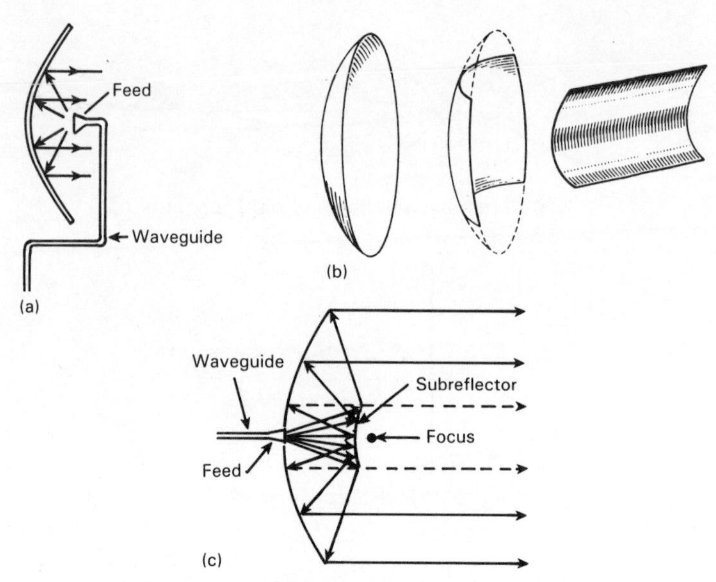

Fig. 26.20 (a) Horn feed of parabolic reflector (b) Types of parabolic antennas (c) Cassegrain feed

the reflector from the rear. The rear feed is often a half-wave dipole which is excited by a waveguide and followed by a reflector.

In the **Cassegrain feed** the source is located in the paraboloid (Fig. 26.20(c)) and the beam is reflected by a small subreflector in such a way that the rays apparently originate from the focus of the paraboloid.

26.4.12 Metal lenses

Some microwave antennas operate without reflectors. A metal lens is used to focus the energy into a directive beam.

Figure 26.21 shows a horn and the curved wavefronts which fall on a number of waveguide sections of different lengths. Since the phase velocity of a wave in a waveguide is greater than that of a wave in free space, waves passing through long waveguide sections are advanced in phase.

A wavefront is a surface of equal phase and the passage of the waves through the sections can be made so that the emerging wavefront is a plane. The advantage of a metal lens is the simple construction.

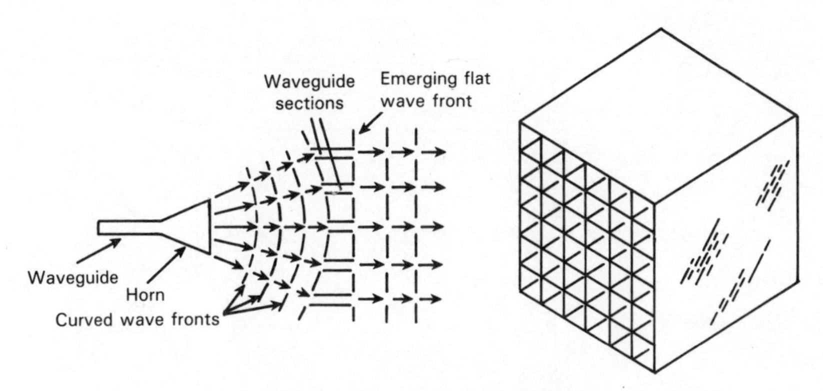

Fig. 26.21 Example of metal lens

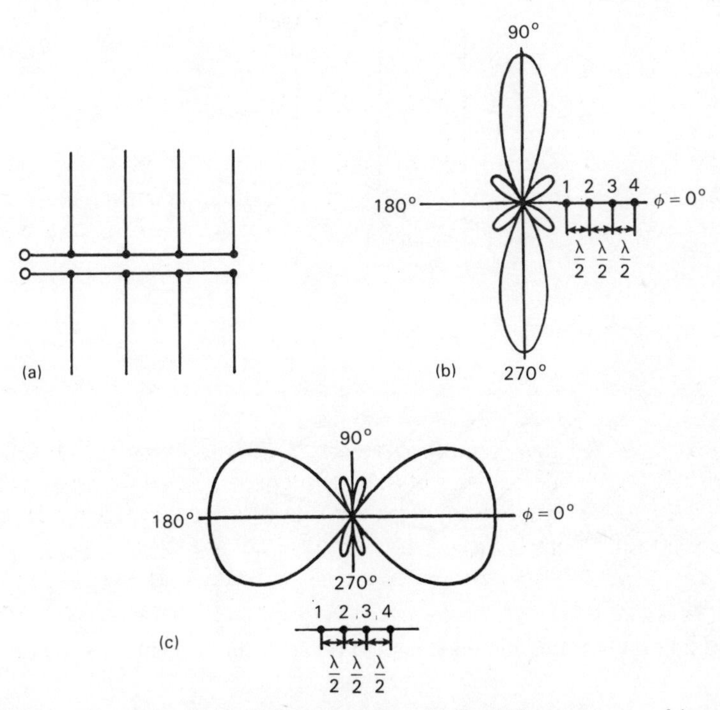

Fig. 26.22 (a) Linear array of four antennas (b) Radiation patterns of broadside array (c) Radiation pattern of endfire array

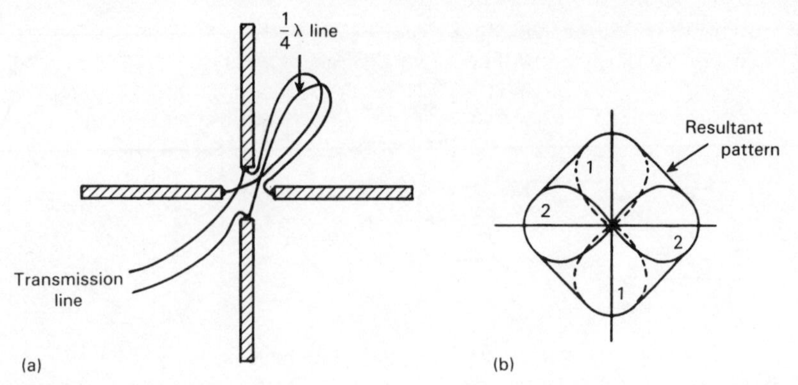

Fig. 26.23 (a) Turnstile construction with crossed dipoles (b) Turnstile radiation pattern

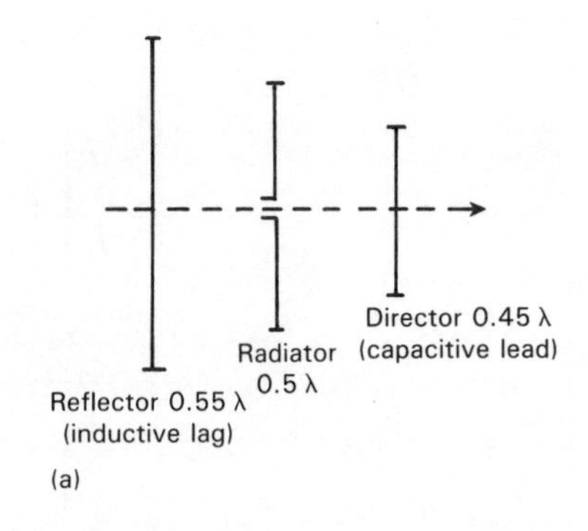

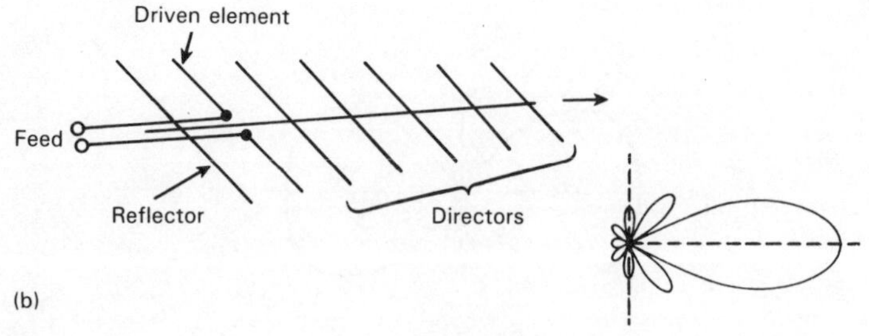

Fig. 26.24 (a) Radiator with director and reflector (b) Yagi antenna and radiation pattern

26.5 ARRAYS

An array is an assembly of radiating antennas, usually $\frac{1}{2}\lambda$ dipoles, in order to obtain a maximum field intensity in a particular direction.

In Fig. 26.22(a) a number of vertical antennas is uniformly spaced at distances d along a straight line. All currents have the same magnitude and a phase difference α exists between adjacent antennas. The resultant field at a distant point P can be derived as

$$E_{\phi} = \frac{\sin n\left(\dfrac{\pi d}{\lambda}\sin\phi - \dfrac{1}{2}\alpha\right)}{n\sin\left(\dfrac{\pi d}{\lambda}\sin\phi - \dfrac{1}{2}\alpha\right)}$$

In a **broadside array** $\alpha = 0$ and pattern maxima occur at $\phi = 0^{\circ}$ and $\phi = 180^{\circ}$ (Fig. 26.22(b)).

In the **endfire array** $\alpha = 2\pi d/\lambda$ and a maximum occurs at $\phi = 90^{\circ}$ (Fig. 26.22(c)). A **turnstile** is constructed with two cross-connected dipoles which are fed so that the currents have a 90° phase difference. This is obtained by connecting two dipole centers with a $\frac{1}{4}\lambda$ loop as shown in Fig. 26.23(a).

The radiation pattern is omnidirectional and linearly polarized in the plane of the two dipoles while circular polarization occurs in the two directions perpendicular to the two dipole axes (Fig. 26.23(b)).

26.6 PARASITIC ARRAYS

The field of an excited element or dipole will induce currents in an adjoining element (parasitic excitation). These elements are tuned to cause a leading or lagging phase shift which causes a change in the directional pattern. Dipole elements slightly shorter than $\frac{1}{2}\lambda$ act as **directors** and reinforce the field of the excited dipole in the direction from dipole to director (Fig. 26.24(a)). In a line of directors, each one will excite the next one.

Slightly longer elements act as **reflectors** and reinforce the field of the excited dipole in the direction from reflector to dipole. A well-known construction is the Yagi–Uda endfire antenna array which consists of a half-wave driven folded dipole, a reflector and usually several directors (Fig. 26.24(b)) of such lengths as if the previous director were the exicted one. Directive gains of 10–16 dB can be realized produced in one lobe. Typical resistance values are 200–300 Ω.

27
Microwaves; Radar Principles

27.1 MICROWAVE SOURCES

27.1.1 Magnetron

The range of microwaves extends from 0.3–300 GHz and is located between the radio waves and the optical waves.

One of the earliest devices for the generation of microwaves is the high-power oscillator, called **magnetron**. Basically, it is a high-vacuum cylindrical diode (Fig. 27.1(a)) with a high voltage (10–50 kV) applied between anode and cathode so that electrons travel from cathode to anode. If now a uniform magnetic field is applied at the anode, the electron paths will be curved. At a certain critical field strength electrons will be unable to reach the anode and they follow a circular path back to the cathode.

In practical constructions resonant cavities are made in the wall of the cylinder (Fig. 27.1(b)) and oscillations will be produced.

Most multicavity magnetrons are used in a pulsed mode where several MW can be produced at frequencies of several GHz with small duty cycles (<0.001). Magnetrons are used for feeding the reflector antennas of radar systems and in domestic applications.

27.1.2 Klystron

The operation of a **klystron** is based on electron velocity modulation. Electrons are emitted by the cathode (Fig. 27.2(a)) and travel to a control and focusing grid g_1 and an acceleration grid g_2. The cathode, g_1 and g_2 are termed the **beam source**. The electrons reach g_2 with equal velocities. Then the electrons move into a drift space with a constant potential.

If an alternating field is applied between g_2 and g_3, electrons will pass g_2 with velocities which depend on the instantaneous value of the voltage between g_2 and g_3. Therefore, in the drift space the electron beam is velocity modulated as fast electrons catch up with electrons which were accelerated in the preceding half cycle. Thus local current densities called **bunching** result and density maxima occur at distances equalling half the

4 1 c a 3 2

1 = no magnetic field
2 = weak magnetic field
3 = stronger magnetic field
4 = critical magnetic field

(a)

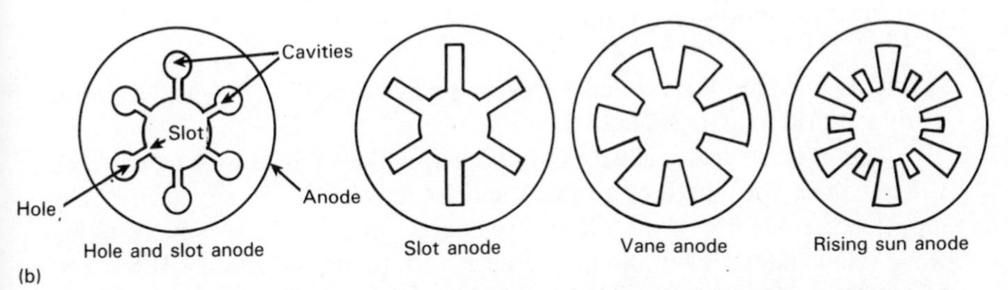

Fig. 27.1 (a) Illustration of the principle of magnetron operation (b) Examples of magnetron constructions

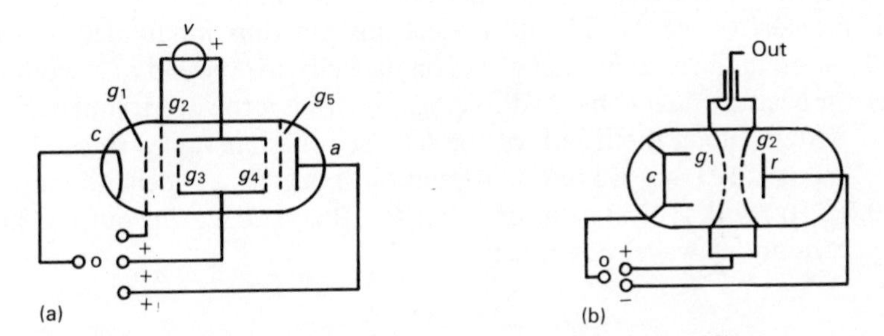

Fig. 27.2 (a) Principle of the klystron (b) The reflex klystron

wavelength of the alternating field. Different numbers of electrons arrive at grids g_4 and g_5 (the catcher grids). The buncher grids g_2, g_3 form part of a resonant cavity excited by a microwave frequency source, as well as the catcher grids g_4, g_5.

As the intensity of the maxima increases along the tube, the microwave signal is amplified. Oscillations are stimulated by providing a feedback path between catcher and buncher which behave as a pair of coupled tuned circuits.

A different oscillator construction is the **reflex klystron** (Fig. 27.2(b)) where one cavity acts as both buncher and catcher. Grid g_1 serves as control and acceleration grid, grid g_2 modulates the beam. Following g_2 is a reflector r which is negative with respect to the cathode.

A electron passing through the grids when these are at the same potential and g_2 going negative, is retarded by reflector r and brought to rest, after which it returns to the cathode. An electron passing the grids a little earlier, when g_2 is positive, approaches the reflector at a higher velocity so that it takes longer to stop. Thus after reflection it takes longer to return to the grids and the result is a bunching process.

27.1.3 Traveling wave tube

A microwave amplifier, called a **traveling wave tube** (TWT) was developed during World War II (Fig. 27.3).

The basic structure consists of two waveguides joined by a helix which is fixed along the length of a tube, the helix terminating in a collector. The input waveguide impresses microwave energy upon the helix. The resulting wave propagates slowly along the helix due to its inductance. Electrons which are shot along the axis of the helix by an electron gun are not retarded by the inductance and have a higher velocity. The field caused by the helix will accelerate some axial electrons and decelerate others. Thus an interaction (velocity modulation) occurs between the applied microwave energy and the electron beam. The axial electrons are thus modulated and this modulation induces additional waves on the helix. At the end of the tube the electron beam is then converted to a beam with microwave frequency. The wave pattern along the length of the tube supplies energy to the collector.

Since a TWT has no resonant cavities, it acts as a wideband amplifier ($\simeq$0.5 GHz) with typical gains of about 40 dB at 4 GHz. The output can be CW (continuous wave) or pulsed.

27.2 RADAR

27.2.1 General

Radar is an acronym for RAdio Direction And Range (also RAdio Detecting And Ranging).

In the early days of World War II the first cavity magnetrons were developed and used in radar equipment on battleships, submarines and aircraft. Radar uses microwaves to detect the presence of objects as well as their direction, range and velocity.

Basically, microwave energy is transmitted and, when it strikes an object, some of the energy is reflected back to the radar antenna where it is

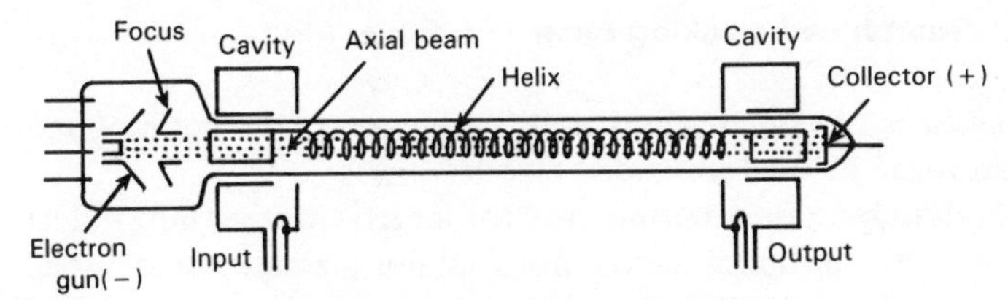

Fig. 27.3 The traveling wave tube

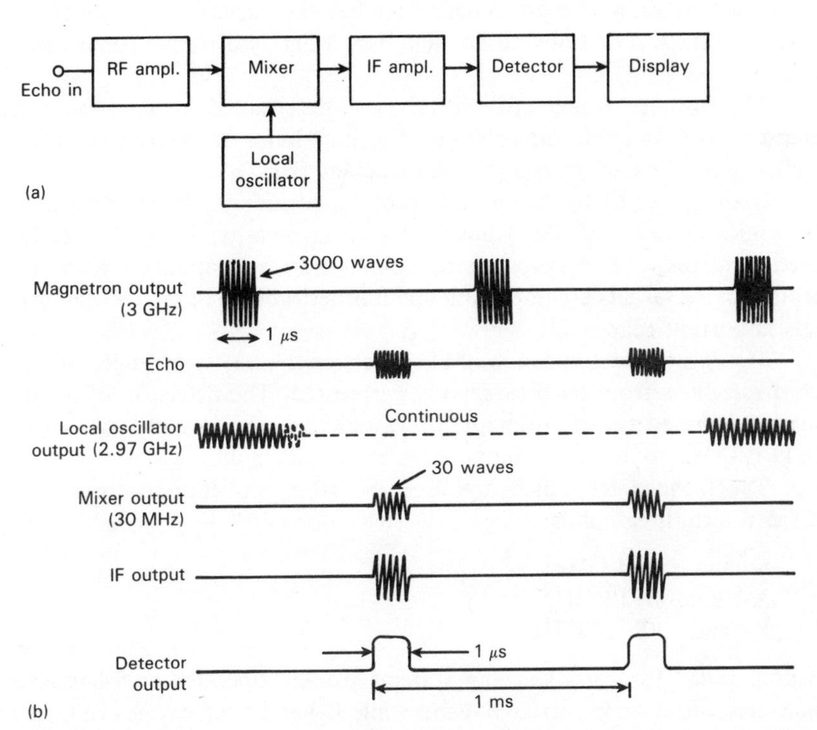

Fig. 27.4 (a) Block diagram of superheterodyne radar receiver (b) Waveform example of pulsed radar

processed in a receiver and finally displayed on a video screen. The block diagram of a radar receiver (Fig. 27.4(a)) is a conventional superheterodyne system (Section 24.1).

Most radar systems operate in the pulsed mode. In the circuit of Fig. 27.4(a) a wave train of 3 GHz is transmitted at intervals of 1.0 ms during 1 μs. The received echoes are mixed with the signal from a local oscillator operating at 2970 MHz and converted to the intermediate frequency of 30 MHz. The resulting wave trains are amplified and converted to pulses (Fig. 27.4(b)) which are displayed on an oscilloscope screen.

27.2.2 Search and tracking radar

Radar systems can be ground based (harbor, airport, highway patrol), on ships, airborne or used in guided missiles.

Search radar looks continuously for targets or other important echoes. The transmitter beam is narrow horizontally (azimuth) and broad in the vertical direction (elevation). The search antenna rotates constantly at a speed of about one revolution per second.

Tracking radar is used to determine bearing, elevation and range of a detected target. The tracking antenna uses a highly directive beam (pencil beam).

The distance (range) of the target is determined by measuring the elapsed time after the transmission of a short burst of microwave energy. This system is called **pulsed Doppler tracking** (PDT).

Both modes of operation are based on scanning: the antenna moves continuously so that the transmitted beam sweeps through a certain predetermined area of space (Section 27.2.4). In most applications, echoes from moving targets are important and it is desirable to ignore and suppress less important echoes.

Moving target indication (MTI) (Section 27.2.3) is arranged so that **clutter** (echoes from fixed targets) is suppressed. The detection of moving targets is based on the principle of Doppler frequency shift. Obviously, MTI is sensitive to radial components of velocity only.

The frequencies which are used in radar applications are roughly divided into three bands:

S-band: 2–4 GHz
X-band: 8–12 GHz
K-band: 18–27 GHz

Search radar for early-warning systems usually operates at S-band frequencies, airborne radars at X-band while K-band frequencies are used in guided missiles. The dimensions of radar antennas decrease drastically with increasing operating frequency.

27.2.3 Moving target indication

The objective of MTI radar is to suppress clutter from fixed objects such as hills and buildings. Less fixed but insignificant objects are sea waves, moving leaves on trees, rain drops, etc.

Part of this problem is solved by subtracting the output of the receiver from the output of the succeeding scan. This way only difference signals are detected. These difference signals are obtained by using a delay line in conjunction with a subtraction circuit, called a **canceler**. Echoes from sea

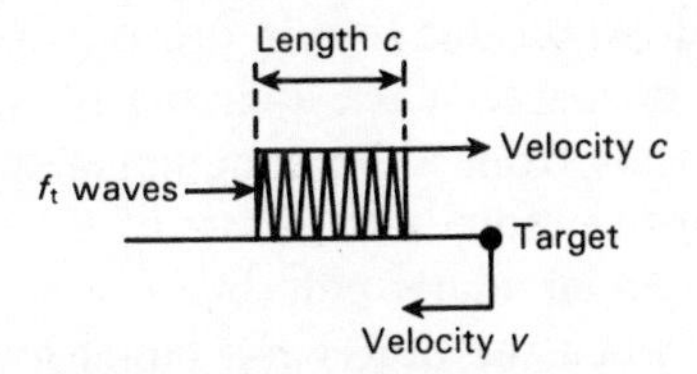

Fig. 27.5 Derivation of the Doppler frequency

waves, vegetation and rain are not suppressed by this method. The radial velocity of a moving target is determined by measuring the Doppler frequency.

The principle of Doppler frequency shift can be explained as follows. The transmitter (Fig. 27.5) fires a wave train with frequency f_t during one second; the radial target velocity is v. The wave train, which moves with the speed of light c, thus has a length c (m). When this wave train meets the target a time $t = c/(v + c)$ has elapsed between the first and last received wave. The total number of waves is f_t and the apparent frequency seen by the target is thus

$$f_{a,t} = \frac{f_t}{t} = f_t\left(1 + \frac{v}{c}\right) = f_t + f_{d,t}$$

where

$$f_{d,t} = f_t \frac{v}{c}$$

is the Doppler frequency at the target. When the waves are reflected back to the antenna, the same effect occurs again and the Doppler frequency at the antenna is

$$f_{d,a} = 2f_t \frac{v}{c}$$

The target velocity is thus

$$v = \tfrac{1}{2}c \frac{f_{d,a}}{f_t}$$

The maximum detectable range of a target can be easily determined. When r is the target range, the wave trains travel twice this range at the speed of light so that $r = \frac{1}{2}ct$. When T is the time interval between two consecutive pulses (receiving time), the maximum detectable range is

$$r_{max} = \tfrac{1}{2}cT$$

Beyond this range, a returned echo may be confused with a short range echo of the next cycle.

The suppression of clutter due to rain drops is based on properties of electromagnetic waves in space. When a circularly polarized wave with $\boldsymbol{E}$ and $\boldsymbol{H}$ components reflects from a flat conducting surface at right angles with the direction of propagation, both phases of $\boldsymbol{E}$ and $\boldsymbol{H}$ are reversed and the reflected wave is again circularly polarized.

This remains true when the object has equal horizontal and vertical dimensions like rain drops. This situation is unlikely for other targets where horizontal and vertical dimensions differ so that the reflected wave will be circularly polarized as well as linearly polarized. This difference in behavior is utilized to discriminate between rain drops and other moving targets.

In spite of all these measures, some clutter will remain. The **subclutter visibility** is defined as the attenuation between target echo and clutter echo beyond which detection becomes impossible.

In MTI–CW a continuous wave (CW) transmitter is used for speed measurements. When the phase of the transmitted signal is $\phi_t = \omega_t t$, the phase of a returning echo of a target at range r is

$$\phi_e = \omega_t \left[t\left(1 + \frac{2v}{c}\right) - \frac{2r}{c} \right]$$

where v is the target speed. Hence the phase difference is

$$\phi_d = \omega_t \frac{2}{c} (vt - r)$$

and the Doppler frequency is

$$f_d = \frac{1}{2\pi} \frac{d\phi_d}{dt} = \frac{2v}{c} f_t$$

In a phase sensitive detector (Section 24.4.6) the Doppler frequency can be detected and the speed determined.

Although ϕ_d contains the range term $\omega_t 2r/c$, range measurements are not possible since it is unknown how many times the phase has increased by $360°$.

In later developments CW radar is used in tropospheric measurements. The carrier is triangularly modulated in frequency and the target range is determined from measurements of the beat frequency obtained from the frequency of the reflected wave and the instantaneous frequency of the transmitted wave. These developments are progressing towards detection of air turbulence, the presence of dust clouds, air pollution and measurements of the structure of tropospheric layers.

In pulsed radar systems phase discrimination is unambiguous. If the phase difference obtained from the first echo is

$$\phi_{d1} = \omega_t \frac{2}{c} (vt - r)$$

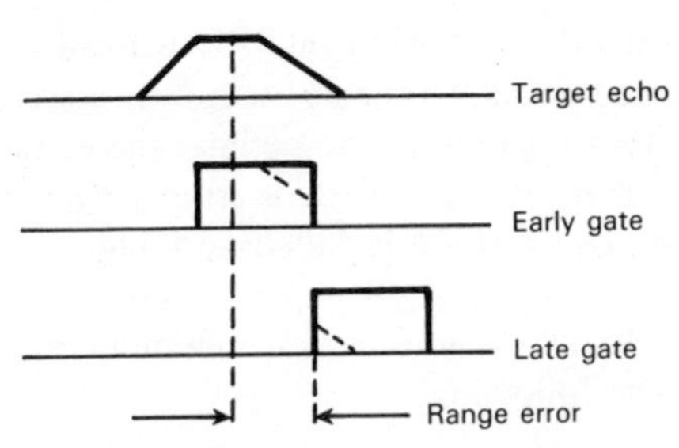

Fig. 27.6 Early gate and late gate

that of the next echo is

$$\phi_{d2} = \omega_t \frac{2}{c} [v(t - T) - r]$$

where T is the receiving time. Thus the difference of the phase differences is

$$\Delta\phi_d = \omega_t \frac{2v}{c} T$$

Certain speeds of moving targets go undetected. These **blind speeds** arise when the phase difference between two succeeding echoes is 360°. The output of the phase sensitive detector will be identical with the one obtained from the first echo so that no output is produced in the canceler. The solution to this problem is the use of two alternating pulse repetition frequencies.

In **tracking radar** ambiguity may evolve when trying to detect the range of two targets which are in close proximity. This problem is solved by the use of a so-called **split-gate** (Fig. 27.6) in combination with two amplifiers. The **early gate** opens the first amplifier during the gate time, then the **late gate** opens the second amplifier.

The range gates are set by the operator at a range which is slightly greater than the target range. As shown, almost all of the echo is passed by the early gate and very little by the late gate.

The amplifier outputs are compared in a differential amplifier and the difference signal, which represents the range error, is used to correct the range setting so that the outputs of both amplifiers become equal.

27.2.4 Scanning

Plan position indication (PPI). This is a scanning system used in search radar: the display tube is intensity modulated with the sweep moving from the center radially outward (Fig. 27.7). The screen contains range marks as well as angular marks while the antenna position is in the center of the screen. The scanned area is usually the horizontal plane.

Sector scan. This is a variation of PPI. Instead of sweeping over a full circle, the antenna sweeps back and forth over a sector of a circle in a horizontal plane (azimuth scanning). Sometimes the vertical position of the scanning antenna is changed during the scanning cycle so that the beam describes a rectangle. This motion is called **nodding**.

Tracking radar knows a number of scanning methods. The most frequently used are the following:

Helical scan. The antenna rotates continuously while the elevation angle changes slowly to a certain maximum value and then starts again (Fig. 27.8).

Conical scan. The beam is swept around the surface of a narrow cone (Fig. 27.9), e.g. by moving the feed system in a small circle. A target on the cone axis returns a constant echo. Targets off the axis return amplitude modulated echoes, the amount of modulation indicating how far the target is off the axis.

Since the nose of the radiated beam is rather blunt, direction measurements suffer from a certain degree of inaccuracy. The error can be decreased considerably by **monopulse lobing** where two overlapping beams are used simultaneously (Fig. 27.10). Generally, the echo in one beam will be stronger than in the other resulting in a difference signal. This difference pattern is very sharp so that much higher accuracies can be achieved (Fig. 27.11). The antenna is called a **monopulse antenna**.

27.2.5 The radar equation

The **radar equation** expresses the range at which targets can be detected in known parameters.

Assume a pulse of length τ is transmitted with a peak power P. The radiated energy is then

$$W = P\tau$$

and the energy density on the surface of a sphere with radius r is

$$w = \frac{P\tau}{4\pi r^2}$$

If the antenna has a directive gain G, the energy density in the direction of the beam is

$$w_a = \frac{GP\tau}{4\pi r^2}$$

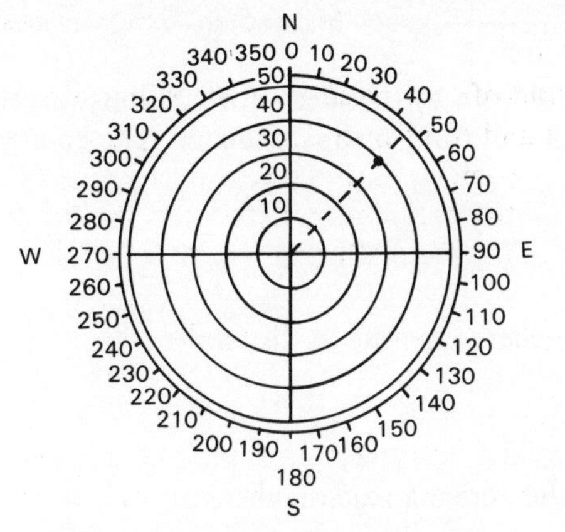

Fig. 27.7 Plan position indication

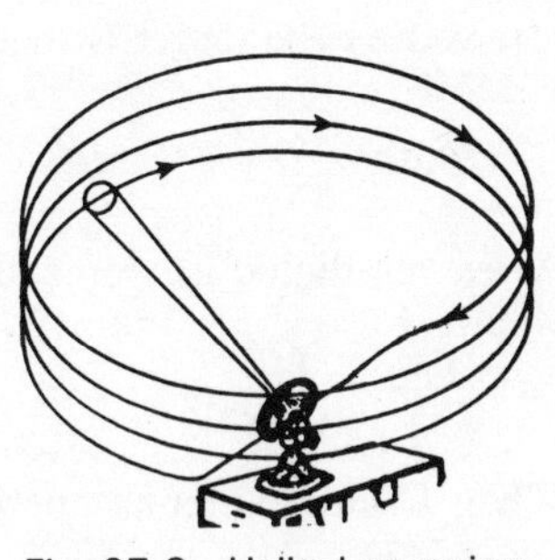
Fig. 27.8 Helical scanning

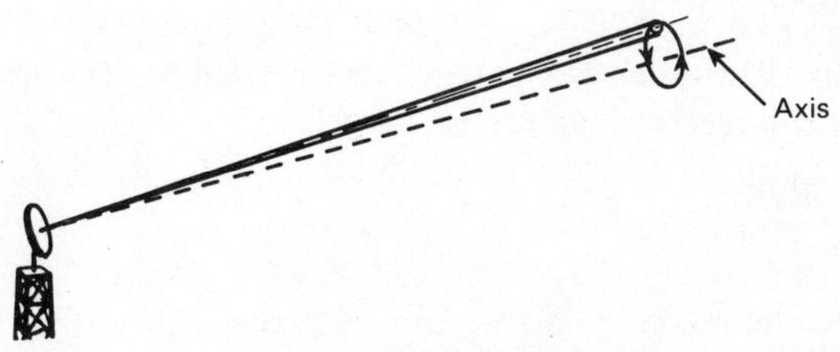

Fig. 27.9 Conical scanning

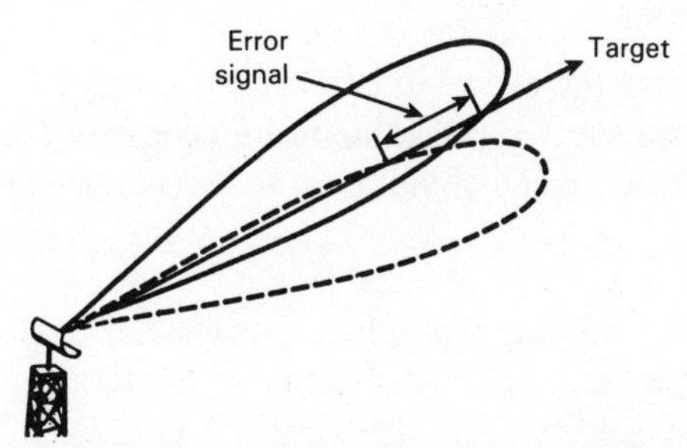

Fig. 27.10 Pattern of lobing antenna

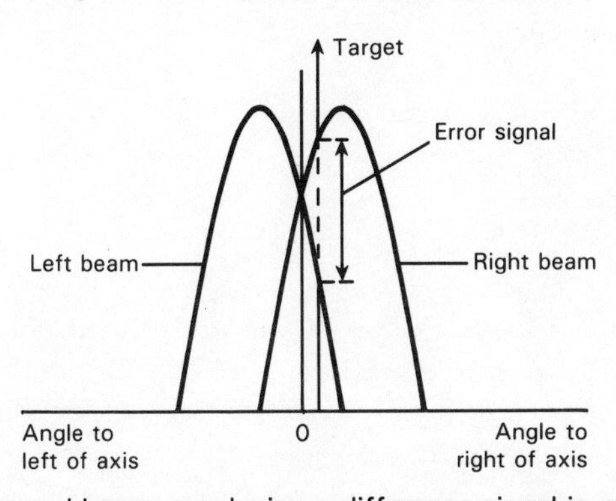

Fig. 27.11 Staggered beams producing a difference signal in monopulse lobing

The **effective target area** σ is defined as the area of a fictitious target reflecting all incident energy as if it were an isotropic radiator. The energy received by the target is then

$$w_a \sigma = \frac{GP\tau\sigma}{4\pi r^2}$$

When reradiated isotropically, the energy density at distance r is

$$\frac{w_a \sigma}{4\pi r^2} = \frac{GP\tau\sigma}{(4\pi r^2)^2}$$

When S is the antenna aperture, the antenna receives the energy

$$\frac{S w_a}{4\pi r^2} = \frac{SGP\tau\sigma}{(4\pi r^2)^2}$$

Directive gain and antenna aperture are related as $G = 4\pi S/\lambda^2$ (Section 26.6) so that the received energy is

$$w_r = \frac{G^2 \lambda^2 P\tau\sigma}{(4\pi)^3 r^4}$$

It depends on the receiver performance whether this energy is sufficient or not for detection. The input noise power of the receiver is

$$P_n = FkTB \text{ (W)}$$

where F is the noise factor.

During the transmitted pulse the noise energy is $P_n\tau$. Assuming that a signal can be detected when signal and noise energies are equal,

$$P_n\tau = w_r$$

it then follows that the range at which targets can be detected at is

$$r = \sqrt[4]{\frac{G^2 \lambda^2 P\sigma}{(4\pi)^3 FkTB}}$$

28
Optoelectronics

28.1 GENERAL

Light is radiant energy which travels in the form of an electromagnetic wave. Although this energy covers a large region of the electromagnetic spectrum, optoelectronics usually implies a smaller region namely the visible part of the spectrum (400–700 nm) as well as the adjacent regions of ultraviolet and infrared.

Light is produced when atoms of a material release energy. This energy is stored in the atoms due to excitation by heat or other energy sources which cause electrons to move to higher energy states.

The elementary unit of light is the **photon**; an uncharged particle having an energy which depends only on its frequency. It is characterized by the equation

$$\Delta W = hf$$

where

h is Planck's constant (4.32×10^{-15} eV s)
f is the transition frequency of the photon (Hz)
ΔW is the energy of the photon (eV)

A change of energy to a higher state implies absorption of the photon, a change to a lower state results in emission of the photon.

For convenience, the equation is often expressed in terms of wavelength:

$$\lambda = \frac{1290}{\Delta W} \text{ (nm)}$$

Optoelectronic devices are grouped into two classes:

- Absorption of electromagnetic radiation and conversion of this radiation into measurable electrical quantities. Such a device is a **detector**.
- Emission of electromagnetic radiation, e.g. due to a current passing through it. The device acts as a **source**.

Examples of detectors are photoresistors (LDRs), photodiodes, phototransistors, photomultipliers, etc. In Chapter 8 a number of these components were discussed. Examples of sources are incandescent lamp, fluorescent lamp, xenon lamp, LED, laser.

28.2 PHOTODETECTORS

28.2.1 Photodiode

Figure 28.1(a) shows a reverse biased *pn* junction. A photon entering the depletion region is readily absorbed thereby generating an electron–hole pair. The electron moves to the cathode, the hole to the anode and a current flow through the diode results. Generation of an electron–hole pair outside the depletion region has no effect since electron and hole will rapidly recombine. To allow light to enter the depletion region, the *p* region is made very thin and the depletion region wide to increase the yield.

If the reverse bias is removed and a resistor connected between cathode and anode, a current proportional to the light intensity flows. Therefore, light energy is converted to electrical energy.

The **solar cell** is a form of photodiode based on this principle. It is used in space probes and satellites to provide the energy for the electronic systems. The efficiency is about 10–15 percent.

A typical characteristic of a photodiode is shown in Fig. 28.1(b). The material is mostly silicon, the spectral sensitivity of which is shown in Fig. 28.1(c). Photodiodes are usually manufactured by the planar process.

28.2.2 Avalanche photodiode

The avalanche photodiode operates in the avalanche breakdown region. This means that a rather large reverse bias (100–400 V) is applied so that breakdown is about to occur. A small additional energy suffices to pull electrons from the atomic structure thereby creating free electrons and holes.

Figure 28.2(a) shows the generation of electron–hole pairs by penetrating photons. The electrons move to the positive side (cathode) with high velocities due to the large reverse bias. The high velocity electrons generate additional electron–hole pairs by collision. These pairs are accelerated as well and produce additional pairs so that avalanche results. In this way, a single photon may produce 100–300 electrons.

A typical characteristic is shown in Fig. 28.2(b). The avalanche photo-

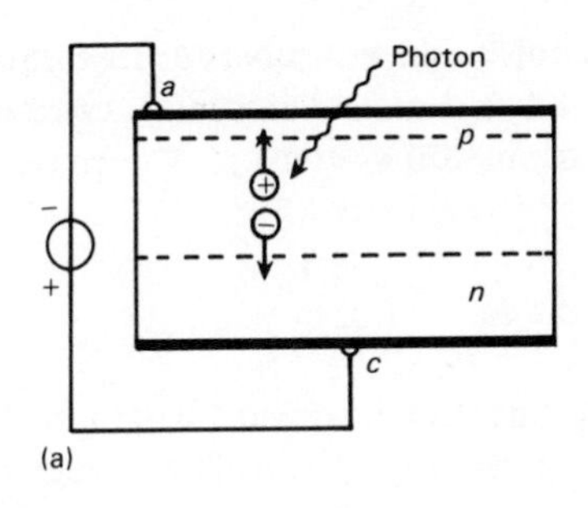

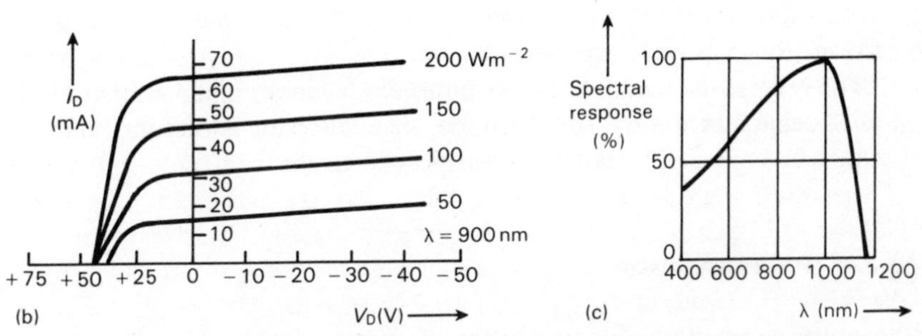

Fig. 28.1 (a) Photocurrent generation in a reverse biased *pn* junction (b) Typical characteristic of a photodiode (c) Spectral sensitivity of silicon

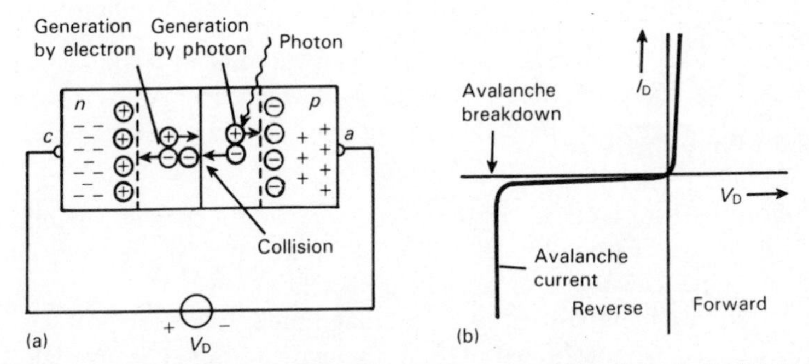

Fig. 28.2 (a) Photocurrent generation in an avalanche photodiode (b) Typical characteristic of an avalanche photodiode

diode has very low noise and is usable up to frequencies of the order of 50 GHz. They are used as detectors in optical communications systems, for example in fiber transmission systems.

28.2.3 PIN photodiode

The PIN photodiode has an intrinsic low-conductance zone sandwiched between *p* and *n* regions (Fig. 9.15). The *p* region is very thin to allow photons to reach the *i* region. When reverse biased, a photon produces an electron–hole pair in the *i* region. These carriers move in opposite directions thus causing a current to flow.

The reverse resistance of a PIN photodiode is very high ($\simeq 10\ \text{G}\Omega$), the parallel capacitance very small (a few pF), allowing switching times of about 10 ns or less. The noise level is very low.

28.2.4 Phototransistor

Phototransistors may have two different constructions:

- A photodiode plus a bipolar transistor, both integrated on a Si crystal (Fig. 28.3(a)).
- A special bipolar transistor construction (Fig. 28.3(b)) with a large base area covered by a transparent window. Light falling on the collector–base junction causes a photocurrent which is multiplied by the current amplification factor h_{fe} of the transistor. The large base area results in a rather poor frequency response in comparison with the photodiode. Often the construction contains a lens to focus the light falling on the base region.

The phototransistor is also manufactured as a Darlington phototransistor to increase the sensitivity of the device.

28.2.5 Photothyristor

The photothyristor is characterized by its large output current capability. In the structure of the photothyristor (Fig. 28.4) the gate is made accessible for incident light. The thyristor acts as a switch so that it is triggered into conduction by a light pulse of sufficient magnitude. It is turned off when the anode current is interrupted. Often the gate terminal is available for additional biasing.

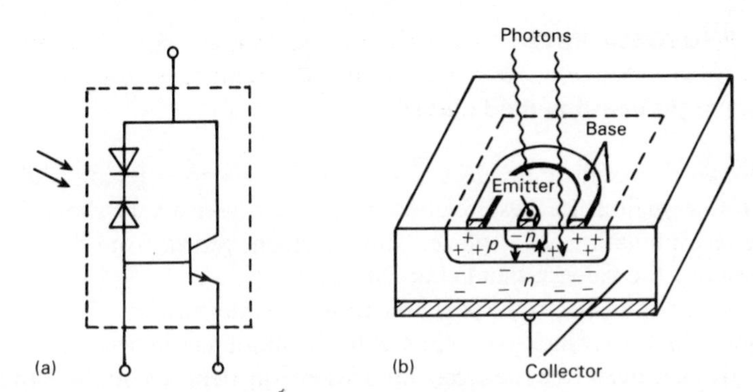

Fig. 28.3 (a) Phototransistor consisting of photodiode plus transistor (b) Bipolar phototransistor construction

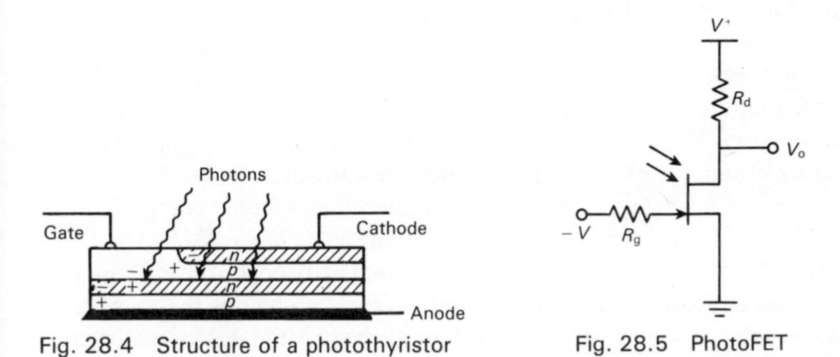

Fig. 28.4 Structure of a photothyristor

Fig. 28.5 PhotoFET

28.2.6 PhotoFET

Similar to the phototransistor the gate of the photoFET is made sensitive to incident light. A gate resistor R_g is connected to a negative supply voltage (Fig. 28.5). The photocurrent causes a voltage change ΔV to develop across R_g which results in a change of drain current according to

$$\Delta I_D = \Delta V_G g_m = \Delta I_G R_g g_m$$

This current change is converted to a voltage change ΔV_D in the drain, since

$$\Delta V_D = \Delta I_G R_g R_d g_m$$

Large R_g values increase the sensitivity, at the same time decreasing the frequency response due to the input capacitance of the FET.

The photoFET turns out to be 10–100 times more sensitive than the phototransistor while its frequency response is superior.

28.3 PHOTOSOURCES

28.3.1 Light emitting diode (LED)

In a forward biased diode electrons travel from *n* to *p* region and holes from *p* to *n* region. In these regions the carriers have a very short lifetime before recombination takes place. Thus electrons return from the conduction band to the valence band (Fig. 28.6).

Meeting of an electron and a hole results in the emission of one photon the energy level of which corresponds to the amount of energy which was originally needed to free the electron. This energy depends on the band gap of the material used.

Recombination can be **direct** recombination which is the transition between conduction band and valence band or **indirect** recombination. This recombination takes place in steps from conduction band to intermediate energy levels which exist as a result of the doping of the material.

In indirect recombination the energy differences are smaller and the released photons have lower energy (larger wavelengths) than in direct recombination. The energy differences in direct recombination of Ge and Si are 0.66 eV and 1.1 eV and these values are too small to produce visible light. For visible light production the required minimum energy level is about 2.3 eV. These levels are available in materials such as GaAs and GaP.

Recombination energy is partly radiation, partly heat. The efficiency of this process is expressed by the **quantum efficiency** which is the number of electron–hole pairs generated per incident photon. A number of materials is listed in Table 17.

28.3.2 Liquid crystal display (LCD)

Although the LCD does not emit light, it performs largely the same function as the LED, namely for display purposes.

A layer of liquid crystals is sandwiched between two glass plates which are covered with two polarizing plates (Fig. 28.7). The molecular structure of the crystals consists of long oval-shaped molecules which align themselves parallel to each other. On the inner surface of the glass plates SiO is evaporated at a certain angle. The crystal molecules in contact with the surface align themselves in parallel to that direction as well as with neighboring molecules. The patterns of the two glass plates are oriented in perpendicular directions so that the crystal molecules will show a 90° twist between top and bottom. Incident light is polarized by the polarizing plates but is gradually twisted by the crystal molecules. Behind the cell is a bright reflecting surface so that the light returns with its original direction of

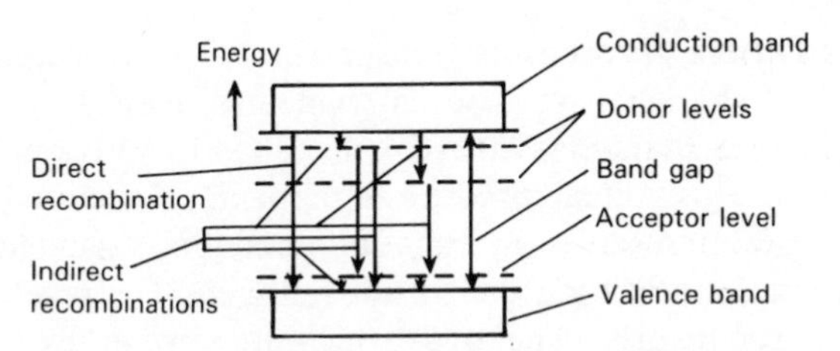

Fig. 28.6 Operation of electroluminescent LED

Table 17

Material		*Wavelength*	*Color*
$GaAs_{0.6}P_{0.4}$		650 nm	red
$GaAs_{0.35}P_{0.65}$	: N	630 nm	red
$GaAs_{0.15}P_{0.85}$	: N	590 nm	yellow
GaP	: N	560 nm	green
GaP	: ZnO	635 nm	red
GaAs		950 nm	infrared

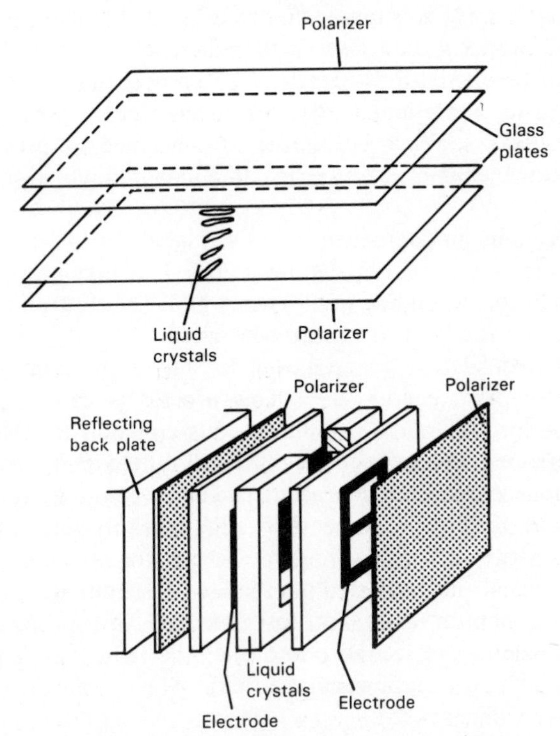

Fig. 28.7 Construction of liquid crystal display

polarization. On the glass covers transparent InSnO is evaporated in the desired shape of characters (e.g. seven-segment display).

To display the characters an AC voltage (5–10 V) is applied between these electrodes. This voltage produces a perpendicular electric field which aligns the crystal molecules in random directions. Incident light thus scatters, loses its polarization and is not reflected. The result is a frosted appearance of the display. The areas which are covered by the electrodes turn dark and are displayed on a bright background.

Since the only requirement for operation is the existence of an electric field, power consumption is extremely low (a few μW). A disadvantage is that ambient light is required to recognize the contrast. LCD displays are abundantly used in watches and pocket calculators.

28.3.3 Laser

A laser (acronym for Light Amplification by Stimulated Emission of Radiation) is a form of optical oscillator, somewhat analogous to a microwave transmitter. Common features are a resonant structure and a means of energy supply to support electromagnetic oscillations.

The basic property of a laser is its coherence of emitted light rays. Coherence can be divided into **temporal coherence**, indicating the monochromaticity of the light, and **spatial coherence** which expresses the phase equality of the waves. When both types of coherence are present, a laser beam will show negligible divergence thus maintaining a high energy density.

An electron in an atom can be in a higher or lower energy state (Fig. 28.8). If the electron is in the lower state 1, it may absorb a photon with energy ΔW corresponding to the energy gap. The electron jumps to the higher state 2 and the atom is said to be **excited**.

If the electron decays from state 2 to state 1 the atom radiates an identical photon with energy ΔW. This process is called **spontaneous emission.** The spectral range of spontaneous emission is relatively wide since no coherence exists between the phases of different photons. Thus the radiation is noncoherent and characteristic for all common light sources.

However, instead of spontaneously falling back to state 1, an incident photon with energy ΔW can stimulate the electron to return to state 1 whereby the original and generated photon are both emitted. In this process the stimulating photon does not lose energy. This process is called **stimulated emission**. The second photon has the same wavelength, phase and direction of propagation as the first. This means that coherent amplification has taken place.

A requirement for this amplification is that sufficient numbers of excited electrons are available in the material. This is achieved by **pumping**

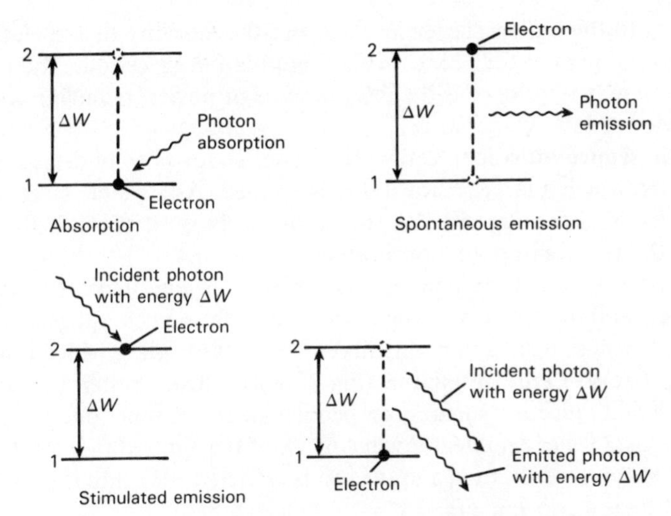

Fig. 28.8 Principle of spontaneous and stimulated emission

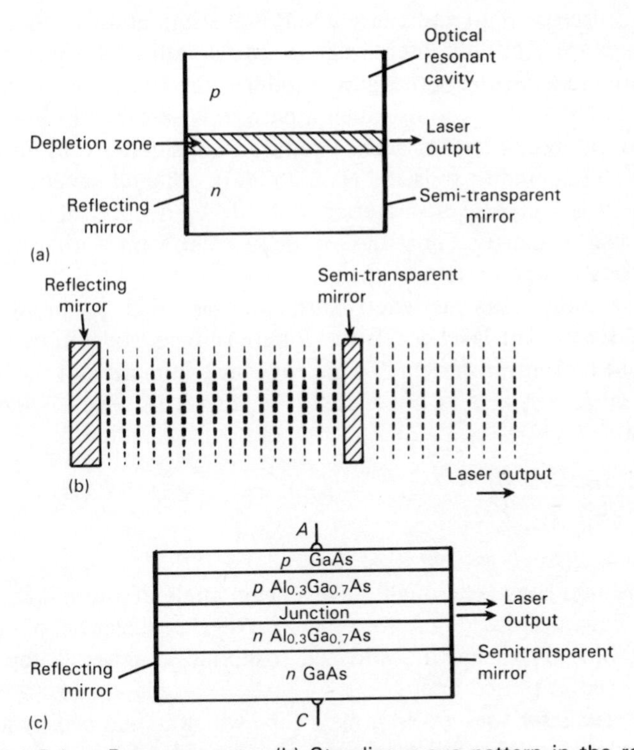

Fig. 28.9 (a) Fabry–Perot resonator (b) Standing wave pattern in the resonator (c) Heterostructure semiconductor laser

electrons to the higher energy level so that the majority of the electrons in the material are excited. This is called **population inversion**. The pumping operation may require considerable amounts of power, requiring cooling of the device to low temperatures.

The **semiconductor injection laser** consists of heavily doped p and n regions to which a large forward bias is applied. As a result, large numbers of electrons and holes are created in the vicinity of the junction which causes the required population inversion.

If the forward current is insufficient, recombinations will occur randomly and the result is luminescence like in fluorescent tubes. Therefore, a second requirement is that optical resonance takes place which is achieved by the **Fabry–Perot resonator** (Fig.28.9(a)). The crystal contains two polished and reflecting surfaces perpendicular to the junction. The surfaces are at a distance of an integral value of $\frac{1}{2}\lambda$ of the emitted radiation. Part of the radiation is transmitted and the rest is reflected back into the laser cavity formed by the two mirrors.

Above a certain threshold current some photons will cause emission of more photons which reach the reflecting surface and are reflected, thereby stimulating more photons before reaching the opposite surface. A number will be reflected again and cause additional stimulation so that oscillation will take place. The light intensity grows to a stabilized value where the light losses are in balance with the light amplification. In operation, a pattern of standing plane waves results. Such a pattern is sketched in Fig. 28.9(b).

Current requirements can be reduced considerably by using heterostructure semiconductor lasers (Fig. 28.9(c)). The pn junction is sandwiched between layers of material with different optical and dielectric properties. At current densities of about $500\ \mathrm{A\,cm^{-2}}$ they can operate continuously at room temperature.

In a **ruby laser** a xenon flash tube is used to pump the laser (Fig. 28.10(a)). The laser consists of a pure ruby crystal and two mirrors of which one is semitransparent. To improve the collimation of the laser beam, an optical lens system is often used (Fig. 28.10(b)) which is based on the formula

$$\frac{\tan \phi_1}{\tan \phi_2} = \frac{f_2}{f_1} = \frac{d_2}{d_1}$$

As shown, $d_2 > d_1$ so that $\phi_2 < \phi_1$.

A 12 mm ruby laser might have a beam angle of about 0.06°. The lens system can decrease this to 0.006°. At the distance of the moon (384 000 km) this beam has widened to about 25 miles. Ruby lasers are usually used in pulsed operations.

The **gas laser** uses gas in a glass tube which is pumped by a source of microwave energy by DC discharge. Gases used are helium-neon, argon, krypton and CO_2.

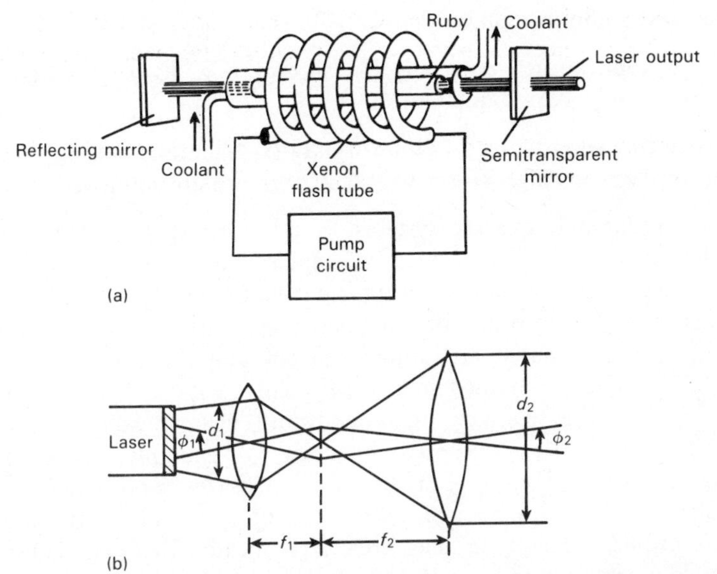

Fig. 28.10 (a) Ruby laser (b) Lens system to improve beam collimation

In a helium–neon laser the energy states are excited states of neon. A helium atom may go to a lower state while a neon atom absorbs this energy and is excited, causing population inversion. Helium–neon lasers usually operate at several wavelengths around 1153 nm. The proper wavelength is selected by design of the mirrors for maximum reflectivity.

Some applications of lasers are:

- Plasma physics: by laser radiation materials can be evaporated and pressurized to values comparable to those in the interior of stars.
- Nuclear fusion: gases in a spherical enclosure are hit by two or more laser beams at different angles. The gas compresses to such densities that it becomes a solid. When using this procedure on isotopes of hydrogen, deuterium, tritium, etc. cores can be melted to release energy.
- Holography: infrared lasers are used to design X-ray lasers which in turn can be used to construct holograms of single molecules.
- Distance measurements: during the first mission to the moon a reflector was left behind to allow measurement of the moon distance. A pulsed laser was used (just like radar) and the distance was measured with an accuracy of about 3 mm.
- Data communication: Semiconductor lasers are used in combination with glass fibers in telephone systems.

- Video and audio disc players.
- Surgical applications, especially eye surgery (e.g. to weld a detached retina) or for sealing blood vessels.
- Microcutting and welding of tools for industry, diamond cutting, etc. For these applications high power CO_2 lasers are usually employed.
- Optical radar and weapon systems, e.g. for tracking and ranging of missiles in a manner as explained in Chapter 27 on radar. The advantage of such a system is the narrow beam and the immunity to electromagnetic interference. The information carrying capacity of laser beams is illustrated by the fact that all radio frequency and radar bands through 300 GHz can be contained in a 0.1 percent bandwidth of the carrier frequency. At this relatively small bandwidth, channel capacity is about 1000 times that of the present frequencies. For example, in the optical range of 5–1000 THz a total of 10^5 TV channels can be placed. The main problem is the requirement of a direct line of sight and deterioration in the atmosphere. These problems are solved in cases where glass fiber can be used as the transmission medium.

Many of the radio and microwave techniques have been converted to optical frequencies and realized with lasers, for example superheterodyne detection, frequency doubling, optical rectification, etc.

28.3.4 Opto-couplers

The function of an opto-coupler is to transfer information from one point to another by optical means. A requirement is that source and detector are compatible with respect to spectral response.

If both responses overlap, the matching factor is said to be 100 percent. For example, a GaAs LED having a wavelength of 900 nm is not detectable by the eye but will be completely detected by a Si *pn* junction.

A second requirement is that the transmission medium passes the signal with minimum attenuation. Often a lens system is located in the transmission medium. An example of such a system is the bar code reader, used in department stores.

Some other applications are punched card readers, position sensing to detect the presence or absence of objects, speed sensing, fluid level detection, guidance systems, chopper amplifiers.

A special opto-coupler is the opto-isolator where source and detector are isolated from each other allowing large voltage differences between the two. These devices are used in medical applications, the oil industry, explosive environments, etc. A few opto-coupler circuits are shown in Fig. 28.11.

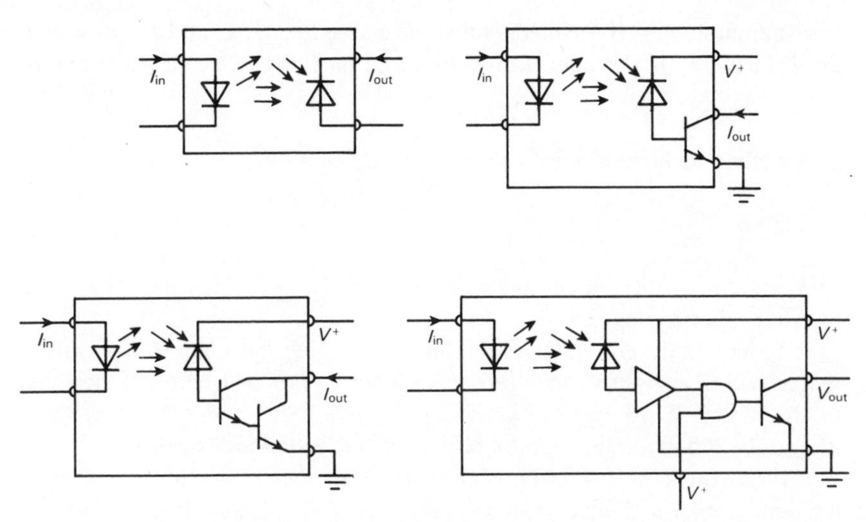

Fig. 28.11 Examples of opto-isolator circuits

28.4 PRINCIPLES OF FIBER OPTICS

In the field of communications (which employs wires, coaxial cables, radio, waveguides, etc.) fiber optics is one of the fastest growing developments. Fibers are gradually replacing coaxial cables in systems such as telephone trunk lines, cable television, ship and airplane wiring, etc.

A few major advantages of fiber versus coaxial cable are:

- Weight: fiber cable weight per kilometer is about 1 percent of that of coaxial cable.
- Interference: fiber optics is immune to electromagnetic interference and crosstalk; fiber cables have no conductivity.
- Low-line attenuation.
- Large transmission capacity ($\simeq 10$ Gbit s^{-1}).

A typical fiber optic telephone link is shown in Fig. 28.12. Sound waves are successively converted to electrical and optical signals and at the receiver the inverse conversions take place. As shown, optical transducers are needed at both ends.

Sources for optical links can be infrared luminescent LEDs or laser-diodes. A comparison of the spectral distributions of the two types is shown in Fig. 28.13.

A basic law in light transmission is Snell's refraction law (Fig. 28.14) concerning the behavior of light rays crossing the boundary between two transparent media. If the media have refractive indices n_1 and n_2 ($n_2 < n_1$), an incident ray is partially transmitted and partially reflected according to

$$n_1 \sin \phi_i = n_2 \sin \phi_t$$

If the angle of incidence $\phi_i = \phi_c$ is such that $\phi_t = 90°$,

$$\sin \phi_c = \frac{n_2}{n_1}$$

The angle ϕ_c is the **critical angle** and for angles of incidence $\phi_i \geqslant \phi_c$ total internal reflection occurs.

To eliminate crosstalk in a bundle of fibers, the fiber is clad with a material with a slightly lower refractive index n_s (about 1 percent) than that of the core (Fig. 28.15(a)). The core refractive index is n_c, the diameter is about 50–200 μm. This type of fiber is called a **step-index fiber**.

If an incident ray is launched into the fiber a maximum value of incident angle is defined whereby the ray will be conducted along the interior of the fiber by multiple reflections. This maximum value is the **acceptance angle** ϕ_a:

$$\sin \phi_a = \sqrt{n_c{}^2 - n_s{}^2} = \text{NA}$$

where NA is called the **numerical aperture**.

When rotating the acceptance angle about the fiber axis, the **acceptance cone** is described.

A disadvantage of step-index fiber is the small transmission bandwidth. Light rays propagate through the fiber at different angles between 0 and ϕ_c which results in different path lengths. A short light pulse gradually widens along the fiber requiring larger bandwidths. This is avoided when the refractive index of the core fiber is made to change gradually from the center outwards. These fibers are called **graded-index fibers**. A light ray follows an undulating course (Fig. 28.15(b)). The longer path lengths are offset by the higher velocity of the rays due to the lower refractive index. Thus phase delay can be largely eliminated.

Since core diameters of the fibers are large compared with the optical wavelengths, interference of light waves will occur. A method to reduce this effect is to reduce the core diameter. Such small diameter fibers are called **single-mode fibers;** core diameters are between 2 and 4 μm.

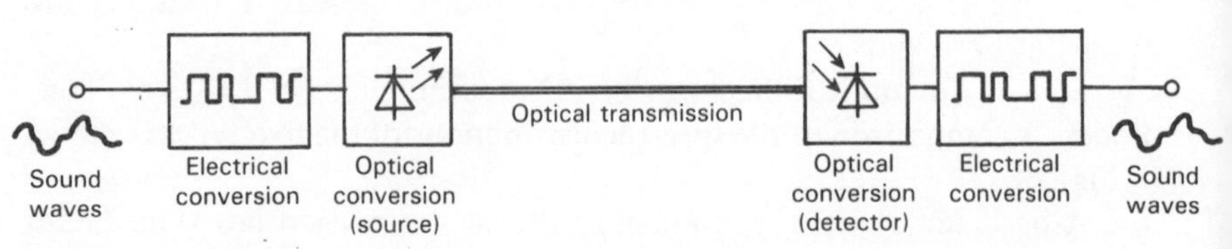

Fig. 28.12 Example of fiber optic used in telephone link

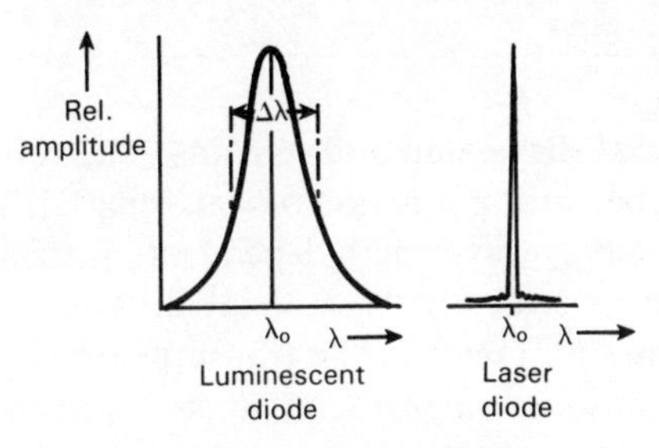

Fig. 28.13 Spectral distributions of luminescent and laser diode

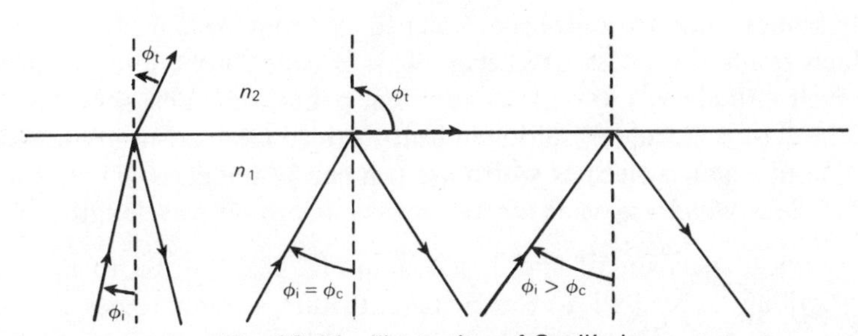

Fig. 28.14 Illustration of Snell's law

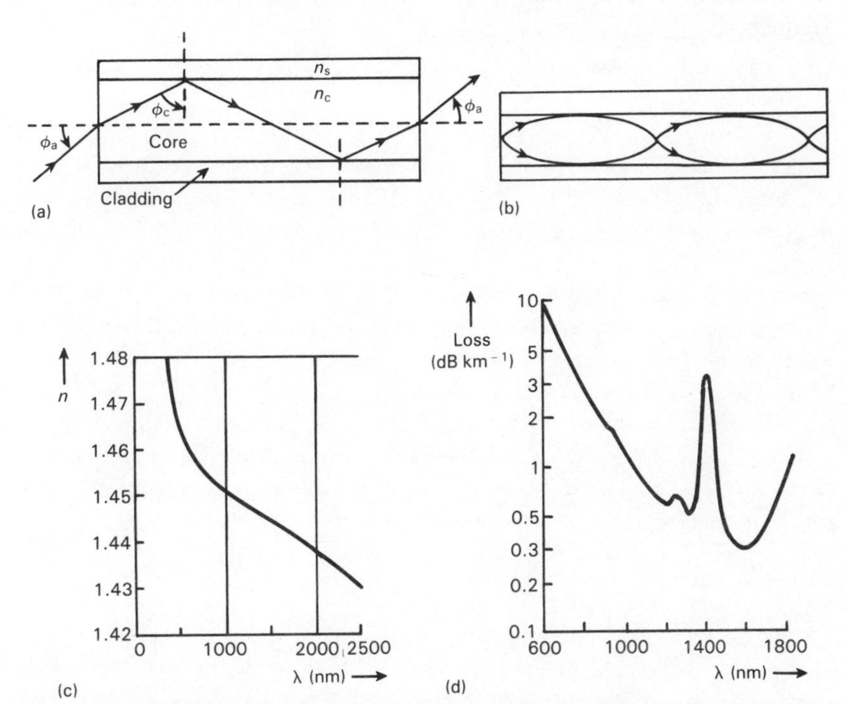

Fig. 28.15 (a) Step-index fiber with cladding (b) Propagation of light rays in graded-index fiber (c) Refractive index of silica glass fiber as a function of wavelength (d) Typical loss spectrum of silica glass fiber

An effect called **dispersion** still remains. The refractive index is not a constant but a rather strong function of wavelength (Fig. 28.15(c)) implying that the light velocity is wavelength dependent. Since an optical transmitter occupies a certain spectral width $\Delta\lambda$, delay differences still remain, leaving dispersion as a limiting factor in the transmission rate.

As in any transmission medium, losses in fibers are inevitable. The main causes of these losses are as follows:

- Rayleigh scattering: due to impurities, density variations, etc.
- Absorption: due to ionization, valence electrons become free electrons which results in a loss of energy. This is called ultraviolet absorption (wavelength about 140 nm). In addition, photons are absorbed by atoms of the glass molecules resulting in heat. A third mechanism of absorption is due to water molecules which are trapped in the glass. These contain OH^- ions which cause absorption peaks at certain wavelengths.

A typical loss spectrum of fiber is shown in Fig. 28.15(d). Further developments will undoubtedly lead to new manufacturing technologies and novel devices, e.g. optical switches. It is conceivable that a significant portion of conventional electronics will shift in the spectrum towards optical frequencies.

Index